Occupational Toxicology
2nd Edition

About the 1st Edition –

'This book flows smoothly . . . the professor and students utilising this text would feel
well-informed and satisfied with the information gained from the contents.'

American Journal of Pharmaceutical Education

'This is a useful book that offers an overview of toxicology with a special emphasis on
the workplace and occupational hazards . . . It could easily serve as a price-worthy text
for introductory courses in toxicology at the upperclass and graduate levels.'

Veterinary Human Toxicology

Hazardous agents are an ongoing concern and there are now many examples of
workers being severely affected by chemicals as a result of both acute and chronic
exposure.

This new edition of **Occupational Toxicology** introduces the basic aspects of
the science of toxicology which underpin the application of toxicological information to the workplace environment.

This revised book contains updated chapters on the most important hazardous
agents in the workplace, such as metals, pesticides, solvents and inhaled materials.
The importance of a multidisciplinary approach to the field is highlighted by
covering inter-related aspects of hygiene, epidemiology and occupational medicine.

The types of toxicity seen in the workplace environment, and the organs affected
are also discussed. The lung and skin are given individual attention and this new
edition also contains additional chapters on the liver, kidneys and nervous system.

As the cancer-related effects of chemicals continue to cause concern, genotoxicology and carcinogenesis are singled out for particular attention.

The final section covers the management of chemicals in the workplace and the
practical use of toxicological information; this includes a new chapter on the workplace assessment of toxic chemicals.

Occupational Toxicology retains the lucid, practical approach that won the
first edition such praise and will be of interest to students of occupational and
industrial toxicology and practitioners in industry.

Dr Neill Stacey has been, until recently, Project Safety and Environmental Affairs
Manager in Sydney, Australia. **Dr Chris Winder** is Head of the School of Safety
Science at the University of New South Wales, Australia.

Occupational Toxicology

2nd EDITION

Edited by

Chris Winder and Neill Stacey

CRC PRESS

Boca Raton London New York Washington, D.C.

Library of Congress Cataloging-in-Publication Data

Winder, Christopher
    Occupational toxicology / Chris Winder and Neill H. Stacey. — 2nd ed.
      p. cm.
    ISBN 0-7484-0918-1 (alk. paper)
    Previously published under title: Occupational toxicology.
    Includes bibliographical references and index
    1. Industrial toxicology. I. Stacey, Neill H. (Neill Hubert), 1950-II. Title.
    RA1229.O28 2003
    615.9'02—dc21
                                               2003053133

Visit the CRC Press Web site at www.crcpress.com

© 2004 by CRC Press LLC

No claim to original U.S. Government works
International Standard Book Number 0-7484-0918-1
Library of Congress Card Number 2003053133
Printed in the United States of America 1 2 3 4 5 6 7 8 9 0
Printed on acid-free paper

Contents

Figures

Tables

Contributors

J. Barter
Manager, Environmental Affairs
Chemicals Group
PPG Industries Inc.
Pittsburgh, PA
USA

T. Driscoll
(formerly) Epidemiology and
Biostatistics Unit
National Occupational Health and
Safety Commission
PO Box 58
Sydney, NSW 2001
Australia

C. Gray
Associate Professor of Occupational
Hygiene
School of Biological and Chemical
Sciences
Deakin University
Geelong, VIC 3217
Australia

W.M. Haschek
Professor of Toxicology
Department of Veterinary Pathobiology
University of Illinois
2001 South Lincoln
Urbana, IL 61801
USA
Email: whaschek@cvm.uiuc.edu

M. Lotti
Instituto di Medicina del Lavoro
Universita degli Studi di Padova
Via Facciolati 71
I-35127 Padova
Italy

A.D. Mitchell (deceased)
Genesys Research Incorporated
PO Box 14165
Research Triangle Park
North Carolina, NC 27709
USA

A. Moretto
Instituto di Medicina del Lavoro
Universita degli Studi di Padova
Via Facciolati 71
I-35127 Padova
Italy

W.-O. Phoon
Pymble Medical Consultants
PO Box 818
Pymble, NSW 2073
Australia

N.H. Stacey
Southern Cross Pharma Pty Ltd
18a Ellison Road
Springwood, NSW 2777
Australia
Email: nstacey@bigpond.net.au

H. Vainio
Chief, Unit of Carcinogen
Identification and Evaluation
WHO International Agency for
Research on Cancer
150 Cours Albert Thomas
69372 Lyon Cedex 08
France
Email: vainio@iarc.fr

C. Winder
Associate Professor in Chemical Safety
and Head
School of Safety Science
University of New South Wales
Sydney, NSW 2052
Australia
Email: C.Winder@unsw.edu.au

Preface

Occupational health and safety (OHS) is a cooperative activity that involves people from various disciplines. Indeed, OHS practitioners often come from a range of backgrounds, bringing their skills and expertise to the field. Interaction amongst these fields is essential in dealing with workplace problems, making it a truly multidisciplinary endeavour.

Chemicals are found in most workplaces, and where such chemicals have hazards they require careful attention from the OHS practitioner to avoid overt acute poisoning from single or short-term repeated exposures, as well as the more insidious adverse health consequences from longer term repeated exposures. Chemicals may also damage the workplace or harm the environment.

Toxicology is a speciality in its own right with its own approaches, terminology and viewpoints. Many people involved in occupational health and safety must deal with chemical issues from time to time, but do not necessarily have a strong background in toxicology, so it is an aim of this book to provide such people with the opportunity to understand the basis of this science and how toxicological information is used. Thus the book is directed towards OHS practitioners at different levels, to allow them to attend to chemical-related issues with greater understanding and confidence.

Indeed, occupational toxicology is emerging as an important part of workplace health and safety, with contributions from the health and chemical sciences. Some of the relationships to occupational toxicology are shown below.

In addition to an understanding of what toxicology is about, this book is structured to provide an overview of the effects of chemicals on target tissues, with emphasis on those more often affected by workplace exposures. The agents most often associated with occupational diseases are then dealt with so that the reader can gain an appreciation of the types of substances of most concern in the workplace. It is not intended that the reader is able to use this as a reference source to look up the toxic effects of various chemicals; there are many texts and databases devoted to this sort of information. Rather, it is intended to give an overview of

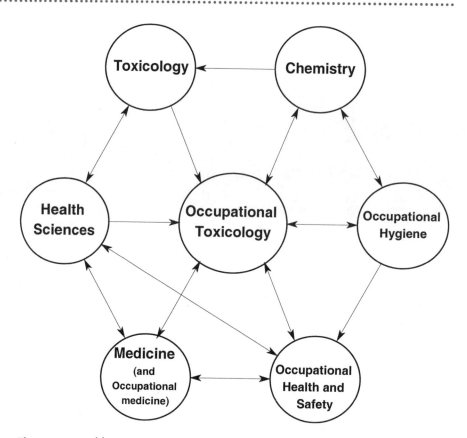

Figure P.1 Fields interacting with toxicology.

the classes of agents with one or two appropriate examples in more detail. The important fields interacting with toxicology are then covered to help the reader understand just what these fields involve and how they all inter-relate. Finally the book covers aspects of chemical use, and systems for managing workplace chemicals, and some working examples requiring the use of toxicological information are provided. This is meant to allow the reader to appreciate how information is used and some of the difficulties and uncertainties involved in the interpretation and decision-making with toxicological data.

Overall this book covers the field of occupational toxicology and is intended to serve as a valuable guide for those involved with dealing with workplace chemicals.

Understanding occupational toxicology

Chapter 1

Introduction

N.H. Stacey

The science of toxicology has many applications. One of these, which is the topic of this book, relates to exposure of people to noxious or hazardous agents during the course of their work. In order that one may understand and use toxicity information on workplace chemicals it is necessary to obtain a grounding in the basic principles and concepts of toxicology. Chapter 2 of this book covers these aspects in some detail, while more comprehensive coverage may be found in the fundamental textbook of toxicology *Casarett and Doull's Toxicology. The Basic Science of Poisons* (Klaassen *et al.* 2001).

Although the basic principles and concepts of toxicology relating to particular organs and/or groups of chemicals are similar across different fields of toxicology, many of the details are deserving of different emphasis with regard to the occupational setting – and this is a focus of this book. Furthermore, there are specific uses of toxicological data particularly related to managing workplace chemicals which are dealt with in detail in our final chapters.

What is occupational toxicology?

Occupational toxicology is the study of the adverse effects of agents that may be encountered by workers during the course of their employment. The adverse effects may be in the workers themselves, or in experimental animals, or other test systems used to define and/or understand the toxicity of the agent of interest. The term 'occupational' is used in preference to 'industrial' because the latter may have the connotation of chemical exposure in factories; this would not necessarily include work such as farming, with potential exposure to pesticides – or office work, with issues such as photocopiers in enclosed spaces.

Why is occupational toxicology required?

There is considerable public awareness of the health effects of chemicals due to events such as the thalidomide tragedy and environmental contamination with chemicals. Recognition of workplace exposures resulting in health effects has been part of this. A cursory glance at the Group 1 carcinogen list of the International Agency for Research on Cancer (IARC) should be evidence enough from this perspective. The epidemic of asbestos-related cancer that continues in some countries, and which will become evident in others in the coming years, indicates that there is no room for complacency in our efforts to curb substance-related ill health.

Since the 1940s the production of synthetic organic chemicals has risen dramatically such that annual production is in the hundreds of millions of tonnes. The increases in production continued steadily through the mid-1990s, as shown for some chemical industry sectors in Table 1.1.

Some figures for individual organic chemicals are provided in Table 1.2 to

Table 1.1 Industrial production for some chemical-related industry sectors in the USA from 1987 to 1997

Industry sector	% Increase
Total	31
Chemicals and products	28
Chemicals and synthetic materials	25
Basic chemicals	7
Alkalis and chlorine	22
Industrial organic chemicals	30
Plastic materials	40
Chemical products	32
Rubber and plastic products	40

Source: Adapted from Chemical and Engineering News (1998).

Table 1.2 Industrial production (millions of pounds) of some organic chemicals in the USA

Chemical	1987	1992	1997
Acrylonitrile	2182	2829	3291
1,3-Butadiene	2931	3232	4107
Ethylene	34,951	40,924	51,078
Isopropyl alcohol	1371	1463	1478
Propylene	19,019	23,421	27,533
Styrene	8014	9000	11,366
Urea	14,866	17,532	15,530
p-Xylene	5155	5656	7789

Source: Adapted from Chemical and Engineering News (1998).

indicate the large volumes, as well as growth, in production of some chemicals. Not all chemicals grew in production continuously over the entire 10-year period, as exemplified by urea, which fell between 1992 and 1997.

As well as the continuing production of familiar chemicals, there are many new chemicals being introduced to which workers may be exposed. Of the 20 million or so chemical entities that are in existence it has been estimated that about 80,000 have reasonable potential for human exposure.

The scope of the matters that occupational toxicology is confronted by may be appreciated by considering the numbers of chemical entities available. This is not too difficult, and it is possible to get a good idea of the numbers of chemical substances available in the world today. For instance, the Chemicals Abstract Service (CAS) allocates a unique identifying number to each new structure notified. The CAS number system consists of three numbers divided by hyphens. For example, the CAS numbers for some common chemicals are:

- formaldehyde   50-00-0
- aspirin   50-78-2
- ascorbic acid (vitamin C) 50-81-7
- ethanol   64-17-5
- toluene   108-88-3
- glutamic acid   617-65-2
- asbestos   1332-21-4
- salt (sodium chloride)   7647-14-5
- water   7732-18-5
- sulfuric acid   7664-93-9
- polyvinyl chloride   9002-86-2
- quartz   14808-60-7
- ethoxylated olive oil   103819-46-1

It can be seen that the three-number CAS system involves anything from two to six digits and a hyphen, then two digits and another hyphen, and a single digit. This is the basis of the CAS Registry number system.

Numbers of substances

The last major review of the health and safety information of substances for which there is at least a reasonable possibility of human exposure was a survey carried out by the US National Toxicology Program (NTP) in 1984. At the time there were 5–6 million chemicals with CAS numbers, and it was estimated that about 70,000 of these chemicals were in commercial use. When this is broken down into categories of chemicals, the numbers in Table 1.3 below are obtained.

Although additional materials have entered commerce since 1984, many of the chemicals listed by the NTP are no longer in use because of technological change, regulation and environmental concerns.

By February 2002, the CAS had allocated CAS numbers for 19,357,053 organic and inorganic substances and 17,255,658 sequences, making a total of 36,612,711

Table 1.3 Chemical categories and numbers of chemicals

Chemical category	Total numbers
Therapeutic substances	1815
Cosmetics	3410
Agricultural chemicals	3350
Food additives	8620
Industrial chemicals	48,523

Source: NTP 1984.

chemical substance registrations. At the time of this analysis, the most recent CAS number is 393780-00-2. This therefore defines a 'universe' of known chemical entities.

The available information on these substances

In the 1984 study, the NTP also estimated the proportion of those chemicals for which full, partial, minimal, below minimal, or no toxicity information was available. A complete hazard assessment could be made for only about 2% of the chemicals produced commercially; even for a partial hazard assessment the figure was only about 14%. These data are shown in Table 1.4.

These are low figures. In 1990 the OECD produced a list of 1338 high-volume chemicals (defined as more than 10,000 tonnes per year in any member country). These accounted for about 90–95% of the global chemical production. Even so, there was sufficient information for a detailed hazard assessment of only 434 of 948 organic chemicals on the list; no information was available for 147 organic chemicals (Bryden *et al.* 1990).

In recent years there has been a significant increase in the availability and development of toxicological and environmental information as a result of both voluntary and regulatory required testing programs and hazard communication initiatives in the USA and Europe. Most importantly, industrial chemicals are the category of chemicals to which people at work have the greatest potential for exposure. It would seem that the growing requirements for material safety data sheets have resulted in additional basic toxicity information, such as acute

Table 1.4 Chemical categories availability of information (NTP 1984)

Chemical category	Full	Partial	Minimal	Below minimal	None
Therapeutic substances (1815)	18%	18%	3%	36%	25%
Cosmetics (3410)	2%	14%	10%	18%	56%
Agricultural chemicals (3350)	10%	24%	2%	26%	38%
Food additives (8620)	5%	14%	1%	34%	46%
Industrial chemicals (48,523)	0%	10%	10%	0%	80%

lethality and irritant effects, becoming available for industrial chemicals – although this is anecdotal.

Ill health caused by chemicals

The true extent of work-related chemical associated ill health in both human and economic terms is unknown and has proven very difficult to define accurately. Estimates for occupational diseases have given figures like 5000–7000 deaths per year from occupational cancer, pneumoconioses, cardiovascular disease, chronic respiratory disease, renal disease and neurologic disorders in New York State, which had a workforce of 7–8 million at about that time (Markowitz and Landrigan 1989). The same report estimated that 35,000 new cases of occupational disease develop each year in New York State. Their estimate for the cost of occupational cancer alone in New York State amounted to approximately 500 million US dollars. More recently, these estimates were extended to the whole of the USA (Leigh *et al.* 1997) with a figure of 60,300 deaths from disease in 1992 being calculated. The cost of these deaths plus illness was estimated to be 26 billion US dollars. Another recent report from the US has reiterated the difficulties in providing reliable figures for substance-related death and illness (NIOSH 2000). For pneumoconioses, which are deemed entirely attributable to occupation, there have been over 114,000 deaths in the US since 1968. Most of these have been associated with coal mining, although these have decreased in recent years, while asbestos-related deaths are on the increase (NIOSH 2000).

The difficult task of estimating mortality and associated outcomes from exposure to hazardous substances was taken up by Morrell *et al.* (1998). In Australia – a country of about 17 million people in the period covered by the study – it was estimated that 2290 deaths per year were attributable to occupational exposure to hazardous substances. Clearly, much of the exposure would have occurred in earlier years, but that does not mean that the study results should be dismissed as irrelevant to contemporary times. The figures were similar to those for road deaths and suicides, and were substantially higher than deaths from AIDS. The prime cause of the deaths was cancer, followed by renal, cardiovascular, neurological, and chronic respiratory disease. Acute toxic episodes accounted for only 1–2% of the deaths. Estimates of productive years of life lost indicated that the yearly figure was nearly 6000. While the methodology used in such studies is subject to criticism because of the assumptions that have to be made, this study does provide an indication, based on a scientific approach, that there is a significant issue with exposure to hazardous substances in the workplace.

Thus there remains much work to be done in the area of workplace chemicals and other agents of toxicological concern. A reliable and accurate estimate of the true extent of substance-related ill health and death is still a contentious issue. Furthermore, while there is still a lack of comprehensive data on existing industrial chemicals, new ones are continually becoming available. In addition, the methods used for carrying out risk assessment are by no means perfect and there is great scope for advances in this field.

How is occupational toxicology used?

The primary goal of occupational toxicology is to protect the health of workers. Application of toxicological information in this way is but one component of the overall strategy for achieving this end. Other components can be organised in an hierarchical order, the 'heirarchy of controls', as shown in Table 1.5.

Role in decision making within hierarchical control

Toxicology plays an important role in deciding the appropriateness of some of these steps. For example, a hazardous chemical may be eliminated for a less hazardous process (use of staples instead of adhesives). Or a substitute for a particularly toxic chemical currently used in a workplace may be found, but the toxicity of the replacement still requires evaluation, for example, synthetic fibres for asbestos. It is not satisfactory to replace one chemical with a second of unknown toxicity just because the first is of toxicological concern. All those involved in occupational toxicology (and, indeed, occupational health and safety) need to be aware of this. It is not unheard of that the sole reason to request registration of one product for use is on the basis that it will replace one known to be responsible for a particular toxic effect.

Role in setting exposure standards for workplace chemicals

Apart from contributing to the above strategy for controlling exposure to workplace chemicals, toxicology is intimately involved in decision making on what levels of human exposure are acceptable for chemicals that are in use; that is, the

Table 1.5 Strategy for controlling exposures to hazardous substances in the workplace

Control measures	Examples
Elimination	Removal
Substitution	Replacement with a low hazard substance
Containment	Process enclosure
Engineering controls	Local exhaust ventilation
Safe work practices	Specific procedures on specific work tasks where hazardous exposures occur
Nonspecific procedures	Hazard communication Training Supervision
Personal protective equipment	Gloves Aprons Respirators

setting of workplace exposure standards. Investigations into the toxicity of a chemical seek to define a level at which there are no toxic effects. Terms used to describe this include NEL (no effect level), NOEL (no observable effect level) and NOAEL (no observable adverse effect level).

These terms warrant some explanations. NEL describes the dose level used in an animal experiment where no effect was noted. NOEL goes a step further by including the word 'observable' to indicate that only a finite array of possible toxic endpoints was investigated. It therefore acknowledges that some form of toxicity may have been present but that the appropriate test was not employed to detect it. It also allows for other uncertainties such as differences among species and so on. NOAEL is designed to accommodate the possibility that an effect may have been observable but it may be one that is not deemed to have toxicological significance for the organism. An example of such an effect, considered by some to be unimportant in setting acceptable exposure limits, is an increase in liver weight of the exposed animals. Many chemicals are known to do this by increasing the amounts of enzymes responsible for biotransformation, and this is often considered to be an 'adaptive response'.

Once the level at which no effect of toxicological significance has been derived, it can be used to decide what constitutes an acceptable level to which humans may be exposed. Often this is done by incorporating a *safety factor* or *uncertainty factor*, the latter being a more contemporary term which better reflects why this factor is used. It is used because of our uncertainty in this process due to aspects such as extrapolation across species, differing sensitivities among individuals, and so on. It cannot, however, guarantee that the level of exposure so chosen will protect every individual to each and every possible toxicological manifestation of that chemical. Nevertheless, it may be expected that below the chosen level it is most unlikely that any adverse effects of the chemical will be observed. Different factors have been employed under differing circumstances. While there is no set rule for deciding on the factor to be used there are guidelines that may be followed as outlined in Table 1.6.

Apart from the uncertainty factor approach, mathematical models have been used as a means by which to arrive at an exposure standard. They can be used to predict what level of exposure would result in an incidence of, say, one in a million. However, the use of such models is contentious as they all depend on particular assumptions and can give final values that vary by orders of magnitude. Therefore, their use remains limited today.

It is important to recognise that toxicological data from experimental animals are not the only source of information used in setting exposure standards. Clearly,

Table 1.6 Guidelines for choice of uncertainty factors in setting exposure standards

Factor	Rationale
10	Suitable human data
100	Chronic data in animals
At least 1000	Limited animal data

any human data that are available must be included and are of great importance. Often, however, we find that these data are of limited use for a variety of reasons, such as co-exposures in the workplace, or the reports being of unsatisfactory scientific rigor. Nevertheless, well-conducted positive epidemiological studies in humans will always outweigh analogous studies from experimental animals. Furthermore, other data may be used, such as those from cells relating to the genotoxic effects of the chemical in question. It is also important that data relating to the mechanism of toxicity are considered when making conclusions. The appropriateness of the study and quality of the data are other important factors that must not be overlooked. Overall, all the available information must be considered in order to arrive at an acceptable value for the exposure standard.

The process for setting exposure standards for workplace chemicals has been somewhat different to that used for chemicals found in other settings over the last 50 years or so. Many workplace chemicals had been in use when the standards were first set, so information was already available on their effects in exposed humans. Thus the setting of exposure limits often relied heavily on this information. Furthermore, little toxicological data from experimental animals was available for consideration because detailed animal testing has only been required in recent years. By comparison, animal data have been a requirement for many years before registration and marketing of pesticide residues. Thus, standard's setting for pesticide residues has relied more heavily on animal data gathered before any human exposure. It should be the end result of toxicological testing, of course, that positive human toxicity data for that chemical are never obtained. This view sometimes seems to arouse condescension in committee colleagues during meetings to decide on acceptable practices for using chemicals in the workplace. For example, should not a genotoxic aromatic amine with clear and strong evidence of causing urinary bladder cancer in dogs be treated as if it were a human carcinogen? In asserting that it should be, one often hears that there is no supporting human data. The author's response is 'Aren't we fortunate that this is so?'. It is interesting to consider the IARC list of Group 1 carcinogens (proven human carcinogens) in this context where many are related to occupational (as compared to environmental) exposures (IARC 1987).

Over recent years the process of setting standards for exposure to workplace chemicals has received considerable criticism (Castleman and Ziem 1988; Halton 1988; Roach and Rappaport 1990). One reason for this has been the use of uncertainty factors to a much lesser extent than has been applied to other chemical groups. Nevertheless, it should be appreciated that the process began because there was a clear void that required attention, and this gap was filled due to the efforts particularly of the American Conference of Governmental Industrial Hygienists (ACGIH). This does not mean that the process should remain unchanged, as more and newer chemicals are brought into the workplace, or existing ones are re-evaluated. There are now increasing requirements for toxicological information to be provided about new chemicals before they may be brought into the workplace. However, it should also be recognised that the extent of this information is not as great as that required for pesticides, for example.

Role in biological monitoring for chemical exposure

Biological monitoring is the measurement of a substance, its metabolites, or its effects in body tissues, fluids or exhaled air of exposed persons. While biological monitoring itself is not a preventive measure, consideration of biological monitoring is an interesting exercise because it is a topic that clearly exemplifies the interdisciplinary nature of protecting workers from the deleterious effects of exposure to chemicals.

There has been increasing attention to biological monitoring in recent years, as evidenced by the publications of world bodies (Elinder *et al.* 1994; WHO 1996) and journals dedicated to the topic (such as *Biomarkers*). The WHO documents are useful reference materials for those wanting further general information and for specific details on measurements of particular substances. The textbook by Lauwerys and Hoet (1993) also contains information on the biological monitoring of a number of individual substances. Advances with newer, more generally applicable techniques may also provide assistance to the monitoring, and therefore management, of exposed workers in the future. Examples from the author's (former) laboratory include detection of DNA adducts, especially by ^{32}P-postlabelling (Qu *et al.* 1997a) and use of serum bile acids as a sensitive indicator of exposure to organic solvents (Neghab and Stacey 2000).

There is usually a relationship between workplace air levels of substances and the actual levels in the workers. Determination of the air levels is largely the domain of the occupational hygienist, while information about levels in the body requires the biological component. Information from experimental animals is very useful, if not essential, with regard to the latter, especially in the initial considerations of what may be the most suitable entity to determine. Thus the involvement of the occupational toxicologist is required. Then we have the issue of biological exposure and biological effect. A metabolite of a chemical in the urine can be considered an index of exposure, but is cholinesterase depression in serum or red blood cells an indicator of exposure or of effect? It is a biological effect, but it is used as a marker of exposure at the early stages of its depression. DNA adducts are certainly related to exposure and provide a measure of internal dose at a critical site, but do we have sufficient information at this time to consider them as an effect as such? The three disciplines of occupational toxicology, medicine and hygiene meet here. Lastly, when there is evidence of organ dysfunction related to chemical exposure we move into the domain of the occupational physician. These inter-relationships are represented schematically in Figure 1.1.

When resolving problems such as evidence of an adverse health effect, it is appropriate for the occupational physician to consult a hygienist for advice on ventilation controls and/or a toxicologist for advice on the toxicity of alternatives, for example. Thus a team approach is very much called for to obtain the best end result. The same view is promulgated by Thorne (2001).

Comparison with chemicals in the general environment

Having referred to pesticides above it is interesting to compare exposure to chemicals in the environment to that in the workplace. Exposure in a workplace is often

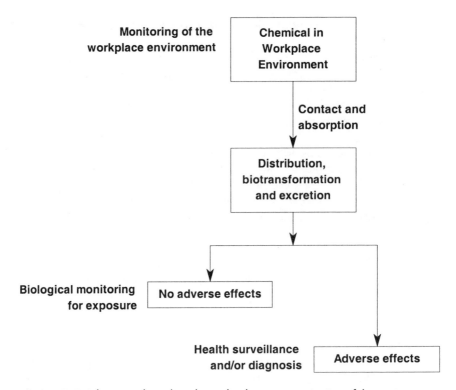

Figure 1.1 Scheme to show the relationship between monitoring of the environment, biological monitoring, health surveillance and/or diagnosis.

to higher concentrations of the chemicals than in the environment, although the duration of the exposure will be limited to the time at work while exposure in the environment could be for 24 hours a day, 7 days a week. Another difference is the size of the exposed population; it is limited to the particular workforce in the occupational setting, while environmental exposure includes larger numbers depending on the spread of the chemical in question. It is perhaps easier to investigate occupational toxicological problems because of the generally higher concentrations and a more easily defined exposed population. Then it must be considered that there may be exposure to the chemical in both the workplace and the environment as well as to multiple chemicals. Perhaps passive smoking or exposure to diesel emissions may be suitable examples. It is of interest to consider recent studies of DNA adducts in this regard. Hemminki *et al.* (1990) found that workers at the site and people in the surrounding community had similar levels of adducts while those in a rural population had much lower levels. However, Qu *et al.* (1997b) found higher levels of DNA adducts associated with work practices involving higher exposures to diesel engine emissions.

Relationship of occupational toxicology to other disciplines relating to workplace chemicals

As is apparent from the preceding sections, the overall management of workplace chemicals requires input from various disciplines. These include occupational toxicology, hygiene, medicine and epidemiology. All are specialities in their own right but each is required for satisfactory attention to this important endeavour. For this reason individual chapters have been included on each of these aspects. The practising occupational toxicologist must have an appreciation of these other areas to make the most meaningful input into the overall deliberations on issues involving occupational exposure to chemicals. As alluded to above, the need to consider information emanating from each of these fields is particularly apparent during the standard's setting process for chemical exposures and during resolution of workplace problems involving chemicals.

Future directions and challenges for occupational toxicology

Continued vigilance and improvement in the safety of chemical products should be a goal for those of us working in the area of occupational toxicology. There are some particular issues that are deserving of and will receive increasing attention over the coming years. Some are issues relating to toxicology in general while others are more specific to occupational toxicology.

The ethical issues associated with animal testing and the dilemma faced in trying to balance this with protecting human health continue today and will continue for some time yet. This matter was addressed by Purchase (1999) whereby he used a recent development in toxicology – endocrine disrupters – to illustrate this dichotomy.

The continued development of *in vitro* tests should be a priority because this will have both humane (as in the previous paragraph) and economic benefits once successfully established. It will also allow a more rapid procedure for assessing the toxicity of chemicals. While there have already been some considerable achievements in the area of *in vitro* toxicology (Tyson and Stacey 1992) it must be appreciated that there is much work to be done and that human health must not be compromised. In the years since the first edition of this textbook, it is the opinion of the authors that progress has been slow in this important area, especially with regard to the use of *in vitro* data in risk assessment.

The toxicity of mixtures remains an issue as it is appreciated that people are exposed to many chemicals simultaneously and that the vast majority of toxicological data are on the individual chemicals. This presents a massive problem, as it would not be possible to test all permutations and combinations of chemicals to which humans may be exposed. Here there may be a role for *in vitro* systems as a screen for likely problem mixtures, although this is still to be realised. It does seem, however, that calculation of the toxicity of mixtures based on their individual ingredients is becoming used and accepted more often and that, usually, the calculated value has been found to be consistent with the toxicity when this has actually been determined.

The greater application of molecular biological tools to aid toxicological assessment continues to emerge. One example is the use of transgenic mice so that genotoxic chemicals can be identified in multiple organs, with all body systems interacting so that a more reliable indication of likely applicability to the human can be obtained. A second application, which has had an impact on toxicological matters relating to the workplace, is found with the identification of adducts to cellular macromolecules as indicators of carcinogen exposure. These adducts, especially those to DNA – the genetic material – are thought to reflect exposure at the critical site of action of the chemical. The promise continues to emerge of being able to monitor workers for exposures at a site where differences in disposition and interaction at the molecular target and repair mechanisms in that individual are no longer an unaccounted for entity.

There will also be increasing attention to the detection of toxicity that may have been unappreciated in the past due to inadequate measurement techniques. Examples of these can be found with investigations into neurobehavioural effects of solvents, and perhaps insecticides, and greater emphasis on immunological alterations induced by chemicals and the relevance of such changes to the organism as a whole. For example, improvement of cognitive test regimens may well assist in more reliably predicting the neurotoxic effects of chemicals in humans (Slikker *et al.* 2000). Another example is the improvement in monitoring in recent years, such as the measurement of serum bile acids to reflect possible subtle hepatic effects of solvents (Driscoll *et al.* 1992; Franco 1991; Neghab and Stacey 2000).

It might also be anticipated that with the passage of time there will be an improvement in data used for hazard assessment for occupational chemicals. Toxicology in general would also benefit substantially from improvements in models used for hazard assessment. This is exemplified by the increasing use of the pharmacokinetic modelling approach in recent times (Andersen 1995). However, it must be remembered that substantial amounts of data are required in the first instance to allow the successful application of such models.

Three final issues deserve special mention because of their importance, and because they are not really the province of the occupational toxicologist but impinge greatly on our efforts.

- Firstly, it is apparent that even the existing toxicological knowledge is not always transmitted and disseminated as well as it should be. Efforts to improve this should be continued.
- Secondly, it would seem that chemical-related injury in the workplace may be reduced by invoking programs to change attitudes with regard to exposure to occupational chemicals. It seems as if even when the knowledge is there it is sometimes ignored for questionable reasons.
- Thirdly, it will be difficult to evaluate the benefits of such programs because our baseline knowledge of the extent of workplace chemical-related injury is unknown.

While it should be appreciated that this is a difficult problem, it will be essential to have such knowledge in the future to determine the areas where resources should be concentrated. Finally, while there have been some advances since the first edition of this textbook, it is the opinion of the author that it has been slow moving. While the reasons for this would be manifold, the expression of will by government, industry and the scientific community has to be regarded as a contributing factor.

References

Anderson, M.E. (1995) Development of physiologically based pharmacokinetic and physiologically based pharmacodynamic models for applications in toxicology and risk assessment. *Toxicol. Lett.* 79: 35–44.

Bryden, J.E., Morgengroth, V.H., Smith, A. and Visser, R. (1990) OECD's work on investigation of high production chemicals. *Int. Environ. Report.* June, pp. 263–270.

Castleman, B.I. and Ziem, G.E. (1988) Corporate influence on threshold limit values. *Am. J. Ind. Med.* 13: 531–559.

Chemical and Engineering News, North East News Bureau (1998) World chemical outlook. *Chem. Eng. News* 76(50): 1–80.

Driscoll, T., Hamdan, H., Wang, G.F., Wright, P.F.A. and Stacey, N.H. (1992) Concentrations of individual serum or plasma bile acids in workers exposed to chlorinated aliphatic hydrocarbons. *Br. J. Ind. Med.* 49: 700–705.

Elinder, C.G., Friberg, L., Kjellstrom, T., Nordberg, G. and Oberdoerster, G. (1994) *Biological Monitoring of Metals.* WHO International Program for Chemical Safety. Geneva: World Health Organization.

Franco, G. (1991) New perspectives in biomonitoring liver function by means of serum bile acids: experimental and hypothetical biochemical basis. *Br. J. Ind. Med.* 48: 557–561.

Halton, D.M. (1988) A comparison of the concepts used to develop and apply occupational exposure limits for ionizing radiation and hazardous chemical substances. *Reg. Toxicol. Pharmacol.* 8: 343–355.

Hemminki, K., Grzybowska, E., Chorazy, M. *et al.* (1990) DNA adducts in humans related to occupational and environmental exposure to aromatic compounds. In: Vainio, H., Sorsa, M. and McMichael, A.J. (eds) *Complex Mixtures and Cancer Risk.* Lyon: IARC.

IARC (1987) *IARC Monographs on the Evaluation of Carcinogenic Risks to Humans Supplement 7.* Lyon: International Agency for Research on Cancer.

Klaassen, C.D., Amdur, M.O. and Doull, J. (eds) (2001) *Casarett and Doull's Toxicology. The Basic Science of Poisons,* 6th edn. New York: McGraw-Hill.

Lauwerys, R.R. and Hoet, P. (1993) *Industrial Chemical Exposure. Guidelines for Biological Monitoring,* 2nd edn. Boca Raton: CRC Press.

Leigh, J.P., Markowitz, S.B., Fahs, M., Shin, C. and Landrigan, P.J. (1997) Occupational injury and illness in the United States. *Arch. Intern. Med.* 157: 1557–1568.

Markowitz, S. and Landrigan, P. (1989) The magnitude of the occupational disease problem: An investigation in New York State. *Toxicol. Ind. Health* 5: 9–30.

Morrell, S., Kerr, C., Driscoll, T., Taylor, R., Salkeld, G. and Corbett, S. (1998) Best estimate of the magnitude of mortality due to occupational exposure to hazardous substances. *Occupat. Environ. Med.* 55: 634–641.

National Research Council (1984) *Toxicity Testing. Strategies to Determine Needs and Priorities.* Washington, DC: National Academy Press.

Neghab, M. and Stacey, N.H. (2000) Serum bile acids as a sensitive biological marker for evaluating hepatic effects of organic solvents. *Biomarkers* 5: 81–107.

NIOSH (2000) *Worker Health Chartbook, 2000.* Washington, DC: National Institute of Occupational Safety and Health/US Department of Health and Human Services.

Purchase, I.F.H. (1999) Ethical review of regulatory toxicology guidelines involving experiments on animals: the example of endocrine disruptors. *Toxicol. Sci.* 52: 141–147.

Qu, S.X., Bai, C.L. and Stacey, N.H. (1997) Determination of bulky DNA adducts in biomonitoring of carcinogenic chemical exposures: features and comparison of current techniques. *Biomarkers* 2: 3–16.

Qu, S.X., Leigh, J., Koelmeyer, H. and Stacey, N.H. (1997) DNA adducts in coal miners: association with exposures to diesel engine emissions. *Biomarkers* 2: 95–102.

Roach, S.A. and Rappaport, S.M. (1990) But they are not thresholds: A critical analysis of the documentation of threshold limit values. *Am. J. Ind. Med.* 17: 727–753.

Slikker, W. Jr, Beck, B.D., Cory-Slechta, D.A., Paule, M.G., Anger, W.K. and Bellinger, D. (2000) Cognitive tests: interpretation for neurotoxicity? *Toxicol. Sci.* 58: 222–234.

Thorne, P.S. (2001) Occupational toxicology. In: Klaassen, C.D., Amdur, M.O. and Doull, J. (eds). *Casarett and Doull's Toxicology. The Basic Science of Poisons*, 6th edn. New York: McGraw-Hill, pp. 1123–1140.

Tyson, C.A. and Stacey, N.H. (1992) *In vitro* technology, trends and issues. In: Frazier, J.M. (ed.). *In vitro Toxicity Testing. Applications to Safety Evaluation*, New York: Marcel Dekker, pp. 13–43.

WHO (1996) *Biological Monitoring of Exposure in the Workplace*, volumes 1 and 2. Geneva: World Health Organization.

Chapter 2

Basics of toxicology

N.H. Stacey

Introduction

The intention of this chapter is to provide an outline of the basic aspects of toxicology to enable the reader to appreciate and understand the principles of the science when using toxicological information in a workplace situation. Furthermore, areas that are pertinent to toxicology in general will be covered in a broad sense so that the reader is aware of their existence and how they are integrated with toxicology as a whole. Thus a framework for the field of toxicology will be provided for future application by the reader.

It is perhaps of interest to mention at this introductory stage some of the difficulties experienced by those working in the field of toxicology. These difficulties relate primarily to the perceptions of others about toxicology and the general concern about chemicals amongst the public at large. For example, there are difficulties in science due to the uncertainties in particular areas such as the interpretation of some experimental animal carcinogenicity data. It is often inappropriate, if not incorrect, to make unequivocal pronouncements on such matters but this is often interpreted by others as evasiveness. Dayan (1981) has raised similar issues in pointing out that the toxicologist is often lauded for preventing harm, but attacked for failing to prevent all undesired effects of chemicals, and disparaged as a source of delay and cost to the industry concerned. He goes on to indicate that many industrial toxicologists are troubled by the impossibility of their work ever meeting public and managerial expectations. The media also plays an important role in the perception and appreciation of toxicology by the general public. One example that illustrates the point very clearly to the author concerns the issue of beef meat contaminated with pesticide residues. This became an issue and many media communications over several days addressed the matter expressing great concern over the impact on the export market. For weeks nothing was published on the possible health effects on local consumers. If there had not been a significant export market at stake, and the story had been followed up by the media, the slant would have been different and could well have focused on the long-term health implications of the local consumer. The point is that the media play an important role in the public perception of toxicological matters by the manner in which they present material.

Terms and definitions

Toxicology and other associated terms have been defined in various ways by different experts. The definitions used here are believed to be the most appropriate but the reader should be aware that others would present slightly different views. Where it is felt necessary and/or beneficial, explanations relating to differences or confusion in terminology will be provided.

Toxicology is the study of the adverse effects of agents on living organisms. This definition will benefit from some expansion. Study refers to various aspects such as experimentation, collection, collation, evaluation and review of toxicological material. 'Adverse' refers to unwanted effects – these are usually overt but may also include more subtle effects. The adverse nature of an effect may be contentious as referred to in Chapter 1, in that an effect may be considered adverse by some but not by others. Then there will be other effects about which there may be uncertainty as to whether they are truly adverse to the homeostasis of the organism as a whole. The term 'Agent' is used because even though chemicals are most often dealt with, other entities such as radiation, dusts, and endo- or exotoxins produced by microorganisms fall within the bounds of toxicology. 'Living organisms' indicates that the target of concern spans plants and lower animals through to humans. In different branches of toxicology any one of these may be the target of concern.

It is now worth defining a **toxicologist**, who may be considered as an individual with expertise in the nature of the adverse effects of agents on living organisms, and the assessment of the probability of their occurrence. While this may appear somewhat repetitive it highlights one very important aspect. That is, once the process of assessment involving humans is being undertaken there is the opportunity for subjectivity to become a factor in any final decisions. This relates very much to what some have described as the art and the science of toxicology (Doull 1984). That is, the science is the hard data upon which assessments are made, while the art is the less certain process of decision making which will often involve subjective and/or arbitrary components. One example of this is the choice of a safety or uncertainty factor when deciding on an acceptable exposure standard.

Now that a definition of toxicology has been given and the extra dimension that the toxicologist brings to the science has been explained it is pertinent to split toxicology into different divisions to better understand what toxicology involves. Table 2.1 lists many of the disciplines and subdisciplines in which a toxicologist may be involved.

However, while there is a significant diversity in the different types of toxicology, ultimately the professional activities of toxicologists fall into five main areas, as shown in Figure 2.1.

It may be seen that the first two areas can be collectively viewed as generator areas of toxicology, while the other three can be regarded as evaluator divisions. That is, descriptive and mechanistic toxicology are involved in actually generating

Table 2.1 Disciplines, applications and functions of toxicology

Type of toxicology	Type of toxicology
Analytical toxicology	Information toxicology
Behavioural toxicology	Mutagenesis
Carcinogenesis	Neurotoxicology
Clinical toxicology	Occupational toxicology
Dermatoxicology	Ocular toxicology
Developmental toxicology	Pesticide toxicology
Drug toxicology	Pulmonary toxicology
Ecotoxicology	Regulatory toxicology
Environmental toxicology	Renal toxicology
Experimental toxicology	Reproductive toxicology
Food toxicology	Respiratory toxicology
Forensic toxicology	Risk analysis toxicology
Genetic toxicology	Teratology
Hepatotoxicology	Toxicokinetics
Immunotoxicology	Toxinology

Descriptive

Mechanistic
} Generator

Informational

Risk assessment

Regulatory
} Evaluator

Figure 2.1 Breakdown of toxicology into divisions.

the required information about the toxic mechanisms and effects of chemicals, while the other three are primarily involved in using and evaluating the information for particular purposes. A toxicologist may work in any of these divisions or, as is often the case, across these divisions. Needless to say a mix of expertise brings advantages to activities in any particular area. It should also be appreciated that others have used different terms, such as testing, research and safety evaluation (Malmfors 1981), but the meaning is essentially the same.

Descriptive toxicology involves the undertaking of experiments or test procedures to provide standard toxicological data for the assessment of the safety of a chemical. This includes various protocols from acute to chronic toxicity testing (which will be detailed later). Thus the outcome is data that describe the toxicity of the chemical under test.

Mechanistic toxicology involves the study of mechanisms by which chemicals exert their toxic effects. While it seems this has been regarded by some in the past as an academic pursuit it is becoming increasingly realised just how important it is to understand the mechanism of action if a meaningful risk assessment for humans is to be undertaken.

Informational toxicology involves the collection, collation and dissemination of toxicological information. It increasingly includes some interpretation and summarising of data as seen, for example, in the provision of material safety data sheets for chemicals. This area of information provision has been overlooked by others, but it is one of developing importance and visibility particularly in the application of toxicological knowledge. It is crucial that those involved in the area have an adequate background in toxicology so that their efforts are not thwarted by misinterpretation of often complex material. This is becoming more so as the ready availability of databases over the Internet continues to develop.

Risk assessment toxicology involves review of the results of descriptive and mechanistic studies, combined with study of potential exposures to make probabilistic estimations of risk in exposed populations. It provides for a better understanding of the impact of toxic materials on exposed populations.

Regulatory toxicology involves the evaluation of the available data for an agent and decision making about its application. A clear example of this is the setting of workplace exposure standards. As outlined earlier, tasks like this bring another dimension to toxicology in that there is much more scope for subjectivity and opinion to have an influence on outcome (IPCS 1994). The final decision may often be based on factors in addition to the toxicological information. This is not to say that the toxicological information is disregarded, but that recognition of socioeconomic factors as well as differing opinions among experts have been involved in the final decision on how a chemical may or may not be used.

Other terms that are intimately involved in toxicology and the application of data also require attention, as follows:

- **Toxicity** is the intrinsic property of an agent to adversely affect an organism
- **Hazard** is the potential for the toxicity of an agent to be realised in a particular situation
- **Risk** is the probability that a hazard will be realised
- **Safety** is the probability that a hazard will not be realised.

Again one may find that others define these terms a little differently, so it is worth explaining them further by way of example. Consider a chemical that has a known acute toxicity in that y mg/kg will result in death. It is to be used in a factory for the production of a certain material in a fully enclosed process. Before it is delivered to the factory it presents no hazard to the workers at that plant. Once it is delivered there is potential for exposure during procedures such as handling, for accident during operation, during maintenance procedures, and so on. Thus there is a certain hazard now present. However, the toxicity of the chemical (its inherent properties) belong to the chemical itself and remain unchanged irrespective of whether it is in the supply house, on the delivery truck, or at the plant. Risk is when one quantifies the likelihood of a hazard being realised – that is, an estimate of the likelihood of an event occurring is given in numerical terms.

Consideration of these terms highlights a very important aspect of toxicology whereby for a toxic response to be manifest exposure must occur. That is, not only is the toxicity of the chemical of importance but so are the conditions under which it is used, which relate to the exposure that occurs. If there is no exposure there can be no toxicological effects.

Empirical truisms in toxicology

Two fundamental aspects of toxicology are critical to its functionality and application. They are usually described as basic assumptions, or tenets, but empirical truisms may better represent their origins. These are the relationship between dose and response, or effect, and the validity of extrapolating of data from experimental animals to humans.

Dose response and dose effect

As the dose of a chemical is increased, the response or effect also increases. The first formal recognition of this relationship is attributed to Paracelsus (also known as Philippus Theophrastus Aureolus von Hohenheim) from the 16th century. Interestingly, his reason for espousing the relationship was to justify his use of a remedy of recognised toxicity for treatment of the Great Pox. As justification for using this poisonous remedy he informed those concerned about its use that it was the dose that was important in determining whether or not a toxic reaction occurred. Thus he was indicating that if the appropriate dose was given the therapeutic benefit could be obtained, without the toxicity being expressed. The obvious relationship to concerns over exposures to workplace chemicals is that it should be possible to determine an exposure level below which a response or effect would not be seen. Thus an acceptable exposure level should be attainable. It is worth noting at this stage that this is an area of uncertainty and debate for carcinogenic effects of chemicals.

The terms 'response' and 'effect' require some clarification because they are used differently in different areas of occupational health. 'Response' is used to mean the proportion of a population showing a specified change. 'Effect' is the actual magnitude of the change in the parameter. An example will help to explain the difference: if there is evidence of liver damage in response to exposure to different levels of a particular agent, the data could be presented as the number of individuals as compared to the total with a serum alanine aminotransferase level over, say, 100 IU/l at the different exposure levels – this would be the response. Alternatively, the data could be presented as the actual enzyme measurements at the different exposure levels, say, 95 or 870 IU/l – this would be the effect. These terms will be used differently and often interchangeably by others. This is most probably related to the use in pharmacology of the terms quantal response and graded response which are equivalent to response and effect, respectively, as defined above.

The relationship between dose and response or effect is often depicted graphically using log (dose) against response or effect arithmetically, or using a probit scale. Return to arithmetic versus arithmetic is also useful in dealing with extrapolation from high to low dose levels. Examples of plots using these scales are shown in Figure 2.2. Use of the log scales has facilitated calculation of values such as the dose calculated to cause lethality in half the exposed population (LD_{50}), the dose calculated to cause a certain toxic outcome in 1% of those exposed (TD_1) and so on. However, they have not proved useful when it is necessary to extrapolate from the high doses used experimentally in small numbers of animals to the low doses experienced by humans, where it is desirable to be able to predict response rates of down to 0.0001%. This remains a contentious issue in contemporary toxicology and is of primary concern when dealing with endpoints such as carcinogenicity as there is debate about the existence of a threshold for this event. The existence of a threshold means that there is a dose below which there will be no effect. However, when there may be no threshold there is a need to extrapolate to very low exposure levels and the shape of the curve is unknown at this point. Figure 2.3 illustrates some possible alternatives. In order that experimental data can be accurate from a statistical point of view huge numbers of experimental animals would be required – tens or hundreds of thousands for each chemical. The only alternative that seems to be available currently is to continue to use the high-dose protocol and extrapolate cautiously due to the added uncertainty. If known, factors relating to nonlinear kinetics due to saturation of metabolic pathways must be incorporated into the evaluation process as these may be important in allowing a more accurate determination of appropriate exposure levels.

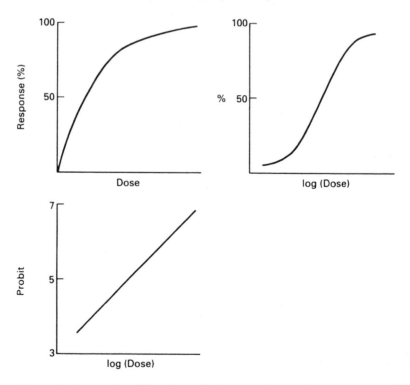

Figure 2.2 Examples of the relationship between dose and response using different scales.

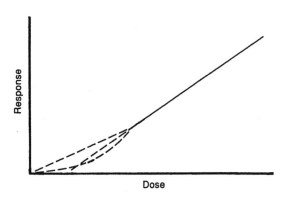

Figure 2.3 Possible alternative ways of extrapolating to very low levels of response.

Extrapolation across species

Extrapolation of data from experimental animals such as rats to humans is central to the use of toxicological data. However, there are undoubtedly many exceptions to this association and one must be circumspect in extrapolating without due thought being given to the undertaking. In fact there have been so many differences documented that there is serious reservation by some about accepting any animal data for extrapolation to humans. While, as noted above, caution is required in undertaking the process, one should not lose the overall perspective. In this regard the author challenges the sceptic to think of general groups of chemicals, pick one or two chemicals at random and compare toxicity in rats and humans along the lines in Table 2.2.

It is interesting to do this, and to illustrate the point some of the examples in Table 2.2 will be used. More details on these will be found in subsequent relevant chapters. Therefore, only general aspects will be raised here to draw out the similarities.

Cadmium is recognised as causing nephrotoxicity as one of its main toxic endpoints on long-term exposure of humans. In rats, the toxic response is also found to involve the kidneys. In both rats and humans, evidence of kidney damage is first

Table 2.2 Extrapolation across species	
Consider the cross-species sensitivity of	
Metals	Cadmium
Solvents	Trichloroethane
	n-Hexane
Pesticides	Organophosphate compounds
Plastic monomers	Vinyl chloride
Contaminants	Benzo-a-pyrene

seen from protein in the urine once the kidney cortex cadmium concentrations reach around 180 μg/g wet weight.

With 1,1,1-trichloroethane, as with most organic solvents, the toxic response of primary concern in humans relates to an effect on the central nervous system, especially for high levels of acute exposure where deaths have occurred. Similar observations are made in experimental animals where at high doses the narcotic effects of 1,1,1-trichloroethane are evident. With *n*-hexane the toxicity of concern is one of peripheral neuropathy, which has been found in exposed workers. Similar effects occur in exposed rats with clear biochemical and morphological correlates.

The mechanism by which organophosphates kill their target species – insects – is by inhibition of cholinesterases. In humans, decreases in the blood levels of cholinesterase in workers are used to determine when workers should be removed from exposure to these chemicals. Thus the same biochemical effect is being observed across species.

Vinyl chloride is the monomer used to make the very common plastic PVC (polyvinyl chloride). In the early 1970s it was found to cause haemangiosarcoma in the livers of rats. About 3 years later the same toxic endpoint was found in workers exposed to vinyl chloride during PVC manufacture. Thus the same very rare tumour was detected in rats and humans. Species specificity has now been seen with other plastic monomers, such as 1,3-butadiene (Himmelstein *et al.* 1996).

It should be reiterated that there are many examples of important differences across species and the possibility of such differences must be considered when applying animal data. Nevertheless, these differences should not detract from the continued recognition of the predictive ability of data from experimental animals to humans. It should also be appreciated that while there are often qualitative similarities across species for chemical toxicity data there are often quantitative differences. This is one sound reason for the inclusion of safety or uncertainty factors when using animal data to determine acceptable exposure levels for humans.

How the body affects chemicals and influences toxicity

As previously mentioned, in order that a toxic response be manifest there must first be exposure to the chemical. Then, unless the toxic response is local in nature (that is, at the point of first contact) the chemical will need to move to the site within the body where the toxic response occurs. Once in the body further events may be required for the toxic response, such as biotransformation, depletion of critical protective body molecules, interaction with key components in the cells, and so on. Elimination of the chemical will be necessary otherwise it will build up to toxic levels eventually. Then there will be other variables that may influence the toxic response. This section will deal briefly with how the body affects chemicals, especially with respect to the toxicity that they may incur. Figure 2.4 schematically represents the major relevant influences of the body on chemicals.

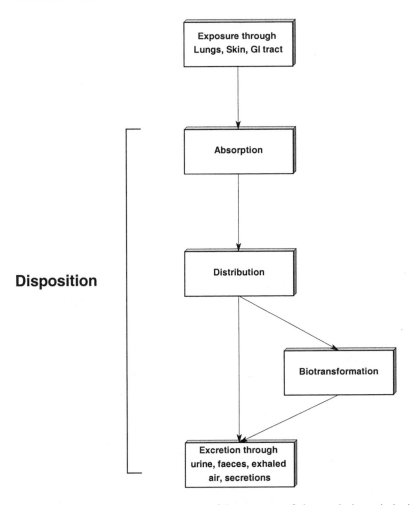

Figure 2.4 Schematic representation of the passage of chemicals through the body (GI: gastrointestinal tract).

Exposure

Both the route and duration of exposure are important with regard to toxicity. Major routes of exposure are shown in Figure 2.4. Those of primary concern in the workplace are the inhalational and dermal routes. Ingestion can be of importance if hands are contaminated, as can swallowing respiratory secretions after inhalational exposure. Inhalation of dusts, gases, vapours and aerosols from the workplace air is an obvious way of exposure by breathing, while the dermal route is less obvious. There are examples of situations where repeated air monitoring shows undetectable levels of a chemical, while urinary monitoring continually presents positive results (see the MOCA exercise in Chapter 17). Not all chemicals cross the skin, however. Indeed, the skin is generally a very efficient barrier to chemicals but there are some that cross readily and those people responsible for workplace health and safety need to be aware of this possibility.

When testing chemicals for their toxic properties the route of exposure chosen should reflect the most likely route by which humans would be exposed. This is not always the case as can be appreciated from the earlier section on comparison of routes. In experimental toxicology it is often convenient to use routes that are quite unusual such as the intravenous, subcutaneous or intraperitoneal routes. The reader should especially be familiar with the last of these (the intraperitoneal route is by injection into the peritoneal cavity) as it is quite often used as a convenient way of administering a chemical that will result in high absorption.

Duration of exposure needs to be considered from two aspects. Firstly, whether the exposure is one that occurs on a single occasion or whether there is an ongoing day by day exposure to (presumably) low levels of the chemical. Table 2.3 presents these types of exposure with examples. It should be noted that the terms used are similar to those for the test methods outlined in the section on Toxicity testing – descriptive toxicology.

Secondly, duration of exposure in an acute situation is particularly important in some workplace situations. For example, it may be possible to be exposed to a certain concentration of a chemical in the inspired air for, say, 15 minutes without serious health effects, but exposure to that concentration for, say, 2 hours may have serious consequences. Thus in such situations the duration of exposure becomes critical. This is, of course, related to the actual dose that the individual is exposed to. It should be noted that one way of expressing toxic or lethal concentrations in air is by using toxic time (TT) or lethal time (LT) values. For example, an LT_{50} of 60 minutes means that at the designated concentration of that chemical in air, exposure for 1 hour would be expected to result in death of 50% of those exposed.

When gathering toxicity information it becomes clear that much more acute toxicity data exist for oral exposures than for other routes, such as inhalational or dermal. This can be limiting when preparing information about chemicals for which the likely exposure route is via the lungs or the skin. Some extrapolation across routes can be undertaken, however, as determined by Winder and Agrawal (1995) who comprehensively compared acute toxicity by different routes of administration. Table 2.4 shows comparative classifications for acute toxicity for the three major routes of exposure.

Table 2.3 Duration of exposure

Nature of exposure	Example
Acute or single	Accident Major spill Maintenance in enclosed space
Subchronic or short-term repeated	Worker on process for e.g. >1 year
Chronic or long-term repeated	Worker on process for e.g. 20 years Exposure to pollution Pesticide in food Contaminant in drinking water

Table 2.4 Classification system for acute toxicity

Classification	Oral LD_{50} mg/kg	Dermal LD_{50} mg/kg	Inhalational LC_{50} mg/m³
Supertoxic	<5	<250	<250
Extremely toxic	5–50	250–1000	250–1000
Very toxic	50–500	1000–3000	1000–10,000
Moderately toxic	500–5000	3000–10,000	10,000–30,000
Slightly toxic	>5000	>10,000	>30,000

Disposition

Once there is exposure to a chemical other factors become involved in the processes that relate to the eventual expression of a toxic response. Collectively these are often referred to as 'disposition', which includes absorption, distribution, biotransformation and excretion of chemicals. These are included in Figure 2.4 as a general outline of the interaction between the body and chemicals. For each of the processes of absorption, distribution and excretion there are certain common aspects that govern the movement of the chemical molecules. They relate directly to the passage of the molecules across cellular membranes, which relates to the composition of the membranes consisting of a lipid bilayer containing protein molecules (see Figure 2.5). Both components play a role in the passage of molecules across the membrane to the inside of cells which is the site of the most crucial interactions for the actual toxic event.

Those chemicals that are readily soluble in oils (reflected in their oil/water partition coefficients) are able to easily dissolve in the lipid of the membrane which is in a fluid state. The chemical molecules can then move in the lipid and cross to the inside of the cell. This process is generally referred to as 'simple diffusion'. Factors affecting this simple diffusion apart from the oil/water partition coefficient of the chemical in question, include the concentration gradient and the area and thickness of the membrane. This process will only occur down a concentration gradient – that is, from high to low concentration. This raises another important concept for the process, however, in that the overall movement is dynamic and will occur in both directions across the membrane. Thus it is net movement down a concentration gradient that is being referred to. It should be appreciated that many chemicals exist in both ionised and nonionised forms and it is the latter that are more lipid soluble. Therefore the degree of ionisation under body conditions will be important for the amount that will cross membranes. There is also some passage by aqueous channels and by endocytosis but these are relatively minor pathways. Some protein components of the cell membrane are also involved in the movement of molecules to the interior (or exterior) of the cell. They act as carriers for the passage of particular chemical molecules across the membrane. The process is subject to saturation and to competition by like molecules (competitive inhibition). Both facilitated diffusion and active transport have these characteristics, being differentiated by the requirement of metabolic energy for active transport and the ability of this process to result in net movement against a concentration gradient.

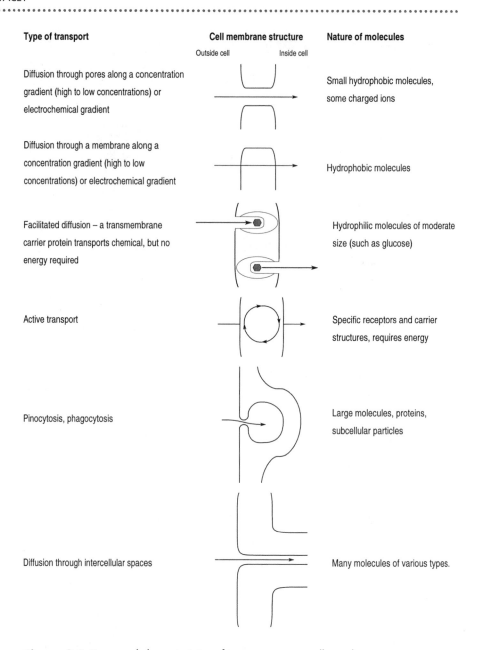

Figure 2.5 Types and characteristics of transport across cell membranes.

Absorption

Absorption is essentially the passage of chemical molecules across the membranes of the cells that separate internal and external aspects of the body. Once across these cell membranes they enter the blood that will then distribute the chemical to other parts of the body (see Figure 2.5). While there are similarities for absorption at the major sites – gastrointestinal tract, lungs and skin – there are also some important differences. Absorption from the gastrointestinal tract occurs along its entire length although mostly in the upper parts. As there are varying pH values from, for example, 1–2 in the stomach to about 8 in the small intestine, chemicals that exist in an ionised/nonionised equilibrium will be differentially absorbed. Other factors include the amount of surface area available and the effects of digestive enzymes and intestinal flora on chemicals. As mentioned already, the lungs are the most important site for absorption of chemicals from workplace exposures. Factors that predispose to rapid absorption in the lungs include their large surface area, the high blood flow to them, and the close proximity of the alveolar air to the blood. Other factors such as size and dimensions of dusts and aerosols are important with regard to how far into the respiratory system they penetrate, which affects their absorption substantially. These aspects in particular will be covered in greater detail in the chapters covering the respiratory system (Chapter 4) and particulate matter (Chapter 15). Similarly, further information on dermal absorption is to be found in the chapter on occupational skin diseases (Chapter 5). Suffice to reiterate here that the skin presents a considerably greater barrier to absorption than either of the other two routes outlined above. The chemical molecules need to pass a large number of cells to reach the blood that inhibits the process.

Distribution

Distribution is the movement of the chemical molecules around the body from the site of absorption to the various tissues. Factors influencing this process include the blood supply to the tissue, movement across capillary beds and, of course, movement across the cell membranes. Concentration in different tissues will depend on factors such as binding to cellular constituents (for example, metallothionein), exchange into the tissue (for example accumulation of lead and fluoride in bone), active transport in that tissue, and high fat solubility. It should be appreciated that the brain is protected by extra barriers that inhibit access by many chemicals. Once an initial distribution has occurred there will be ongoing redistribution back to the blood to other tissues and so on as the situation is dynamic with some movement or exchange always occurring. Redistribution is readily apparent in the kinetics of some chemicals such as lead, which has an initial distribution to red blood cells but later is concentrated in bone.

Excretion

Excretion serves to remove chemicals from the body. This is important because if this did not occur chemicals would accumulate in the body until toxic and/or lethal levels were reached. The main routes of excretion are via the kidney in the urine, via the gastrointestinal tract in the faeces (from nonabsorbed material or the bile), via the expired air, or to a minor extent via other bodily secretions such as sweat, saliva, semen and milk. These are included in Figure 2.4. The urine is the most important route of excretion as many water-soluble chemicals and/or their metabolites are eliminated in this way. The blood is filtered at the glomerulus of the nephron with the plasma (minus high molecular weight proteins) forming the filtrate. This fluid passes to the tubules where components are subject to reabsorption across the membranes of the cells lining the tubules. If this happens, as it will with liposoluble compounds, the chemical re-enters the blood and will not be excreted. If the molecules are not reabsorbed they will remain in the urinary fluid, pass to the bladder and eventually be voided with the urine and hence excreted. Thus the more water soluble the molecule is at this stage the more likely it is that it will be excreted. This is important when we consider biotransformation (in the next section) as one overall outcome of biotransformation is to make metabolites more water soluble and thereby promote excretion from the body. Some molecules are actively secreted by the cells of the kidney tubules but this is not an important aspect of excretion in general. It should be noted that the pH of urine is variable and that the excretion of chemicals that are subject to an equilibrium of ionised to nonionised will thus be affected. The other routes of excretion are of lesser importance, although for some metals the biliary route is important and for some volatile chemicals excretion across the lungs is a major elimination pathway.

Biotransformation

This process, which is also sometimes called metabolism, is rather distinct from the other components of disposition. While molecules must attain certain intracellular sites to undergo biotransformation the process is biochemically quite different. However, it should be noted that biotransformation is an important component in the handling of chemicals by the body because it can eliminate the parent molecule by changing its structure. This performs a similar function to excretion because that parent molecule no longer exists in the body. Nevertheless, the metabolite is still present and therefore needs to be dealt with.

Further to the elimination of a molecule by changing its structure to a metabolite, the advantage to the body of the process of biotransformation is to change the parent molecules to forms that are more readily excreted. Overall, biotransformation can be viewed as turning liposoluble compounds into more water-soluble forms. As noted above this promotes the excretion of the chemicals from the body, which is necessary so that they do not build up to toxic levels. Thus in general terms biotransformation may be regarded as a detoxifying process.

Biotransformation is usually separated into two parts – phase I and phase II. They are generally regarded as being linked in that the products of phase I often provide substrates for phase II (see Figure 2.6). A representative example of this process is shown in Figure 2.7.

Phase I biotransformation involves oxidation, reduction and hydrolysis reactions on the parent chemical. This often adds or exposes functional groups which then may assist in more phase I biotransformation reactions, or provides a substrate for phase II biotransformation reactions.

The enzyme systems responsible for biotransformation are found primarily in the liver although lower amounts of these enzymes exist in several other organs including the lung, small intestine, adrenal cortex and kidney. They not only biotransform exogenous chemicals but also various endogenous moieties such as steroidal hormones, bile acids and fat-soluble vitamins. The main site and disparate function of these enzyme systems may relate (in an evolutionary sense) to the wide range of chemical structures found in the foods human beings eat, remembering that blood containing absorbed foods goes through the liver before reaching the general circulation.

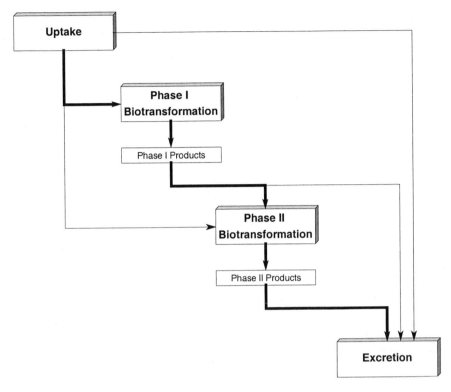

Figure 2.6 Biotransformation of exogenous compounds.

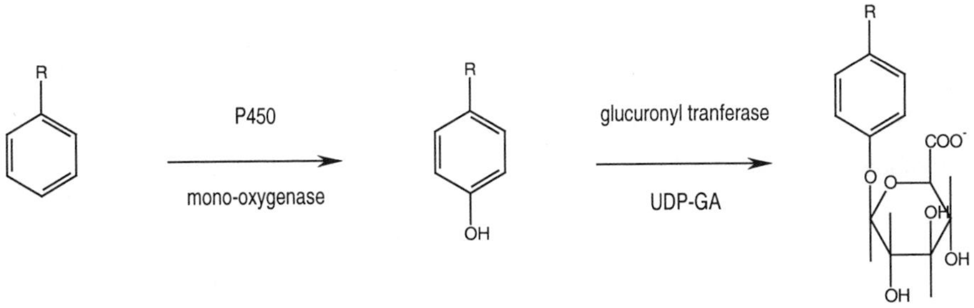

Figure 2.7 Representative examples of phase I and phase II biotransformation.

Phase I biotransformation is mainly carried out by a family of enzymes known as the cytochrome P450 system which is also referred to as the P450 mono-oxygenases, or the mixed-function oxidases. They are found mainly in the endoplasmic reticulum of the liver and in lower amounts in other organs. They consist of many similar proteins with varying substrate specificities and, as the name implies, catalyse largely oxidative reactions. This is not always the case, however, as some reactions are reductive. The enzymes are subject to induction by a variety of chemicals and also to inhibition.

Examples of the different sorts of phase I biotransformation reactions are shown in Table 2.5.

Table 2.5 Reactions catalysed by cytochrome P450

Reaction	Example	
	Reaction	Product(s)

Oxidative reactions
Hydroxylation

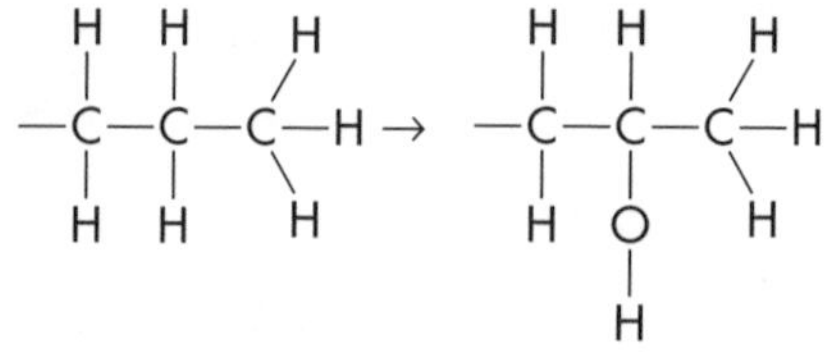

Epoxidation

Deamination

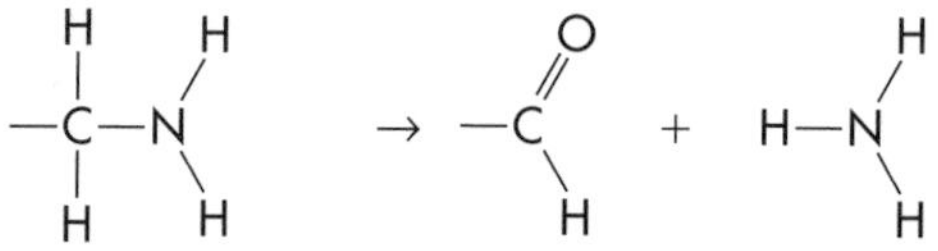

N-Dealkylation

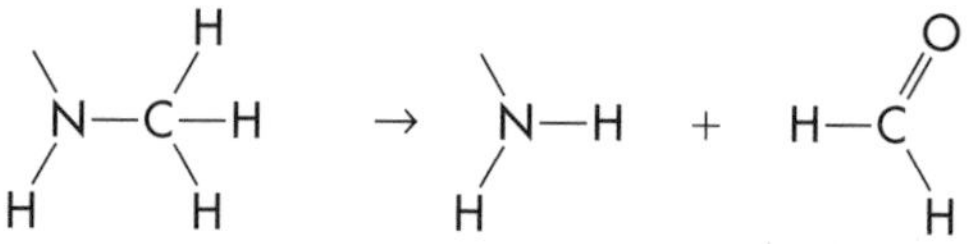

N-Hydroxylation

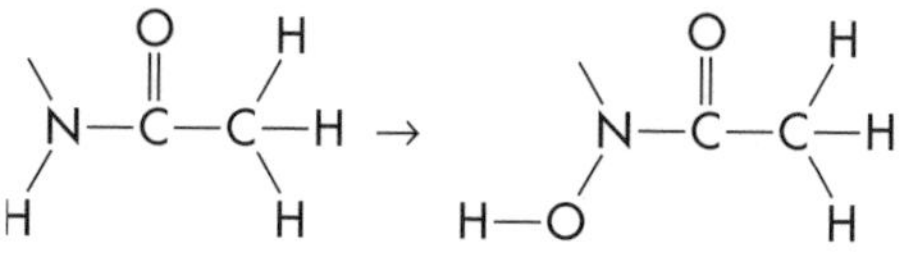

Dehalogenation (oxidative)*

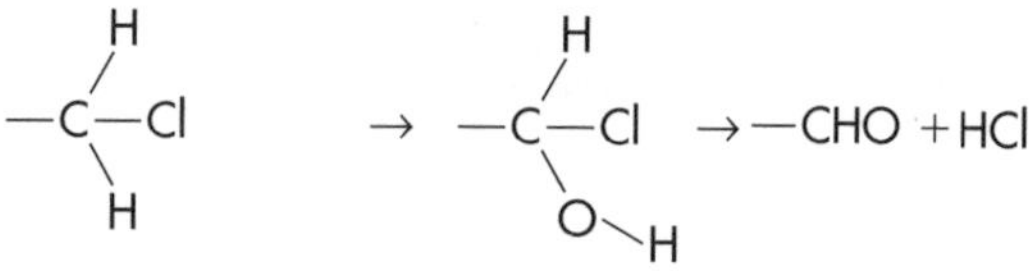

O-Dealkylation

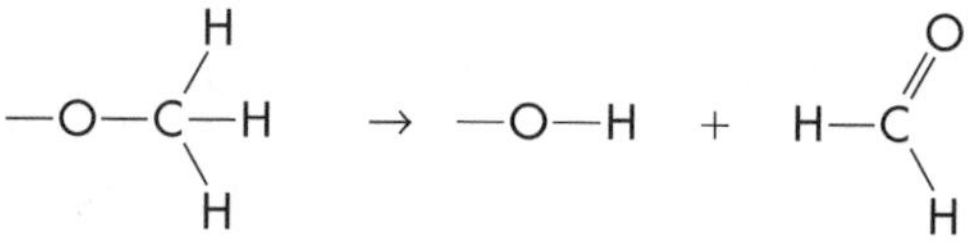

Desulfuration

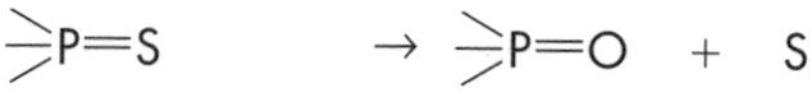

Sulfoxidation

S-Dealkylation

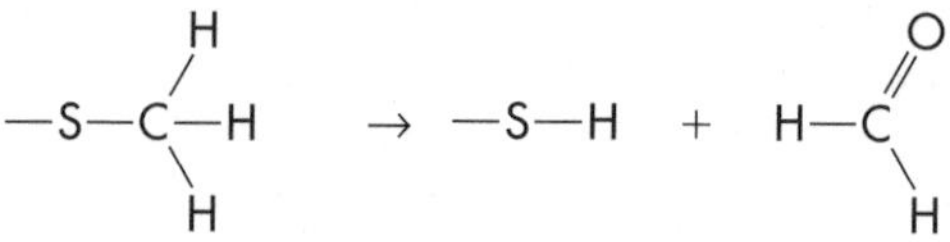

Table 2.5 continued		
Reaction	**Example**	
	Reaction	Product(s)
Reductive reactions		
Dehalogenation (reductive)	$-\overset{\displaystyle Cl}{\underset{\displaystyle Cl}{C}}-Cl$	$\rightarrow$ $-\overset{\displaystyle Cl}{\underset{\displaystyle Cl}{C}}-H$ $+$ $H-Cl$
Azo-reduction	$-N{=}N-$	$\rightarrow$ $-\overset{\displaystyle H}{N}$ $+$ $\overset{\displaystyle H}{N}-$
		$ \underset{\displaystyle H}{} \underset{\displaystyle H}{}$
Nitro-reduction	$-NO_2$	$\rightarrow$ $-NH_2$

*The chlorine can be replaced by any halogen.

While the process of biotransformation is generally one of advantage to the organism from the toxicity point of view there are also many examples of toxicity where it is the metabolite that is the offending moiety rather than the parent molecule. One example of this is with the formation of epoxides (an oxygen bridge across two carbons) which are very reactive and have been implicated as important in the process of carcinogenicity. The genotoxicity of 1,3-butadiene, for example, involves the formation of epoxide metabolites. Other phase I biotransformations are carried out by enzymes other than those of the P450 system, such as the flavin-containing monooxygenase, epoxide hydrolase, esterases, amidases, and alcohol and aldehyde dehydrogenases. Some substrates are also metabolised by a peroxidase-dependent co-oxidation. To extend the above example with epoxides, the epoxide hydrolase enzyme may biotransform the epoxide to a dihydrodiol that may be further biotransformed to the diol epoxide, which may play a critical role in carcinogenesis. As can be appreciated from this one example, biotransformation can be a complex process with many steps to either detoxification or, in some cases, to toxic moieties.

Phase II biotransformation generally results in the addition of a relatively large moiety to the substrate, which is often the product of a phase I reaction. Examples of the different sorts of phase II biotransformation reactions include:

- Glucuronyl transferase
- Glutathione S-transferase
- Sulfotransferase
- Amino acid conjugase
- Methyl transferase
- N-Acetyl transferase.

An example of the most common of these reactions, glucuronidation, is shown in Figure 2.7. The products of phase II reactions are often far more water soluble than the substrates, thereby promoting the overall excretion of the chemical. The glucuronidation process uses uridine-5′-diphospho-α-D-glucuronic acid (UDP-GA) as a cofactor and is carried out mainly in the endoplasmic reticulum of the liver.

Another common phase II reaction is via the enzyme glutathione S-transferase, which is actually a family of enzymes responsible for the eventual formation of mercapturic acid or thioether metabolites. These have been determined in urine in biological monitoring for exposure to some chemicals. As the name implies the cofactor for the enzyme is the tripeptide, reduced glutathione. This is a widely studied reactant in the detoxification of a variety of chemicals. If its supply is limited increased toxicity may be observed.

While it has been stated that the biotransformation of chemicals overall is a detoxification process, it must be remembered that many chemicals are toxic because they are biotransformed; that is the metabolites of some chemicals actually cause the toxicity. This has been extensively studied and the reader is referred to other texts for further information. As it is important to understand how biotransformation can result in a toxic product, and as it is also important that the reader appreciates the value of such knowledge, a salient example is to be found with n-hexane in Chapter 14 on solvents.

Toxicokinetics

The quantitative study and mathematical description of the disposition of chemicals related to their toxic effects is termed toxicokinetics. Thus the amount absorbed, the relative distribution to different tissues, the rate of removal by biotransformation and/or excretion are all included in a toxicokinetic appraisal of the movement of a chemical through the body. While there will be no attempt to detail the field of toxicokinetics in this text it should be appreciated by the reader that important information is attainable from understanding the kinetics of chemicals in the body. For example, the amount absorbed will be a factor in the toxicity of a chemical – if none is absorbed then there will be no systemic toxicity at a site removed from the portal of entry. The reader should be warned not to oversimplify this, however, as lack of absorption of particles like nickel subsulfide from the lung may in fact be a major factor in the toxicity expressed at that site. That is, no absorption does not simply equate to no toxicity.

Knowledge of the toxicokinetics of a chemical also allows prediction of body burdens of a chemical, both temporally and quantitatively, even after exposure has stopped. It is also crucial to know about the kinetics (and biotransformation) of a chemical when planning or implementing urinary monitoring programs. This can be clearly appreciated by considering this aspect in the chlordimeform exercise in Chapter 19.

The application of toxicokinetics is also important in the risk assessment process both in extrapolating from high dose to low dose and across species. Knowledge of saturation of kinetic processes at high dose as compared to low dose must be integrated into the risk assessment process if it is known to occur. Similarly, differences in the quantitative handling of chemicals across species can be crucial to a meaningful risk assessment for a chemical. It should be appreciated, however, that often critical pieces of data, such as the kinetics in humans, may not be readily available or obtainable for ethical reasons.

How chemicals affect the body

There are different types of toxic effects that chemicals have on the body as listed in Table 2.6. **Local** effects, as opposed to **systemic** effects, are discussed elsewhere in this text (Chapter 3). The differences between **acute** and **chronic** effects have been addressed earlier in this chapter. This generally relates to effects as well where there may be a reaction due to a single one-off exposure, perhaps to a high amount of a chemical. On the other hand a toxic response to another chemical may be seen after repeated long-term, low-level exposures as is usually the case with carcinogenic effects. Most chemicals will result in **immediate** effects but for some there will be a delay before the toxicity is manifest. An example of both can be found with paraquat where at a high enough dose death will be immediate, but at a moderate dose death is delayed until about 2 weeks after intoxication. The delayed death is due to pulmonary insufficiency related to fibrosis in the lungs.

Many toxic effects, like death, are **irreversible** while others may be **reversed** – for example, by regeneration of liver tissue after toxic insult to that organ. While ultimately all toxicity is related to interactions at the **molecular** level, the result may be visible only at the **cellular** level. For example, liver toxicity may be seen as death of the parenchymal cells but may be related to destruction of cellular lipids or covalent binding of chemical molecules to cellular macromolecules.

There are also specific types of effects that require some attention. **Allergic** reactions are not uncommon and are of concern with workplace chemicals such as toluene diisocyanate (TDI). Here one individual may become sensitive to TDI, reacting to very low levels, while a workmate remains unaffected. **Hypersensitivity** or **idiosyncratic** responses are those seen in particular individuals and are related to some underlying metabolic abnormality. They occur at exposure levels well below those at which the 'normal' population would respond. The term is often used to account for otherwise inexplicable observations in particular people. **Carcinogenic**, **genotoxic** and **developmental** effects are specific types of effects that are dealt with in separate chapters.

A number of other variables can influence the toxic response to a chemical

---

Table 2.6 Types of effects of chemicals

Local/systemic
Acute/chronic
Immediate/delayed
Reversible/irreversible
Cellular/molecular
Allergic
Hypersensitivity/idiosyncratic
Carcinogenic/genotoxic
Developmental

which should be kept in mind. These include impurities in the chemical in question, especially if it is a commercial formulation, the vehicle that may be used in the formulation, characteristics of the individual like sex, age, disease and immunological status, and other environmental factors.

Toxicity testing – descriptive toxicology

In order that information on the toxicity of chemicals is available for the risk assessment process tests need to be carried out (Hayes 2001). They will be discussed only superficially here and the reader is referred to other sources for further information – in particular to the OECD guidelines for testing of chemicals (OECD 1993). These are detailed procedures covering the tests required for general marketing of chemicals. As well as these, there will be requirements in individual countries that may vary among jurisdictions (covered in more detail in Chapter 19). Readers should also become familiar with any special requirements that their country may have. The information that is usually required includes:

- Physical and chemical properties
- Production, use and disposal
- Toxicological (mammalian) data
- Ecotoxicological data.

It is the third of these which is of concern here – that is, a description of the toxicity of the chemical in mammals with the most common test species being the rat.

The major general testing protocols are related to the duration of exposure as shown in Table 2.7 with duration given in terms as related to rats.

It should be noted that there is some confusion in the use of these terms, although they are consistent with their use by the OECD. 'Subacute' has also been used – in some instances as the equivalent to subchronic, and by others to mean the same as short-term repeated dose (STRD). Thus care is required when considering data of this nature. The safest approach is to refer to the actual duration of exposure to the chemical.

The general information sought under the acute toxicity testing protocols includes that listed in Table 2.8. While the first six of these constitute the 'acute toxicity package' or 'six pack', it is surprising how few chemicals with exposure standards (that is, those encountered commonly in many workplaces) have data for each of these endpoints (Agrawal and Winder 1996). It should also be noted that

Table 2.7 Testing protocols for mammalian toxicity

Name	Duration (for rats)
Acute	Single
Short-term repeated dose (STRD)	14–28 days
Subchronic	Up to 90 days
Chronic	6–30 months

Table 2.8 Acute toxicity testing

Toxic endpoint	Route of administration
Lethality	Oral
	Inhalational
	Dermal
Irritation	Skin
	Eye
Sensitisation	Skin
	Respiratory

there is no generally accepted toxicity test method for determining respiratory sensitisation even when there are examples where this has been a significant issue in the workplace (for example, toluene diisocyanate).

Thus the information will include estimates of the amount of chemical that might result in death of exposed individuals by various routes, an indication as to how damaging skin or eye contact of the chemical would be, and determination of the possibility that a chemical may result in sensitisation of individuals. It should be noted that these last tests are not well developed and it is not common to test specifically for the potential for respiratory sensitisation. However, it is included here because of its importance to the workplace.

While details of the tests are not provided it should be appreciated that there is a single administration of chemical for the acute lethality studies (for a stated duration for inhalation) with subsequent observation of the animals for 7–14 days.

The STRD, subchronic and chronic protocols all involve administration of the test chemical repeatedly for the time specified. The route of exposure should be that likely for humans and there should be a control group plus groups exposed to increasing doses. There should be several animals per group with larger numbers in the chronic protocols. Animals are assessed for a number of parameters, including:

- Mortality
- Physical examination
- Body weight
- Diet consumption
- Hematology
- Clinical chemistry
- Gross and microscopic pathology (including organ weight).

It should also be noted that it is a prudent use of resources to combine chronic toxicity studies with studies to determine the oncogenicity of chemicals. The tests cost millions of dollars so it is economically rational to do this. However, it must also be appreciated that the aims of the two are somewhat different. The protocol for chronic toxicity seeks primarily to determine a level of exposure at which no toxic effects are observed – that is, a no observed effect level (NOEL), while that for oncogenicity seeks to push the dose as high as practicable without other toxic effects to provide the best chance of detecting an oncogenic effect.

As well as the standard test protocols that have been outlined above there are some speciality tests that require individual test procedures. These are listed in Table 2.9.

There will be still more information on certain chemicals used for particular purposes that may be appropriate. Thus it should be remembered that the assessment of the toxicity of chemicals is more than a routine checklist approach. Flexibility should be retained to allow for the provision of additional information as appropriate. Similarly there may be occasions when information for one or other aspect is unnecessary.

Interactions between chemicals

As well as the effects of single effects on the body, consideration in toxicology also needs to be given to the problems of the effects of exposure to combinations of chemicals, and the effects such interactions may cause. The interaction can cause an increase in effect, in that the effect is more than the individual effects added together, or it can be decreased, where the effect of two chemicals is less. These can be expressed mathematically (see Table 2.10).

Table 2.9 Speciality tests

Reproduction and fertility
Developmental toxicity (including teratogenicity)
Genotoxicity
Neurotoxicity
Behavioural toxicity
Immunotoxicity
Toxicokinetics
Antidote studies

Table 2.10 Chemical interactions

Type of interaction	Numerical example
Independent interactions*	$2 + 4 = 4$
Subtractive or divisive interactions (antagonism)[†]	$0 + 3 = 1$
	$2 + 4 = 3$
	$-4 + 4 = 0$
Additive interactions	$2 + 4 = 6$
Potentiation[†]	$0 + 3 = 9$
Multiplicative interactions (synergism)	$2 + 4 = 25$

*Note: The interaction is equal to the largest single effect.
[†]Note: The number 0 indicates absorption, but no toxic effect.

The most dangerous interaction of combinations of chemicals is where the effects multiply. This interaction is called 'synergism' (see Table 2.10) and the effect may be many more times than can be predicted by the individual effects. For example, the synergistic interaction between asbestos exposure and tobacco smoking is well established: if it assumed that the background rate of lung cancer in nonsmokers is equal to one, then smokers have eleven times the background rate of lung cancer, asbestos workers five times the background rate, and smoking asbestos workers (that is, they have both exposures) up to one hundred times the background rate. This is a huge interaction. In other cases, a strong synergistic effect may be observed where the effect of the individual chemical is usually so small it is not noticeable.

Of what is known about interactions between chemical exposures, it seems that many interactions, where they have been reported, are of the additive type. Indeed, it is common practice in occupational hygiene to assume that the interaction of multiple exposures to chemicals at work can be estimated additively by summing exposures as a fraction of their exposure standards.

Toxicological information

One aspect of toxicology that everyone involved in the science needs is information. The researcher must be aware of what is already known, the regulator of the extent of information available for assessment, and so on.

Several levels of toxicity information are available. These include the primary literature, reviews, books and computerised databases. Less available, for reasons of confidentiality, are data in government or company files. The primary literature consists of reports of original research published in scientific journals. Reviews are generally found in journals both as covering articles and in publications dedicated to reviewing information in particular fields. There are now several textbooks in the field of toxicology – some covering the science in general, others focusing on one or other aspect of toxicology. The most comprehensive text for toxicology in general remains *Casarett and Doull's Toxicology* (Klaassen *et al.* 2001) although *Principles and Methods of Toxicology* (Hayes 2001), *Toxicology: Principles and Applications* (Niesink *et al.* 1996), *Handbook of Human Toxicology* (Massaro 1997), *Encyclopedia of Toxicology* (Wexler and Gad 1998) are readable and authoritative texts. *Patty's Industrial Hygiene and Toxicology* (Harris 2000) consists of several volumes and contains a large amount of toxicological information on chemicals. In the medical toxicology area *Ellenhorn's Medical Toxicology: Diagnosis and Treatment of Human Poisoning* (Ellenhorn *et al.* 1997) remains an outstanding text.

Several computerised databases are also available for seeking toxicological data on chemicals. While they were once mostly available through substantial libraries they are now being marketed for more personal use. Many are now freely available over the Internet. Some of the useful databases include RTECS, TOXNET, TOXLINE, HSELINE, NIOSHTIC, CISDOC and CCINFO.

For a comprehensive listing of sources of toxicological information the reader is referred to a publication by Wexler *et al.* (2000) dedicated to this topic.

References

Agrawal, M.R. and Winder, C. (1996) The frequency and occurrence of LD_{50} values for materials in the workplace. *J. Appl. Toxicol.* 16: 407–422.

Dayan, A.D. (1981) The troubled toxicologist. *Trends Pharmacol. Sci.* 2: 1–4.

Doull, J. (1984) The past, present, and future of toxicology. *Pharmacol. Rev.* 36: 15S–18S.

Ellenhorn, M.J., Schonwald, S., Ordog, G. and Wasserberger, J. (eds) (1997) *Ellenhorn's Medical Toxicology: Diagnosis and Treatment of Human Poisoning*, 2nd edn. New York: Williams and Wilkins.

Harris, R.L. (ed.) (2000) *Patty's Industrial Hygiene*, 5th edn. New York: John Wiley and Sons.

Hayes, A.W. (ed.) (2001) *Principles and Methods of Toxicology*, 4th edn. New York: Taylor and Francis.

Himmelstein, M.W., Turner, M.J., Asgharian., B. and Bond, J.A. (1996) Metabolism of 1,3-butadiene: inhalation pharmacokinetics and tissue dosimetry of butadiene epoxides in rats and mice. *Toxicology* 113: 306–309.

IPCS (1994) *Assessing Human Health Risks of Chemicals: Derivation of Guidance Values for Health Based Exposure Limits*. In: Environmental Health Criteria 170: World Health Organisation International Program for Chemical Safety, Geneva.

Klaassen, C.D., Amdur, M.O. and Doull, J. (eds) (1996) *Casarett and Doull's Toxicology. The Basic Science of Poisons*, 6th edn. New York: McGraw-Hill.

Malmfors, T. (1981) Toxicology as science. *Trends Pharmacol. Sci.* 2: I.

Massaro, E.J. (ed.) (1997) *Handbook of Human Toxicology*. Boca Raton: CRC Press.

OECD (1993) *OECD Guidelines for the Testing of Chemicals*. Paris: Organisation for Economic Cooperation and Development.

Niesink, R.J.M., de Vries, J. and Hiollinger, M.A. (eds) (1996) *Toxicology: Principles and Applications*. Boca Raton: CRC Press.

Schlosser, P.M., Bond, J.A. and Medinsky, M.A. (1993) Benzene and phenol metabolism by mouse and rat liver microsomes. *Carcinogenesis* 14: 2477–2486.

Wexler, P., Hakkinen, P.J., Kennedy, G. and Stoss, F. (eds) (2000) *Information Resources in Toxicology*, 3rd edn. New York: Elsevier.

Wexler, P. and Gad, S.C. (eds) (1998) *Encyclopedia of Toxicology*. New York: Academic Press.

Winder, C. and Agrawal, M.R. (1995) Oral, dermal and inhalational toxicities: Classification of hazard. *Chem. Health Safety* 2: 31–34.

Targets of chemicals

Chapter 3

Systemic toxicology

W.M. Haschek, N.H. Stacey and C. Winder

Introduction

This section deals with the important effects of chemicals on various organs or tissues within the body not considered elsewhere in this text. The range of potential health effects from chemical hazards is extensive. They can be divided into specific or nonspecific effects, and local or systemic effects. While this chapter will concentrate on specific and systemic effects on internal organ systems and tissues, a short description of nonspecific and local effects will be provided.

Nonspecific versus specific effects

Most cells have very similar molecular and biochemical processes like protein synthesis, cellular respiration, and anabolic and catabolic processes. Some chemicals affect these general processes, and will therefore affect many different cell types should they reach effective concentrations in such cells. For example, cyanide will block energy metabolism in any cell into which it is absorbed, and the toxicity cannot be described as specific to a particular organ or tissue type. When this occurs, the toxicity is considered 'nonspecific' (Chaisuksant *et al.* 1999). On the other hand, specific toxicity refers to injury to cell types with characteristics that make them more susceptible to injury than the general population. For example, bronchiolar nonciliated epithelial cells (Clara cells) in rodents contain high concentrations of certain P450 metabolising enzymes that allow metabolism of compounds such as 3-methylfuran to the proximate toxicant, resulting in death of these cells while the adjacent ciliated cells are unaffected.

Local toxicity

When a chemical reaches the body, it interacts with the tissue that it first comes in contact with (Stoughton 1974). In the occupational environment, these tissues or organs are the lungs, the skin, the digestive tract and, in some cases, the eye. If this interaction is adverse, local toxicity will occur. Toxic effects may be slight, such as skin redness or an increase in nasal discharge of mucous, or severe, such as acid-based skin corrosion or asphyxiation. Local effects are described in Table 3.1.

Local toxicity commonly occurs with workplace chemicals due to the pattern of usage and likely routes of exposure. Absence of suitable control measures can increase the occurrence of local toxicity. Both nonspecific and target organ toxicity are important toxicological concepts.

Systemic toxicity

Toxic chemicals may affect specific organs, systems or tissues by virtue of some property of the chemical or some property of the organ/system/tissue. Unlike local toxicity, for systemic toxicity to occur, absorption and distribution of the agent are required since the effect is expressed at a site distant from exposure or contact. Most chemical substances cause systemic toxicity and some can cause both local and systemic toxicity. Most chemicals that produce systemic toxicity elicit their major effect on one or two organs. These sites of toxicity are called target organs.

Each organ or tissue in the body, or each function, can be individually affected. Some organ toxicities are common, such as hepatotoxicity (liver damage), nephrotoxicity (kidney damage), haematotoxicity (damage to the blood-forming tissues) or immunotoxicity (damage to the immune system). For example, the kidney is high in the protein metallothionein, which binds cadmium to allow cadmium to accumulate in the kidney. Once cadmium reaches a certain concentration it causes renal damage. The term 'primary' target organ refers to the organ directly or most severely affected by the toxin or toxicant, while 'secondary' target organs are those less severely or indirectly affected.

Chemicals are sometimes grouped with respect to their primary target organ for classification purposes. For example:

- cardiotoxicants are chemicals that affect the heart
- dermal toxicants are chemicals that affect the skin
- haematopoietic toxicants are chemicals that affect the blood-forming tissues
- hepatotoxicants are chemicals that affect the liver
- immunotoxicants are chemicals that affect the immune system
- nephrotoxicants are chemicals that affect the kidney
- neurotoxicants are chemicals that affect the nervous system
- ocular toxicants are chemicals that affect the eye
- respiratory or pulmonary toxicants are chemicals that affect the respiratory system or, more specifically, the lung
- reproductive toxicants are chemicals that affect the male or female reproductive system.

In addition, chemicals can be grouped according to other specific targets. For example:

Table 3.1 Types of local effects

Effects	Signs and symptoms	Toxicants
Irritation	Inflammation of the skin and mucous membranes of nose, eyes, mouth and upper respiratory tract. Related to the solubility of the substance onto skin or moist surfaces	Solvents (skin) Chlorine, ammonia, nitrogen dioxide and phosgene (mucous membranes)
Corrosiveness	Irreversible damage to tissue on contact	Hydrochloric, sulfuric and nitric acids and sodium hydroxide
Asphyxiation	A reduction in the concentration of oxygen in inspired air by physical displacement (simple asphyxiation) leading to hypoxia	Nitrogen and hydrogen
	A reduction in oxygen transport in the body by chemical reaction (chemical asphyxiation)	Carbon monoxide, hydrogen sulfide and cyanides
Skin damage (dermal toxicity)	Irritation, corrosive or allergic reactions The skin contains fairly high levels of fatty tissue and therefore chemicals which can dissolve fat will cause defatting, which can lead to drying of the skin, cracking and possibly dermatitis	Any corrosive, irritating or skin allergenic chemicals Organic solvents
Damage to the respiratory system (respiratory or pulmonary toxicity)	Irritation of the airways, sneezing, nose bleeds, wheezing, coughing, obstruction of airway passages and sinusitis	Sulfur dioxide
	Bronchoconstriction and epithelial injury Asthma (reversible bronchoconstriction)	Ammonia Isocyanates, formaldehyde
	Increase in secretion of mucous Damage to cellular components of airways and alveoli	Sulfur dioxide, tobacco smoke Hydrogen chloride, phosgene
Eye damage (ocular toxicity)	Ocular irritation, lacrimation, conjunctival irritation/ conjunctivitis, corneal damage and iritis	Bleaches, detergents, acids and alkalis
Damage to the digestive system (gastrointestinal toxicity)	Gastric irritation, nausea, diarrhoea, abdominal pain, colic	Ingestion of pesticides, chlorinated hydrocarbons, food allergens and metal salts

- teratogens are chemicals that affect the development of the organism
- carcinogens are chemicals that cause cancer
- genotoxicants are chemicals that affect genetic material (also called mutagens).

This is one convenient way of splitting the toxic properties of chemicals into manageable and meaningful subunits. Furthermore, this has a useful purpose clinically in that specific organ dysfunction is most likely to be seen at clinical presentation. In conjunction with a good occupational history, it should be possible to determine if the injury is related to workplace exposure to a chemical. However, chemical-induced organ dysfunction must be differentiated from that due to other causes because each organ has a limited manner in which it can respond to injury. Therefore, detection of the chemical or its biotransformation products in body fluids or tissues at levels associated with toxicity, as well as evidence of toxicity compatible with exposure to the particular xenobiotic, are needed to make a diagnosis of occupational or environmental disease.

The ultimate expression of toxicity at any particular target depends on a variety of factors as discussed in some detail in Chapter 2. These include the nature of the chemical itself, the route of exposure, the disposition of the chemical, including any biotransformation, and the sensitivity of the exposed organ or tissue.

Any organ of the body is a potential target for injurious effects from chemicals, but some are more susceptible to adverse effects than others. In the workplace setting, the skin and respiratory tract are the two systems most commonly affected since exposure is often by direct contact with the skin or by inhalation. Since the liver is an important organ for metabolism and excretion of absorbed chemicals, and the kidney is the major organ of excretion, individual chapters have been devoted to these organs. The occupational toxicology of the nervous and reproductive systems and developmental toxicity are also addressed in separate chapters as these systems have specific issues to be covered. Another potential target for chemicals is genetic material. Interaction of xenobiotics with DNA may result in cancer, a major concern with workplace chemicals. Therefore, two chapters are devoted to this very important aspect of occupational toxicology.

This chapter provides an overview of the systemic toxicology of the haematopoietic, immune, cardiovascular and gastrointestinal systems. Clearly, each of these is a topic in its own right. For a more detailed coverage of these (and other) target organs the reader is referred to texts such as the *Target Organ Toxicology Series* (Dixon), *Casarett and Doull's Toxicology* (Klaassen 2001), *Handbook of Toxicologic Pathology* (Haschek *et al.* 2002), or specific references used in subsequent sections.

Gastrointestinal system (W.M. Haschek)

Introduction

Although many industrial chemicals cause gastrointestinal symptoms and even tissue injury, little information is available regarding their specific toxic effects and even less regarding their mechanism of action. This may be due to the generally held view that the gastrointestinal tract is not an important site of action for toxic chemicals except when exposure is to corrosive agents and carcinogens.

The gastrointestinal tract is the site of entry for ingested chemicals, for inhaled chemicals that are removed from the lung by coughing or expectoration and then swallowed, and those excreted in the bile (Rozman and Klaassen 2001). Gastrointestinal toxicity may occur from ingested chemicals themselves, or secondarily by chemicals reaching the tract through the circulation or from the bile. Susceptibility of the gastrointestinal tract to injury is due to its barrier role, allowing direct contact with ingested chemicals; its marked absorptive capacity; and its high mucosal proliferative and metabolic rate. Protective mechanisms against chemicals include the low pH in the stomach, the mucous surface layer, and the presence of biotransforming enzymes. These enzymes are present both within the lining epithelium and within endogenous lumenal bacteria; they may metabolise chemicals to either reactive or less toxic intermediates. Reactive metabolites may be directly toxic to the gastrointestinal tract or may be absorbed into the circulation to reach other target organs. The intestines also contain large amounts of inert binding materials that can act as protective adsorbents.

Structure and function of the gastrointestinal tract

The gastrointestinal tract is a tubular organ whose wall consists of a mucosa, a submucosa, smooth muscle, and serosa. The mucosa is the primary site of chemical-induced injury. It consists of an epithelial lining, a lamina propria of vascularised connective tissue, and the muscularis mucosae, a thin layer of muscle that delineates the mucosal boundary. The epithelium that covers the lumenal surface of the mucosa functions as a selectively permeable barrier to substances within the lumen, and facilitates the transport and digestion of food, as well as the absorption of water and digestive products. Abundant lymphoid tissue is also present in the mucosa and submucosa; protection against infection is primarily through immunoglobulin A (IgA) production.

In the stomach, the mucosa is folded into rugae. The major types of epithelial cells are the mucous, parietal, and chief cells, which secrete large amounts of mucous, hydrochloric acid, and pepsinogen and lipase, respectively. The mucous protects the surface from the secreted acid as well as from ingested materials. In the small intestine, the mucosa is folded into villi with crypts present at the base. The crypt epithelial cells are the progenitor cell with a high mitotic rate. They differentiate as they move up the villus into nondividing absorptive cells which line the lumenal surface. The renewal rate for the mucosal epithelium is less than 5 days in humans. The crypt epithelial cells are very sensitive to compounds that interfere with cell division such as radiation and chemotherapeutic agents. Damage to these cells ultimately decreases the availability of absorptive cells, resulting in malabsorption as well as diarrhoea. The absorptive cells are rich in biotransforming enzymes, similar to those present in the liver, and therefore can metabolise chemicals. Resident bacteria can also metabolise ingested chemicals. The large intestine does not have villi, only crypts, with many mucous cells. It is the major site for water absorption from the intestinal contents and for mucous production.

Chemicals that enter the digestive tract may pass straight through and be excreted unchanged in the faeces; may be directly toxic to the mucosa; be absorbed, primarily by passive diffusion, into the epithelial cells; or may be metabolised by bacteria in the lumen. Once within the cell, a chemical may be directly toxic to the cell, metabolised to reactive or less toxic intermediates, or bound to protein before passing into the lymphatics or capillary bed. After passing through the liver, the absorbed chemical can then be excreted in the urine or the bile. Enterohepatic circulation occurs when the chemical excreted in the bile is absorbed from the intestines, re-enters the circulation, and is once more excreted by the liver into the bile. This process may be repeated several times resulting in concentration of the chemical and increased toxicity.

Classification of gastrointestinal toxicity

Classification of gastrointestinal toxicity can be based on the type of injury, as shown in Table 3.2, or on the pathophysiological or clinical response, as shown in Table 3.3. There are four major types of tissue responses to injury (see section on mechanisms below for additional information):

- ulceration (Rozman and Hanninen 1986)
- necrosis and inflammation (which often occur together)
- proliferation, including cancer.

Table 3.2 Classification of gastrointestinal toxicity based on tissue response

Ulceration
Necrosis
Inflammation
Proliferation

Table 3.3 Clinical manifestations of chemical-induced gastrointestinal injury

Clinical signs	Examples
Diarrhoea	Mercury
	Solvents
Vomiting	Arsenic
Constipation	Lead
Change in motility	Lead
	Organophosphorus compounds
Abdominal pain	Arsine
	Lead
	Bromochloromethane

Classification based on pathophysiological responses (see Table 3.3) is useful in the detection of injury, as discussed below. Diarrhoea is the most common presenting sign associated with toxicity and can result from increased mucosal permeability and exudation, hypersecretion, malabsorption, or abnormal motility of the gastrointestinal tract (Bertram 1991). The consequences include dehydration from loss of water, acidosis from loss of bicarbonate, and electrolyte imbalance. Vomiting can result from direct gastric irritation, an effect on the central nervous system, or liver damage. Constipation can result from functional abnormalities, including neurological effects, or a result of structural changes, such as the presence of a tumour. Changes in motility due to chemicals can result in spasms, diarrhoea or constipation. These can be due to neurological effects, either regional or central, as well as direct toxicity.

Detection and evaluation of gastrointestinal injury

Gastrointestinal symptoms, especially nausea and vomiting, are very widespread after exposure to chemicals. Most of the time these effects are regarded as secondary, with the intestine not considered a target organ, with injury to other organs, such as the liver (for example, chloroform-induced hepatotoxicity) overshadowing intestinal toxicity. Gastrointestinal injury can be detected based on clinical signs, as shown in Table 3.3; however, determination of a role for chemical exposure in the clinical disease syndrome relies on an accurate history of exposure and exclusion of other possible causes. Evaluation of chemical-induced gastrointestinal injury is similar to that used to diagnose gastrointestinal disease from other causes. Endoscopic examination, radiography, and biopsy may be useful. Measures of the rate of absorption and propulsive activity can also be obtained.

Mechanisms of gastrointestinal toxicity

The mechanisms of chemical-induced gastrointestinal injury are listed in Table 3.4. Insecticides, such as carbamate and organophosphorus compounds, and toxic nerve gases, such as sarin and soman, induce a pharmacologic effect by inhibiting acetylcholinesterase. Muscarinic cholinergic actions are the predominant effects

Table 3.4 Mechanisms of chemical-induced gastrointestinal injury

Pharmacologic effects
Alteration of absorptive functions
Decreased blood supply or hypoxia
Indirect damage to the mucosa
Direct cytotoxicity
    Parent compound
    Reactive metabolite (binding to macromolecules)
    Block prostaglandins/alter mucous
Hypersensitivity reactions
DNA damage

exhibited. These include anorexia, nausea, vomiting, abdominal cramps and diarrhoea. These chemicals may also increase motility and tone, and induce vomiting
due to central neurogenic effects (Scheufler and Rozman 1986). Other chemicals
can alter absorptive functions resulting in malabsorption. Abnormal absorption
can occur due to alteration in epithelial transport mechanisms, reduction of surface
area, or binding of chemicals to unabsorbed intestinal contents (Bertram 1991).
Increased absorption of allergens can also occur. Decreased blood supply and resultant hypoxia, such as that occurring in shock and hypotension, increase the susceptibility of the mucosa to injury and are frequently manifested as ulceration.
Alteration of the local protective mechanisms, such as mucous production and
inhibition of certain prostaglandins, is important in gastric ulceration induced by
steroidal and nonsteroidal anti-inflammatory drugs.

Direct toxicity to epithelial cells by the parent chemical or its metabolite can
result in necrosis and inflammation (gastritis, enteritis, colitis) with subsequent
malabsorption and maldigestion. Crypt epithelial cells are highly susceptible to
cytostatic chemicals such as epoxides, which are DNA alkylating agents, and radiation, which also damages DNA. Absorptive cells, on the other hand, are frequently injured by corrosive chemicals due to direct contact or chemicals that
require metabolic activation for toxicity. Aldehydes, for example, are potent local
irritants and phenols are highly corrosive agents on ingestion. Arsenic is cytotoxic
to the vasculature and results in increased capillary permeability. Immune-mediated hypersensitivity reactions, characterised by an inflammatory process, may also
occur; trinitrobenzene is one such example. DNA damage, such as that induced by
some nitrosamines, can lead to cancer. Since colon cancer is the second most
common cancer in humans, the contribution of chemicals to this process is under
active investigation.

Examples of gastrointestinal injury from workplace exposure

Ingestion of corrosive chemicals typically causes severe gastrointestinal damage.
Gastrointestinal damage can occur following ingestion of other compounds;
however, workplace exposure to chemicals more commonly occurs by inhalation
and dermal exposure. Table 3.5 lists chemicals known to cause gastrointestinal
injury.

Epidemiological studies show that workers in plants producing polyvinyl chloride and those exposed to asbestos have a higher incidence of intestinal cancers
than the control populations (Pfeiffer 1985). Lead poisoning can cause gastric
ulcers and gastric hypoacidity, while chronic mercury poisoning has been associated with diarrhoea and recurrent bouts of abdominal pain (ATSDR 1999a,
1999b). In addition, stomatitis (inflammation of the oral mucosa) is a classical sign
reported in both acute and chronic occupational metallic mercury poisoning. Colic
is a consistent early symptom of lead poisoning in occupationally exposed cases
(ATSDR 1999a). The rapidly dividing small intestinal crypt epithelial cells are
very susceptible to the cytotoxic effects of ionising radiation (Potter and Grant
1998). This leads to lack of replacement of absorptive cells with acute effects
including diarrhoea, and more chronic effects consisting of susceptibility to malabsorption syndromes, secondary bacterial infections, and fluid and electrolyte imbal-

Table 3.5 Examples of gastrointestinal injury from workplace exposure

Asbestos (peritoneal mesothelioma only)
Corrosive chemicals
Aliphatic chlorinated hydrocarbons
  Chloroform
  Carbon tetrachloride
Metals
  Lead
  Mercury
  Arsenic
  Cadmium
Vinyl chloride
Ionising radiation

ances. Nausea and vomiting are common signs of inhalation exposure of humans to chloroform and carbon tetrachloride.

Overall, there is very little information on the gastrointestinal toxicology of occupational exposures (DiPalma *et al.* 1991; Schiller 1979). Occupational and environmental diseases of the gastrointestinal tract are underemphasised (Banwell 1979), although associations between workplace exposure and peptic ulcer disease, pancreatitis, gastroenteritis, coeliac disease, and pneumatosis cystoides intestinalis have been noted (Fleming 1992). Cancers of the gastrointestinal tract due to occupational exposure have also been reported (Neugut and Wylie 1987), for example rectal cancer following exposure of workers to tetraethyl lead (ATSDR 1999a).

Haematopoietic system (W.M. Haschek)

Introduction

The incidence of chemically-induced injury to the haematopoietic system is low, even in populations chronically exposed to the well-documented haematotoxicant, benzene (Irons 1992). However, when injury does occur, it can be life-threatening because of the critical functions of the haematopoietic system. These include transport of oxygen, host resistance to infectious agents, and haemostasis. Chemicals reach the haematopoietic system through the circulation. Reasons for susceptibility to injury include heavy metabolic demands to meet the need for rapid production of cells and the dependence of this system on small numbers of stem cells with limited proliferative activity. In addition to direct toxicity, where one or more blood components can be directly affected, secondary toxicity may occur as a consequence of other organ injury or systemic effects. In most cases, toxicity depends on the exposure dose and is predictable. However, in some cases toxicity may occur in only a small number of susceptible individuals. This is termed an idiosyncratic response and is often immune-mediated.

There are two major classes of haematotoxic compounds, drugs and occupationally important chemicals. In the case of drugs, blood dyscrasias (abnormalities) other than cancer have been extensively studied. However, for occupationally important chemicals the major emphasis has been on their ability to cause cancer. In spite of extensive studies, benzene is the only major industrial chemical that is known to cause cancer of haematopoietic tissue in humans (Leventhal and Khan 1985). Radiation can also cause similar cancers.

Structure and function

Haematopoiesis occurs in the bone marrow with differentiated cells entering the circulation to replace senescent cells (Leventhal and Khan 1985; Testa and Gale 1988). Within the bone marrow, haematopoietic cells are located extravascularly and are supported and regulated by stromal cells that produce a variety of growth factors. In addition, erythropoietin, produced by the kidney, is essential for erythropoiesis. The microenvironment of the haematopoietic cells is critical for their proliferation and differentiation. The pluripotent stem cell gives rise to the progenitor cells of all lineages; it has little proliferative capacity in spite of unlimited self-renewal. The progenitor cells have greater proliferative capacity but are committed to a limited number of lineages. The proliferative precursors of each cell lineage give rise to the well-differentiated end-stage cells that enter the circulation. These end-stage cells cannot proliferate, except for monocytes and lymphocytes, and all have a finite lifespan (Young 1988). The two basic cell lineages are the lymphoid (discussed later in the section on the immune system) and haematopoietic, the latter differentiating into the myeloid, erythroid and megakaryocytic series, as shown in Figure 3.1. Differentiated cells released into the circulation consist of granulocytes and monocytes (called macrophages in tissues) from the myeloid lineage, erythrocytes (red blood cells) from the erythroid lineage, and platelets from the megakaryocytic lineage (Irons 1991).

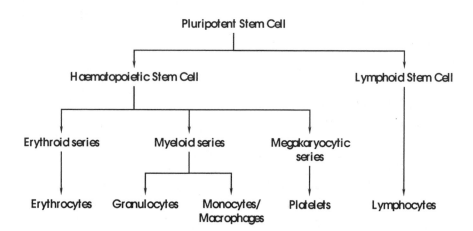

Figure 3.1 Haematopoiesis. Bone-marrow differentiation into mature elements of peripheral blood.

Granulocytes play a major role in the nonspecific host defence mechanisms due to their capacity for phagocytosis and release of inflammatory mediators. Monocytes/macrophages also play a role in nonspecific host defences as well as acting as regulatory cells in haematopoiesis and the immune response. Monocytes from the bone marrow differentiate into a variety of mononuclear phagocytic system cells located in the liver (Kupffer cells), spleen, lymph nodes and other tissues. Erythrocytes provide haemoglobin for the transport and delivery of oxygen to tissues and facilitate the removal of carbon dioxide. Platelets have an important role in haemostasis.

Classification and mechanisms of chemical-induced haematotoxicity

Haematotoxicity can be targeted to cells of the bone marrow or those in the circulation (Irons 1985). Injury to the pluripotent stem cells or their microenvironment can result in underproduction of all cell types. Once differentiation begins, individual cell types can be affected, either in the bone marrow or in the blood, leading to a decrease in cell numbers of a particular cell type or abnormal function of that cell type. Direct toxicity to circulating cells results in an increased demand on the bone marrow resulting in increased cell proliferation (hyperplasia) and decreased differentiation time. Abnormal cell function will not be discussed in detail, since the target organs lie outside the haematopoietic system (see Bloom and Brandt (2002) for an in-depth discussion). A decrease in erythrocyte numbers, mean corpuscular volume (MCV), mean corpuscular haemoglobin content (MCH) or any combination thereof, is termed anaemia. A decrease in white blood cells is termed cytopenia (granulocytopenia, lymphocytopenia or thrombocytopenia depending on the cell type affected) or pancytopenia if all white cells are affected. Effects will vary in severity and reversibility, depending on the chemical and exposure conditions. Possible outcomes are recovery following cessation of exposure, persistence of changes, or progression to aplastic anaemia, myelodysplastic syndrome, or leukaemia (Rosner and Grunwald 1990).

Classification of haematotoxicity is generally based on the type of injury or the compartment affected, as shown in Table 3.6, rather than the mechanism of injury.

Bone-marrow suppression can be general or selective. General suppression at the stem cell level results in pancytopenia with decreased numbers of all cell types. Pancytopenia has been identified following exposure to hydrochlorofluorocarbons and 2-bromopropane, recently introduced substitutes for chlorofluorocarbons. If the marrow fails to respond, the condition is called nonregenerative or aplastic anaemia.

Of the industrial chemicals, benzene is the best known haematotoxicant; other causes include chronic mercury poisoning and radiation exposure. Selective suppression affects one of the more differentiated cellular elements resulting in a specific cytopenia. Mechanisms by which chemicals can induce bone-marrow suppression include selective toxicity to proliferating cells, as seen with chemotherapeutic drugs and ionising radiation; and interference with RNA or DNA synthesis or with cell division, as seen with folate antagonists.

Myelotoxic agents, such as radiation and certain chemicals, can dysregulate hematopoiesis and induce myelodysplastic syndrome (MDS) and leukaemias in

Table 3.6 Classification of chemical-induced haematopoietic injury according to site and type of injury

Bone marrow
Suppression
Red blood cells – anaemia (e.g. arsenic, lead)

Other cell types – cytopenia
All cell types – pancytopenia, aplastic anaemia (e.g. benzene, trinitrotoluene, chemotherapeutic drugs, radiation)

Carcinogenesis
Any cell type – leukaemia (e.g. benzene, chemotherapeutic drugs, radiation)

Blood
Direct toxicity leading to decreased numbers of circulating cells
Red blood cells – haemolytic anaemia (e.g. arsine)
White blood cells – cytopenia

Functional impairment
Red blood cells – methaemoglobinaemia (e.g. analine, nitrobenzene) – carboxyhemoglobinaemia (e.g. carbon monoxide)
Platelets – aggregation
White blood cells – phagocytosis

some circumstances. Leukaemias are monoclonal in origin and both gene alterations and epigenetic factors appear to play a role. The degree of cell differentiation loosely correlates with rate or degree of progression, with poorly differentiated types designated as 'acute' and well-differentiated types as 'chronic'. Although leukaemias can arise from any cell lineage, leukaemias associated with chemical and drug exposure are predominantly of the acute myelogenous type (Irons 1985). This is well documented for most alkylating chemotherapeutic drugs, radiation, and benzene.

As in other organs, metabolic activation can play a role in some of the chemical-induced bone-marrow toxicities. Metabolism is a requirement for benzene-induced bone-marrow toxicity. Both the liver and the bone marrow can metabolise benzene to benzene oxide which then can be further metabolised to either of two hematotoxins, hydroquinone via a phenol, or muconic dialdehyde (ATSDR 1997). Benzene cytotoxicity is cycle specific and phase specific, with dividing cells in G_2 and M phases of mitosis being targets. Progression to leukaemia is thought to be a continuum of a single toxic response and involves cellular growth factors, proto-oncogenes and growth-promoting genes (Irons 1985).

Another mechanism of toxicity is exemplified by lead which causes abnormal erythropoiesis by its effects on haeme biosynthesis so that the differentiated erythrocyte is abnormal. Lead inhibits ferrochelatase and γ-aminolevulinic acid dehydrase (ALAD), resulting in an increase in aminolevulinic acid synthetase (ALAS) and therefore aminolevulinic acid (ALA). The activity of ALAD is inhibited at very low blood lead levels and a threshold has not been identified. It also binds directly to the red blood cell membrane resulting in increased fragility (Irons 1985). On cytological examination, erythrocytes may exhibit basophilic stippling which represents altered ribosomes. A mild, normocytic hypochromic anaemia with elevation of reticulocytes (erythrocyte precursors) and basophilic stippling is an indicator of lead exposure.

Toxicity to circulating cells that results in decreased cell numbers increases the demand for formation of new cells in the bone marrow. For example, a decrease in red blood cells in the circulation due to haemolysis results in increased haematopoiesis in the bone marrow. Haemolysis may be caused by direct toxic damage to the erythrocyte membrane, oxidative injury to haemoglobin leading to its precipitation (Heinz body), or by immune-mediated mechanisms. The best known example of an industrial chemical causing haemolytic anaemia is arsine gas (AsH_3). However, pharmaceutical drugs are the most common cause, with people deficient in 6-phosphate dehydrogenase being especially susceptible to erythrocyte damage. Toxicity to circulating granulocytes can be caused by drug-induced immune-mediated mechanisms.

Functional impairment of specific cell types without an actual decrease in cell numbers is also possible. This is especially common for red blood cells since many agents can affect oxygen transport or utilisation. For example, cyanide increases the affinity of haemoglobin for oxygen, while carbon monoxide causes carboxyhaemoglobinaemia (replacement of oxygen by carbon monoxide) with a decrease in the total oxygen-carrying capacity of blood. As discussed above, lead also causes erythrocyte abnormalities. Abnormalities in the function of white blood cells and platelets can also occur, such as inhibition of platelet aggregation by acetylsalicylic acid.

Detection and evaluation of chemical-induced haematopoietic injury

Most patients with acute blood dyscrasias tend to present with overt clinical sign and symptoms. In such cases, routine haematological examination and determination of haemoglobin concentration are adequate for diagnosis of a problem. To determine a role for occupational exposure, both the clinical course of the disease and the type and level of exposure should be considered. Measurement of blood levels of specific chemicals (for example, lead) or their metabolites may be warranted. Most chemically induced blood dyscrasias are transient and resolve following termination of exposure. In chronic poisoning, evaluation for Heinz bodies, basophilic stippling, and methaemoglobin are indicated.

Another parameter that can be used as a diagnostic tool is the red cell distribution width index (RDW), or coefficient of variation associated with the mean corpuscular volume (MCV) (Irons 1985). Additional tools include bone marrow biopsy, with quantitation of cellularity, and flow cytometry, which is beginning to be used in evaluation of leukaemias and lymphomas.

Examples of haematopoietic injury from workplace exposure

Although a wide variety of industrial chemicals have been implicated epidemiologically in haematopoietic toxicity, careful evaluation of the available data has narrowed the list of definite haematotoxicants to those listed in Table 3.7

Table 3.7 Examples of haematopoietic injury from workplace exposure

Definite haematotoxicants
Lead (anaemia)
Radium, radiation (leukaemia)
Trinitrotoluene (anaemia)
Benzene (anaemia, leukaemia)

Possible haematotoxicants
Arsenic
Glycol ethers
1,3-Butadiene
Chlordane and heptachlor
Phenoxy herbicides, especially 2,4-dichlorophenoxyacetic acid

Haemoglobin toxicants
Aromatic amino and nitro compounds (methaemoglobinaemia)
Carbon monoxide (carboxyhaemoglobinaemia)

(Irons 1985). While the results of studies in animals provides some information, the difficulty in identifying haematotoxicants in humans is mainly due to the difficulty in obtaining a reliable history of chemical exposure, simultaneous exposure to multiple potentially toxic chemicals, and the marked variability in individual responses.

Chronic lead poisoning, with anaemia as a late manifestation, is still an occupational hazard in lead workers and was a hazard previously in painters. Occupational exposures to radium and trinitrotoluene (TNT) are now of historical interest only. Benzene is the only primary aromatic solvent with haematotoxic properties. It is an important industrial compound, with its production ranking among the top ten commodity chemicals in the USA (Irons 1985). Haematopoietic toxicity, affecting all (pancytopenia) or any cell type (specific cytopenia), occurs with repeated exposure to high concentrations of benzene. In fact, workplace concentrations of benzene above 1 ppm. may induce haematological abnormalities. Severe damage can result in aplastic anaemia and can progress to acute myelogenous leukaemia.

Some recent well-documented studies show an association between the agricultural use of the herbicide 2,4-dichlorophenoxyacetic acid (2,4-D) and the risk of non-Hodgkin's lymphoma (Irons 1985). In addition, associations between certain occupations, such as the rubber and tyre industry, and increased risk of leukaemia or lymphoma are well known; however, the nature of the causative agent remains obscure.

Immune system (N.H. Stacey and C. Winder)

Introduction

Exposure to chemicals, pharmaceuticals or food constituents may result in impaired resistance to infections and tumours, autoimmune disorders and allergic reactions (Descotes and Bernard 1999). Although the immune system has only recently been studied substantially from a toxicological perspective, immunotoxicity has now emerged as a formal field of toxicology. Immunotoxicity encompasses both reduced and heightened immune function (Karol 1998).

The end result of suppression of the immune system ranges from viral infections such as the common cold lasting a little longer than usual, to increased incidence of tumour development. Chemicals can also initiate immune-mediated disorders such as allergic reactions and the development of autoimmune conditions. As these immune-mediated effects are most commonly manifested in the skin and the respiratory system, they will not be discussed further in any detail in this section.

It is worth noting that toxicity testing is normally conducted under specific-pathogen free (SPF) conditions, which actually means that animals are free of specific infectious agents, and indeed that animals are protected from exposure to infectious agents. Thus effects on a system that normally protects against infectious organisms will not be observed unless specific functional assays or challenges by specific infectious agents are undertaken. The immune system differs from other organs in its accessibility; white blood cells from peripheral blood can be separated and studied, while tissues from other organs are far less accessible. This easy accessibility improves the investigators ability to understand and interpret chemical-induced immune changes in humans.

The reader is referred to other sources for detailed and comprehensive information on the effects of chemicals on the immune system (Burns-Naas *et al.* 2001). The purpose of this section is to provide only a broad outline of this area.

Structure and function

The immune system is a little different to other major organs from a structural point of view. It is rather diffuse in that it has a circulating component as well as a presence in other parts of the body. The circulating component originates from the pluripotent stem cells (see previous section). Then there are the primary lymphoid organs which include the thymus where T cells mature and the bone marrow where B cells mature. The secondary lymphoid organs include the spleen, lymph nodes and mucosal-associated lymphoid tissue.

From a functional aspect, the immune system plays a crucial role in the protection of the body against invading and potentially infectious organisms. It is also involved in a surveillance function for cancerous cells with the apparent ability to eliminate certain neoplastic cells as they arise. The immune system consists of numerous different cell types, many of which are interactive with one another. Thus interference with one component of the immune system can affect other components.

The major cell types and their functions are shown in Table 3.8. Macrophages, granulocytes and lymphocytes can be distinguished from one another by their appearance under the microscope. Further morphological distinction, among the lymphocytes in particular, can be made by examination of surface markers using specific antibodies which are often attached to a fluorescent marker for identification by flow cytometry. This approach is applicable to other cell types as well and has become widely used.

Table 3.8 Cellular components of the immune system

Cell type	Function
Macrophage/monocyte	Phagocytose foreign material
Polymorphonuclear cells or granulocytes	Phagocytose foreign material
Lymphocytes	
T	Cell-mediated cytotoxicity – major histocompatibility complex restricted (MHCr)
T_H or T_S	Helper or suppressor
B	Antibody production
Natural killer (NK)	Cell-mediated cytotoxicity (non-MHCr)

Mechanisms of sensitisation and allergy

The immune system defends the body against invading organisms, exogenous antigens, and host cells that have become neoplastic. In addition, the immune system is an active participant in autoimmune disease, hypersensitivity reactions and transplant tissue rejections.

Most of the blood cells, first described in the previous section on the haematopoietic system, have immunological functions. The most significant of these are the lymphocytes, which are involved in the two main phases of the immune response – the induction phase and the sensitisation phase. This second phase then consists of two components – an antibody-mediated component and a cell-mediated component.

The lymphocytes are divided into two main groups:

- B cells, which are involved in the production of antibodies.
- T cells, which are important in the induction phase of the immune response and are responsible for cell-mediated reactions. There are helper/inducer, suppressor and cytotoxic T cells.

Specific immunological response

The specific immunological response increases the efficiency of the organism's defence response. The response is both more efficient, and can be specific to an individual pathogen or toxicant.

The antigen

The initial step in the immune response is recognition by the organism that something is present which is foreign. This is called the antigen. The antigen may be:

- a bacteria or microorganism
- products of organisms, such as excreta of dust mites or bacterial proteins
- a large chemical molecule

- a small chemical molecule called a hapten (these are chemicals that are too small to be recognised as antigens in their own right, and must bind to proteins to evoke a response)
- proteins
- some drugs.

T cell–antigen–B cell complex

The principal cells in antigen recognition are the T and B lymphocytes; more specifically, the T helper cells and B cells. These bind cooperatively or simultaneously to the antigen at different sites, and a T helper cell–antigen–B cell complex forms. The T helper cell also releases soluble factors that stimulate the proliferation of B cells. These cloned B cells mature into antibody-producing cells, the plasma cells.

The antibody

The antibody is a Y-shaped protein structure called gammaglobulin or immunoglobulin (Ig), with the arms of the Y being specific recognition sites for specific antigens. There are a range of different categories of antibody, which differ from each other in certain structural respects. These are IgG, IgM, IgE, IgA and IgD.

Antibodies improve the organism's capacity to respond to toxic insult. Apart from their ability to interact directly with pathogens or toxins, they can also multiply many times, making the defence reaction specific and effective.

Systemic responses in inflammatory reactions

In addition to the local changes in an area where an inflammatory response is occurring, there are sometimes various general responses which may affect the whole organism. For example:

- rise in temperature, due to increased biological activity
- increase in blood cell populations, such as leukocytes (leukocytosis), neutrophils (neutrophilia) or monocytes
- increase in concentration of circulating proteins, enzymes and mediators (such as fibrinogen, α_2-macroglobulin and components of complement)
- change in immune function.

All of these reactions are based on and are dependent of the scope and intensity of the inflammatory response.

Unwanted inflammatory and immune responses

The inflammatory and immune responses outlined above are complex and elaborate. They are also not without problems. The immune response to an antigen is potentially a 'double-edged sword' that can either protect or harm the host.

Of course, the paramount biological effects of the immune response are the humoral (B cell) and cellular (T cell) recognition and elimination of infectious agents and other foreign antigens. Nevertheless, re-exposure of some previously sensitised individuals to the same antigen may bring about an exaggerated ('hypersensitive') or misdirected immune response that results in local tissue injury or in systemic manifestations, including shock or death. Immune responses that result in tissue injury or other pathophysiological changes are called hypersensitivity (allergic/immunopathological) reactions.

The ensuing tissue injury may involve:

- release of vasoactive substances (primary and secondary mediators)
- phagocytosis or lysis of cells
- activation of the inflammatory and cytolytic components of the complement system
- release from recruited inflammatory cells of proteolytic enzymes, cytokines, and other mediators of tissue injury and inflammation.

The antibody-induced inflammatory response can, in certain circumstances, be inappropriately triggered, or triggered to induce exaggerated reactions. Unwanted immune responses are termed allergic or hypersensitivity reactions, and have been classified into four types: Type I (anaphylactic), Type II (antibody-dependent/cytotoxic), Type III (immune complex) and Type IV (cell-mediated/delayed) hypersensitivity (see Table 3.9) (Gallin 1993).

Type I Immediate or anaphylactic hypersensitivity

This occurs when the antigenic material, which is not of itself noxious, evokes the production of IgE-type antibodies. Antigens which can cause this reaction include grass pollen, house-dust mite products, some foods, chemicals and drugs. IgE antibodies bind to mast cells and on subsequent exposure cause the release of histamine, platelet-activating factors, eicosanoids and cytokines. Effects may be localised to the nose (hay fever), skin (urticaria), bronchus (asthma) or gastrointestinal system. In some cases, the reaction is more generalised and can produce anaphylactic shock which can be immediately life-threatening. This process is basically an inappropriate deployment of antibody–mast cell interactions.

Type II Antibody-dependent cytotoxic hypersensitivity

All cells within an organism have properties which allow the immune system to recognise that they belong to the host organism ('self'), and should not be targeted for an immune reaction. If the immune system fails to recognise cells as being part of the host organism, it will initiate an immune response. This Type II response can arise spontaneously in autoimmune conditions (where antibody is directed against host antigen or 'self') or can occur, for example, after incompatible blood transfusions or alteration of cell properties to remove 'host identity' by drugs or chemicals. The basic mechanism is an initiation of the complement sequence by the antigen–antibody reaction.

Table 3.9 Immunological mechanisms of inflammation and tissue injury

Reaction type	Type I	Type II	Type III	Type IV
Description	Anaphylaxis (immediate)	Cytotoxic (antibody-dependent)	Immune-complex mediated	Cell mediated
Time elapsed	Seconds to minutes	Hours to a day	Hours to days	2–4 days
Immune mechanism	IgE	IgG, IgM	IgG, IgM	T cells
Mediators of injury and inflammation	Mast cell vasoactive factors, basophils (histamine, arachidonate derivatives)	Complement	Cytolytic, chemotactic, vasoactive factors	Lymphokines/monokines
Pathology	Accumulation of neutrophils, eosinophils. Smooth muscle contraction	Phagocytosis, lysis of target cells, receptor dysfunction	Accumulation of neutrophils, macrophages. Release of lytic lysosomal enzymes	Lymphocytes and macrophages, granulomas
Examples	Anaphylaxis, atopic disorder (allergic rhinitis, bronchial asthma, hay fever, allergic dermatitis, urticaria, angioedema)	Autoimmune haemolytic anaemia, transfusion reaction, haemolytic disease of the newborn, immune thrombocytopenia, Goodpasture's syndrome, myasthenia gravis, Graves' disease	Systemic lupus erythematosus (nephritis, vasculolitis), immune complex glomerulonephritis, serum sickness	Contact dermatitis, granulomatous diseases (tuberculosis, sarcoidosis, leprosy). Hashimoto's thyroiditis, organ rejection

Type III Immune-complex hypersensitivity

This type of hypersensitivity occurs when antibody reacts with a *soluble* antigen. The antigen–antibody complex can then stimulate the release of mediators (as in Type I) and/or activate complement (as in Type II). Type III hypersensitivity is implicated in farmer's lung and lupus erythematosus (an autoimmune inflammatory disease of connective tissue).

Type IV Cell-mediated hypersensitivity

In type IV hypersensitivity, cytotoxic T cells and helper T cells recognise either intracellular or extracellular synthesised antigen when it is complexed, respectively, with major histocompatibility complex (MHC) molecules. Macrophages function as antigen-presenting cells and release interleukin-1 which promotes the proliferation of helper T cells. Helper T cells release interferon and interleukin, which together regulate delayed-hypersensitivity reactions centred around macrophage activation and T-cell mediated immunity. Activated cytotoxic T cells destroy target cells on contact. Natural killer (NK) cells can kill target cells directly, in the absence of prior immunisation and without MHC restriction, and by antibody-dependent cellular cytotoxicity. Activated macrophages express increased phagocytic, bactericidal and cytocidal activity and, when activated, transform into epithelioid cells. Granulomas form and frequently contain multinucleated giant cells that result from macrophage fusion.

Classification and detection of chemical-induced injury with examples

Studies in humans designed to detect immunotoxicity from exposure to xenobiotics present a challenge (Burrell *et al.* 1997; Luster and Kimber 1996). Exposed and control groups must be carefully selected, exposure to the xenobiotic must be sufficiently high and well documented, and the reference group should be as similar as possible to the exposed (IPCS 1997). Immune markers/functional tests in an individual may be influenced by sunlight exposure, medication, illness and use of recreational drugs; all of these potential confounding factors must be addressed. Sample acquisition is usually performed at sites geographically distant from the controlled environment of an investigator's laboratory, yielding an assortment of new problems that would not occur in clinical or hospital situations (Trizio *et al.* 1988).

Classification and detection will be considered together for the immune system, as one is dependent upon the other.

Table 3.10 lists the major types of chemical-induced interference with immune function. Chemicals can be further classified on the basis of the function with which they interfere. For example, a chemical (trichloroethylene) may be shown to inhibit NK cell function, but not T or B cell function.

Many chemicals of occupational relevance have been shown to cause immunosuppression of one type or another in experimental animals. Often more than one function is affected. Considerable caution is required, however, in this emerging field because there are many conflicting reports for many chemicals. One particular

Table 3.10 Types of immunotoxic reactions

Reaction	Example
Immunosuppression	Benzene, asbestos
Carcinogenicity	Cyclophosphamide*
Autoimmunity	Trichloroethylene, epoxy resins
Hypersensitivity	Nickel, di-isocyanates

Note: *Cyclophosphamide is a cytotoxic drug used in chemotherapy. Workers preparing this and similar drugs require a high level of protection.

example with cadmium highlights this point. In 1984, this heavy metal was reported to increase, to decrease, and to have no effect on one parameter of immune cell function. Interestingly all three reports came from the same laboratory. This brings out the issue of interpretation of interference with immune cell function in studies of this nature. At this stage it is apparent that not all aspects of this interference can be ascribed the role of necessarily resulting in a clear compromise of the organism's homeostasis. Factors such as reserve capacity for parts of immune system function need to be borne in mind.

Functional assays performed after treatment, either *in vivo* or *in vitro*, with chemicals are a major way to detect an adverse effect on the immune system. The numbers of the various cell types can also be determined as an indication of the effects of a chemical. Other functions may also be evaluated as indicated in Table 3.11, which displays the suggested tiered approach to screening for immunotoxic effects (IPCS 1997).

Since the application of immunotoxicological techniques to populations exposed to xenobiotics is relatively new, and the ability to measure an increasing number of immune biomarkers of activation, suppression, autoimmunity or hypersensitivity is rapidly expanding, there are difficulties in the interpretation of statistically positive results (which sometimes lie within the normal range) and their potential health significance (Biagini 1998).

The evaluation of immunological function in routine toxicological screening is still not commonplace and the appropriate level of testing remains to be determined. The effects of chemicals on the immune system are potentially wide

Table 3.11 Detection of chemical-induced immune dysfunction

Tier I	Tier II
Immunopathology	Immunopathology (extended)
Haematology	Host resistance
Organ weights	Cell-mediated immunity (extended)
Histology	Humoral immunity (extended)
Cell-mediated immunity	Macrophage function
Lymphocyte blastogenesis	Granulocyte function
Natural killer cell activity	Bone-marrow cell quantitation
Humoral immunity	

ranging and vary from the obvious to the subtle, which requires careful investigation so that the effects are recognised (Burns-Naas *et al.* 2001). Effects on immune system function are known to be caused by occupational exposures to (among other things) asbestos (Rosenthal *et al.* 1998), benzene (Bogadi-Sare *et al.* 2000), dithiocarbamates (Irons and Pyatt 1998), isocyanates (Karol and Jin 1991), jet fuel (Harris *et al.* 2001), lead (Goyer 1990), mercury vapour (Moszczynski 1997), polychlorinated biphenyls (Tryphonas 1995), pesticides (Vial *et al.* 1996) and welding (Tuschl *et al.* 1997).

Finally, both biological and methodological factors complicate the assessment of dose–response/concentration effect relationships in human immunotoxicity studies, and traditional dose–response relationships may not always be present (Luster and Rosenthal 1993).

Cardiovascular system (W.M. Haschek)

Introduction

Cardiovascular diseases, especially atherosclerotic vascular disease, are the major cause of death in the USA. Major risk factors for atherosclerosis include cigarette smoking, diabetes, obesity, hypertension, hyperlipidaemia, and genetic factors (Benowitz 1992). Occupational exposure has been linked to cardiovascular diseases such as arrhythmias, atherosclerosis, problems with coronary blood supply, systemic hypotension, and cor pulmonale (right ventricular hypertrophy usually due to pulmonary hypertension). Although many chemicals have been shown experimentally to be potentially cardiotoxic, the role of low-level chronic exposure to chemicals in the workplace in the aetiology of chronic cardiovascular disease in humans is still largely unknown.

The cardiovascular system can be a primary target for chemicals or may be affected secondarily by damage to other organs. An example of a secondary effect is the cor pulmonale induced by severe, asbestos-induced lung fibrosis. Cardiac dysfunction may result from abnormalities in ion movements, membrane functions, and energy-producing systems. Susceptibility to injury is due to factors such as exposure to chemicals at maximal concentrations, often for prolonged periods; presence of low concentrations of protective/detoxification enzymes in the heart; and exquisite sensitivity of the heart to hypoxia due to its high metabolic rate. Some chemicals produce structural changes with secondary functional alterations, and others produce solely functional changes, such as cardiac arrhythmias. In either case, these alterations can have serious and often lethal consequences.

Structure and function of the cardiovascular system

The cardiovascular system consists of the heart, which pumps blood by rhythmic contraction of its chambers; the arteries (muscular vessels that carry oxygenated blood away from the heart to the tissues); the capillaries (tiny vessels that transport oxygen and nutrients from blood to the tissues, and collect carbon dioxide and waste products from the tissues); and the veins (large, thin-walled vessels that carry blood from the capillaries back to the heart, and then to the lung for reoxygenation). The entire circulatory system is lined by a single cell layer called the endothelium. Veins and arteries have a subendothelial layer consisting of connective tissue (the intima). Beneath this layer lies smooth muscle (the media), which is well developed and highly elastic in the larger arteries, and an outer layer of loose connective tissue (the adventitia). The most common vascular disease, atherosclerosis, is characterised by plaque formation due to intimal degeneration and lipid deposition. This can result in partial to complete occlusion of the vascular lumen. Contraction of the smooth muscle in the media can result in vasospasm or vasoconstriction.

The heart muscle or myocardium consists of cardiac muscle cells whose main functions are electrical impulse generation and conduction, and contraction. These cells cannot regenerate; therefore injury leading to cell death (apoptosis or necrosis) can result in significant problems. Both the sympathetic and parasympathetic divisions of the autonomic nervous system innervate the heart. Blood supply to the myocardium is through the coronary arteries. Because of the high energy demand of the myocardium, any interference with blood supply (such as coronary atherosclerosis) leading to hypoxia or ischaemia can have serious consequences. These can be manifested clinically as angina pectoris (severe chest pain), heart attack, or sudden death. Decreased blood supply due to cardiovascular dysfunction can affect other organs, especially the brain and kidneys.

Classification of cardiovascular toxicity

There are many ways to classify cardiovascular toxicity (Acosta 1992; Balazs 1992). Classification can be based on the type of injury, as shown in Table 3.12, clinical manifestations, mechanism of toxicity (see Table 3.13), and organellar localisation of injury. Direct cardiovascular injury is discussed under mechanisms below.

Detection and evaluation of cardiovascular injury

Evaluation of workplace exposure as a factor in cardiovascular disease is very difficult since cardiovascular disease from other causes is very common in the general population and the clinical signs are usually nonspecific as to cause. Major clinical

---

Table 3.12 Types of cardiovascular injury with examples

Heart
Disturbances in electrical impulse formation or conduction (leading to arrhythmia) (e.g. halogenated hydrocarbons, organophosphates, arsine)
Direct myocardial injury (e.g. arsenic, arsine)
Ischaemia due to coronary vasospasm (e.g. organic nitrates)
Asphyxia (e.g. carbon monoxide, cyanide, hydrogen sulfide)

Vessels
Vasoconstriction (functional injury leading to hypertension) (e.g. lead, carbon disulfide)
Atherosclerosis/arteriosclerosis (e.g. carbon disulfide)
Vasculitis (e.g. arsenic, gold, many drugs)

Table 3.13 Mechanisms of cardiovascular toxicity

Pharmacological (functional)
Cardiotonic agents increase force of myocardial contraction by
   inhibiting Na^+/K^+ ATPase (e.g. digitalis)
   increasing Na^+ influx (e.g. ciguatoxin)
   increasing Ca^{2+} influx (e.g. catecholamines)

Myocardial depressants acting by
   decreasing Na^+ permeability (e.g. polyethylene glycol, higher alcohols)
   replacing sarcolemmal Ca^{2+} and increasing Ca^{2+} influx (e.g. cadmium)
   altering contractility (e.g. organic solvents)

Direct cellular injury
reaction with functionally or structurally important molecules (e.g. cobalt)
reactive metabolite binding to macromolecules (e.g. allylamine)
reactive oxygen radical formation (e.g. adriamycin)

Hypersensitivity responses
immune-mediated (e.g. penicillin)

Aggravation of pre-existing disease
coronary heart disease (e.g. carbon disulfide)

---

features of acute toxicosis are electrocardiographic changes, arrhythmias, acute cardiac failure, and death. Knowledge of potentially cardiotoxic chemicals and their effects on other organ systems are helpful in defining the role of chemical exposure in the cardiovascular disease of a specific patient since chemical exposure may aggravate pre-existing conditions. Evaluation of chemical-induced cardiovascular injury is identical to that for other causes of cardiovascular disease.

Mechanisms of cardiovascular toxicity

Cardiovascular toxicity may be due to direct or indirect injury or due to an interaction with exogenous or endogenous chemicals as shown in Table 3.13. In occupationally induced cardiovascular injury, pharmacological or functional effects arise from alteration of electrical or contractile properties of the heart, without detectable structural injury (Magos 1981). This type of toxic effect is frequently manifested as arrhythmia and sometimes as sudden death.

Reaction with functionally or structurally important molecules by a chemical or its metabolite can cause a direct effect on the heart, frequently resulting in degeneration and necrosis of cardiac muscle, sometimes accompanied by inflammation. Allylamine, an industrial chemical, is metabolised by amine oxidase to acrolein and hydrogen peroxide, which conjugates to and depletes glutathione. Selective vasculotoxicity occurs with damage to smooth muscle. Chronic exposure results in atherosclerosis. In addition, acrolein conjugates to and depletes glutathione, putting cardiac muscle at risk for damage due to free radicals. Chemotherapeutic agents such as the anthracyclines (for example, adriamycin) form reactive oxygen radicals which can directly injure cardiac muscle. Cadmium, lead and cobalt exhibit negative inotropic and dromotropic effects and produce structural changes in the heart. Immune-mediated hypersensitivity reactions can also affect the cardiovascular system. These are not dose-related, are primarily induced by drugs, and characterised by inflammation, especially vasculitis.

Solvents can produce cardiotoxicity due to multiple mechanisms, including effects through the nervous system on cardiac electrical activity and directly on the heart affecting contraction and energy production, as well as release of circulating hormones such as catecholamines, vasopressin, and serotonin. Effects such as these have been reported following short-term occupational exposure to high concentrations of organic solvents or fluorocarbons (Benowitz 1992). At lower levels of exposure, these solvents sensitise the heart to the effects of adrenergic agonists such as endogenous catecholamines, which cause increased contractility, accelerated heart rate and even arrhythmias leading to hypoxia (Balazs and Ferrans 1978). Endogenous catecholamines may be released due to central nervous stimulation induced by the solvents themselves, or due to physiological responses such as those following exercise or excitement (Benowitz 1992). Halogenated hydrocarbons and ketones have similar effects as they cause a decrease in parasympathetic activity and increased adrenergic sensitivity. There is altered ion haemostasis with all three types of agents.

Several chemicals have been found to aggravate pre-existing cardiovascular disease (Ramos et al. 2001). Carbon disulfide, combustion products, and arsenic irreversibly accelerate coronary heart disease (Benowitz 1992). A two- to threefold increase in coronary heart disease has been reported following carbon disulfide exposure. Carbon monoxide and nitroglycerine may also aggravate coronary artery disease but the association is not as strong. Carbon monoxide can lead to decreased oxygen availability (myocardial asphyxiation) manifesting as increased heart rate (tachycardia) and other electrocardiographic changes suggestive of hypoxia. Nitroglycerine and other organic nitrates can cause vasospasm of the coronary artery and thus lead to ischaemic heart disease, even in the absence of atherosclerosis (Benowitz 1992).

Examples of cardiovascular injury from workplace exposure

Many chemicals are known to be potentially cardiotoxic; however, few have been shown to be of unequivocal importance in the workplace. Those workplace chemicals which have been identified as significant include halogenated hydrocarbons, such as chloroform, fluorocarbons, and methylene chloride; heavy metals such as cadmium, lead, and nickel; and carbon monoxide (CO) and carbon disulfide. Table 3.14 lists those workplace chemicals that are accepted, on the basis of epidemiological studies, to have a role in cardiovascular disease; also listed is the type of injury produced (Benowitz 1992; Kristensen 1989; van Vleet et al. 2002).

Hypertension, increased mortality due to cardiovascular and cerebrovascular disease, and Reynaud's phenomenon have been reported in vinyl chloride workers (ATSDR 1997). Fingers in workers with Reynaud's syndrome become cold, numb and painful following cold exposure (Noel 2000). This is due to structural changes in the digital arteries which include wall thickening with narrowing of the lumen due to vasculitis.

Table 3.14 Workplace exposures implicated in cardiovascular disease

Disease	Causation	
	Definite	Probable
Atherosclerotic ischaemic heart disease	Carbon disulfide	Tobacco smoke
Nonatheromatous ischaemic heart disease	Organic nitrates	
Myocardial asphyxiant	Carbon monoxide, cyanide, hydrogen sulfide	
Direct myocardial injury	Arsenic, arsine	Antimony
Arrhythmias	Halogenated hydrocarbons, organophosphates, arsine	Antimony, arsenic, lead
Hypertension	Carbon disulfide	
Peripheral arterial occlusive disease	Arsenic, lead, carbon disulfide, vinyl chloride	

Source: Adapted from Benowitz (1992).

References

Acosta, D. (ed.) (1982) *Cardiovascular Toxicology*. New York: Raven Press.

ATSDR (1977a) *Toxicological Profile for Benzene*. Agency for Toxic Substances and Disease Registry/US Department of Health Human Services, Atlanta.

ATSDR (1997b) *Toxicological Profile for Vinyl chloride*. Agency for Toxic Substances and Disease Registry/US Department of Health Human Services, Atlanta.

ATSDR (1999a) *Toxicological Profile for Lead*. Agency for Toxic Substances and Disease Registry/US Department of Health Human Services, Atlanta.

ATSDR (1999b) *Toxicological Profile for Mercury*. Agency for Toxic Substances and Disease Registry/US Department of Health Human Services, Atlanta.

Balazs, T. (ed.) (1992) *Cardiac Toxicology*, 2nd edn. Boca Raton: CRC Press, Inc.

Balazs, T. and Ferrans, V.J. (1978) Cardiac lesions induced by chemicals. *Environ. Health Persp.* 26: 181–191.

Banwell, J.G. (1979) Environmental contaminants and intestinal function. *Environ. Health Persp.* 33: 61–69.

Benowitz, N.L. (1992) Cardiac toxicity. In: Sullivan, J.B. and Krieger, G.R. (eds) *Hazardous Materials Toxicology: Clinical Principles of Environmental Health*. Baltimore: Williams and Wilkins, pp. 168–178.

Bertram, T.A. (1991) Gastrointestinal tract. In: Haschek, W.M. and Rousseaux, C.G. (eds) *Handbook of Toxicologic Pathology*. San Diego: Academic Press Inc, pp. 195–251.

Biagini, R.E. (1998) Epidemiology studies in immunotoxicity evaluations. *Toxicology* 129: 37–54.

Bloom, J.C. and Brandt, J.T. (2001) Toxic responses of the blood. In: Klaassen, C.D. (ed.) *Casarett and Doull's Toxicology. The Basic Science of Poisons*, 6th edn. New York: McGraw-Hill, pp. 389–417.

Bogadi-Sare, A., Zavalic, M., Trosi, I., Turk, R., Kontosi, I. and Jelci, I. (2000) Study of some immunological parameters in workers occupationally exposed to benzene. *Int. Arch. Occup. Environ. Health* 73: 397–400.

Burns-Naas, L.A., Meade, B.J. and Munson, A.E. (2001) Toxic responses of the immune system. In: Klaassen, C.D. (ed.) *Casarett and Doull's Toxicology. The Basic Science of Poisons*, 6th edn. New York: McGraw-Hill, pp. 419–470.

Burrell, R., Flaherty, D.K. and Sauers, L. (1997) *Toxicology of the Immune System: A Human Approach*. New York: Wiley.

Chaisuksant, Y., Yu, Q. and Connell, D.W. (1999) The internal critical level concept of nonspecific toxicity. *Rev. Environ. Contam. Toxicol.* 162: 1–41.

Descotes, J. and Bernard, C. (1998) *Introduction to Immunotoxicology*. New York: Taylor and Francis.

DiPalma, J.A., Cunningham, J.T., Herrera, J.L., McCaffery, T.D. and Wolf, D.C. (1991) Occupational and industrial toxin exposures and the gastrointestinal tract. Gastrointestinal Toxicology Subcommittee of the American College of Gastroenterology Patient Care Committee. *Am. J. Gastroenterol.* 86: 1107–1117.

Fleming, L.E. (1992) Unusual occupational gastrointestinal and hepatic disorders. *Occup. Med.* 7: 433–448.

Gallin, J.I. (1993) Inflammation. In: Paul, W.E. (ed.) *Fundamental Immunology*, 3rd edn. New York: Raven Press, p. 1017.

Goyer, R.A. (1990) Lead toxicity: from overt to subclinical to subtle health effects. *Environ. Health Persp.* 86: 177–181.

Harris, D.T., Sakiestewa, D., Titone, D., Robledo, R.F., Young, R.S. and Witten, M. (2001) Jet fuel-induced immunotoxicity. *Toxicol. Ind. Health* 16: 261–265.

Haschek, W.M., Rousseaux, C.G. and Wallig, M.A. (eds) (2002) *Handbook of Toxicologic Pathology*, 2nd edn. San Diego: Academic Press.

IPCS (1997) Principles and Methods for Assessing Direct Immunotoxicity associated with Exposure to Chemicals. *Environmental Health Criteria* 180: WHO International Program on Chemical Safety, Geneva.

Irons, R.D. (1991) Blood and bone marrow. In: Haschek, W.M. and Rousseaux, C.G. (eds) *Handbook of Toxicologic Pathology*. San Diego: Academic Press Inc, pp. 389–420.

Irons, R.D. (1992) Benzene and other haematotoxins. In: Sullivan, J.B. and Krieger, G.R. (eds) *Hazardous Materials Toxicology: Clinical Principles of Environmental Health*. Baltimore: Williams and Wilkins, pp. 718–731.

Irons, R.D. (ed.) (1985) *Toxicology of the Blood and Bone Marrow*. New York: Raven Press.

Irons, R.D. and Pyatt, D.W. (1998) Dithiocarbamates as potential confounders in butadiene epidemiology. *Carcinogenesis* 19: 539–542.

Karol, M.H. (1998) Target organs and systems: methodologies to assess immune system function. *Environ. Health Persp.* 106 (Suppl. 2): 533–540.

Karol, M.H. and Jin., R.Z. (1991) Mechanisms of immunotoxicity to isocyanates. *Chem. Res. Toxicol.* 4: 503–509.

Klaassen, C.D. (ed.) (2001) *Casarett and Doull's Toxicology. The Basic Science of Poisons*, 6th edn. New York: McGraw-Hill.

Kristensen, T.F. (1989) Cardiovascular disease and the work environment. A critical review of the epidemiologic literature on chemical factors. *Scand. J. Work Environ. Health* 15: 245–264.

Leventhal, B.G. and Khan, A.B. (1985) Haematopoietic system. In: Mottet, N.K. (ed.) *Environmental Pathology*. New York: Oxford University Press, pp. 344–355.

Luster, M.I. and Kimber, I. (1996) Immunotoxicity: hazard identification and risk assessment. *Human Exp. Toxicol.* 15: 947–948.

Luster, M.I. and Rosenthal, G.J. (1993) Chemical agents and the immune response. *Environ. Health Persp.* 100: 219–226.

Magos, L. (1981) The effects of industrial chemicals on the heart. In: Balazs, T. (ed.) *Cardiac Toxicology*, Volume II, Boca Raton: CRC Press, Inc, pp. 203–211.

Moszczynski, P. (1997) Mercury compounds and the immune system: a review. *Int. J. Occup. Med. Environ. Health* 10: 247–258.

Neugut, A.I. and Wylie, P. (1987) Occupational cancers of the gastrointestinal tract. I. Colon, stomach, and esophagus. *Occup. Med.* 2: 109–135.

Noel, B. (2000) Pathophysiology and classification of the vibration white finger. *Int. Arch. Occup. Environ. Health* 73: 150–155.

Pfeiffer, C.J. (1985) Gastrointestinal tract. In: Mottet, N.K. (ed.) *Environmental Pathology*. New York: Oxford University Press, pp. 230–247.

Potten, C.S. and Grant, H.K. (1998) The relationship between radiation-induced apoptosis and stem cells in the small and large intestine. *Br. J. Cancer* 78: 993–503.

Ramos, K.S., Melchert, R.B., Chacon, E. and Acosta, D. (2001) Toxic responses of the heart and vascular systems. In: Klaassen, C.D. (ed.) *Casarett and Doull's Toxicology. The Basic Science of Poisons*, 6th edn. New York: McGraw-Hill, pp. 597–651.

Rosenthal, G.J., Corsini, E. and Simeonova, P. (1998) Selected new developments in asbestos immunotoxicity. *Environ. Health Persp.* 106 (Suppl. 1): 159–169.

Rosner, F. and Grunwald, H.W. (1990) Chemicals and leukemia. In: Henderson, E.S. and Lister, T.A. (eds) *Leukemia*. Philadelphia: WB Saunders, pp. 271–287.

Rozman, K. and Hanninen, O. (eds) (1986) *Gastrointestinal Toxicology*. Amsterdam: Elsevier.

Rozman, K. and Klaassen, C.D. (2001) Absorption, distribution, and excretion of toxicants. In: Klaassen, C.D. (ed.) *Casarett and Doull's Toxicology: The Basic Science of Poisons*, 6th edn, New York: McGraw-Hill, pp. 107–132.

Scheufler, E. and Rozman, K. (1991) Industrial and environmental chemicals. In: Rozman, K. and Hanninen, O. (eds) *Gastrointestinal Toxicology*. Amsterdam: Elsevier, pp. 397–415.

Schiller, C.M. (1979) Chemical exposure and intestinal function. *Environ. Health Persp.* 33: 99–100.

Stoughton, R.B. (1974) Local toxicity. *Clin. Pharmacol. Ther.* 16: 964–969.

Testa, N.G. and Gale, R.P. (eds) (1988) *Haematopoiesis*. New York: Marcel Dekker.

Trizio, D., Basketter, D.A., Botham, P.A., Graepel, P.H., Lambre, C., Magda, S.J., Pal, T.M., Riley, A.J., Ronneberger, H. and Van Sittert, N.J. (1988) Identification of immunotoxic effects of chemicals and assessment of their relevance to man. *Food Chem. Toxicol.* 26: 527–539.

Tryphonas, H. (1995) Immunotoxicity of PCBs (Aroclors) in relation to Great Lakes. *Environ. Health Persp.* 103 (Suppl. 9): 35–46.

Tuschl, H., Weber, E. and Kovac, R. (1997) Investigations on immune parameters in welders. *J. Appl. Toxicol.* 17: 377–383.

Van Vleet, J.F., Ferrans, V.J. and Herman, E. (2002) Cardiovascular and skeletal muscle systems. In: Haschek, W.M., Rousseaux, C.G. and Wallig, M.A (eds) *Handbook of Toxicologic Pathology*, 2nd edn, Volume II. San Diego: Academic Press Inc, pp. 363–455.

Vial, T., Nicolas, B. and Descotes, J. (1996) Clinical immunotoxicity of pesticides. *J. Toxicol. Environ. Health* 48: 215–229.

Young, N.S. (1988) Drugs and chemicals as agents of bone marrow failure. In: Testa, N.G. and Gale, R.P. (eds) *Haematopoiesis*. New York: Marcel Dekker, pp. 131–157.

Chapter 4

Occupational respiratory diseases

C. Winder

Introduction

The respiratory system is the organ involved in supplying oxygen (O_2) to the body and removing the waste product carbon dioxide (CO_2) (Haldane and Priestley 1935). To do this, the lungs have evolved into a system which brings the atmosphere from the external environment into the body, humidifies this air, brings it to body temperature, and delivers it to the gas exchange area in the deep portions of the lungs. It has been estimated that the total area of the gas exchange surface within the lungs is 140–150 m². While this gas exchange area is very thin (in most cases from 0.4–2.5 μm thick), this surface area is about 70 times that of the skin (West 1985).

A number of fairly basic problems are important to the problems arising in the respiratory system:

- The human lung comes into contact with about 10–12 m³ of air every day, with its accompanying chemical, particulate and microbial load.
- There are no membrane barriers between the external environment and the physiological surfaces of the respiratory system, and therefore the only obstacles that inhaled gases, vapours and particles have before they can be absorbed into the body are based on physical properties, such as particle size and solubility. By necessity, the mammalian lung has developed an elaborate defence system to preserve its structure and function.
- From an evolutionary perspective, the respiratory system has evolved to deal with the kinds of airborne materials found in the natural environment. The respiratory system is poorly equipped to deal with the range of pollutants and contaminants of the modern environment, particularly those found in work-places. For this reason, sometimes the conventional responses of the respiratory system are not adequate to cope with such exposures.

■ There is wide variability in individual responses to inhaled materials. This variability may be observed as a genetic predisposition to particular contaminants (atopy), or as an acquired trait from previous exposures (for example, impairment in lung function induced by cigarette smoking).

■ The likelihood of respiratory problems from inhalation of airborne contaminants depends on a range of factors, including how much is inhaled, how much is retained, how much is cleared out of the lungs, how much is biotransformed, whether critical doses are reached in target tissues, and the pathological sequence of molecular, biochemical, and cellular events, as well as tissue effects and defence responses that can lead to clinical disease (see Figure 4.1).

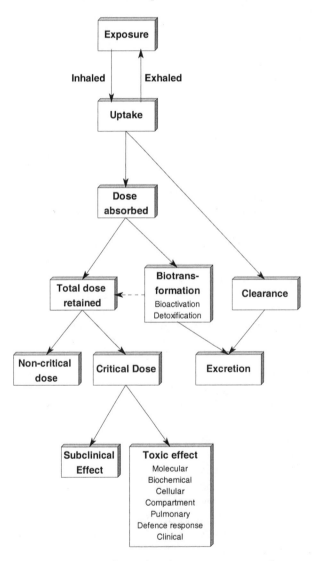

Figure 4.1 Inter-relationships between exposure, absorption, excretion and effects from inhalation of airborne contaminants.

Occupational lung diseases are among the earliest occupational diseases reported. Coal miner's pneumoconiosis, cereal handler's lung diseases, silicosis and asbestosis are examples of occupational respiratory disease that have well-established histories. Occupational lung diseases are preventable (Sahu 1992).

The respiratory system

The function of the respiratory system

The main target organ of entry into the body for gases and vapours is the respiratory system. The primary function of the respiratory system is to provide a means for the efficient exchange of oxygen and carbon dioxide (see Figure 4.2). To do this, the respiratory system must be in continuous contact with atmosphere in the external environment; and it is indeed a unique and direct link between the blood supply and the external environment. The functional structure of the contact

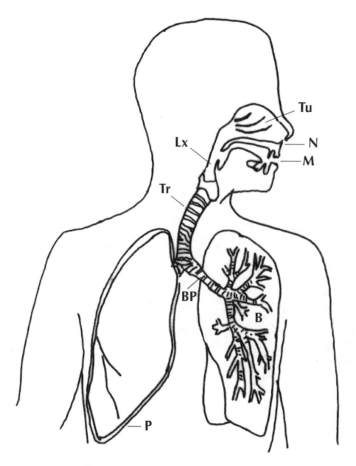

Figure 4.2 The respiratory system. B, bronchus; BP, primary bronchus; Lx, larynx; M, mouth; N, nostrils (nares); P, pleura; Tr, trachea; Tu, turbinates (conchae).

between the external environment and the body is the air sac, or alveolus. There are many millions of alveoli in the lungs, as shown in Figure 4.3.

All cardiac output travels through the lungs with each cardiac cycle through the body. This makes the lung particularly suited for the uptake of gases and volatile compounds, and an efficient organ for the uptake of nonvolatile materials as well. Materials that can inflict direct injury to lung tissue can lead to respiratory malfunction. This can cause a spectrum of pathological changes up to and including death.

The importance of inhalation as a route of exposure at the workplace cannot be overemphasised.

The anatomy of the respiratory system

Regional structures of the respiratory system

The lung has marked regional differences. It is useful to consider the respiratory system as having four compartments (see Figure 4.2):

- the nasopharyngeal compartment, from the nostrils to the vocal cords
- the tracheobronchial compartment, from the vocal cords to the respiratory bronchioles

Figure 4.3 The alveoli. A, alveolus; Ad, alveolar duct; As, alveolar sac; Br, respiratory bronchiole; Bt, terminal bronchiole; C, alveolar sac with pulmonary network. (Only one alveolus shown.).

- the pulmonary/parenchymal compartment, from the respiratory bronchioles to the alveoli
- the pleural space (the space between the lining of the lungs, and the lining of the ribs).

Of these, the first three are important in occupational lung diseases, and the last one has importance for the occupational lung disease mesothelioma.

The anatomical features of the nasopharyngeal compartment are:

- it is the entry for inspired air
- it consists of nasal turbinates, epiglottis, glottis, pharynx, larynx
- it is rich in blood vessels.

The anatomical features of the tracheobronchial compartment are:

- it is lined with ciliated cells and mucous-secreting goblet cells
- it is the delivery system for inspired air
- it consists of branching ducts, from trachea to terminal bronchioles
- it is covered with mucous-secreting goblet cells and ciliated cells forming a mucociliary blanket.

The anatomical features of the pulmonary compartment are:

- it is the area of gas exchange
- it consists of alveolar ducts and alveoli
- alveolar cells are very thin, and bring inspired air into intimate contact with blood and lymph
- cells of the alveoli secrete surfactant, which assists in gas exchange and alveolar integrity.

Further subdivisions of these areas are possible. For example, the nasopharyngeal region contains the olfactory epithelium, where the olfactory and other sensory cells are located; and the nasal epithelium, which contains nerves and a venous plexus which can lead to nosebleeds (epistaxis) following intense exposures.

Lung cell types

There are over 40 cell types required to perform the various functions of the respiratory tract:

- seventeen types of epithelium
- nine types of connective tissue
- two types of bone and cartilage
- seven types of blood cells
- two types of muscle cells
- five types associated with pleural and nervous tissue elements.

The cells of most interest are those that are unique to the respiratory tract, such as:

- ciliated epithelium
- nonciliated bronchiolar epithelium (Clara cells)
- Type I (squamous alveolar) pneumocytes
- Type II (great alveolar) pneumocytes
- alveolar macrophages.

The mucociliary blanket

The conducting airways from the trachea to the terminal bronchioles are lined with epithelial cells that differ in type and function at the different levels. Ciliated epithelial cells are the predominant cells in the trachea, bronchi, and bronchioles of airways greater than 1 mm in diameter. Cilia are fine hair-like extrusions that project from the top of the epithelial cell into the lumen of the airway.

The other predominant cell type is the goblet cell, so called because it is shaped like a wine goblet. The goblet cells secrete mucous into the lumen of the airway, where is acts to keep the surfaces moist and to collect inhaled particles (see Figure 4.4a).

Other cell types include support cells in the epithelial cell layer, and basement membrane and connective tissue cells (see Figures 4.4a and 4.4b).

The cilia are organised to beat in a coordinated fashion to propel the mucous from the terminal bronchioles to the bronchi, to the trachea. The mechanism of action of the cilia is shown in Figure 4.5. This mucous blanket is a means of

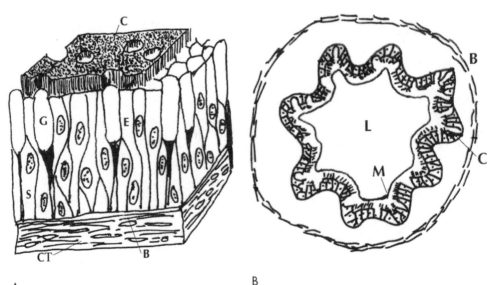

A   B

Figure 4.4 Ciliated epithelium. A) Section of ciliated epithelium: B, basement membrane; C, cilia; CT, connective tissue; E, epithelial cell; G, goblet cell; S, supporting cell. B) Section through a bronchus: C, cilia; L, lumen of bronchus; M, layer of mucous (not to scale); B, muscle lining of bronchus.

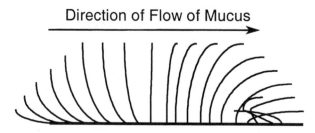

Direction of Flow of Mucus

Figure 4.5 Action of the cilia.

defence, and the 'mucociliary blanket' is a means of clearance from the airways. These diagrams give an idea of the way in which the ciliated epithelial and goblet cells work.

A number of structural components of the mucociliary blanket are important for normal function. These features, and the sorts of diseases or airborne contaminants that can damage them, are shown in Table 4.1.

Airway smooth muscle

Airway smooth muscle has an important role in lending rigidity to the luminal walls and in regulating air flow and ventilation by changing airway diameter. The majority of airway smooth muscle is located in the conducting airways, and lies beneath the epithelial cells. The relaxation and contraction of airway smooth muscle is controlled by nerve stimulation and by cellular mediators released locally. Nervous control is by cholinergic (contraction) and adrenergic (relaxation) stimulation. Cellular mediators such as the eicosanoids participate in the regulation of airway smooth muscle tone. During inflammation or hypersensitivity, mediators such as histamine, tachykinins and platelet-activating factor will also contribute to the contractile state of airway smooth muscle.

As the terminal bronchiole diminishes in diameter and terminates in the respiratory bronchiole, the cilia-bearing cells eventually disappear. Nonciliated bronchiolar cells (Clara cells) are typically present in only the small bronchioles. These are major sites of damage from xenobiotic compounds, and may be involved in biotransformation process by a lung cytochrome P450 system.

The alveolar surface

The bronchioles eventually terminate into the alveoli. These are small air sacs where gas exchange occurs. There are literally millions of these air sacs in the lung, providing a huge surface area for gas exchange. The outside walls of the alveoli (that is, the side away from the air supply) are highly vascular, with a huge network of capillaries that bring blood close to the alveolar surface.

The alveolar surface on the luminal (air) side of basement membrane is lined by squamous alveolar type I pneumocytes and by Type II alveolar pneumocytes. The Type I cell has few cellular organelles and is believed to function as a barrier. The Type II cell has significant cellular contents, including endoplasmic reticulum,

Table 4.1 Structural components of the mucociliary blanket and the diseases or contaminants that can cause damage

Structural component	Mechanism of damage	Disease or toxicant
Membrane	Altered cellular surface tension	–
Mucous production	Altered viscosity	Tobacco smoke Chronic bronchitis Formaldehyde
	Excessive production	Chronic bronchitis Asthma Cystic fibrosis
	Tethering to goblet cells	Dehydration
	Altered composition	Cystic fibrosis
	Altered pH	Sulfuric acid mist
Cilia	Decreased or absent	Irritants Tobacco smoke
	Reduced beat (frequency)	Acetaldehyde Cadmium salts
	Reduced beat (amplitude)	Dimethylamine
	Uncoordinated beating	Tobacco smoke
	Beating in reversed direction	Tobacco smoke
Goblet cell microvilli	Impaired function	—
Cell junctions	Leakage of ions and materials	Tobacco smoke Diethyl ether Antigen challenge
Epithelial cells	Defective ion transport	Physical trauma Cystic fibrosis
	Altered intracellular pH	Ammonium ions Sulfate ions
	Defective energy metabolism	Cadmium
	Cell death	Irritants (at high exposures)
Blood vessels	Increased permeability	Allergy
Mucoserous glands	Altered secretions	—
Nerves	Increased goblet cell and other secretions	Some drugs
	Modified ciliary activity	—

Golgi apparatus, and other cytosomes. Type I cells are very thin (0.1–0.3 µm) with a large surface area covering 93% of the alveolar surface area. Type II cells are more numerous but due to their cuboidal shape make up about 7% of the surface area. These cells synthesise and release surfactant that is vital in regulating surface tension for parenchymal support.

Alveolar macrophages

Alveolar macrophages are resident phagocytic cells that are found free in the alveoli. They ingest inhaled particulate material and are the main clearance mechanism for the pulmonary region. Infectious particles are usually engulfed by the macrophages except in some chronic diseases, such as tuberculosis, and in some viral diseases. The macrophages are cells which can move through and across the surface of the alveoli, and contain enzymes that can break down the materials they engulf (such as bacteria). Following injury to the alveolar epithelium, macrophages can also release factors that promote growth of the Type II cells and thus play a role in the repair process.

Physiology of the respiratory system

Gas exchange

The specialised anatomy of the pulmonary system ensures the efficient and effective gas exchange of oxygen and carbon dioxide. The movement of these gases across alveoli cell walls to (or from) the capillary occurs through diffusion down a concentration gradient and depends on the different partial pressures of the gases. Table 4.2 shows the concentrations of oxygen and carbon dioxide in atmospheric air, alveolar air, and blood.

With an alveolar oxygen concentration of 105 mm Hg and a blood concentration of 40 mm Hg oxygen will diffuse into blood. With an alveolar oxygen concentration of 40 mm Hg and a blood concentration of 45 mm Hg carbon dioxide will diffuse into the alveoli.

Gas transport

Oxygen is transported by the blood either in combination with haemoglobin (Hb), oxyhaemoglobin (HbO_2), or a tiny amount is dissolved in plasma. During rest,

Table 4.2 Oxygen and carbon dioxide levels in air and blood

	Air		Blood	
	Atmospheric	Alveolar	Deoxygenated	Oxygenated
O_2	20.9%	14%	—	—
pO_2	160 mm Hg	105 mm Hg	40 mm Hg	105 mm Hg
pCO_2	0.3 mm Hg	40 mm Hg	45 mm Hg	40 mm Hg

95% of all oxygen delivered to tissues is transported in combination with haemoglobin, while during exercise this value may exceed 99%. The amount of oxygen combined with haemoglobin depends on the partial pressure of oxygen in the blood to which the haemoglobin is exposed (that is $O_2 \Leftrightarrow pO_2$). When haemoglobin is fully converted to oxyhaemoglobin it is fully saturated. When both forms exist, haemoglobin is partially saturated. The percentage saturation of haemoglobin and oxygen is shown by the oxygen–haemoglobin dissociation curve as shown in Figure 4.6.

When pO_2 is high, oxyhaemoglobin is high and, conversely, when pO_2 is low, oxyhaemoglobin is low. The rate of saturation differs, depending on the pO_2; at higher levels the rate of increase is less. This explains why people can still perform well at low oxygen concentrations (at high altitudes or with cardiac and pulmonary diseases); for example, at a pO_2 of 40 mmHg, haemoglobin is 75% saturated.

Damage to the respiratory system can displace the oxygen–haemoglobin dissociation curve to the left. This means that the respiratory system has to work harder to maintain body oxygen levels. In turn, this means that the lungs may be made more vulnerable to further adverse exposures.

Lung volumes and capacities

A number of physiological parameters are used in measuring respiratory function as shown in Figure 4.7.

These parameters include:

- Tidal volume (TV): the volume of air that enters the lungs in a breath.
- Minute volume (MV): the volume of air that enters the lungs in a minute. A product of tidal volume (TV) and respiratory rate.

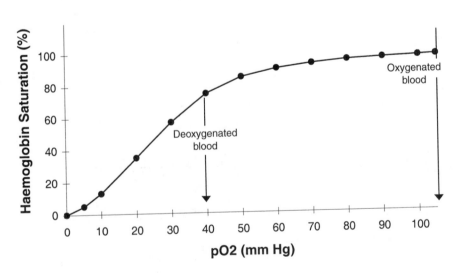

Figure 4.6 The oxygen–haemoglobin dissociation curve.

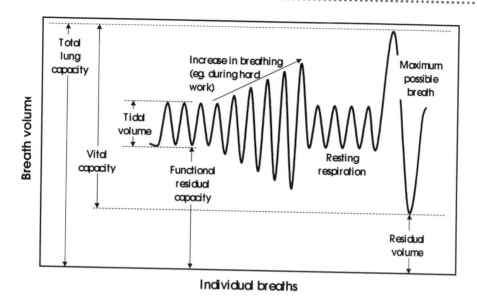

Figure 4.7 Respiratory volumes and capacities.

- Functional residual capacity (FRC): the volume of air remaining in the lungs at the end of expiration.
- Expiratory reserve volume (ERV): one part of the FRC. This is the maximal additional volume of air that can be exhaled following a normal exhalation.
- Residual volume (RV): the other part of the FRC. This is the volume of air remaining in the lungs at the end of maximal exhalation.
- Inspiratory reserve volume (IRV): the maximal volume of gas that can be inhaled following spontaneous inspiration.
- Inspiratory capacity: this is made up from VT and IRV. It is the maximal volume of air that can be inhaled from the FRC.
- Total lung capacity (TLC): the total volume of air in the lungs following a maximum inspiration (about 4.5–6 L).
- Vital capacity (VC): the maximum volume of air that can be exhaled following a maximal inspiration.

Other parameters exist. For example, the alveolar minute volume is the tidal volume minus the dead space multiplied by the respiratory rate.

The maximum rate of air flow that can be achieved during expiration is called the peak expiratory flow rate (PEFR). This can be measured using a peak flow meter and healthy subjects have a PEFR of 400 to 600 L/min.

Two further important physiological parameters are the forced expiratory volume in one second (FEV_1) which is the volume expired in the first second of a forced expiration, and the forced vital capacity (FVC).

Respiration and work

To appreciate the amount of air that can be inhaled by people at work, it is first necessary to calculate the amount a worker might breathe over one day at work. At a conservative estimate of 500 ml per breath and 15 breaths per minute, a worker might inhale 3.6 m^3 of air over an 8-hour shift (see Table 4.3). Over a full day, a person would breathe three times this amount (10.8 m^3). Over a 70-year lifetime a person would breath about 275,000 m^3 of air. Therefore, the human lung comes into contact with approximately 20 kg of air each day. Along with the air is a chemical and microbial load.

Parameters affecting deposition of inhaled materials

Many factors can affect the deposition of inhaled materials, once they reach the respiratory system. However, many of these factors depend on what is happening in the relevant region of the respiratory system (see Figure 4.8).

The process of respiration moves large volumes of air from the external environment to the gas exchange areas. It follows that as each breath is inhaled or exhaled, significant changes are occurring in the direction and rate of air flow. These will affect *absorption* (which can occur in all regions of the respiratory system) and deposition (see Figure 4.8a).

In the upper airways, as the rate of the air flow is so high, many inhaled particles are propelled into the lining of the airways where they can be absorbed (in the size range 5–30 μm). This is called *inertial impaction* (see Figure 4.8b).

Further down the system, in the bronchi and bronchioles, the speed and directional change is less traumatic, and smaller particles (in the range 1–5 μm), which were carried along by the force of respiration, begin to drop out of the air flow. This is called *sedimentation* (see Figure 4.8c).

Finally, at the end of the airways, the terminal bronchioles, and in the gas exchange areas, there is hardly any change of air movement at all, and gases, vapours and small particles (less than 1 μm) tend to diffuse across a concentration gradient (for example, oxygen in and carbon dioxide out). This is the process of *diffusion* (see Figure 4.8d).

Table 4.3 Respiratory parameters in workers

Breath volume	0.5 L per breath
Respiratory rate	15 breaths per minute
Minute volume	7.5 L per minute
Hour volume	450 L per hour
Shift volume	3600 L per 8-hour shift
Shift volume	3.6 m^3 per 8-hour shift

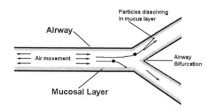

a. **Absorption**. Dependent on solubility in mucus (important for soluble gases, vapours and particulates.

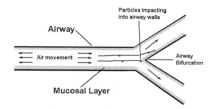

b. **Inertial impaction**. Dependent on air speed. More particles would collide with the airway walls if the air is turbulent or is moving fast.

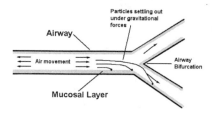

c. **Sedimentation**. Dependent on gravity, such as particulates or fibres.

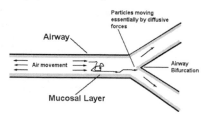

d. **Diffusion**. Requires low air speeds (for example, in terminal bronchioles or alveoli).

Figure 4.8 Absorption on inhaled materials in airways.

The role of the nasal turbinates

One other factor is important when looking at the interactions between work and respiration. Human beings are nose-breathers when the demand for oxygen is not great. This is a good thing, because air inhaled through the nasal passages and turbinates passes it over the mucous lining, humidifying it and removing any particulates or soluble gases and vapours. As such, this region acts like a filter (see Figure 4.9).

Once oxygen demand increases so that inhaling through the nose is not sufficient to supply the body's needs, the human becomes a mouth-breather. This means that the nasal passages are bypassed in the need to get more air into the lungs. Under normal physiological conditions, for example during exercise, this is a perfectly acceptable thing to do.

However, the possible toxic exposure of workers carrying out hard work in a contaminated environment is likely to be increased because they are breathing more contaminated air, and the air they are breathing is going straight into the lungs without the benefit of filtering through the nasal passages. Therefore, careful consideration should be given to inhalational exposure for workers with heavy workloads.

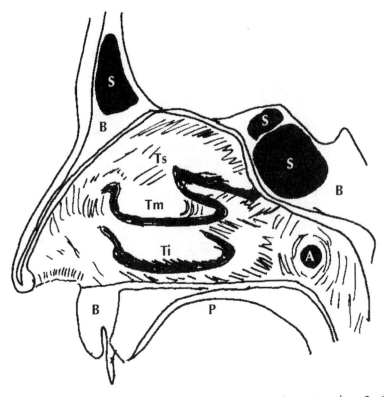

Figure 4.9 The nasal turbinates. A, auditory tube; B, bone; P, palate; S, sinus; Ti, inferior turbinate; Tm, medial turbinate; Ts, superior turbinate.

Pathology of the respiratory system

Defences, clearance and retention of inhaled materials

Again, the mechanisms of clearance or retention of inhaled materials is dependent on the region of the lung.

The nasopharyngeal compartment has the following mechanisms:

- smell/avoidance
- coarse nose hairs
- impaction and collection in nasal turbinates
- irritation of the nasal epithelium evoking sneeze
- some production of mucous.

The tracheobroncheal compartment has the following mechanisms:

- cough
- irritation of epithelial lining
- bronchoconstriction

- mucociliary clearance moves dissolved or suspended material through the respiratory system to the back of the mouth where it can be expectorated or swallowed – the layer of mucous varies in thickness from 5–10 μm (thinner at the bronchial end) and the layer of mucous flows faster at the top of the mucociliary escalator than at the bronchiolar end.

The pulmonary compartment has the following mechanisms:

- phagocytosis primarily by alveolar macrophage activity
- some drainage of material into lymph
- inevitably some material will be retained by lung tissue.

The status of lung defences are also dependent on a number of factors: they may not be developed; they might be impaired (for example, by prior exposure, smoking, immunosuppression); they might be exaggerated (for example, in asthma); or they might be aged.

Pathological responses of inspired materials in the respiratory system

The respiratory system has a range of responses to inhaled toxicants; these range from immediate reactions to highly toxic or irritant materials, to longer term effects from long-term repeated dose exposures. The ultimate response of the respiratory system largely depends on the site of deposition and action of the inhaled agent.

In the nasopharyngeal compartment the response includes:

- rhinitis
- nasal ulceration and septal perforation (acid mists)
- nasal cancer (formaldehyde).

In the tracheobroncheal compartment the response includes:

- bronchoconstriction
- bronchitis
- lung cancer (radon, nickel carbonyl).

In the pulmonary compartment the response includes:

- alveolitis and bronchiolitis
- diffuse alveolar damage (ammonia, nitrogen dioxide)
- pulmonary oedema
- emphysema.

A full range of pathological responses in the respiratory system to inhaled materials (whether gases, vapours or particulates) is summarised in Table 4.4.

Table 4.4 Pathological responses of the respiratory system to inspired materials

Response	Mechanism
Absorption	Absorption of agents can have toxic effects, either in the lung, or elsewhere in the body
Asphyxiation	Either by a reduction in the concentration of oxygen in inspired air by physical displacement (simple asphyxiation) Or by a reduction in oxygen transport in the body by chemical reaction (chemical asphyxiation)
Local irritation	Related to the solubility of the substance on moist surfaces and mucous membranes of nose, eyes, mouth and upper respiratory tract
Irritation of airways/ bronchoconstriction	Irritation of the airways leads to bronchoconstriction. More extensive and smaller airway constriction occurs on exercise or exertion than at rest
Increase in the secretion of mucous	Increase in secretion of mucous will slow down ciliary movement, and may block smaller airways
Cell damage/oedema	Damage to cellular components of airways and alveoli results in increased permeability, loss of compliance, necrosis and intraluminal (within the airways) rather than interstitial (within cells of the airways) oedema. Pulmonary oedema may, in turn, be compounded by secondary infection
Macrophage cytotoxicity	Alteration in function or destruction of alveolar macrophages will alter clearance processes, which can lead to collection of respired particles in a given area
Sensitisation and allergy	Dependent on immunological status and disposition to asthma
Lung overload by particles	When lung burdens of particulates are sufficient to exceed physiological clearance mechanisms, such as macrophage phagocytosis, lung burdens of such particulates will persist, and completely nonphysiological mechanisms of disease pathogenesis may occur
Emphysema	Abnormal presence of air. In the lungs, emphysema is an overdistension of the alveoli, and destruction of their walls in parts, giving rise to the formation of large sacs from the rupture and running together of a number of contiguous air vesicles. Another form, acute interstitial emphysema, involves the infiltration of air beneath the pleura and between the pulmonary air cells

continued

Table 4.4 continued	
Response	Mechanism
Granulomatous reactions	Granuloma is a new growth made up of granular cells, caused by chronic inflammation
Fibrogenesis	The growth of fibrous tissues, comprising fibres of collagen and elastin. Between these cells lie star-shaped cells, or fibroblasts, from which collagen or elastin is formed. Elastin has elastic properties and is used in the walls of arteries, and so forth. Normally, collagen is grouped into bundles which are held together by other fibres, used to make ligaments, tendons and sinews. It is also the substance laid down in the repair of wounds, or as a result of inflammation, and forms scar tissue. Fibrogenesis is the growth of collagen fibres in response to cellular inflammation and damage
Cancer	Oncogenesis leading to primary lung tumours

The toxicity of atmospheric contaminants

Types of contaminants

There is a huge range of airborne contaminants in workplace air, including vapours, fumes, mists, dusts and fibres (Amdur 1991). In common usage, these terms require definition (Kennedy and Trochimowicz 1982):

- **Gas** is usually applied to a substance that is in the gaseous state at room temperature and pressure.
- **Vapour** is applied to the gaseous phase of a material that is ordinarily a solid or liquid at room temperature and pressure.
- **Aerosol** is applied to a relatively stable suspension of solid particles in air, liquid droplets in air, or solid particles dissolved or suspended in liquid droplets in air.
- **Mists** and fogs are aerosols of liquid droplets formed by condensation of liquid droplets on particulate nuclei in the air.
- **Fumes** are solid particles formed by combustion, sublimation or condensation of vapourised material. Conventionally, the size of fume particles are considered to be less that $0.1\,\mu m$, though the size can increase by aggregation or flocculation as the fume ages
- **Dusts** are solid particles in air formed by grinding, milling or blasting.
- **Fibres** are solid particles with an increased aspect ratio (the ratio of length to width). Fibres have special properties because of their ability to be suspended in air for longer periods than dusts and other aerosols.

These can be divided into two main types of contaminant:

- those that are dissolved in air, such as gases and vapours
- those that are suspended in air, such as fumes, dusts, mists, aerosols, fibres (also called particulates).

Dissolved contaminants will reach, and can have effects in, all parts of the respiratory system. Suspended particles may be inspired into the respiratory system, although the depth that they can reach is a function of their size, measured as the mean aerodynamic diameter of the particle.

There are a number of factors that can effect the severity of inhaled gases and vapours, fumes, mists and particles:

- size of the inhaled contaminant (for particulates)
- solubility (ability to dissolve in tissue fluids)
- reactivity (ability to react with tissue components); included in reactivity is the general property of oxidative burden, caused by the generation of free radicals)
- conditions of exposure, such as concentration and duration of exposure
- lung defences, genetic or acquired
- exercise or rest state
- immunological status
- tissue water content.

Of these, size and solubility are the most important intrinsic properties, and conditions of exposure relate to risk.

Figure 4.10 shows the size of various atmospheric contaminants. Only those below a size of 5 μm are considered respirable, that is, they can reach into the deep portions of the respiratory system where the alveoli are located.

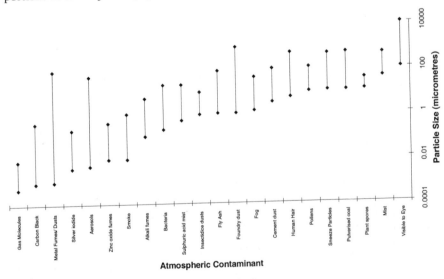

Figure 4.10 The sizes of various airborne contaminants.

Measurement of inhalational toxicity in animals

The toxicity studies

Rats and mice are the most common choices for most single-dose toxicity tests, for reasons of cost and relative ease of handling. Furthermore, factors such as genetic background and disease susceptibility are well known for these species.

With time, toxicity testing methods have become codified, and standard experimental protocols have now been issued by various regulatory agencies worldwide, such as the US Federal Food and Drugs Agency, US Environmental Protection Agency and the Organization for Economic Co-operation and Development (OECD). One of the most well recognised of these are those of the OECD which has issued a number of standard protocols for the conduct of physicochemical, toxicity and ecotoxicity tests since the early 1980s. These protocols include some that have single-dose toxicity values, such as the inhalational LC_{50}, as an endpoint.

A number of features are common to all recommended experimental protocols, including:

- temperature and humidity control
- light
- atmospheric conditions
- noise
- diet
- ventilation
- biological factors such as infections and diseases
- requirements for quality assurance, storage of samples and data and other aspects of Good Laboratory Practice.

Parameters for the standard single-dose inhalational toxicity study are shown in Table 4.5. The LC_{50} provides an estimate of the relative toxicity of a substance by the inhalational route of exposure. Extrapolation of inhalational LC_{50} values in animals to humans is valid only to a limited degree.

Duration of inhalational studies

The variability of duration of inhalation toxicological studies is quite substantial. Review of the duration of inhalational studies for chemicals with exposure standards indicates that they range from 10 minutes to 9 hours, with many being carried out for 4 hours.

From a procedural standpoint, there are a number of reasons why tests of different duration are carried out. For example, arsine is a very toxic compound and is absorbed rapidly by the inhalational route so the study for arsine was carried out for only 10 minutes, whereas the exposure duration for ethyl alcohol was 9 hours, as this chemical has only a moderate vapour pressure and is not particularly acutely toxic.

The OECD guidelines for conducting acute inhalational studies suggest that the duration of exposure should be at least 4 hours. However, this is optional, and the

Table 4.5 Parameters for the single-dose inhalational toxicity study

Test parameter	Skin irritation
Animals	Five animals per dose (rats being the preferred species) of each sex. At the end of the study, one group of five animals of the other sex should be dosed with an appropriate dose to check for sex sensitivity
Dose	At least three dose groups, and enough to produce a dose–response curve. Dosing should be chosen to produce a range of toxic effects and mortality rates to allow estimation of toxicity
Equipment	Designed to enable a dynamic air flow of 12–15 air changes per hour, to ensure an oxygen concentration of at least 19% and an evenly distributed exposure atmosphere. The 'volume' of animals should not exceed 5%, the volume of the test chamber. Alternatively, oronasal, head only, or whole-body individual chambers may be used
Duration of contact	Single uninterrupted exposure by inhalation over a short period of time (24 hours or less) to a substance capable of being inhaled. The preferred exposure period is 4 hours, although the time of exposure depends on the toxicity of the inhalant (see Figure 4.11)
Assessment criteria	Estimation of inhalational LC_{50} and signs of toxicity from pathology
Duration of test	At least 14 days

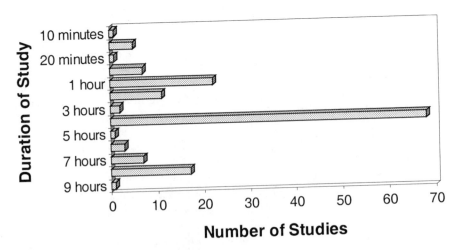

Figure 4.11 Duration of inhalational studies.

OECD also suggest other durations may be needed to meet specific requirements. If a test is carried out for longer than 4 hours, the reason for the change should be given.

Gases and vapours

Properties of gases and vapours

Gases are one of the three fundamental states of matter, with fundamental properties of no fixed form or volume. They share these properties with vapours (the gaseous phase of a liquid). Some vapours have significant potential to cause problems in the workplace, because they can reach high concentrations. It is therefore appropriate to give such vapours the same consideration as gases.

Gases are extremely easy to handle. They can be enclosed, compressed, liquefied, containerised, piped, and so forth, without too much technological difficulty. While contained, they present little risk to users or other potentially exposed individuals.

However, in that there is no physical barrier between the air and the various structures that make up the human respiratory system, gases can enter the body very easily and quickly (Gardner et al. 1988). So risks can arise once gases are released and are available for exposure (whether occupational, domestic or environmental).

Toxicity of gases and vapours

A significant number of gases can be encountered in the workplace environment, including asphyxiants, irritants, sensitisers and toxic gases (Witschi and Last 1996). The toxic effects of gases and vapours are as follows:

- Asphyxiation is the ability of a gas or vapour to displace oxygen from air by dilution (simple asphyxiation) or by interfering with the ability of the body to transport oxygen (toxic asphyxiants).
- Irritation is the ability of a gas or vapour to cause local symptoms of irritation upon exposure or contact with the tissues of the respiratory system.
- Sensitisation is the ability of a gas or vapour to cause an immune response upon exposure or contact with the tissues of the respiratory system leading to respiratory sensitisation and asthma.
- Toxicity is the ability of a gas or a vapour to produce a toxic effect either at the site of contact (that is, the respiratory system) or following absorption and distribution to other tissues or organ systems.

These effects may be observed in the nasal passages (Barrow 1986), the mucous membrane lining the airway (Jeffery and Reid 1977), in the alveoli in respiratory function (Mauderly 1989), or in clearance mechanisms (Schlesinger 1990).

While the target organ is the respiratory system, these gases can cause a range of health effects in a number of body systems. These include effects in the nervous system, blood-forming tissues, and the liver (see Table 4.6).

Table 4.6 Gases and vapours, and the body systems they affect

Organ systems affected			Examples
Respiratory system	Irritants		Chlorine (Cl$_2$)
			Ammonia (NH$_3$)
			Oxides of nitrogen (NO$_x$)
			Sulfur dioxide (SO$_2$)
			Sulfur trioxide (SO$_3$)
			Fluorine (F$_2$)
			Phosphine (PH$_3$)
			Phosgene (COCl$_2$)
			Ozone (O$_3$)
			Formaldehyde (CH$_2$O)
			Acrolein
	Corrosives		Acid mists
			Caustic mists
	Asphyxiants	Simple	Nitrogen (N$_2$)
			Hydrogen (H$_2$)
			Methane (CH$_4$)
			Helium (He$_2$)
			Argon (Ar$_2$)
			Neon (Ne$_2$)
			Ethylene (C$_2$H$_4$)
			Ethane (C$_2$H$_6$)
		Toxic	Carbon monoxide (CO)
			Hydrogen cyanide (HCN)
			Hydrogen sulfide (H$_2$S)
	Sensitisers		Isocyanates (—N=C=O)
			Amines (—C—NH$_2$)
Central nervous system			Carbon disulfide (CS$_2$)
			Aliphatic hydrocarbons
			Solvent vapours
Blood-forming system			Arsine (AsH$_3$)
Carcinogens			Vinyl chloride (C$_2$H$_3$Cl)
			Nickel carbonyl (NiCO)
			Formaldehyde (CH$_2$O)
			Sulfuric acid mists (H$_2$SO$_4$)

Gases have both local effects in the respiratory tract and systemic effects after inhalation. Therefore, the respiratory tract is both a point of entry and a target organ of toxicity. Local adverse effects relate mainly to irritation, although increased susceptibility to allergy, infection, structural disease and cancer also occur in the respiratory tract in response to gas, vapour or fume exposure. There are a number of systemic effects, although the main effect is asphyxiation.

At the level of exposure to gases, the main controls are by somehow placing a barrier between the gas and the respiratory system. As with the control of all hazards, the hierarchy of controls should be applied, and elimination, substitution and isolation are preferable to measures that dilute gas concentrations (such as ventilation). The final control that should be considered is the use of respiratory protection (Terenski and Cheremisinoff 1983).

Particulates

Classification of particulates

In this chapter, particulate is a term used to describe all materials (whether liquid or solid) suspended in air. There are many different types of particulates found in workplaces, and they can be broadly classified as either organic or inorganic. Organic dusts originate from plant and animal materials, but also from synthetic material of an organic nature (for example, chemical intermediates or plastics). Inorganic particulates can be further subdivided into metallic or nonmetallic, and whether or not they have silica present. A fuller classification is shown in Figure 4.12.

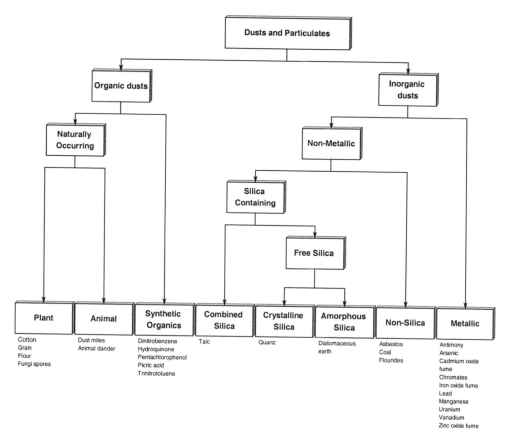

Figure 4.12 Classification of particulates.

The toxicity of particulates

While the main risk of exposure to particulates is to the respiratory system, some particulates (for example, nickel) can also cause skin problems on contact. Still other dusts can have soluble components that can cause systemic toxicity (such as kidney damage from cadmium or nervous system damage from lead).

The size of an airborne contaminant is important in inhalational toxicology (Witschi and Brain 1985). Airborne particles can have a range of masses, densities, shapes and sizes, and these will bear on the amount of time the particle will be airborne (Oberdorster 1995). From the time it is generated, the particle will be subject to forces of:

- intrinsic kinetic energy from any forces that arose during the process of particle generation (for example, particles from a grinding wheel)
- air movement at the source of particle generation (such as air flow from ventilation systems)
- diffusion
- air resistance
- gravity.

A fuller discussion of these factors is given in Chapter 15 on atmospheric contaminants.

The critical factor for determining the time that a particle is airborne is the terminal velocity. Large particles fall out of the air quite quickly, becoming dusts. Small particles fall under the force of gravity, but are subject to air resistance. The smaller the particle, the greater the effect of air resistance, and the longer a particle stays in air. Eventually, the forces of gravity and air resistance reach an equilibrium, and terminal velocity is reached.

Experiments with 'standard spherical' particles show that terminal velocity is fairly linear, until the particle diameter falls below 20–30 μm, where air resistance has a greater effect.

The terminal velocity becomes very low with particles of very small diameter. For example, a particle with a diameter of 300 μm will reach terminal velocity (at 1 m/s) very quickly, and fall 1 metre in 1 second. Such particles do not represent a risk to exposed individuals as they are not generally available to be inhaled. Because of the experimental basis of this data (using spherical objects unlikely to be encountered in the occupational environment), it is likely that terminal velocities would be lower for workplace dusts, fumes and other particles. Indeed, only particles with a diameter below 50–100 μm are relevant to inhalational toxicology.

Further, it is generally accepted that particles with an aerodynamic diameter below 50 μm, with a terminal velocity of less than 7 cm/s (or 0.07 m/s), do not remain airborne for too long. This should be contrasted with a dust particle with a diameter of 1 μm, which has a terminal velocity of 0.03 mm/s (or 0.00003 m/s) where settling due to gravity is negligible for practical purposes, and movement in air is more important than settling through it.

Inhaled aerosols are rarely of constant dimensions – most contain particles of varying sizes and shapes. The size distribution of many aerosols encountered in workplaces approximates to a log normal distribution, which can be statistically characterised by the median or geometric mean and the geometric standard deviation. The relative mass of small particles to large particles is dependent on the cube of the radius of the particle, and a toxic effect caused by inhalation of a particulate may be due to a very small proportion of the total particles present in the aerosol.

For this reason, dusts and particulates are divided into size-specific fractions, based on their ability to enter and penetrate into the respiratory system. The main fractions of a total aerosol are shown in Figure 4.13.

Total dust is divided into *noninspirable* and *inspirable* (or inhalable) fractions. For a particulate to have a toxic effect in the respiratory system, it must be inhaled.

The largest inhaled particles, with aerodynamic diameters greater than 30 μm, are deposited in the upper airways of the respiratory system, that is the air passages from the point of entry at the nostrils or lips to the larynx (the *nasopharyngeal* fraction). During nasal breathing, particles are captured by nasal hairs and by impaction onto the mucosal lining of the turbinates and upper airways. Of the particles that fail to deposit in the nasopharyngeal fraction, the larger particles will deposit in the tracheobronchial airway region and can either be absorbed (if soluble or capable of dissolution) or be cleared by mucociliary clearance; this is the *tracheobronchial* fraction. The remaining particles not deposited in the nasopharyngeal and tracheobronchial regions can penetrate to the alveolar region, the region of gas exchange with the body; this is the *respirable* fraction.

A great deal of work has gone into establishing the limits of regional and total deposition of particulates in the respiratory system based on size. It is generally considered that the nasopharyngeal region collects particles of above about 7 μm (to 30 μm), that particulates of 2–30 μm are captured in the tracheobronchial region, and particles of less than 7 μm can penetrate to the alveolar regions. However, these values are approximate, and there is a discernible range of particle sizes that fall into nasopharyngeal, tracheobronchial and respiratory fractions (see Figure 4.14).

Following on from this work is the effort given to developing occupational hygiene monitoring methods that allow estimation of total, inspirable and respirable fractions on workplace aerosols. Some of the more common particulates

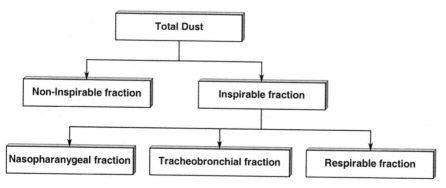

Figure 4.13 Size-selective components of particulates.

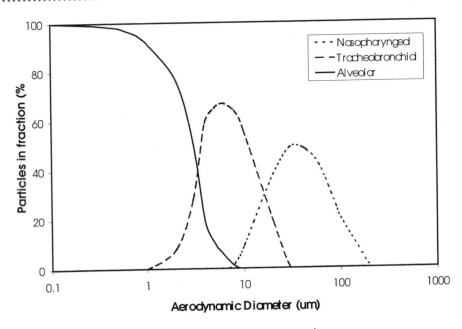

Figure 4.14 Particulate deposition in the lungs by size and region.

encountered in occupational environments are shown in Table 4.7. The main lung diseases from occupational exposure to particulates are shown in the section below (Morgan and Seaton 1987; Beckett and Bascom 1992; Coultas and Samet 1992; Wright 1993; Mitchell 1997; Schwartz and Peterson 1997).

Occupational lung diseases

The pneumoconioses

The word pneumoconiosis literally means dust in the lungs. Since not all dusts deposited in the lungs cause disease, the more widely accepted definition is that of the International Labour Organisation (ILO) which is that pneumoconiosis is the accumulation of dust in the lungs and the tissue reaction to its presence. The term normally means fibrosis of the lung due to accumulation of dust (mostly inorganic dusts).

Some dusts are not known to be toxic following inhalation. The term 'benign pneumoconiosis' has been used to describe this situation, and a range of inorganic materials have been shown not to damage alveolar architecture or give rise to collagenous fibrosis when they are inhaled and retained in the lungs. Materials such as iron ore (siderosis), tin ore (stannosis), barium compounds (baritosis), antimony ore, zirconium compounds, and titanium dioxide are classified as inert dusts (provided they are free of other toxic impurities and contain less than 1% quartz). The most common of these is caused by iron ore (siderosis), which occurs in welders, iron ore workers, and foundry workers.

Table 4.7 Common particulate toxicants

Particulate	Sources	Toxic effects/diseases
Asbestos	Mining, manufacture of asbestos products, construction, ship building	Asbestosis, pleural plaques, lung cancer, mesothelioma
Aluminium dust and abrasives	Manufacture of aluminium products, ceramics, paints, electrical goods, fireworks, abrasives	Aluminosis, alveolar oedema, interstitial fibrosis
Beryllium	Mining and extraction, alloy manufacture, ceramics	Berylliosis, pulmonary oedema, pneumonia, granulomatosis, lung cancer, cor pulmonale
Cadmium (oxide)	Welding, manufacture of electrical goods, pigments	Pneumonia, emphysema, cor pulmonale
Chromium [VI]	Manufacture of chromium compounds, pigment manufacture, tanneries	Bronchitis, fibrosis, lung cancer
Coal dust	Coal mining	Fibrosis, coal miner's pneumoconiosis
Cotton dust	Textile manufacture	Byssinosis
Iron oxides	Hematite mining, iron and steel production, welding, foundry work	Siderosis, diffuse fibrosis-like spneumoconiosis
Kaolin	Pottery manufacture	Kaolinosis, fibrosis
Manganese	Alloy production, chemical industry	Manganism, manganese pneumonia
Nickel	Mining, production, electroplating	Pulmonary oedema, lung cancer, nasal cavity cancer
Silica	Mining and quarrying, stone cutting, construction, sand blasting	Silicosis, fibrosis, silicotuberlosis
Talc	Rubber industry, cosmetics	Talcosis, fibrosis
Tantallum carbide	Manufacture and sharpening of cutting tools	Hard metal disease, hyperplasia of bronchial epithelium, fibrosis
Tin	Mining, tin production	Stanosis
Tungsten carbide	Manufacture and sharpening of cutting tools	Hard metal disease, hyperplasia of bronchial epithelium, fibrosis
Vanadium	Steel manufacture	Irritation, bronchitis

Inhalation of other inorganic dusts will stimulate responses that eventually lead to structural alterations in lung tissue and irreversible fibrosis (Phoon 1988). These are called the collagenous pneumoconioses. There are a range of materials that cause these conditions, including silica (silicosis), quartz, cristobalite, tridymite, diatomaceous earth, beryllium, asbestos (asbestosis), coal dust (coal worker's pneumoconiosis), graphite (graphite lung), carbon black, mica, aluminium and talc (talcosis). Of these, the most important are silica, asbestos and coal dust.

Silicosis

Free silica is silicon dioxide, which usually occurs in a crystalline form where the silicon and oxygen are arranged in a definite regular tetrahedral pattern throughout the crystal, or in an amorphous form where the silicon oxide has no definite pattern. There are a number of forms of silica of which quartz is the commonest. Silica particles can vary in size from 0.1–100 μm, indicating that silica can be deposited throughout the entire respiratory system.

Silicosis is a chronic fibrotic disease resulting from prolonged and intense exposure to free crystalline silica. Industrial sources of silica include foundry work, mining, quarrying and tunnelling, abrasive sand blasting, stone cutting and masonry, glass manufacture, ceramics and enamelling. Both acute and chronic forms of silicosis exist, although the acute form (silicoproteinosis) is now quite rare. The pathological changes are produced by the penetration of silica particles into the alveoli, where they are ingested by alveolar macrophages and, initially, lie in the engulfed vesicle (the phagosome). The phagosome usually interacts with the lysosome, a cellular organelle which contains proteolytic enzymes. These enzymes are released into the phagosomes and are involved in the digestion of the normal phagosome contents (e.g. bacteria) and breakdown of the phagosome wall, thereby releasing the phagosome contents into the cytoplasm of the cell (including the silica particles). This causes further release of lysosomal enzymes and eventually leads to the death of the cell. The liberated silica particles are then ingested by another macrophage, thereby perpetuating the cycle of cell destruction. Macrophages are also activated to produce proinflammatory mediators that then initiate the disease process, by causing fibroblast production. With continual exposure to silica, and continual activation of macrophages, the sustained release of these mediators and the production of fibroblasts leads to development of silicosis.

The earliest lesions in silicosis arise in the walls of the alveoli and respiratory bronchioles. Activation and/or death of macrophages leads to fibroblast proliferation, reticulin formation and production of collagen fibres. The fibres are laid down in a concentric fashion and become hyalinised. These eventually develop into nodules. The nodules expand by concentric growth, with transfer of silica to the outer layers (Ng and Chan 1991).

Chronic silicosis occurs with moderate exposure over a period of 20–45 years, usually involving respirable dusts containing less that 30% quartz. Symptoms may not develop for many years. The lesions are usually nodular (node sizes up to about 5 mm) and tend to be prominent in the upper lobes of the lungs, probably due to better clearance of dust from the lower lobes. Impairment of pulmonary function is uncommon in early stages. People with early silicotic nodules and only small opacities on X-ray can survive quite normally. The most common symptom of established silicosis is shortness of breath with exertion, although symptoms may not correlate with the radiographic appearance. A productive cough may also develop as the disease progresses, especially in cigarette smokers. There may be a decrease in lung function (FVC and FEV$_1$), total lung capacity, and lung compliance.

Advanced silicosis is complicated by the formation of massive fibrotic lesions, usually in the upper lobes, resulting from the coalescence of the fibrotic nodules and containing blood vessels and bronchi. Ventilatory capacity continues to decrease (this appears to be due to indirect effects, such as emphysema, rather than direct effects). There is distortion of the lung structure, with contraction of the upper lobes and the development of emphysematous changes in the lower lobes, resulting in airway obstruction. Radiographic findings include shadows in the early stages, followed by larger nodular opacities due to fibrosis. Typically, the mediastinal lymph nodes are enlarged and the pleura and pericardium may also be thickened with fibrosis. Later on, cyanosis and right-heart enlargement with pulmonary hypertension may be present. During the later stages, the disease is usually progressive despite removal from exposure.

Accelerated silicosis occurs with moderately high exposure to dusts containing 40%–80% quartz for 5–15 years, where the nodes tend to be smaller and the massive fibrosis favours the mid zones in the lungs. This disease is seen most frequently in sand blasters.

People with silicosis are also prone to tuberculosis, and silica has been implicated in the production of lung cancer (particularly in smokers).

Asbestosis

This is a fibrosis caused by asbestos. Asbestos is a term given to a group of fibrous silicates most of which contain magnesium. The health effects of asbestos are related to the fact that asbestos fibres can be small enough to be inhaled. Asbestos fibre bundles can break into smaller and smaller fibres until dust particles in the respirable size range are produced. Particles in the respirable range (2–10 μm) can reach the lower regions of the lung (the alveolar region) which is beyond the reach of the normal mucous clearing system.

Fibres retained in lung tissue may be up to 200 μm long and 3 μm or less in diameter. Some of the longer fibres get coated with a iron protein complex, producing drumstick-shaped 'asbestos bodies'. These are considered indicative of occupational exposure.

A range of health effects have been reported following asbestos exposure, depending on the type of asbestos, and the type, duration and intensity of exposure. Some of these effects are related to the amount of exposure (dose-related), some are not. In the expression of health effects, the length of the asbestos fibre has been related to health effects: asbestos fibres less than 5 μm in length may not be pathogenic. The ratio of the fibre's length to its width (called the aspect ratio) is important in defining a fibre, although there is still some debate about the correlation of aspect ratio with health effects.

Asbestosis is defined as a diffuse interstitial fibrosis (scarring) of the lung, the result of exposure to asbestos dust. It was fully recognised as an industrial disease in

1930. The severity of the condition is related to the degree of exposure. However, the scarring of lung tissue and loss of lung elasticity are not sufficiently different from other causes of interstitial fibrosis (such as silicosis), and diagnosis of asbestosis is difficult unless evidence of significant exposure to asbestos is available. Scarring of the pleural lining of the chest wall and diaphragm can be observed on an X-ray (pleural plaques) without any loss of lung function. Bilateral pleural plaques are strongly indicative of asbestos exposure and this supports the belief that such individuals are at greater risk of developing asbestos-related disease than individuals without such evidence of exposure.

Coal miner's pneumoconiosis

Historical cases of respiratory disease in miner's, called 'miner's black lung' or more colloquially 'black lung' or 'the dust', date back to the early 19th century. Coal worker's pneumoconiosis was initially thought to be due to silica exposure, as many coal seams are found in conjunction with silica-containing rock. However, studies in miners extracting coal with a low content of silica refute this idea (Schlueter 1994). Further, there were pathological differences between coal miner's pneumoconiosis and silicosis, and workers exposed to carbon materials in other industries (carbon black, graphite and carbon electrode manufacturing) developed similar lung lesions.

In general, coal miner's pneumoconiosis is more prevalent if the coal being extracted is of a 'hard' type such as anthracite, than of a soft type such as bituminous coal. Initially there may be no signs and symptoms. Clinically, coal dust causes a fibrogenic reaction that histologically does not have the clear-cut nodular picture of silicosis. The features are black pigmentation of lung tissue. Macrophages laden with dark colouration can be observed surrounding focal collections of coal dust, and are located in the vicinity of the respiratory bronchioles. Initially the lesions are discrete, but eventually they tend to coalesce into large areas. This is called progressive massive fibrosis and it most often occurs in miners with extensive exposures and heavy deposits of coal dust in their lungs. It is considered that without progressive massive fibrosis the condition remains relatively static after removal from exposure (unlike silicosis). The presence of progressive massive fibrosis often shortens the life of the miner owing to respiratory failure, pulmonary infection and focal emphysema.

Important features of the more common pneumoconioses are shown in Table 4.8.

Hypersensitivity pneumonitis from organic dusts

Dust diseases from inorganic dusts (the pneumoconioses) are pathologically different from diseases caused by organic dusts, which tend to be of a hypersensitivity pneumonitis type. Many of the organic dusts have biological activity, or are contaminated by bacteria or fungi, which cause reactions when they come into contact with lung tissues. The range of materials that can precipitate such reactions is quite large, and includes chemicals, dusts from cereals, mouldy hay, cotton, bird droppings, cork, bagasse (sugar cane fibres), coconuts, paprika, compost and some medical treatments, as listed in Table 4.9.

Table 4.8 Important features of common pneumoconioses

Disease	Pathology	Clinical features	Complications
Silicosis	Nodules, more in upper lobes Dense collagen-containing foci Emphysema Enlarged lymph nodes Pleural thickening Progressive massive fibrosis	Few symptoms unless widespread fibrosis is present Breathlessness and chest pain in later stages	Tuberculosis Chronic bronchitis Pleural effusion Cor pulmonale
Asbestosis	Nodules, more in lower lobes Alveolitis No dense collagen fibrosis Pleural plaques Asbestos bodies	Crepitations Clubbing	Lung cancer Mesothelioma Cor pulmonale Pleural effusion
Coal miner's pneumoconiosis	Black nodules usually in upper portions of upper and lower nodes Star-shaped lesions Little collagen fibrosis Progressive massive fibrosis	Symptomless unless progressive massive fibrosis present Breathlessness and chest pain may occur Clubbing may occur	Pleural effusion Cor pulmonale

In the workplace, exposure to respirable particles (particles with an aerodynamic diameter below about 5 μm) produces an immunologically mediated inflammatory pulmonary response located in the alveoli. There are acute, subacute, and chronic features.

In about two-thirds of sufferers, the typical acute disease presents as an attack of chills, fever, cough and shortness of breath occurring 4–8 hours after exposure. This can be measured by a reduction in FEV_1 and FVC. This is also associated with malaise, gradual resolution of fever (perhaps to 41°C), a harsh cough, headache, myalgia (muscle pain), and persisting dyspnoea. Acute symptoms usually subside within hours but may persist for days.

The subacute syndrome is characterised by gradually progressive dyspnoea and productive cough often associated with repeated low-grade exposures. The acute symptoms appear to abate, but the dyspnoea is progressive and associated with anorexia and weight loss. X-rays may show interstitial reticulonodular densities up to several millimetres in diameter scattered diffusely. These tend to clear out gradually over a period of weeks to months. Individuals with multiple exposures may have fibrotic densities that will clear slowly, or not at all. X-rays of advanced hypersensitivity pneumonitis show honeycombing similar to that of other end-stage lung diseases.

Table 4.9 Agents and occupations producing hypersensitivity pneumonitis

Source	Occupation/population	Disease
Chemicals		
Isocyanate vapours	Bathtub refinishers	Bathtub refinisher's lung
Epoxy resins containing phthalic anhydride	Foundry workers	Foundry worker's lung
Catalysts containing trimellitic anhydride	Urethane foam workers	
Insecticides, pyrethrum	Insecticide manufacturers	Insecticide worker's hypersensitivity pneumonitis
Hairspray	Hairdressers, beauticians	Thesaurosis
Beryllium dust and fumes	Beryllium workers	Chronic beryllium disease
Bakelite	Bakelite workers	Bakelite worker's lung
Red cedar wood (plicatic acid)	Red cedar workers	Red cedar worker's lung
Ionised cobalt	Tungsten carbide workers	Hard metal lung disease
Bacteria		
Mouldy hay	Agricultural workers	Farmer's lung
Mouldy sugar cane	Bagasse workers	Bagassosis
Compost	Mushroom workers	Mushroom worker's lung
Rope dust	Bag and rope workers	Sisal worker's disease
Coffee bean dust	Coffee workers	Coffee worker's lung
Water/ventilation systems	Office workers	Humidifier fever
Dirt	Fertilisers	Fertiliser worker's lung
Fungi		
Mouldy bark	Maple bark strippers	Wood worker's lung
Redwood	Redwood workers	Sequoiosis
Wood	Loggers	Wood pulp worker's lung
Infected old wood	Inhabitants of old houses in Europe	Dry rot disease
Mouldy cork dust	Cork workers	Suberosis
Wood dust	Inhabitants of Japanese wood houses	Summer-type hypersensitivity pneumonitis
Mouldy malt and barley	Malt workers	Malt worker's lung
Mouldy paprika pods	Paprika splitters	Paprika splitter's lung
Infected wheat flour	Flour workers	Wheat weevil's disease
Cheese mould	Cheese workers	Cheese worker's lung
Mouldy barn straw	Horse riders	Horseback rider's lung
Mouldy lichen	Lichen pickers	Lichen picker's lung
Mouldy wood chips	Paper mill workers	Paper mill worker's lung

continued

Table 4.9 continued

Source	Occupation/population	Disease
Animal proteins		
Bird feather bloom, bird excreta, bird serum	Bird handlers	Avian protein diseases Bird fancier's lung Pigeon breeder's lung Poultry handler's lung
Fox fur	Furriers	Furrier's lung
Snuff	Snuff producers	Snuff taker's lung
Rats, mice, gerbils	Animal laboratory workers	Rodent handler's disease
Therapeutic substances		
Bleomycin	Cancer patients	Bleomycin hypersensitivity
Gold salts and colloids	Arthritis patients	Gold-induced hypersensitivity pneumonitis
Acebutolol (β-blocker)	Heart disease patients	Acebutolol-induced hypersensitivity pneumonitis
Amiodarone	Heart disease patients	Amiodarone-induced hypersensitivity pneumonitis

A smaller percentage of patients have chronic disease that has an insidious onset. This is characterised by progressive dyspnoea, with features of interstitial fibrosis and/or airways obstruction. This occurs most commonly after multiple episodes of symptomatic exposure in the subacute and chronic individual. However, chronic hypersensitivity pneumonitis is indistinguishable from other types of combined obstructive airways disease and diffuse interstitial fibrosis.

Death can occur at any stage and the overall mortality rate has been estimated at between 9% and 17%.

The mechanism of hypersensitivity pneumonitis is not completely understood, but is essentially immunological. The hypersensitivity is classified as a type III or immune-complex mediated reaction. The alveolar macrophages, T lymphocytes, natural killer cells and cytokines are considered key. Further, as symptoms appear with higher incidence in exposed relatives, genetic mechanisms are likely.

Farmer's lung

This is considered the prototypic hypersensitivity pneumonitis. Symptoms occur following the inhalation of the dust of mouldy hay or other mouldy vegetable produce (from spores of actinomycetes fungi). Fever, malaise, rigours or chills, aches and pains may occur. There may be cough with scanty sputum or dyspnoea. Crepitations (sounds and rattles) may be present in the lungs. On X-ray there are extensive nodular shadows in the lungs, especially in the lower zones. The disease is an extrinsic allergic alveolitis, with fibrosis in the chronic stages.

Byssinosis

Probably the most common of these diseases: it is caused by inhalation of dusts of cotton, sisal, hemp and flax. Exposures that cause problems are from the processing of these fibres in the textile industry. There is bronchial constriction followed by symptoms of emphysema and bronchitis. Initially, these effects are transient, but they may become irreversible after prolonged and repeated exposure.

Early symptoms are quite characteristic. On exposure, the worker feels tightness in the centre of the chest and breathlessness. Measures such as FEV_1 can be reduced after only 6 hours of exposure, suggesting a direct effect on bronchoconstriction mechanisms. Within 1–2 days of continued exposure this reaction tends to get better, indicating physiological adaptation. However, the symptoms recur after a break, such as a vacation, or a weekend – hence the colloquial name of the condition, 'Monday fever'. The effects can become persistent throughout the working week and permanent after many months or even years of exposure. At this stage, the effects are permanent even if exposure ceases.

The mechanisms of the effect are not known, but are considered to involve either the presence of a histamine-like factor in the dust, which causes bronchoconstriction, or endotoxins produced by bacterial contamination of the fibres. Smoking, other atmospheric pollutants and concomitant respiratory conditions may contribute to the disease.

Bagassosis

Bagasse is the waste fibre from sugar-cane processing, which is used in the manufacture of construction materials, such as boards and insulation. Bagassosis is an acute attack of fever with cough, dyspnoea, and sometimes haemoptysis (coughing of blood from the lungs). The mechanism of effect is believed to be an immune response to fungi in the bagasse (which may also contain some silica).

Mushroom worker's lung

This condition is found in mushroom growers, who are exposed to spores from the compost in which the mushrooms are grown. The acute signs and symptoms are similar to farmer's lung, with repeated attacks of upper respiratory illness. The progression to chronic illness has not been reported.

Occupational asthma

Asthma is a widespread lung condition which affects 5–15% of the general population. Asthma can be triggered by a variety of environmental agents such as gases, particulates and allergens. However, other exposures, such as cold air or exercise, can precipitate asthmatic reactions.

Occupational asthma refers to bronchial asthma, caused by factors in the workplace (Salvaggio *et al.* 1986). Over 200 chemicals have been identified as causing asthma at work and these are found in a wide variety of occupations (see Table 4.10) (Demeter and Cordasco 1999). Of course many of these substances have also

Table 4.10 Chemicals identified as causing asthma

Compound	Occupation/industry
Metals and their compounds	
Platinum	Refining, jewellery
Nickel and compounds	Refining, electroplating, welders, chemical workers
Chromium and compounds	Electroplating, tanning, welders
Chromates and dichromates	Cement workers
Cobalt	Hard metal industry
Vanadium and vanadium pentoxide	Boiler and turbine cleaners
Aluminium fluoride	Potroom workers
Zinc	Galvanising workers
Isocyanates	
Toluene diisocyanate	Polyurethane, varnish manufacture, laminators
Diphenylmethane diisocyanate	Foundries, polyurethane foam, laminators
Hexamethylene diisocyanate	Paint manufacture, spray painters
Naphthalene diisocyanate	Chemists, rubber workers
Anhydrides	
Phthalic anhydride, tetrachlorophthalic anhydride	Epoxy resins, plastics
Trimellitic anhydride	Epoxy resins, plastics, chemical workers
Hexahydrophthalic anhydrides	Epoxy resins
Himic anhydride	Fire retardants
Dyes	
Anthraquinone	Fabric dying
Carmine	Cosmetics, dyes
Hexafix® yellow, Drimaren® blue, Cibachrome® brilliant scarlet	Dye manufacturers
Paraphenyl diamine	Fur dyeing
Henna extract	Beauticians
Miscellaneous chemicals	
Ammonium thioglycolate, monoethanolamine, hexamethylenamine, persulfate salts	Hairdressers, beauticians
Azodicarboamide	Plastics and rubber workers
Chloramine T	Brewery workers
Diazonium salt	Photocopying and dye workers
Dimethyl aminoethanolamine, dimethyl ethanolamine	Spray painters
Ethylenediamine	Rubber workers
Ethylene oxide	Medical sterilisers

continued

Table 4.10 continued	
Compound	*Occupation/industry*
Formaldehyde	Laboratory workers, funeral workers/embalmers
Fural and furfuryl alcohol	Foundry mould workers
Organophosphate insecticides	Agricultural workers
Persulfate salts	Chemical workers, beauticians
Pyrethrin	Insecticide applicators, fumigators
Triethyl tetramine	Aircraft fitters
Fluxes	
Colophony	Electronics soldering
Aminoethyl ethanolamine	Aluminium soldering
Animal proteins	
Mammalian (hair, dander, urine)	Laboratory workers, veterinarians, meat processing, furriers, wool handlers
Insects (grain, weevils, mites, silkworms, cockroaches, moths)	Laboratory workers, farmers, grain workers, sewage workers, bee keepers, entomologists
Molluscs (prawns, crabs)	Shellfish workers, fish shop workers, shell grinders
Birds	Pigeon breeders, poultry, feather processors
Plant proteins	
Wheat, soy and rye flour	Milling, baking
Coffee	Coffee processing
Castor bean	Oil and food processing
Tobacco	Growers and manufacturers
Gums (acacia, karaya, arabic, tragacanth)	Printing and food processing
Cotton dust and oils, flax hemp	Textile and food manufacturing
Pollens	Horticulturalists, gardeners
Hops	Brewers
Pepper, tamarind seeds, garlic	Millers, spice processors
Moulds	Mushroom workers, cheese workers
Wood dusts	
Western red cedar, redwoods, cedars, iroko, oak, mahogany, maples, walnut, boxwood, mulberry	Carpenters, construction workers, cabinet makers, sawmill workers
Enzymes and related materials	
Bacillus subtilis	Detergent manufacturers
Pepsin, trypsin, chymotripsin, pectinase, bromelain	Pharmaceuticals, plastics and rubber workers
Papain	Pharmaceuticals, plastics, rubber and food processing workers

continued

Table 4.10 continued

Compound	Occupation/industry
Amylases and proteases	Detergents, manufacturing, bakers
Flaviastase	Pharmaceutical industry
Therapeutic substances	
Penicillins, methyl dopa, spiromycin, salbutamol, tetracycline, piperazine, sulfonamides, cimetadine	Pharmaceutical workers
Cephalosporins	Pharmaceutical workers, nurses
Enfluorane anaesthetic	Hospital staff
Psyllium	Laxative manufacture
Sulfone chloramides	Brewery workers

been shown to cause hypersensitivity reactions. This is not too surprising, as most asthma is produced through a sensitisation process.

Asthma consists of attacks of wheezing and breathlessness due to bronchospasm and secretions of thick mucous. There are a number of variables which can effect the rate of asthma including:

- premorbid health state
- nature of occupationally encountered substances
- concentration and duration of exposure
- availability and use of ventilation and/or protective devices in the workplace
- the presence of coexisting asthmogenic factors or agents.

Asthma can be defined as a disorder characterised by an increased response by the airways to irritants. Simply put, the effects of asthma are bronchospasm, mucous production in the airways, and cough. With occupational asthma, the natural functional responses of the lung are exaggerated in response to irritants.

A variety of protective responses occurs in the lung after inhalation of an irritant. By diminishing the diameter of the airways, bronchospasm limits airway diameter and reduces the amount of irritant entry into the lungs. Mucous secretion and cough assist in the removal of air from the lungs. Everyone has these responses, but individuals who possess exaggerated responses are diagnosed as having asthma.

Asthma can be diagnosed historically (episodes of wheezing/shortness of breath, cough and mucous production), physically (by auscultating wheezes) or physiologically. Occupational asthma is defined as a disorder where there is generalised obstruction of the airways, usually reversible and caused by the inhalation of substances or material which the worker uses or which are incidentally present in the workplace.

Asthma can be initiated or provoked by workplace exposures. The asthmatic response can be activated by either intrinsic (exaggerated response) or extrinsic (exaggerated stimulus) factors.

Reactive airways dysfunction syndrome (RADS) is a term coined in 1981 which defines the events leading to asthma following a toxic exposure:

- preceding absence of respiratory complaints (if possible, documented)
- onset of symptoms after a single or short period of exposure
- other types of respiratory disease have been ruled out
- the exposure is to a gas, vapour, smoke or fume present in high concentrations and with irritant qualities in its nature (it has become established that these exposures are gaseous, rather than particulate)
- onset of symptoms occurs within 24 hours after the exposure and persisted for at least 3 months
- symptoms simulate asthma with cough, wheezing and dyspnoea predominating
- pulmonary tests show air flow obstruction
- challenge testing with methacholine is positive.

This is a definition of one type of asthma that is different from, for example, airways obstruction. Obviously, RADS is important from the occupational perspective.

Table 4.10 lists some of the better-known agents that cause asthma. These can be divided into three main types:

- high molecular weight compounds
- low molecular weight compounds
- high levels of irritants (which can be considered the agents that produce RADS).

High molecular weight irritants are frequently biological proteins and their responses are mediated by immunological mechanisms (including production of immunoglobulin E antibodies). Patients with this form of occupational asthma are commonly atopic and their sensitisation can be diagnosed using skin patch tests.

By contrast, low molecular weight antigens are usually inorganic compounds with a molecular weight of less than 1000, are not usually allergically mediated, and are less frequently a cause of occupational asthma than the high molecular weight compounds. There are exceptions to these rules.

Industrial bronchitis is at the other end of the spectrum. It is a disorder characterised by dyspnoea, cough and mucous production occurring in the workplace because of high concentrations of irritants, often particulate in nature. It usually occurs over long periods of time with repeated exposure and is usually not permanent.

Asthma is typically diagnosed by history, physical examination, and by pulmonary function testing. Obtaining information on timing or asthmatic symptoms is crucial. The objective tests for occupational asthma are either of pulmonary function type or alternative tests designed to demonstrate hypersensitivity (such as immunological tests).

The treatment of asthma is quite simple: standard medical treatment for asthma; and avoidance of further exposure. The latter is essential. The majority of individuals with occupational asthma will have chronic persistent symptoms, and these will diminish with time (but in most cases will not resolve). If continued work is to occur, then the options are transfer to a different area, or the use of ventilation or respiratory protection (air-supplied respirators should be used as filter-type respirators increase the effort required for breathing). However, these measures will not guarantee protection. Even discontinuation of exposure can not be assumed to remove the risk if the agent is encountered in the environment.

Chronic obstructive lung disease

Chronic obstructive lung disease (COLD) develops when many factors work in combination (smoking, air pollution, genetic factors, pre-existing respiratory disease). In some ways it is a condition that develops but does not progress to other conditions (hypersensitivity, fibrosis and so on). Workers are more likely to suffer from COLD if they have dusty occupations, working in foundries, mining and milling, brick and tile manufacture, textile manufacturing and so forth.

Occupational lung cancer

Lung cancer is a major disease in the community for nonoccupationally exposed individuals (mainly associated with smoking). However, there are many occupationally induced cancers of the respiratory system in workers (Whitesell and Drage 1993). As with other occupationally induced respiratory diseases, the site of the cancer is often related to the carcinogenic agent.

Cancer can occur in the upper respiratory tract, such as cancers caused by formaldehyde, some wood dusts, leather work and isopropyl alcohol, which cause cancer of the nasal cavity and turbinates. Asbestos is implicated in producing cancer of the larynx.

Several cancers of the lungs have also been reported in association with the following agents (Steenland *et al.* 1996):

- asbestos
- chromium and its salts
- tars, pitches and bitumen
- ionising radiation
- nickel and its salts and nickel carbonyl
- bis-chloromethyl ether and chloromethyl ether
- mustard gas (now not used)
- arsenic (possible lung carcinogen)
- diesel exhaust particulates (these appear to produce lung tumours in rats with long-term exposure to air concentrations 100–1000 times more than realistic workplace concentrations).

Asbestos has been associated with cancer since 1934, and was firmly established as a cause of cancer in 1955. The lung is the main site, though other sites are also implicated.

Bronchogenic cancer

There is an excess of bronchogenic carcinoma of the lung in people exposed to asbestos: smoking greatly increases this risk through a synergistic interaction. This cancer has a long latency (20 years or more) and there is some suggestion that the incidence is dose-related.

Mesothelioma

See Figure 4.15. This is a rare cancer of the lining of the lung and chest cavity (the pleura), though a small proportion of mesotheliomas are found in other tissues, such as the peritoneum or pericardium. A definite link between mesothelioma and asbestos was established in 1960. The disease is always fatal. The incidence of the disease seems to be related to the small size of the inhaled fibres rather than the chemical composition. About 20% of cases of mesothelioma do not appear to be associated with asbestos exposure, which suggests that there is a poor dose–response relationship. However, this is difficult to establish owing to the long latency of the cancer (up to 40 years for asbestos-related cancers). Many mesothelioma cases are associated with exposure to blue asbestos, though all forms of asbestos are now considered carcinogenic to some degree. Smoking plays no role in mesothelioma, though other exposures have been implicated in the development of this cancer, including ionising radiation, farming of sugar cane in India, previous pleuropulmonary disease and exposure to erionite, a fibrous, nonasbestos mineral from Turkey.

Other sites

Other cancers of the respiratory system, such as carcinoma of the larynx, have been reported in asbestos-exposed workers. Also, most inhaled asbestos is eliminated from the respiratory tract by the mucociliary escalator, where it is cleared

Development of Mesothelioma

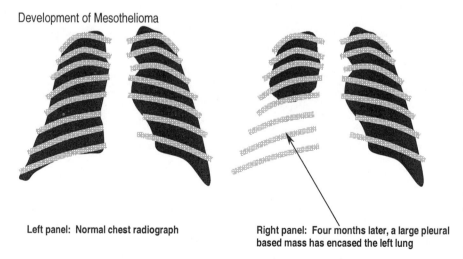

Left panel: Normal chest radiograph

Right panel: Four months later, a large pleural based mass has encased the left lung

Figure 4.15 Development of mesothelioma.

into the throat, swallowed, and then leaves the body through the gastrointestinal tract. Consequently, cancers of the oesophagus, stomach and intestine have been found following substantial industrial exposure.

With the exception of mesothelioma, the carcinogenic potency of asbestos fibres is related to a number of factors, including duration and intensity of exposure, type of asbestos and most importantly the fibre dimensions. Only thin and comparatively long fibres with a diameter of less than 1.5 μm and length of more than 8 μm appear to be critical in carcinogenic potency.

Other factors

Finally, cigarette smoking very appreciably enhances occupational pulmonary carcinogenesis. The role of smoking in cancer and other occupational lung diseases must not be ignored as a significant contributor to the severity of the effects of occupational exposures.

Detection of occupational lung diseases

The approaches for measuring and detecting pathophysiological effects of various lung diseases differ, depending on the agent of effect, the site of the disease and its location in the lung.

The main tools of detection of occupational lung diseases are:

- the occupational history
- structured questionnaires
- spirometry
- exercise activities
- chest X-ray.

Other methods, such as clinical tests, antibody tests, biomarkers or lung biopsy are also available (Schulte 1996). Table 4.11 shows the main methods for measuring respiratory pathophysiology and their usefulness in the different respiratory compartments.

Table 4.11 Methods for detecting respiratory diseases in different compartments

Method	Nasopharyngeal	Tracheobronchial	Pulmonary	Pleural
History	✓✓	✓✓	✓	✓
Questionnaire	✓✓✓	✓✓	✓	
Spirometry		✓✓✓	✓✓	✓
Exercise		✓✓	✓✓✓	✓
Chest X-ray		✓	✓✓	✓✓✓

✓ Some usefulness.
✓✓ Moderately useful.
✓✓✓ Highly useful.

The first approach is obviously the interview and the compilation of occupational and medical history. Further information on specified occupational diseases can be obtained using structured questionnaires. These two approaches are useful for identifying toxic effects in the nasopharyngeal compartment and, to a lesser extent, in the tracheobronchial compartment, as they elicit information on signs and symptoms.

As many of these symptoms are signs of an activated pulmonary defence system, standardised spirometry is useful for detecting disease in the tracheobronchial compartment. Measures such as FEV_1 and FVC have been developed to indicate specific components of lung function, and have been used for many years by occupational physicians to measure lung function in workers. In addition, lung volume measurements can detect disease in the lung parenchyma and pleural space.

Damage to the pulmonary compartment can be assessed using gas exchange tests. The simplest test is to measure respiratory performance following exercise, using the step test, treadmill or bicycle ergometer. Measures of pO_2 and pCO_2 can indicate changes in the cells in the gas exchange area. A combination of spirometry and gas exchange tests can be very useful in defining damage to the respiratory system, especially as blood gases and pH are often normal at rest but become abnormal during exercise.

Finally, diseases of the pleural compartment and the lung parenchyma can be detected with chest X-rays. The X-ray is an indispensable aid in the evaluation of some diseases of the respiratory system, such as pneumoconiosis, but the limitations of the X-ray should be recognised: X-rays may be strikingly abnormal due to the presence of dusts, and there may be significant clinical impairment and extensive pathological changes in the presence of a normal X-ray. The chest X-ray should not be used to estimate the presence or extent of ventilatory impairment. However, workers exposed to respiratory hazards are often required to have an annual X-ray, and often serial X-rays are available to investigate the build up of respiratory disease.

Indeed, all these measures have developed to the stage that they are used as surveillance tools to detect disease from inhaled toxicants.

Summary

The very nature of the respiratory system places it at risk when it is exposed to airborne contaminants. In turn, a range of diseases of the lungs have been described based on properties of the inhaled material, the anatomy of the respiratory system itself, and the specialised defences it possesses.

Consideration of the tissue dose may help to identify the critical sites within the respiratory tract. The physical, chemical and biological properties of contaminants and their action within the lung helps identify the pathophysiological effects that might be anticipated. Knowledge of the pathophysiological consequences allows appropriate endpoints to be recognised and measured, and it can identify appropriate exposure monitoring, health surveillance and health surveillance protocols.

In conducting a risk assessment for a toxic inhaled material, the sequence of events that such materials progress can be used in the assessment:

- How much is available in the air?
- How much is inhaled?
- What is the tissue dose?
- What is the toxic effect?
- What is the pathological effect?
- What is the clinical outcome?

The range of effects of inhaled materials are fairly restricted to:

- inflammatory reactions
- hypersensitivity reactions
- bronchoconstriction
- granuloma formation
- fibrosis
- macrophage activity
- cellular proliferation
- cell death.

Ultimately, impairment or damage to the respiratory system will impede the flow of air to the gas exchange system. The impairment of the flow of gases to the gas exchange regions can be classified into either obstructive or restrictive airway disease.

Restrictive diseases are typically characterised by a decrease in lung volumes (e.g. vital and total lung capacities). Restrictive defects may occur when the elastic properties of the lung tissue are decreased and the lung becomes stiff (in fibrotic diseases such as silicosis, asbestosis and pneumonia).

Obstructive diseases are characterised by an obstruction to air flow. This increase in the resistance to air flow is characterised chiefly by a decrease in expiratory flow rate (e.g. FEV_1).

An increasingly important concept in developing safe exposure to respiratory toxicants is the recognition of the role of protective mechanisms. If the exposure to a toxicant results in a tissue dose that does not overwhelm the protective response, then it is quite unlikely to cause adverse pathophysiological effects. However, disease can develop if these defences are overwhelmed and a range of normal biological responses becomes exaggerated (such as fibrosis, hypersensitivity or neoplasia) and possibly health-damaging or life-threatening.

References and further reading

Amdur, M.D. (1991) Air pollutants. In: Amdur, M.O., Doull, J. and Klaassen, C.D. (eds) *Cassarett and Doull's Toxicology: the Basic Science of Poisons*, 4th edn. New York: Macmillan, pp. 854–871.

Barrow, C.S. (1986) *Toxicology of the Nasal Passages*. Washington: Hemisphere Publishing.

Beckett, W.S. and Bascom, R. (eds) (1992) Occupational lung disease. *Occupat. Med. State Art Rev.* 7(2): 189–370.

Coultas, D.B. and Samet, J.M. (1992) Occupational lung cancer. *Clin. Chest Med.* 13: 341–354.

Demeter, S.L. and Cordasco, E.M. (1994) Occupational asthma. In: Zenz, C., Dickerson, O.B. and Horvath, E.P. (eds) *Occupational Medicine*, 3rd edn. St Louis: Mosby, pp. 213–228.

Gardner, D.E, Crapo, J.D. and Massaro, E.J. (eds) (1988) *Toxicology of the Lung*. New York: Raven Press.

Haldane, J.S. and Priestley, J.G. (1935) *Respiration*. Oxford: Clarendon Press.

Jeffery, P.K. and Reid, L.M. (1977) The respiratory mucous membrane. In: Brain, J.D., Proctor, D.F. and Reid, L.M. (eds) *Respiratory Defence Mechanisms*, Part I. New York: Marcel Dekker.

Kennedy, G.L. and Trochimowicz, H.J. (1982) Inhalation toxicology. In: Hayes, A.W. (ed.) *Principles and Methods of Toxicology*. New York: Raven Press.

Mauderly, J.L. (1989) Effect of inhaled toxicants on pulmonary function. In: McClellan, R.O. and Henderson, R. (eds) *Concepts in Inhalation Toxicology*. New York: Hemisphere, pp. 347–401.

Mitchell, C.A. (1997) Occupational lung disease. *Med. J. Aust.* 167: 498–503.

Morgan, W.K.C. and Seaton, A. (1984) *Occupational Lung Diseases*, 2nd edn. Philadelphia: Saunders.

Ng, T.P. and Chan, S.L. (1991) Factors associated with massive fibrosis in silicosis. *Thorax* 46: 292–232.

Oberdorster, G. (1995) Lung particle overload: implications for occupational exposures to particles. *Regulat. Toxicol. Pharmacol.* 21: 123–135.

Phoon, W.-O. (1988) Dust diseases of the lung. In: *Practical Occupational Health*. Singapore: PG Publishing, pp. 161–190.

Sahu, A.P. (1992) Prevention of occupational and environmental lung disorders. *Toxicol. Indust. Health* 7: 521–529.

Salvaggio, J.E., Butcher, B.T. and O'Neil, C.E. (1986) Occupational asthma due to chemical agents. *J. Allergy Clin. Immunol.* 78: 1053–1058.

Schlesinger, R.B. (1990) The interaction of inhaled toxicants with respiratory clearance mechanisms. *Crit. Rev. Toxicol.* 20: 257–286.

Schlueter, D.P. (1994) Silicosis and coal worker's pneumoconiosis. In: Zenz, C., Dickerson, O.B. and Horvath, E.P. (eds) *Occupational Medicine* 3rd edn. St Louis: Mosby, pp. 171–178.

Schulte, P.A. (1996) Use of biomarkers to investigate occupational and environmental lung disorders. *Chest* 109: (3 Suppl) 9S–12S.

Schwartz, D.A. and Peterson, M.W. (1997) Occupational lung disease. *Adv. Intern. Med.* 42: 269–312.

Steenland, K., Loomis, D., Shy, C. and Simonsen, N. (1996) Review of occupational lung carcinogens. *Am. J. Ind. Med.* 29: 474–490.

Terenski, M.F. and Cheremisinoff, P.N. (1983) *Industrial Respiratory Protection*. Ann Arbor: Ann Arbor Science.

West, J.B. (1985) *Respiratory Physiology – The Essentials*. Baltimore, MD: Williams and Wilkins.

Whitesell, P.L. and Drage, C.W. (1993) Occupational lung cancer. *Mayo Clin. Proc.* 68: 183–188.

Witschi, H.P. and Brain, J.D. (1985) *Toxicology of Inhaled Particles*. Berlin: Springer-Verlag.

Witschi, H.R. and Last, J.A. (1996) Toxic responses of the respiratory system. In: Klaassen, C., Doull, J. and Amdur, M.O. (eds) *Cassarett and Doull's Toxicology: The Basic Science of Poisons*, 5th edn. New York: McGraw-Hill, pp. 443–462.

Wright, P.H. (1993) Occupational lung diseases other than asthma. *Postgrad. Med. J.* 69: 129–135.

Chapter 5

Occupational skin diseases

C. Winder

Introduction

With an area of 20,000–23,000 cm^2 the skin is the body's largest organ. The skin can serve as an avenue of entry for certain toxicants that cause chemical toxicity through skin absorption. The skin can also be damaged by the agents with which it comes into contact, and occupational skin disease is a very common industrial illness (NOHSC 1988).

Historically, occupational diseases of the skin have paralleled industrial development, with mechanical injuries being a common early cause, and the hazards of later developments producing unplanned consequences on skin (and other body systems); for example, fissuring dermatitis of the hands from handling lye, leg ulcerations among salt miners, and grain dermatitis were reported by Ramazzini in the early 1700s, and Sir Percival Pott reported the first known case of occupational cancer (cancer of the scrotum in chimney sweeps) in 1775.

Conditions such as arsenical keratoses, asbestos corns, cement dermatitis, chrome holes, nickel itch, fibreglass itch, oil bumps, rubber rash, tar warts, the dermatitis of cobblers (from shoe wax), of bakers (from sugar and spice), of tobacco handlers, silk winders, hide handlers and hairdressers (shampoos) and dermatitis from coal tar, chromates, mercury, fungi, vaccines, and so on, are all well-established occupational skin disorders, some of which remain causes of worker ill health today. The term occupational dermatosis is used to describe any abnormality of the skin resulting directly from or aggravated by the work environment (Mathias 1994).

Occupational skin disorders represent up to one-third of work-related diseases causing absenteeism of more than 1 week (Mathias 1990). The calculated yearly cost of occupational skin disorders in New South Wales is approximately 12 million Australian dollars (Freeman 1991; Rosen and Freeman 1992). Diagnosis of occupational causes of skin disease is often difficult, but is essential for prevention, treatment and recovery, as well as for workers' compensation and rehabilitation.

The skin

The skin is a major barrier between the internal workings of the body and the outside world. Except for the palms of the hands and the soles of the feet, human skin is quite thin. However, its variable thickness, and its collagen and elastic components allow it to function as a flexible barrier. It is a unique barrier which (within limits) restrains water loss from within, and protects against mechanical trauma, penetration by various chemicals, microorganisms, natural and artificial radiation, and stress of heat and cold. The structure of the skin is shown in Figure 5.1.

The skin is composed of two principal layers – the epidermis and the dermis. Each layer contains structural elements that are important not only with regard to their function, but also with regard to the disease processes which may affect them.

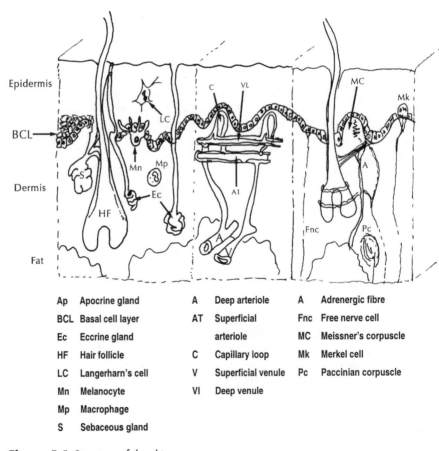

Ap	Apocrine gland	A	Deep arteriole	A	Adrenergic fibre
BCL	Basal cell layer	AT	Superficial	Fnc	Free nerve cell
Ec	Eccrine gland		arteriole	MC	Meissner's corpuscle
HF	Hair follicle	C	Capillary loop	Mk	Merkel cell
LC	Langerharn's cell	V	Superficial venule	Pc	Paccinian corpuscle
Mn	Melanocyte	Vl	Deep venule		
Mp	Macrophage				
S	Sebaceous gland				

Figure 5.1 Structure of the skin.

Structural elements of the skin

Epidermis

The epidermis is between 100 and 200 μm thick. The innermost layer comprises a thin ridged layer of basal cells, which reproduce about every 14 days. These produce squamous cells (also called keratinocytes) that are gradually pushed outwards by dividing cells from beneath. The keratinocytes are metabolically active, and synthesise filaments made of keratin and keratohyaline granules. As the cells continue to be pushed outwards, they migrate through a zone of transition consisting of various layers, described by the activity of the cell at that time. The keratinocytes also become less metabolically active and undergo the process of keratinisation (cornification), in which the cellular contents condense and disintegrate, and membrane coating granules are extruded, where they produce a densely packed layer of relatively impermeable, cornified 'dead' cells. By the time the cells reach the outermost layer, the cells are flattened, dead and cornified. The layer is called the stratum corneum, and is the principal protective barrier. This layer consists of the remains of the keratinocytes, embedded in a matrix containing mucous and lipids and ceramides. It takes a cell up to 12–14 days following its formation in the basal cell layer to move up through the living cell layers to the stratum corneum, and another 14 days or so to reach the outer surface and be sloughed off. The time from growth to sloughing for a skin cell decreases with injury.

The epidermis also contains specialised cell types including melanocytes and Langerhans cells (see below).

Dermis

The epidermis is separated from the dermis by a basal lamina, which is ridged over most of the body surface. In contrast to the epidermis, the dermis is a strong, flexible tissue. This is due to a loose matrix of dermal connective tissue, composed of fibrous proteins (collagen, elastin and reticulin), embedded in an amorphous glycosoaminoglycan ground substance. This provides a slow diffusion medium for tissue materials to disperse, nutrients moving outwards, or absorbed toxicants moving inwards.

Most of the skin structures (hair follicles, sebaceous, apocrine and eccrine glands, blood vessels, lymphatics, nerve endings, sensory receptors and immune processes) are found in the dermis, as are scattered cells such as macrophages, mast cells, lymphocytes and fibroblasts.

Hair follicles

Hair is found on all cutaneous surfaces, except the palms, soles and mucous membranes. Hair roots extend below the level of the sebaceous glands. Hair grows cyclically, with alternating periods of growth and quiescence.

Melanocytes

These cells are located between the dermis and the epidermis (at the level of the basal cell layer), which spread dendritic processes into the layers of squamous cells around and above the basal cell layer. Under suitable stimulation from sunlight and ultraviolet radiation, these cells increase the synthesis of melanin (a protective pigment) which darkens the skin and provides enhanced protection against solar radiation.

Eccrine sweat glands

These are found on all surfaces of the skin, but are most numerous on the palms and soles. The ducts of these coiled glands traverse the epidermis, and deposit their contents on the surface of the skin. They produce sweat which is used to provide evaporative cooling when the body is stressed by heat (some sweating also occurs with emotional stress). The primary control of sweating is through cholinergic nerves.

Sebaceous glands

These are associated with hair follicles, with which they share a common ductal opening. The glands produce sebum, composed primarily of triglycerides, wax esters and squalene.

Nervous innervation

The skin is a primary sensory organ with a range of sensory receptors, including Merkel cells, Vater–Pacini corpuscles (to sense pressure), Meissener's corpuscles (to sense touch), mucocutaneous end organs located in the dermis (Krause bulbs and Ruffini endings). As well as these receptors, the skin is innervated by a dense network of nerve fibres, of both the cholinergic and adrenergic systems.

Blood supply

The dermis has a substantial blood flow, more than required for metabolic activity. Blood is distributed within dermal tissue through a highly developed, interconnected network of superficial and deep blood vessels. Capillary loops extend from dermal arterioles to venules through the dermal papillae close to the basal cell layer. Blood flow through the skin is important for a number of reasons, including heat control (thermoregulation) and tissue nutrition. Cutaneous blood flow is regulated by constriction of meta-arterioles and precapillary sphincters, and a large volume of circulating blood can be brought close to the skin surface. Mast cells are often located around small vessels. These release pharmacologically active substances, such as histamine, heparin, serotonin and leukotriene precursors, which participate in a variety of pathological processes. The dermis also contains a rich supply of lymphatics which drain to regional lymph nodes.

Langerhans cells

These cells are located around and above the basal cell layer. They possess dendritic processes that spread to surrounding cells. They also possess immunochemical receptors and are involved in cutaneous immune regulation and surveillance, and are responsible for antigen processing.

Absorption of materials through the skin

While it is known that the main routes of exposure to occupational hazards and risks are through the skin and respiratory system, more attention has been paid to inhalational exposure.

As noted above, the epidermis is an effective barrier against many agents, including microorganisms, chemicals and radiation. However, once across the epidermis, absorption into the body is largely a matter of diffusing to the closest blood vessel or lymphatic.

Percutaneous absorption occurs mainly through passive transfer. Skin absorption is quite complex and can be affected by many factors. For example:

- physical properties of the hazardous agent (molecular weight, solubility in cornified cells, solubility in water)
- concentration in liquid in contact with skin
- area exposed, either in contact with the toxicant or available for absorption of airborne contaminants (for example, about 20% of exposure to 2-butoxyethanol is considered to be absorption of vapour)
- duration of exposure
- presence of barriers that prevent exposure or occlusion, which may increase exposure
- the role of skin as a storage depot for toxicants
- rate of diffusion across the epidermis
- amount of damage to the skin (intact, as opposed to cracked or damaged skin)
- rate of diffusion across the dermis
- rate of uptake through hair follicles and sweat glands
- rate of biotransformation in the dermis
- blood flow to the skin.

The rate of penetration varies greatly for different substances, and generally fat soluble (lipophilic) substances penetrate more quickly than water soluble (lipophobic or hydrophilic) substances.

Physicochemical studies have shown that the rate limiting step in absorption of chemicals across the skin (in these studies, intact skin is always specified) is getting across the stratum corneum. With this consideration, Fick's law of diffusion provides a reasonable approximation of the absorption of a chemical across the skin. The rate of uptake is given by:

$$\frac{K_m \times D_m \times C}{\delta_m}$$

where K_m is the partition coefficient between the stratum corneum membrane and the chemical or vehicle in which it is dissolved; D_m is the diffusion constant; C is the concentration difference of the toxicant across the skin membrane; and δ_m is the thickness of the skin membrane.

Percutaneous absorption can be measured using a range of *in vivo* or *in vitro* techniques. *In vivo* studies are generally performed using topical application of toxicants (if necessary, dissolved or suspended in a suitable vehicle) to intact skin. *In vitro* studies generally use excised skin, and have the benefit of being able to use human skin. The selection of species is important as rodents generally have higher permeability than humans because they have a thinner epidermis and more hair follicles. The skin of pigs and monkeys is more like that of humans. Radiolabelled compounds can be used in both types of studies, to good effect.

However, for practical purposes, the amount of absorption of a toxicant across skin in practical situations is usually measured using biological monitoring of absorption.

Measurement of skin toxicity, irritation and sensitisation in animals

The toxicity studies

Rats and mice are the most common choices for most single-dose toxicity tests, for reasons of cost and relative ease of handling (Maibach 2000). Furthermore, factors such as genetic background and disease susceptibility are well known for these species. Rabbits are often used for testing dermal (skin) toxicity since their shaved skin is highly sensitive to dermal effects, including irritancy but not sensitisation reactions. Indeed, rabbits are the preferred species for irritancy studies, although the guinea pig is the preferred model for investigation of skin sensitisation reactions (see below).

With time, toxicity testing methods have become codified, and standard experimental protocols have now been issued by various regulatory agencies worldwide, such as the US Federal Food and Drugs Agency, the US Environmental Protection Agency (EPA) and the Organization for Economic Co-operation and Development (OECD). One of the most well recognised of these are those of the OECD which has issued a number of standard protocols for the conduct of physico-chemical, toxicity and ecotoxicity tests since the early 1980s. These protocols include some that have single-dose toxicity values, such as the dermal LD_{50}, as an endpoint.

There are a number of features common to all recommended experimental protocols, including:

- temperature and humidity control
- light
- atmospheric conditions
- noise
- diet
- ventilation
- biological factors such as infections and diseases

- requirements for quality assurance, storage of samples and data and other aspects of Good Laboratory Practice.

Parameters for the standard single-dose skin toxicity include those shown in Table 5.1.

The LD_{50} provides an estimate of the relative toxicity of a substance by the dermal route of exposure. Extrapolation of dermal LD_{50} values in animals to humans is valid only to a limited degree.

The irritation studies

Toxicity studies that focus on irritation as a toxic endpoint are based on the pioneering work of Draize and coworkers from the 1940s (Draize *et al.* 1944). The term irritation is slightly imprecise, as there are a number of ways in which tissues become irritated (for example, contact or allergic irritation). As the Draize tests are a measure of irritation from contact with the test chemical, they are called primary irritation studies.

In these studies a fixed amount of the test material is held against the tissue being evaluated (either skin or eye) for a fixed length of time. The test material is then washed off and the tissue evaluated for irritation at fixed times using a score from 0 (no effect) to 4 (severe effect). Scores are then average for each time point and each animal to provide an overall irritation score.

While test parameters may vary depending the conditions of the test, standard irritation test parameters, as required under standard protocols, are shown in Table 5.2.

Table 5.1 Parameters for the standard single-dose toxicity test

Test parameter	Skin irritation
Animals	Five animals per dose (rats, rabbits or guinea pigs) of the same sex. At the end of the study, one group of five animals of the other sex should be dosed with an appropriate dose to check for sex sensitivity
Dose	At least three dose groups, and enough to produce a dose–response curve (under an occluded dressing). Dosing should be applied uniformly over an area less than 10% of the total body area
Duration of contact	24 hours, held in contact using a porous gauze dressing and nonirritating tape. Restrainers may be used, but complete immobilisation is not recommended
Assessment criteria	Estimation of dermal LD_{50} and signs of toxicity from pathology
Duration of test	At least 14 days

Table 5.2 Standard irritation test parameters

Test parameter	Skin irritation	Eye irritation
Animals	Three rabbits	Three rabbits
Dose	500 mg (under an occluded dressing)	100 mg (instilled under eyelid)
Duration of contact	4 hours	1 hour
Assessment criteria	Skin redness (0–4) Skin swelling (0–4)	Conjunctival redness (0–4) Damage to iris (0–4) Damage to cornea (0–4)
Time of assessment	24, 48, 72 hours	24, 48, 72 hours
Duration of test	10–14 days	10–14 days

The sensitisation study

The production of irritation through immunological mechanisms is termed 'sensitisation'. In sensitisation, an initial exposure produces an immunological change (such as the production of antibodies to the initial exposure) which changes the status of body defences to the exposure (that is, the individual has become sensitised). A sensitisation reaction then occurs on subsequent exposures.

Sensitisation is theoretically possible in any organ or tissue where antibody cells are found. However, the main organs of sensitivity are the skin, the respiratory system and the gastrointestinal tract. Some evidence is also available to indicate sensitisation in the nervous system.

The only mainstream toxicity test for sensitisation is for the skin, and requires an appreciation of the mechanism of sensitisation (of initial exposure developing into a sensitisation reaction). A specialised initiating dosing regimen is required to ensure that the immunological change occurs, followed by a challenge exposure. The initiating doses must be such that primary irritation does not occur (which could complicate the findings of the test).

The species of choice for skin sensitisation is the guinea pig, which has an immunological system more similar to humans than that in rats, mice or rabbits. The initial phase requires dermal exposure to doses of the test chemical that do not induce primary irritation (therefore, the results of the primary skin irritation are necessary in estimating such a dose). Several initiating doses are given, normally every other day for 2–3 weeks. The chemical may be placed on the skin under an occluded dressing, but is more usually administered subcutaneously or intradermally. Skin is evaluated for primary irritation during the initiating phase, and animals showing primary irritation may be removed from the study. Once the initial phase of dosing is complete, there is a short period of a week or so where no chemical administration occurs. If the chemical is a sensitiser, it is during this period that the sensitisation processes will be occurring.

After this period, the animals are tested with a 'challenging' dose, normally lower than the initiating doses. If a sensitisation reaction has occurred, the minor dose of the challenge dose will initiate a skin reaction, showing that the test chemical is a sensitiser.

One important concept to the skin sensitisation study is that animals with compromised immune systems will have impaired sensitisation processes, and therefore might not be able to respond appropriately to test chemicals. If this happens, the possibility of a false-negative result can arise. As this is a distinct possibility it is common practice to include a second test chemical, which is known to be a skin sensitiser in the study, as a positive control. If animals dosed with the positive control show skin effects at the challenge dose, then the study has been conducted properly. If animals dosed with a positive control do not have skin sensitisation, then their immune function is compromised and the results from the study can be rejected.

Structure, function and occupational disorders of the skin

The structure and function of the various components of skin, and its susceptibility to occupational disease, are shown in Table 5.3.

Clinical aspects of occupational skin disorders

Diagnosis of skin disorders

The diagnosis of possibly work-related dermatoses requires some discipline (Freeman 1991). There are three main diagnostic tools:

- the history
- physical examination
- patch testing.

Other forms of identification may also be of use, such as skin prick testing and skin biopsy, although these methods are not used often by dermatologists.

Taking the history

In the course of taking the history from the patient with a work-related skin disease, information should be obtained on personal, family, and related matters (Lushniak 2000). With regard to work factors, information should be sought on at least:

- past and present job and job activities
- a description of the current job, the length of time they have been doing it, job activities and preventive measures employed
- an accurate chronological sequence of events surrounding the onset of disease, its subsequent clinical course and associated work activities
- qualitative and quantitative aspects of exposure

Table 5.3 Structure, function and occupational disorders of the skin

Structure	Function	Occupational disorder
Stratum corneum	Barrier against chemicals, microorganisms, and some radiation	Chapping from low humidity, chemical stains
Squamous cells/basal cells	Cell regeneration, synthesis of stratum corneum, wound repair	Infection, contact dermatitis, neoplasms
Connective tissue	Mechanical protection against trauma, wound repair, development of scar tissue	Infection, granulomatous reactions, scleroderma, solar elastosis, scarring
Hair follicles	Insulation and protection, secondary sensory organs	Folliculitis, traumatic or toxic alopecia
Melanocytes	Absorption of UV radiation	Toxic vitiligo, melanoma, hyperpigmentation
Eccrine sweat glands	Thermoregulation, buffering of skin surface	Miliaria, 'rusting'
Sebaceous glands	Synthesis of skin lipids, chemical barrier against microorganisms	Oil acne, chloracne
Nerve tissue elements, Merkel cells, other sensory receptors	Perception, sensation	Toxic neuropathies
Blood vessels, mast cells	Thermoregulation, nutrition of tissue	Heat stroke, urticaria, flushing reactions, vibration white finger
Langerhans cells, dermal macrophages	Immune regulation and surveillance	Delayed hypersensitivity reactions

- a description of the skin lesions and their initial anatomical locations and (if any) spread to other body sites
- disability caused by the skin condition
- identification of all relevant skin exposures
- identification of skin protection controls
- identification of personal hygiene and skin cleansing methods
- presence of skin disease in coworkers exposed to similar risks
- response to previous medical treatment
- whether the skin problem improves while away from work
- any factors which exacerbate the symptoms

- identification of other risk factors, such as personal or family history of atopic allergies, previous skin problems, medications, potential causative factors in the domestic environment (second job, hobbies, outside activities, and so forth).

These may need to be followed up by contact with the employer, and possibly a workplace assessment.

Physical examination

The physical examination must assist in the differentiation of occupational from nonoccupational skin disorders. For example, some patterns of dermatitis are suggestive of endogenous dermatitis – localised dermatitis spreading towards the heel of the palm is probably always endogenous, as is involvement of elbow and knee creases. Therefore, the entire skin should be examined for the presence of skin lesions.

The patch test

Although virtually any diagnostic procedure can be used to evaluate suspected occupational dermatitis, the patch test is the most frequently employed. This is a simple but useful diagnostic tool.

Firstly, a small amount of the suspect substance is placed on a metal disc or filter paper backed with aluminium foil. This disc (the patch) is placed on adhesive tape with a number of other patches (normally 10–20) containing a range of routine 'screens' (compounds known to cause positive reactions). The adhesive tape is then placed on the upper back or (rarely) the upper, outer arm. Both the patch and the tape are impermeable to water, thus providing total occlusion of the test (and screen) substances against the skin (Freeman 1991).

Test strips are removed about 2 days later, and the sites evaluated. A positive reaction is scored in the presence of numerous papules or vesicles. Weaker reactions (skin redness or swelling without papules or vesicles) cannot be reliably distinguished from false-positive marginal reactions alone. Test sites should be re-evaluated 72–96 hours following initial application, since 30–40% of all positive readings will be equivocal or negative at the 48 hour reading.

The patch test should not be used to diagnose irritant contact dermatitis since the conditions of the test seldom approximate the actual conditions under which exposure occurs.

Common occupational skin disorders

Contact dermatitis

Contact dermatitis (also called contact eczema) is by far the commonest occupational dermatosis (Shama 1988; Stevenson 1989; Adams 1990; Nethercott 1994; Halkier-Sorensen 1996; ATSDR 1996; Rycroft et al. 2001). The term refers to changes in the skin, usually accompanied by inflammation, from direct skin exposure to exogenous physical or chemical agents (Linn Holness 1994). Inflammation is provoked by two mechanisms: irritation or allergy (Lachapelle 1986). A key diagnostic feature is itching. The resulting skin lesions are difficult to differentiate clinically, since each can appear as acute or chronic forms.

The acute form is erythmatous (increased skin redness), may be vesicular (small blisters), bulbous (large blisters), oedematous (swollen) or oozing. This form is usually of short duration, lasting days or weeks.

The chronic form has some of these features, although it is more likely to be lichenified (thickened), scaly, and fissured, which can last for years.

One important clinical feature of occupational contact dermatitis is the distribution of the dermatitic lesions. These generally occur on those areas of the skin which have come into contact with the irritating or allergenic agent (Burry 1982). At work, the most common areas for skin lesions are the hands, wrists and forearms, where they come into contact with materials by dipping or washing. Dusts and mists may affect other areas, such as the forehead, eyelids, face, ears and neck. Areas can be affected where agents accumulate, such as beneath watch straps, collars or waistbands, and in creases and folds of skin (Coenraads et al. 1985).

Physical damage, such as abrasion or erosion, may also predispose a person into developing a contact dermatitis reaction.

Other, less indirect factors include:

- skin thickness
- sweating
- personal or environmental hygiene
- medications
- alcohol, smoking and other social drug use
- season of the year or climatic conditions
- pigmentation.

Finally, individuals may have an innate genetic susceptibility to developing dermatitis. This predisposition is called atopy.

Irritant contact dermatitis

Primary irritants damage the skin because they are irritating. Almost any substance can be a contact irritant.

Irritants are generally divided into strong and mild types. Strong irritants include strong acids, alkalis, aromatic amines, and many metallic salts. They produce an irritating effect within minutes after contact. A feature of many strong irritants is their ability to damage the skin irreparably, causing eschar formation (scarring). An irritant that causes permanent skin damage is also called a corrosive.

On the other hand, mild irritants, such as detergents, solvents, oils, soap and water, can require repeated and prolonged exposures for many days before clinical changes occur in skin. Mild skin irritants cause many cases of occupational irritant dermatitis and are a major problem in the workplace. One reason for this is that the systems for dealing with mild irritants are unlikely to be present in the workplace – while people know to be careful about acids, taking basic preventive measures is not so obvious when, for example, using shampoo in a hairdressing establishment.

Common skin irritants, either in the domestic or occupational environment, are shown in Table 5.4.

Important hazard properties of a primary skin irritant are:

- pH
- solubility
- molecular size (large molecules of molecular weight > 1000 are poor penetrants in general)
- ionisation and polarity
- physical state
- concentration of the irritant within the material.

Important risk properties of a primary skin irritant are:

- duration of contact
- intensity of contact
- occlusion.

Table 5.4 Common irritants in the home and workplace

At home	At work
Bleaches	Acids
Carpet cleaners and shampoos	Alkalis
Copper and metal brighteners	Aluminium powders
Detergents	Cement and concrete
Drain cleaners	Cleaning agents and products
Fertilisers	Coal and petroleum products
Furniture polishers and waxes	Combustion products
Household pesticides	Drugs and medicines
Oven cleaners	Epoxy resins
Pet shampoos	Fibre glass
Scouring pads and powders	Formaldehyde
Soaps	Glutaraldehyde
Toilet bowl cleaners	Insulation foams
Window cleaners	Metallic compounds
	Minerals (some)
	Organic solvents
	Particulates of ores and metal oxides
	Plastics and plasticisers
	Rubber products
	Sawdust
	Soap and detergents
	Wool

The dermatitis caused by contact with irritants is a local cutaneous response by direct action; that is, it is not mediated through an immunological reaction.

The inflammatory cutaneous changes that occur from skin irritation result from a direct, local, toxic effect on cellular elements of the skin, leading to cell death, release of lysosomal enzymes and soluble inflammatory mediators, recruitment of inflammatory cells and further tissue destruction. Obviously, exposure to strong irritants will produce massive damage quickly. However, under the right circumstances, virtually any substance is potentially capable of causing irritant contact dermatitis, and clinical irritant dermatitis generally results from multiple, cumulative exposures to (often) more than one potential skin irritant, rather than just a single substance.

There is a spectrum of degree of severity, depending on the strength of the irritant, the circumstances of exposure and the skin site affected. Since irritation depends on the lower, living layers of the epidermis, those factors that enhance penetration will increase severity of response.

A number of points should be emphasised:

- Contact irritant dermatitis can occur from contact with several mild irritants, in which the effect is cumulative.
- Continual irritant contact dermatitis can produce either a condition where even very mild irritants can produce irritation, or 'hardening' where the skin eventually tolerates repeated exposures.
- Constant exposure to irritants impairs the barrier function of the skin and allows further penetration of other irritants.
- Irritant and allergic contact dermatitis frequently coexist in the same person.

One further form of irritant contact dermatitis is light-induced change to chemicals on the skin surface (phototoxicity). The mechanism is nonimmunological, and some irritants, such as polycyclic aromatic hydrocarbons and some dyes, may be made more irritating by virtue of the action of sunlight. Some protection to this effect occurs in pigmented skin.

Some of the precise mechanisms of irritant contact dermatitis remain undefined. Substances that dissolve skin proteins, such as keratin, or oxidising agents or dehydrating agents may be irritants. Changes in the water content of the skin also appear to be important.

Follow-up studies of people with occupational contact dermatitis indicate that (Cooley and Nethercott 1994):

- about 25% fully resolve (that is, they become symptom free)
- about 50% improve but continue to have periodic bouts of symptoms
- the remaining 25% develop chronic, persistent dermatitis which is as severe (or worse) as the initial dermatitis.

About 30–40% of people with occupational dermatitis change their jobs to get away from the precipitating exposure. Job change is most beneficial (but not strikingly so) for sufferers with allergic contact dermatitis (Vestey et al. 1986). Job change does not always clear up the dermatitis, with perhaps a quarter of those changing jobs experiencing complete resolution of symptoms.

Allergic contact dermatitis

A variety of agents are potential contact allergens, such as:

- metal salts (nickel, chromium, cobalt)
- rubber additives (mercaptobenzothiazole and other benzothiazoles, thiuram, resorcinol)
- ethylenediamine componds
- hydroquinone and *p*-aminophenol photographic developers
- epoxy compounds
- acrylates
- aliphatic and aromatic amines
- formaldehyde
- phenolic compounds
- therapeutics (neomycin, benzocaine, streptomycin antibiotics)
- biocides (formaldehyde-releasing compounds, halogenated germicides, quaternary ammonium compounds).

A variety of natural products also cause allergic contact dermatitis, including grains (barley, oat, rye, wheat), foods (carrot, chicory, coconut, coffee, endive, lettuce, potato, radish), spices (cardamon, tamarind, tumeric, vanilla), plants (rhus, poison ivy, poison oak, sumac), and fragrances (balsam of Peru, cinnamic acid derivatives, citronella derivatives).

Unless they are primary irritants, most agents that cause allergic contact dermatitis do not produce a skin reaction on the first contact (if they did, they would be irritants).

The mechanism of effect of allergic dermatitis is a cell-mediated (or type VI) immune response. There are two main phases – the sensitisation phase and the elicitation phase.

There are a number of steps in sensitisation:

- absorption of the allergic agent or antigen (or hapten)
- binding of the antigen to a carrier protein, which forms the complete antigen
- binding of the complete antigen to the cell surface of macrophages or Langerhans cells within the epidermis
- these cells alter the configuration of the antigen where it then interacts with T lymphocytes
- antigen-bearing lymphocytes then undergo clonal proliferation in the regional lymph node, where two populations of sensitised lymphocytes arise
- effector lymphocytes which are distributed to the skin surface through the blood stream
- memory cells will form new populations of sensitised lymphocytes on repeat contact with the antigen.

Sensitisation usually takes 10–21 days. Following sensitisation, the immune system is in a state of readiness (elicitation phase) should further contact with the allergic agent occur. When this happens, effector T lymphocytes are activated (from formation of the complete antigen). Activated lymphocytes then interact with other skin cells and processes to synthesise a range of substances called cytokines, which mediate the inflammatory response. The elicitation phase can fade with time, as antigen memory is lost, although it can persist for life in many individuals.

Strong allergens have molecular weights of less than about 500 and are quite reactive with proteins. The number of potent allergens is quite small, although several thousand substances have been reported to have caused sensitisation in one or a few individuals.

Other important points to note about allergic contact dermatitis should be noted:

- Differentiation of mild irritants from skin allergens can be quite difficult.
- Allergic contact dermatitis may not develop for months or years after first exposure to an agent.
- Many sensitisers are also irritants (for example, chromates, nickel salts, formaldehyde, wood turpentine).
- Elicitation reactions can be produced or maintained by minute amounts of exposure, often in concentrations insufficient to irritate the nonallergic skin.
- Cross-sensitivity may occur where a worker exposed to one allergen will also react to one or more closely related allergens, even without being exposed to the second allergen before.
- Diagnostic patch testing is the most accurate means of distinguishing between allergic contact from irritant contact dermatitis.

As with phototoxicity and irritant contact dermatitis, photoallergic reactions (an interaction of a chemical and sunlight) are also possible. Photoallergens include phenothiazides and some coumarin derivatives. Photoallergic reactions are quite rare in the occupational environment, but are very serious if they occur.

The factors that should be taken into account in the development of skin irritation are summarised in Table 5.5.

Thermal burns

Skin tissue is not thick and can be susceptible to extremes of temperature. Burns cause problems because the body can loose fluids, thereby leading to dehydration and electrolyte imbalance, and increasing susceptibility to infection.

Thermal burns are classified thus:

- first degree burns: damage to epidermis only
- second degree burns: damage to epidermis and upper dermis (blisters may be seen)
- third degree burns: full thickness burns, usually leading to scarring.

Critical, that is life-threatening, burns are those where there are:

Table 5.5 Factors in the development of skin irritation

Factor	Important aspects
Potential irritant	Chemical properties Physical properties Allergic properties
Exposure	Concentration Duration of exposure Frequency and number of exposures Possibility of occlusion to the skin Temperature
Preventive measures	Protective equipment Barrier creams
Individual	Characteristics (age, gender, ethnicity) Skin site Pre-existing skin damage Susceptibility (atopy)

- second degree burns to over 25% of the skin
- third degree burns to over 10% of the skin surface
- third degree burns to face, hands, feet and joints.

Scalds are a subclass of burns caused by hot liquids or hot, moist vapours like steam.

Pigmentary disorders

Vitiligo is a condition characterised by the destruction of melanocytes. The destruction can be over small or large areas of skin, resulting in patches of depigmentation, often with a hyperpigmented border, and often enlarging slowly. While vitiligo is considered an autoimmune disease, some occupational exposures may produce similar depigmentation. For example, skin absorption of materials that resemble some of the precursors to melanin synthesis (such as phenolic or catecholic derivatives which resemble tyrosine) may inhibit synthesis at low doses, and cause melanocyte death at high exposures. These materials are antioxidants or germicidal disinfectants, which may be found in a range of products, including rubber products, photographic developers, lubricating oils, plastics, adhesives and disinfectant or cleaning solutions.

Compounds known to cause hypopigmentation include:

- antiseptics such as o-benzylchlorophenol and pyrocatechol
- organic compounds such as o-phenylphenol and p-butylphenol
- photographic chemicals such as hydroquinone and its monoethyl and monobenzyl ethers
- disinfectants such as p-cresol
- astringents such as p-tertiary butylcatechol.

Hyperpigmentation is a condition by which the skin darkens. This can follow any episode of skin inflammation, and is due to accumulation of melanin from injured melanocytes of haemosiderin from blood cells present in dermal tissue. This is more common in people with dark complexions.

Discolouration or staining may also be caused by long-term exposure to heavy metals, particularly silver, mercury or arsenic. These exposures produce diffuse slate grey, blue–grey or melanic discolouration of skin.

Other materials encountered in industry may stain the skin. Yellow–orange stains are produced by some nitrogen compounds, such as nitric acid and trinitrotoluene. Coal miner's tattoo is produced by the insertion of coal dust under the skin from abrasive or explosive forces. This type of effect is possible from similar exposures in other industries. Other discolourations are shown in Table 5.6.

Acneform disorders

Acne is an inflammatory disease of the sebaceous gland. The follicle to which the gland is attached becomes plugged with keratin, and the inflamed glands form small pink papules, pustules or cysts. Sometimes these have black centres (comedones or blackheads). Secondary bacterial infection may be present. Common

Table 5.6 Causes of skin discolouration

Colour	Cause	Industry/Occupation
Brown	Arsenic	
	p-Phenylanaline	Dye manufacturers (obviously other colours from other dyes, too)
	Permanganates	Bleach and dye makers
Orange	Tetryl	Explosives manufacture
	Chlorine	Chemical workers
		Laundry workers
		Cleaners
		Swimming pool maintenance workers
Yellow	Picric acid	Explosives manufacture
	Dichromates	Electroplaters
		Tanneries
		Lithographers
	Nitrates and nitrites	
Green	Copper dust	Copper smelting
		Electrical
Blue	Silver nitrate	Photography
	Oxalic acid	Car radiator cleaners
Black	Osmium tetroxide	Electron microscopy
	Mercury	Cosmetologists

acne (*acne vulgaris*) is quite widespread, and not conventionally considered to have an occupational cause. However, pre-existing acne can be aggravated by various occupational stresses, such as heat and high humidity, which cause a swelling of the lining of the duct, with subsequent blocking of the pore ('tropical acne').

Rubbing, pressure or friction may provoke new acne in susceptible areas (*acne mechanica*). Occupational causes of this condition include face masks, respirators, belts and harnesses, tight-fitting work clothing and pressure from seat backs.

Exposure to oil (heavy lubricating greases, oils and pitch fumes) is associated with follicular plugging and pustular folliculitis (**oil acne**). The acne appears on the areas where exposure occurs (fingers, hands, forearms and possibly legs, if covered by oil-soaked clothing). The mechanism is keratinisation of the follicles followed by blocking of the ducts. Scrupulous attention to personal hygiene should help reduce the severity of this problem.

Chloracne is an acneform disorder caused by exposure to polyhalogenated materials. Established exposures known to cause chloracne include polychlorinated biphenyls (PCBs) and polybrominated biphenyls (PBBs), polychlorinated diben-zodioxins and furans, polychlorinated benzenes, azobenzenes and naphthenes, chlorinated phenols, and other polyhalogenated compounds and contaminants. Chloracne is produced by squamous metaplasia of the sebaceous gland ducts, followed by atrophy of the underlying sebaceous gland and subsequent formation of keratin-filled cysts. The basic dominant clinical lesion is a small cystic lesion ranging in size from a pin-head to a small pea (Maibach 2000). Chloracne appears initially on the face, although it may spread to other areas of the body as the severity of the disease increases. The condition does not respond well to treatment, can take 2–3 years to improve, and scarring can occur (Zugerman 1990).

Vascular reactions

Urticaria (also called hives) is a vascular reaction of the skin marked by the transient appearance of smooth, slightly elevated patches (wheals) which are redder or paler than the surrounding skin, and which are accompanied by severe itching. These eruptions usually occur within 15–60 minutes of exposure, and rarely last longer than a couple of days, although chronic forms occur. Urticaria can be caused by certain foods (such as shellfish), drugs (penicillin), infection or emotional stress. Urticaria may be provoked by either immunological or nonimmunological mechanisms, but is not always associated with allergy.

Occupational causes of urticaria are fairly rare, but where they occur they are often associated with inhalational exposure (and are invariably associated with other symptoms of inhalant allergy, such as asthma, rhinitis, conjunctivitis). Causative agents include dusts, moulds, pollen, castor bean pomace, coffee bean dust, platinum salts, penicillin, formaldehyde, some pesticides and irritant gases. Other exposures may cause urticaria by contact. This group includes a wide variety of foodstuffs, medical rubifacients, plant materials and solvents (Maibach 1987).

Compounds known to cause urticaria include those shown in Table 5.7.

Table 5.7 Chemicals that cause allergic contact urticaria	
Animal products	*Medications*
Dander	Bacitracin
Hair	Cephalosporins
Saliva	Chloramphenicol
Seminal fluid	Gentamicin
Serum	Neomycin
	Salicylic acid
Common chemicals	
Ammonia	*Plant products*
Alcohol	Henna
Epoxy resin	Latex rubber
Formaldehyde	Papain
Parabens	Strawberries
Polyethylene glycol	Woods
Cosmetics	*Textiles*
Acrylic monomer	Silk
Hair products	Nylon
Nail polish	Wool
Perfumes	
Food	
Eggs	
Flour	
Fruits	
Meat	
Milk	
Nuts	
Seafood	
Spices	
Vegetables	

Flushing reactions are primarily a transient redness of the skin of the face, but also sometimes the neck and upper chest. Rubber workers exposed to some curing compounds may experience severe flushing reactions, headaches, nausea and vomiting. 'Degreaser's flush' is associated with exposure to trichlororethylene, which may be exacerbated by alcohol consumption.

Connective tissue disorders

Vibration white-finger disease is an occupationally induced condition caused by vasospasm of dermal blood vessels through reflex enervation of blood vessels from the Pacinian corpuscle. This responds to vibration frequencies in the range of 60–125 Hz. Vibration at these frequencies is common from a range of equipment and mechanical plants, including construction machinery, chainsaws, pneumatic drills, rivet guns and so forth.

The first symptoms are persistent numbness and tingling, usually after a few minutes of operating a vibratory tool, followed by fingertip swelling. Whitening of the fingers (blanching) occurs with further exposure and is generally most severe in the finger closest to the most intense vibration. At first the symptoms subside quickly, but soon take some time (perhaps subsiding at the weekend). Reactive hyperaemia and cyanosis follow the blanching (ischaemia), which can be persistent in advanced cases. Later, the symptoms take much longer to subside, or may be precipitated by other exposures, such as the cold. Necrotic ulcerations of the fingertips (from ischaemia) may occur, although this is extremely rare. The most usual symptoms are paraesthesias and neurosensory defects. Unless the condition is in an early or mild form it is generally irreversible.

Vibration white-finger disease needs to be differentiated from Raynaud's phenomenon, with which it shares some common features.

Skin cancer

Among light-skinned people, the overwhelming cause of skin cancer is the ultraviolet (UV) component of sunlight (Emmett 1987). Other causes are shown in Table 5.8.

The evidence implicating exposure to chemicals with the subsequent development of skin cancer comes principally from studies carried out over 30 years ago. Chronic exposure to chemicals was much more common because of elevated exposures, poor control, poor hygiene, a lack of information on risk, mobility of workers, latencies of disease, and difficulties in making the association between exposure and disease (many skin cancers can be treated on an outpatient basis). Risks today are likely to be much lower.

The commonest skin cancers among fair-skinned people are from exposure to sunlight (or more accurately, ultraviolet solar radiation). The nonionising spectrum of sunlight ranges from 200–400 nm, with the higher energies at the lower wavelengths. This spectrum is divided into three regions: (i) UVA from 315–400 nm; (ii) UVB from 280–315 nm; and (iii) UVC radiation below 280 nm. UVA carries less energy than UVB or UVC, although UVA penetrates more deeply into the skin and underlying tissues. While shorter wavelengths carry more energy than the longer wave UVA, they do not penetrate as deeply into the skin as UVA. Skin burn is associated with the 300 nm band. Wavelengths greater than 300 nm are transmitted by the cornea and are absorbed by the lens (indicating that eye protection in sunlight is at least as important as skin protection).

The common skin cancers include basal cell carcinomas with an incidence rate of about 70%, followed by squamous cell carcinomas (about 25%), malignant melanomas (3–5%) and 'others' (cancers of the sweat glands, sebaceous glands and hair follicles).

Squamous cell carcinoma is a tumour of epidermal squamous cells which occurs most frequently on areas of the body that receive repeated and prolonged sunlight exposure (head, neck, lips, top of the ears, hands, forearms). These tumours seldom arise spontaneously on normal-looking skin and involve sites that generally show other signs of damage from sunlight. These changes include freckles (solar-induced lentigines), thickened yellowish furrowed skin (solar elastosis), skin wrinkles,

Table 5.8 Causes of occupational skin cancer

Carcinogen	Industry or exposure	Example(s)
Ultraviolet light	Outdoor workers	Farmers
		Fisherfolk
		Construction workers
		Open-cut miners
	Industrial UV	Welders
		Germicidal lamps
		UV curing processes
Ionising radiation	Nuclear industry	Reactor workers
	Health care	Radiologists
Pitches and tars (polyaromatic hydrocarbons)	Coal, coke, gas industries	Coke oven workers
		Tar distillers
		Coal gas manufacture
	Fuel manufacture	Briquette making
	Creosote users	Tile workers
		Timber proofers
	Soot	Chimney sweeps
		Rubber manufacture
Oils	Shale oil refining	
	Paraffin wax workers	
	Oil refining	Still cleaners
	Engineering	Tool setters
		Lathe operators
	Cotton industry	Mule spinners
Arsenic	Agriculture	Insecticide application
		Animal dips

abundant dilated skin vessels close to the surface of the skin (telangiectasiae) and flat, reddened, scaling, premalignant sun keratoses.

Basal cell carcinoma is a tumour of the epidermal epithelial cells which also demonstrates a close association between site of tumour and exposure to sunlight. However, unlike squamous cell tumours up to a third of basal cell carcinomas may develop on skin that is not chronically exposed to sunlight. Further, many of these tumours have a tendency to cluster around the skin of the eyes, nose and mouth.

Most nonmelanoma skin cancers (80–90%) occur on the head and trunk. These tumours rarely metastasise and a large proportion respond well to treatment, indicating that they have a high incidence, but a low mortality. On the other hand, the majority of melanomas are found on the trunk and limbs, even though the causative agent is invariably sunlight. Conversely, rates of melanoma are lower in outdoor workers than in indoor workers, suggesting that gradual exposure to sunlight during spring and early summer, and the build-up of skin tan may be protective factors. The higher rates of melanoma in indoor workers may be related to the massive (but short-lived) exposures to sunlight at weekends or on annual vacations. Indeed, melanoma may be related more to previous skin damage through sunburn and blistering than to direct sunlight exposure itself.

Melanoma is a tumour of melanocytes. It is a cancer of white people, and is very rare in black people. The cancer tends to be pigmented in an initially flat lesion. As the cancer spreads out irregularly and superficially in the skin the pigmentation can become irregular (Fritschi and Siemiatycki 1996).

The only satisfactory curative treatment for melanoma is surgical excision. The likely outcome depends on the depth of the tumour, the site of the tumour, the gender of the individual, and other factors. The likely success of this treatment (surgery) is much greater when the cancer is thin, in its early stages. Cure is possible in 90% of patients if the cancer is removed before it is less than 0.75 mm thick. The cancer is lethal in 60% of cases where it has been allowed to invade vertically into the skin to a thickness of 4 mm or more by the time it is treated.

Epidemiological surveys show an increased risk for the development of squamous cell, basal cell or melanoma-type tumours among fair-skinned populations as geographic location approaches the equator. The migration of fair-skinned populations during recent centuries to those regions of the world with high sunlight exposure has meant that those individuals (and their families) are more likely to develop skin cancer, a factor confirmed in the medical and scientific literature. Indeed, Australia, now with a predominantly European-based population, has one of the highest incidences of melanoma, particularly in those states with tropical climates.

Prevention of occupational skin disorders

Determination of the agents and risk factors that cause occupational skin disorders is crucial to their prevention (Mathias 1990). However, this can be problematic, especially for skin allergens. As with many workplace risks, the process of occupational risk management (identify, assess, control) are necessary in prevention. Obviously prevention of dermatoses in the workplace would ideally be accomplished through total elimination of skin exposure to dermatitic risk factors (Cooley and Nethercott 1994). However, this is not always possible. Therefore a multidimensional approach to occupational skin disorders is needed, involving:

- recognition of potential cutaneous irritants and allergens
- recognition that some individuals may be more susceptible to exposure than others (atopy)
- elimination, substitution, or engineering controls to control skin exposure
- personal protection, appropriate clothing and barrier creams to prevent skin exposure
- appropriate information and training of workers and managers
- monitoring and health surveillance (including pre-employment screening)
- personal hygiene
- better systems for diagnosis, treatment and reporting of occupational skin disorders
- regulation of established allergens and irritants.

A comprehensive preventive program of this type requires cooperation of workers, employers, product manufacturers and suppliers, safety and occupational health practitioners, the medical profession and government agencies.

Finally, it is important to promptly and accurately diagnose the cause of the dermatitis for the speedy rehabilitation of the worker. A worker who has been incorrectly diagnosed with work-related dermatitis often finds it difficult to accept a later diagnosis of constitutional dermatitis. The confusion, and sometimes hostility, such conflicting diagnoses produce can result in greatly delayed healing of the skin disease.

Summary

A wide variety of agents encountered in the workplace may cause injury, cell death, irritation, sensitisation, infection, discolouration or other changes to the skin of workers. Further, some agents induce cancerous changes to the skin. The term occupational dermatoses is commonly used for all of these abnormalities resulting from, or aggravated by, the work environment.

In addition to being a target organ, the skin may also serve as an entry route to the body for toxic chemicals by percutaneous absorption.

Causal agents in the development of occupational skin disorders may be categorised as:

- chemical: a wide range of corrosives, acids, alkalis, organic and inorganic compounds
- physical: mechanical effects, heat and cold, humidity, sunlight and other radiation
- biological: bacterial, viral, fungal infections, parasite infestations, plants and plant products.

Chemical agents are the greatest cause of occupational skin disorders. Sometimes a combination of several chemical agents, or of chemical and other agents (for example, sunlight) is responsible. The multicausal nature of occupational dermatoses significantly complicates its diagnosis and control in the workplace. While the symptoms of a skin disorder are visibly apparent, accurate diagnosis linking the disorder to occupation requires a high level of clinical skill and expertise in taking occupational histories and in physical examination. Establishing the relative contributions of exposures outside work can also prove difficult and complex.

Hot humid conditions and sun exposures associated with tropical and subtropical climates, in particular combined with the use of personal protective equipment, can exacerbate skin disorders.

As with exposure to all occupational risks, assessment of risks and application of the hierarchy of controls will assist in eliminating or reducing risk. These measures include:

- removal of known dermatoxic agents from the workplace
- substitution of hazardous materials for less hazardous materials (for example, allergen replacement in epoxy resins by high molecular weight chemicals)

- control at source such as enclosed dipping instead of spray painting
- safe working procedures such as reuse of cutting oils in lathe operations
- personal administration, such as adequate supervision, training and information (labels and material safety data sheets), provision of personal hygiene controls (for example, ensuring that skin is washed when contaminated)
- personal protective equipment, including hats, eye protection, work overalls, gloves, aprons, footwear, barrier creams.

Other approaches, such as monitoring and health surveillance, will identify problems as they arise and assist in control. Epidemiological studies may sometimes be helpful in establishing causal relationships with occupational skin diseases (for example, skin cancer).

References

Adams, R.M. (ed.) (1990) *Occupational Skin Diseases*, 2nd edn. Philadelphia: W.B. Saunders.

ATSDR (1996) Skin lesions and environmental exposures. Agency for toxic substances and disease registry. *Aust. Ass. Occup. Health Nursing* 44: 529–540.

Burry, J.N. (1982) Clinical epidemiology and preventive dermatology: A sample of 1,000 cases in South Australia. *Aust. J. Dermatol.* 23: 14–17.

Coenraads, P.J., Foo, S.C., Phoon, W.O. and Lun, K.C. (1985) Dermatitis in small scale metal industries. *Cont. Derm.* 12: 155–160.

Cooley, J.E. and Nethercott, J.R. (1994) Prognosis of occupational skin disease. *Occup. Med. State Art Rev.* 9: 19–24.

Draize, J.H., Woodard, G. and Calvery, H.O. (1944) Methods for the study of irritation and toxicity of substances applied topically to the skin and mucous membranes. *J. Pharmacol. Exp. Ther.* 82: 377–390.

Emmett, E.A. (1987) Occupational skin cancers. *Occup. Med. State Art Rev.* 2: 165–177.

Freeman, S. (1991) Occupational skin diseases: Diagnosis and differential diagnosis. *J. Occup. Health Safety – Aust. NZ* 7: 229–235.

Fritschi, L. and Siemiatycki, J. (1996) Melanoma and occupation. Results of a case–control study. *Occupat. Environ. Med.* 53: 168–173.

Halkier-Sorensen, L. (1996) Occupational skin diseases. *Cont. Derm.* 35: (Suppl) 1–120.

Lachapelle, J.M. (1986) Industrial airborne irritant or allergic contact dermatitis. *Cont. Derm.* 14: 137–145.

Linn Holness, D. (1994) Characteristic features of occupational dermatitis: Epidemiologic studies of occupational skin disease reported by contact dermatitis clinics. Occupational Skin Disease. In: Nethercott, J.R. (ed.) *Occup. Med. State Art Rev.* 9: 45–52.

Lushniak, B.D. (2000) Occupational skin diseases. *Prim. Care* 27: 895–916.

Maibach, H.I. (ed.) (1987) *Occupational and Industrial Dermatology*, 2nd edn. Chicago: Year Book Medical.

Maibach, H.I. (2000) *Toxicology of the Skin*. London: Taylor and Francis.

Mathias, C.G.T. (1990) Prevention of occupational contact dermatitis. *J. Am. Acad. Dermatol.* 23: 742–748.

Mathias, C.G.T. (1994) Occupational dermatoses. In: Zenz, C., Dickerson, O.B. and Horvath, E.P. (eds) *Occupational Medicine*, 3rd edn. St Louis: Mosby, pp. 93–131.

Nethercott, J.R. (ed.) (1994) Occupational skin disease. *Occup. Med. State Art Rev.* 9: 1–125.

NOHSC (1989) *National Strategy for the Prevention of Skin Disorders*. National Occupational Health and Safety Commission/AGPS, Canberra.

Rosen, R.H. and Freeman, S. (1992) Occupational contact dermatitis in New South Wales. *Aust. J. Dermatol.* 33: 1–10.

Rycroft, R.J.G., Menne, T., Frosch, P.J. and Lepoittevin, J.-P. (eds) (2001) *Textbook of Contact Dermatitis*. New York: Springer-Verlag.

Shama, S.K. (1988) Occupational skin disorders. In: McCunney, R.J. (ed.) *Handbook of Occupational Toxicology*. Boston: Little, Brown and Co, pp. 216–235.

Stevenson, C.J. (1989) Occupational skin disease. *Postgrad. Med. J.* 65: 374–380.

Vestey, J.P., Gawkrodger, D.J., Wong, W.-K. and Buxton, P.K. (1986) An analysis of 501 consecutive contact clinic consultations. *Cont. Derm.* 15: 119–125.

Zugerman, C. (1990) Chloracne: Clinical manifestations and etiology. *Dermatol. Clin.* 8: 209–213.

Chapter 6

Occupational toxicology of the liver

N.H. Stacey

Introduction

The liver is the largest internal organ in the body. From a simple operational perspective, the liver is often referred to as the body's chemical purification plant.

The liver is a very important organ for chemical-induced toxicity as it is rather vulnerable to attack by chemicals (MacSween *et al.* 2001). Reasons for this include its anatomical location and its ability to concentrate, biotransform and excrete some chemicals through the bile. The first of these, anatomical location, relates especially to ingested chemicals and is therefore less relevant to workplace exposures which occur mainly by inhalation or skin (Sherlock and Dooley 2001). Nevertheless, it should be appreciated that chemicals absorbed across the gastrointestinal mucosa will then be delivered directly to the liver through the portal blood supply. Thus the liver will be exposed to the highest concentrations of these chemicals and on this basis alone could be expected to exhibit toxicity more often than other organs (Plaa 1986).

Chemicals encountered by other routes of exposure may also reach the liver, through its blood supply from the hepatic artery as well as the portal vein. The second reason listed for susceptibility of the liver to chemical-induced injury is its ability to concentrate, biotransform and excrete chemicals, irrespective of the route of exposure (Zimmerman 1982). Chemicals may be toxic as the parent compound or may be biotransformed to reactive metabolites for expression of toxicity. When the metabolite is formed in the liver this organ will again be exposed to the highest concentrations of the toxic moiety with the associated likelihood of being injured. Furthermore, the liver has the greatest amounts of the biotransformation enzymes in the body, and as these will be involved in the biotransformation of endogenous and exogenously absorbed chemicals, may also be an organ at high risk of injury through the production of reactive metabolites (Bircher *et al.* 1999).

Anatomy of the liver

The liver is situated slightly below the diaphragm and in front of the stomach. It consists of two lobes that are wedge-shaped. The right lobe of the liver is larger than the left. Each lobe is further divided into many small lobules, each one about the size of a pin-head, and consisting of many liver cells, with bile channels and blood channels between them (Bircher *et al.* 1999).

Two blood vessels enter the liver: the hepatic artery, with oxygenated blood from the lungs through the heart and aorta; and the hepatic portal vein with chemicals derived from dissolved food substances from the small intestine. Permeating the entire liver structure is a system of blood capillaries, bile capillaries and lymph capillaries.

The liver cells secrete the bile, and this collects in the bile canaliculi, which then unite to form bile ducts. These bile ducts all eventually unite, forming the common hepatic duct, which is joined by a branch, the cystic duct, which leads from the gall bladder. This is on the inferior surface of the liver, and is the storage site for bile. The common bile duct leads from the gall bladder to the pancreatic duct, forming a common duct that opens into the duodenum (the first section of the small intestine) (O'Grady *et al.* 2000).

Within the liver, the hepatic artery and portal vein branches, together with the bile-collecting system, form the portal triad. The blood entering the liver travels from the portal triad through the sinusoids to the terminal hepatic venule (or central vein) and is thence returned to the systemic circulation via the hepatic vein. Figure 6.1 depicts the structure of the liver.

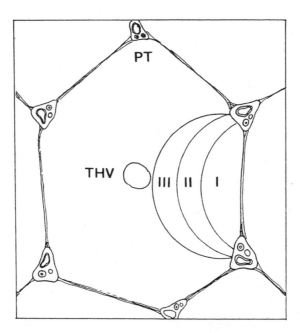

Figure 6.1 Main structural elements of the liver (schematic representation). PT, portal triad; THV, terminal hepatic venule (or central vein).

The functional unit of the liver is the acinus, which is divided into zones 1, 2 and 3 (Rappaport 1980). The area around the portal triads is called zone 1 (or peri-portal according to an old classification) while that around the central vein is zone 3 (or centrilobular). Zone 2 (or midzonal) lies between zones 1 and 3. When a section of liver is viewed under the microscope the classical hexagonal lobule can be seen as shown in Figure 6.1. The structure can be easily recognised by locating the central vein and portal triads.

Knowledge of the structure of the liver is important because a widely used classification of hepatotoxic reactions is based on histological location and appearance. Injury such as necrosis around the central vein is termed centri-lobular (sometimes pericentral) while that in the area of the portal space is called periportal (or periacinar). Liver cell damage in between is described as midzonal.

At the cellular level, the liver consists of four systems:

- the hepatocyte (liver cell) system
- the biliary tract system, involved in the production of bile from the breakdown of haemoglobin from blood cells
- the blood circulatory system. The blood in the liver is supplied by two sets of blood circulatory systems: the systemic circulation (about one-quarter of the cardiac output flows into the liver through the hepatic artery) and the entire blood flow in the portal vein, flowing from the intestinal system. Blood from these two systems mixes together and flows through a latticework of tiny chan-nels inside the liver. Blood from the liver flows back to the heart through the hepatic vein and vena cava.
- the reticuloendothelial system consists of a range of cell types including Kupffer cells (highly mobile macrophages), fat-storing lipocytes, highly mobile, natural killer lymphocytes and endothelial cells.

Physiology of the liver

The functions of the liver are varied, working closely with nearly every fundamen-tal system and process in the human body, in particular biotransformation, homeo-stasis and the regulation of blood sugar.

- **Biotransformation of chemicals.** The liver biotransforms the chemicals absorbed from a number of sources, including food and drink, the environment and the workplace. These include endogenous chemicals circulating in the body (for example, hormones, waste products and so on) or absorbed exoge-nous chemicals. The most important reaction is to convert what are generally fat-soluble substances into more water-soluble substances. By doing this, absorbed chemicals can be more readily excreted in urine. Other materials may also be excreted in the bile.
- **Regulation of blood sugar.** The level of blood sugar stays at around 0.1%, and excess coming from the gastrointestinal tract is stored as glycogen. A hormone excreted by the pancreas called insulin causes the excess glucose to turn into glycogen.

- **Regulation of lipids.** Lipids are extracted from the blood and, depending on body requirements, may be converted to carbohydrates or similar structures, or sent to fat storage sites if not needed immediately.
- **Regulation of amino acids.** Levels of amino acids in the blood are kept at a physiological level. It is not possible to store essential amino acids in the body, so the amine groups of any spare circulating amino acids are converted to urea, redistributed in the blood system, and excreted by the kidneys in urine. The remainder of the amino acid molecule is not wasted; it is changed into a carbohydrate structure that can be used again.
- **Formation of bile.** Bile consists of bile salts and the excretory bile pigments. These assist in the digestion of lipids.
- **Metabolism of cholesterol.** This fatty substance is used in the cells. An excess in the blood is linked with blood vessels becoming blocked, which may lead to heart attacks and other disorders.
- **Formation of plasma proteins.** Some plasma proteins are used for blood clotting and for keeping the blood plasma constant. The main blood proteins include fibrinogen, prothrombin, albumin and globulins.
- **Making heparin.** This is an anticoagulant substance which prevents the blood from clotting as it travels through the body.
- **Storage of blood.** The liver can swell to hold huge amounts of blood which can be released into the circulation if the body suddenly needs more, in injury or disease.
- **Removal of haemoglobin molecules.** When red blood cells become fragile, they are broken down, with the haemoglobin they contain being converted into bile pigments, and the iron atoms being saved for future use.
- **Storage of vitamins.** Vitamin A is also made in the liver from carotene, the orange–red pigment in plants. Vitamin B_{12} is also stored in the liver.
- **Production of heat.** The liver is one of the hardest working regions of the body and produces a lot of heat. This is carried round the body in the blood to warm less active regions.

In general terms the liver is a vital metabolic organ which is essential to the organism's survival. It is singly responsible for the synthesis of many proteins, for energy storage in the form of glycogen, for the majority of biotransformation of exogenous chemicals by the body and the excretion of some chemicals through the bile to name a few. For details on the functions of the liver the reader is referred to other texts (Arias *et al.* 1988; O'Grady *et al.* 2000).

Classification of chemical-induced injury

Liver injury can be classified in various ways. Other more authoritative texts may be consulted for greater detail in this field (Lawrence *et al.* 1998; Zimmerman 1978, 1982; Zimmerman and Lewis 1995). Of course, the evaluation of workers exposed to agents that may impact on liver function also needs to consider the likely agents that may cause liver problems (Fleming 1992; Hodgson *et al.* 1990; Klockars 1986; Zimmerman 1999): Firstly, on the basis of its histological change or morphologic or microscopic appearance with respect to the site within the lobule, and with respect to the nature of the alteration; secondly, on the basis of pre-dictability of the lesion; and thirdly, with regard to the organelle primarily affected (Zimmerman 1993).

Histological change

Because of its unique metabolism and relationship to the gastrointestinal tract, the liver is an important target for the toxicity of drugs, xenobiotics, and oxidative stress (Jaeschke *et al.* 2002). Drug metabolism, lipid peroxidation and thiol oxida-tion are frequently involved in liver toxicity (Batt and Ferrari 1995).

Chemicals can be classified according to the region within the lobule that is damaged. It may be centrilobular as with the classical hepatotoxicant carbon tetrachloride, midzonal as with beryllium, or periportal as exemplified by phosphorus.

The appearance of changes induced in the liver on microscopic examination is varied, as shown in Table 6.1.

- **Necrosis**, which is one type of cell death, is a common effect of acute expo-sure to a hepatotoxic agent. It may be zonal or diffuse. If the insult is not lethal the liver is usually able to undergo complete repair after an episode of acute necrosis.
- **Steatosis**, or intracellular fat accumulation, is also a common change that is generally reversible on removal of the offending chemical.
- **Cholestasis**, which means cessation or slowing of bile flow, is another type of injury often encountered. Bile stasis can be observed histologically. More recently the term cholestasis has been used to describe any interference with biliary function as this may occur with some chemicals without necessarily a clear histological equivalent.
- **Apoptosis** is a more recently described type of cellular event that is important in physiological cell deletion as well as some forms of tissue injury, such as chronic active hepatitis. It is usually associated with immune cell-induced cytotoxicity and remains to be convincingly ascribed as a sole cause of chem-ical-induced damage.

Table 6.1 Histological classification of liver injury

Lesion type	Examples
Necrosis	Carbon tetrachloride, beryllium
Steatosis	Carbon tetrachloride, ethanol
Cholestasis	Organic arsenicals
Apoptosis	Uncertain
Immune cell infiltrate	See text
Fibrosis/cirrhosis	Carbon tetrachloride, trinitrotoluene
Neoplasia	Carbon tetrachloride, vinyl chloride
Mixed	Carbon tetrachloride

- **Immune cell infiltrate** is observed in conjunction with chemical-induced cell injury but it can be unclear if some immune cells are present as a causative factor or as a response to tissue injury to remove cellular debris. Nevertheless, there are a number of chemicals (mainly therapeutic agents) that have a definite association with the presence of immune cells other than in a debris removal function (see section on predictability).
- **Fibrosis** and **cirrhosis** are the end result of ongoing liver injury, generally following long-term exposure. Collagen is deposited in the liver leading to a disruption of normal architecture and function. Occupational exposure to carbon tetrachloride and trinitrotoluene has been implicated in this liver condition (see Table 6.1).
- **Neoplasia** may also ensue from such ongoing occupational exposures with hepatic carcinoma being the most common type of cancer. Other types of cancer have also been recorded with the outstanding example being angiosarcoma following exposure to vinyl chloride (Neugut *et al*. 1987).

As is apparent from Table 6.1 some chemicals produce more than one type of histological alteration so that the pattern is one of a **mixed** response.

Predictability

Many chemicals cause liver injury in a manner that is normally expected – that is, in a predictable way with clear evidence of a dose–effect relationship. However, other chemicals do not, and in recognition of this Zimmerman has proposed a classification scheme (Zimmerman 1978, 1982). The features of the two types of reaction are shown in Table 6.2.

This nonpredictable hepatotoxicity has also been referred to as an 'idiosyncratic response' that may have either an immunological basis resulting in a hypersensitivity type reaction, or it may be due to a metabolic abnormality.

Organelle damage

A less commonly used method of classifying hepatotoxicants is according to the organelle primarily affected. Chemicals have been termed as 'primarily injurious' to mitochondria, endoplasmic reticulum, plasma membrane, lysosomes or the cytoskeleton. If serious injury does occur to one or other of these organelles then

Table 6.2 Features of predictable and nonpredictable hepatotoxicity

	Predictable	Nonpredictable
Histology	Distinct	Diffuse/immune cells
Extrahepatic allergy	No	Yes
Dose–effect	Yes	No
Latent period	Regular	Varied
Incidence	High	Low
Reproducible in test animals	Yes	No

death of the cell may result. Again, there may be a mixture of effects with carbon tetrachloride being a good example, having been associated with damage to all of the above except the cytoskeleton.

Detection and evaluation of liver injury

Liver damage can be detected with a range of procedures, from external palpation to enlargement and tenderness, blood enzyme tests, through to needle biopsy.

Acute hepatotoxicity can be detected with commonly used noninvasive tests; detection of subacute and chronic hepatic damage is more difficult. The methods for testing for liver damage in response to chemical exposure are listed in Table 6.3.

Blood sampling followed by analysis of various serum constituents is employed in the first instance as the sampling of liver tissue (biopsy) for microscopic examination is an invasive procedure which can have complications.

Although the term liver function test is a misnomer, it is still commonly used to describe tests of liver integrity, such as serum enzymes. Hepatocytes contain many enzymes that may be released into the blood if the cell membranes are damaged. An elevation of serum enzymes such as alanine aminotransferase and aspartate aminotransferase reflect damage to the hepatic parenchymal cells, while elevated levels of alkaline phosphatase indicate injury to the biliary apparatus of the liver. The enzyme γ-glutamyl transpeptidase is often found to be increased with excessive alcohol consumption. Bilirubin, bile salts and proteins such as albumin and clotting factors are also serum indicators of liver dysfunction. Thus they can be regarded as tests of liver function using endogenous markers.

In testing for liver damage, the activity of a range of enzymes may be measured (RCPA 1990) as described below.

- Alanine transaminase (ALT), formerly known as SGPT (serum-GPT), is used in the investigation of hepatocellular (liver cell) damage.

Table 6.3 Methods for detecting liver injury

Physical examination

Clinical chemistry
Enzymes
Bilirubin
Bile salts
Proteins

Histopathology
Light microscopy
Electron microscopy

Organ function tests
Dye excretion
Drug biotransformation

- Aspartate transaminase (AST), formerly known as SGOT (serum-GOT), is a specific liver enzyme used in the investigation of hepatocellular (liver cell) disease.
- Alkaline phosphatase (ALP) is used in the investigation of hepatobiliary (liver and bile ducts) disease. ALP is not one enzyme but a group of enzymes with broad substrate specificity – it may be raised in other conditions not related to liver injury, such as bone or gastrointestinal tract conditions.
- γ-Glutamyl transaminase (GGT), used in the investigation of liver disease, particularly cholestasis (suppression of the flow of bile). This enzyme is often increased with chronic intake of alcohol, some therapeutic drugs and some occupational exposures (for example trinitrotoluene).
- Sorbitol dehydrogenase (SDH) is a liver enzyme specific for hepatocytes, but it is unstable and usually elevated for only very short periods.
- Ornithine carbamoyltransferase (OCT) is a hepatocellular mitochondrial enzyme that is a sensitive marker of hepatic damage, although analysis of this enzyme is difficult and is not used routinely.
- Once the specificity of liver injury is established, lactate dehydrogenase (LD) is used to investigate the progress of liver disease.

Elevated levels of one or more of these enzymes may indicate liver damage (Meeks *et al.* 1991). Other markers include bile acids and bilirubin. Other measures of liver function include levels of uric acid, cholesterol or triglycerides, although these are more likely to indicate metabolic dysfunction, of which liver dysfunction is only of one type.

However, the histological appearance of the liver remains the most definitive indicator of liver injury. The nature of such changes has been outlined above. Biopsy in humans for confirmation of diagnosis is usually employed when there is uncertainty.

Organ function tests

The organ function tests listed in Table 6.3 require administration of exogenous chemicals such as the dye indocyanine green or the drug antipyrine. Neither of these are used routinely in animal experimentation but they do have (or have had) some application in the evaluation of liver injury in humans and in experimental animal investigations.

The use of individual serum bile acids, which seem to provide a very sensitive index of liver injury, may be useful indicators of chemical-induced damage as well as of organ dysfunction (Azer *et al.* 1997; Bai *et al.* 1992). Such dysfunction may be detected prior to release of cytoplasmic enzymes into the blood.

Mechanisms of hepatic injury

The mechanisms by which chemicals cause liver injury are deserving of mention because this knowledge is important in prevention and treatment of injury. The liver has received more attention than other organs in this regard, probably due to the number of chemicals that have the liver as their primary target. This in turn may be related to the high levels of biotransformation enzymes located in the endoplasmic reticulum. Another factor may be the greater availability of hepatic *in vitro* preparations that retain *in vivo* characteristics for studies relating to mechanism of action of the toxic agent. The various mechanisms that have been implicated in chemical-induced liver injury are listed in Table 6.4.

Although thousands of papers have been published on these proposed mechanisms over the last 40 years, there is still no comprehensive unified understanding of the steps involved in cell death after exposure to the toxic agent. Each of the mechanisms has considerable evidence supporting its role but there is also conflicting evidence which is difficult to reconcile with that mechanism being causally linked to the ensuing cell death. For example, the theory that lipid peroxidation is the cause of liver injury due to some chemicals has a lot of appeal because of the autocatalytic nature of the process once it is initiated. However, some experimental evidence shows that the peroxidation of the lipids can be inhibited without stopping cell death. This clearly suggests that the peroxidation is not directly responsible for the death of the cells induced by that chemical. Similarly, covalent binding of reactive metabolites to cellular macromolecules and disturbance in the homeostasis of the calcium ion have much experimental support. Some critical evidence also disallows unequivocal acceptance of these as the key event in cell death. Inhibition of protein synthesis is an older hypothesis that lost support when the timing of the inhibition did not coincide well with other observations in the process leading to cell destruction. The involvement of mitochondria as a key and causative event also lost favour because of the timing of events.

However, more recently there has been renewed interest in the role of mitochondrial damage in cell death. The immunological component relates to the hypersensitivity reactions referred to earlier. It is possible that some chemicals act as haptens by binding to tissue components, thereby presenting an antigenic stimulus which elicits an immune response. This may lead to cell destruction. It is possible that apoptosis could be involved in this process as there are now several reports associating apoptosis with lymphocytotoxicity. It is also possible that some chemicals may activate endonuclease(s) that seem to be involved in the process of apoptosis, which could result in the cell being destroyed. The inhibition of triglyceride secretion is listed last because, although it is one of the mechanisms

Table 6.4 Mechanisms of hepatic injury

Lipid peroxidation
Covalent binding
Alteration of calcium homeostasis
Alteration of protein synthesis
Depletion of protein thiols
Mitochondrial dysfunction
Immune-mediated
Endonuclease and apoptosis
Interference with triglyceride secretion
DNA damage

responsible for steatosis, the cellular accumulation of fat does not necessarily pre-dispose to hepatocyte death.

Overall then, there is yet to be identified a common component in the pathway leading to liver cell death. Each of the above proposed mechanisms is clearly involved in some way. With further investigation the current anomalies may be explained, or perhaps the critical event that is the irreversible step leading to cell death is yet to be identified. It may also be possible that there is no single step common to all chemicals and that there are different key effects with different chemicals.

Examples of hepatotoxicity from workplace exposures

Many chemicals have the potential to cause liver injury in workers (Dossing and Skinhoj 1985; Pond 1982). In the past, chemicals such as carbon tetrachloride have been associated with both acute and chronic liver injury in exposed workers. With increased knowledge and improved workplace conditions such examples are now less frequent. Nevertheless, they do continue to occur as evidenced by case reports of workers experiencing liver damage after using various chemicals such as dimethyl formamide and methylene dianiline.

Although some examples of hepatotoxicants have already been listed in Table 6.1 it is worth considering such chemicals in greater detail and from the viewpoint of duration of exposure as shown in Table 6.5. In some cases, groups of chemicals (such as organic solvents) are listed because more than one chemical in the group has been associated with that response.

Summary

The liver contains many biotransformation pathways, and absorbed chemicals may be changed to less toxic or more toxic products. Sufficient exposure to toxic chemicals and their biotransformation products can produce effects in the liver. This damage may impair the capacity of the liver to do its job properly. While a number of chemicals have been shown to induce liver damage (hepatotoxicity) such chemicals are being gradually withdrawn from use (for example, carbon tetrachloride, chloroform).

Table 6.5 Examples of workplace exposures that have resulted in hepatic toxicity

Exposure	Agent	Type of lesion
Acute	Organic solvents	Mixed
Short-term repeated dose	Tetrachloroethane, trinitrotoluene	Acute or ongoing necrosis, may progress to cirrhosis
Chronic	Arsenicals, vinyl chloride, organic solvents	Fibrosis, cirrhosis, cancer

Source: adapted from Dossing and Skinhoj 1985.

Many chemicals cause liver damage in a manner consistent with the dose–response relationship (that is, the higher the dose or exposure, the greater the effect). However, other chemicals do not. This is referred to as an idiosyncratic response, and may be due to immunological or metabolic abnormalities.

Occupational exposures to some chemicals are known to produce significant liver toxicity. Also, the interaction of chemicals with each other, from interoccupational exposure or nonoccupational exposure to solvents (such as alcohol consumption), may cause liver damage in chemical-exposed workers.

References

Arias, I.M., Jakoby, W.B., Popper, H., Schachter, D. and Shafritz, D.A. (eds) (1988) *The Liver – Biology and Pathobiology*, 2nd edn. New York: Raven Press.

Azer, S., Klaassen, C.D. and Stacey, N.H. (1997) Biochemical assay of serum bile acids: Methods and applications. *Br. J. Biomed. Sci.* 54: 118–132.

Bai, C., Canfield, P.J. and Stacey, N.H. (1992) Individual serum bile acids as early indicators of carbon tetrachloride- and chloroform-induced liver injury. *Toxicology* 75: 221–234.

Batt, A.M. and Ferrari, L. (1995) Manifestations of chemically induced liver damage. *Clin. Chem.* 41: 1882–1887.

Bircher, J., Benhamou, J.-P., McIntyre, N., Rizetto, M. and Rodes, J. (eds) (1999) *Oxford Textbook of Clinical Hepatology*, 2nd edn. Oxford: Oxford University Press.

Dossing, M. and Skinhoj, P. (1985) Occupational liver injury. Present state of knowledge and future perspective. *Int. Arch. Occup. Environ. Health* 56: 1–21.

Fleming, L.E. (1992) Unusual occupational gastrointestinal and hepatic disorders. *Occup. Med.* 7: 433–448.

Friedman, L.S. and Keeffe, E.B. (eds) (1998) *Handbook of Liver Disease*. London: Churchill Livingstone.

Hodgson, M.J., Goodman-Klein, B.M. and van Thiel, D.H. (1990) Evaluating the liver in hazardous waste workers. *Occup. Med.* 5: 67–78.

Jaeschke, H., Gores, G.J., Cederbaum, A.I., Hinson, A., Pessayre, D. and Lemasters, J.J. (2002) Mechanisms of hepatotoxicity. *Toxicol. Sci.* 65: 166–176.

Klockars, M. (1986) Solvents and the liver. *Prog. Clin. Biol. Res.* 220: 139–154.

MacSween, R., Burt, A.D. and Portmann, B.C. (2001) *Pathology of the Liver*. Edinburgh: Churchill Livingstone.

Meeks, R.G., Harrison, S.D. and Bull, R.J. (1991) *Hepatotoxicity*. London: CRC Press.

Neugut, A.I., Wylie, P. and Brandt-Rauf, P.W. (1987) Occupational cancers of the gastrointestinal tract. II. Pancreas, liver, and biliary tract. *Occup. Med.* 2: 137–153.

O'Grady, J., Lake, J. and Howdle, P. (2000) *Comprehensive Clinical Hepatology*. London: Mosby.

Plaa, G.L. (1986) Toxic responses of the liver. In: *Casarett and Doull's Toxicology: the Basic Science of Poisons*, 3rd edn. New York: Macmillan, pp. 286–309.

Plaa, G.L. and Hewitt, W.R. (eds) (1997) *Toxicology of the Liver*, 2nd edn. New York: Taylor and Francis.

Pond, S.M. (1982) Effects on the liver of chemicals encountered in the workplace. *West J. Med.* 137: 506–514.

Rappaport, A.M. (1980) Hepatic blood flow: morphologic aspects and physiologic regulation. *Int. Rev. Physiol.* 21: 1–63.

RCPA (1990) *Manual of Use and Interpretation of Pathology Tests*. Royal College of Pathologists of Australasia, Sydney.

Sherlock, S. and Dooley, J. (2001) *Diseases of the Liver and Biliary System*. Oxford: Blackwell Science.

Zimmerman, H.J. (1978) *Hepatotoxicity*. New York: Appleton-Century-Crofts.

Zimmerman, H.J. (1993) Hepatotoxicity. *Dis. Mondiale* 39: 675–787.

Zimmerman, H.J. (ed.) (1999) *The Adverse Effects of Drugs and Other Chemicals on the Liver*, 2nd edn. New York: Raven Press.

Zimmerman, H.J. and Lewis, J.H. (1995) Chemical- and toxin-induced hepatotoxicity. *Gastroenterol. Clin. N. Am.* 24: 1027–1045.

Chapter 7

Occupational toxicology of the kidney

W.M. Haschek and C. Winder

Introduction

The kidney plays a principal role in the regulation of extracellular fluid volume, acid–base balance and electrolyte composition, as well as excretion of metabolic waste products (Brenner 2000). The kidney has a high blood flow–mass ratio and is thus typically exposed to higher concentrations of blood-borne chemicals than most other organs. It also has a unique function of concentrating urine and its constituents, which can include exogenous compounds such as chemicals and drugs. These properties make the kidney uniquely susceptible to injury. Occupational renal injury can occur through trauma, biochemical, haemodynamic or immunological mechanisms (Bennett and Porter 1993). An understanding of the structure and normal functional processes of the kidney is required to fully understand renal injury.

Kidney structure and function

The kidneys are paired organs located retroperitoneally in the lumbar region, ventral and lateral to the vertebral column (O'Callaghan and Brenner 2000). In an adult human, each kidney is 10–13 cm long, 5–6 cm wide and 2.5–3 cm thick, and weighs about 150 g (less in females). The kidneys are reddish-brown, bean-shaped and covered by a thin fibrous capsule (Brenner 2000). The renal artery and vein and the ureter enter the kidney at the hilus, an indentation on the inner border (Brenner 2000). Blood is filtered by the kidney and the filtrate, which contains waste products, is excreted in the form of urine. Urine flows from the renal pelvis of the kidney through the ureters into the urinary bladder and is voided through the urethra to the exterior.

On gross dissection, the kidneys are clearly divided into an outer cortex and an inner medulla with papillae extending into the renal pelvis. The kidneys receive 25% of the cardiac output, with approximately 85% of the total renal blood flow distributed to the cortical region and the remainder to the medulla. Therefore, most blood-borne chemicals enter the cortex, and the cortex is exposed to the greatest concentration of any blood-borne compound. However, the medulla and papillae will also be exposed to high compound concentrations in the tubular filtrate due to the concentrating function of the cortex and will be exposed for longer periods due to the slow flow of filtrate and blood in this region (Schnellmann 2001). Injury to specific regions is dependent on the properties of the toxicant and the constituent tissue factors in the region.

The functional and structural unit of the kidney is the nephron. There are over one million nephrons in each human kidney. The nephron is divided into the vascular and glomerular portions located in the cortex, and the epithelial tubular segment that extends from the glomerulus into the medulla and opens into the pelvis (Figure 7.1). These elements are separated by a sparse but highly vascularised connective tissue, the interstitium.

Vascular and glomerular components

The renal artery enters the hilus of the kidney and branches into the interlobular, arcuate and interlobular arteries. The interlobular artery gives rise to the afferent arterioles which supply blood to the glomerulus. The glomerulus consists of a glomerular tuft separated from the surrounding Bowman's capsule by Bowman's space. The glomerular tuft is composed of capillaries separated by mesangium and lined by visceral epithelium that consists of specialised epithelial cells. This epithelium continues as the parietal epithelium which lines the inner aspect of Bowman's capsule and then as tubular epithelium. Blood passes through the afferent arteriole into the glomerulus, where it is filtered by selective passage of water, electrolytes, low molecular weight proteins and other compounds, including waste products. Glomerular filtration is based mainly on molecular size, with retention of plasma proteins with a molecular weight higher than 40,000; net charge and shape also are factors in retention, with filtration of anionic molecules being somewhat restricted.

Glomerular blood pressure and blood flow is regulated by the afferent and efferent arterioles which are innervated by the sympathetic nervous system and also respond to angiotensin II, vasopressin, endothelin, prostanoids and cytokines. Glomerular filtration pressure is about 50 mmHg (75 mmHg of blood pressure minus an osmotic pressure of 25) which is sufficient to force a considerable fraction of blood plasma through the glomerulus with selective passage of water, electrolytes, low molecular weight proteins and other compounds, including waste products, into the filtrate.

Toxicants can alter the glomerular filtration rate (GFR) and the size and/or charge selective properties of the glomerulus. These alterations can be induced by changes in blood pressure as well as by structural or functional changes to the components of the glomerular capillary tuft. In addition, chemically induced immunological injury (see Chapter 3) can lead to glomerular injury due to trapping of circulating immune complexes in the glomerulus (for example, gold, HgCl, cadmium, silica) (Fenwick and Main 2000; Schnellmann 2001; Stratta et al. 2001). Immune complex deposition may also occur following chemically induced autoimmune disease that affects components of the glomerulus such as the base-

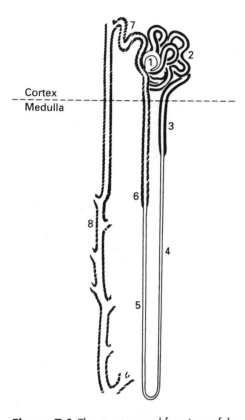

Figure 7.1 The structure and functions of the nephron. The dotted line represents the demarcation between the cortex and medulla regions of the kidney. (1) Glomerulus: water and lower molecular weight components of blood are filtered through the capillary tuft. (2) Proximal convoluted tubule: most reabsorption of substances such as glucose, chlorine, calcium, amino acids, phosphates, bicarbonate occurs in (2) and (3). Proximal straight tubule: reabsorption of organic anions and metabolism of glutathione. (4) Descending limb of the loop of Henle: begins the process of reabsorption of water, chloride ions are pumped out of the loop (4)–(6), which becomes progressively impermeable to water. The resulting filtrate is quite dilute. (5) Thin ascending loop of Henle: uptake of sodium ions, secretion of potassium and hydrogen ions. (6) Thick ascending loop of Henle: same as 5. (7) Distal convoluted tubule: same as 5. (8) Collecting duct: continues the process of regulation of urine, water reabsorption. (From Hewitt *et al.* 1991, with permission.)

ment membrane (for example, volatile hydrocarbons, solvents) (Schnellmann 2001). This type of injury to the glomerulus is called glomerulonephritis. Glomerular disease that affects blood flow through the glomerular capillaries has an adverse effect on the tubules since their blood supply is from the efferent arteriole.

The filtrate (urine) passes into the tubule while filtered blood leaves the glomerulus by the efferent arteriole, either entering the capillary network surrounding the adjacent nephron or the vasculature supplying the medulla. Therefore, chemicals that reach the glomerulus can be filtered into the urine which is in direct contact with the luminal side of the tubular epithelium (Figure 7.2). Alternatively, the chemical can remain in the blood, and reach the basal side of the tubular epithelium through the capillary network or enter the medulla

Tubular component

The filtrate from the glomerulus enters the proximal convoluted tubule (pars convoluta, S1 segment). After several convolutions, the tubule straightens (transition, S2 segment) into the proximal straight tubule (pars recta, S3 segment) and eventually enters the medulla; it has a total length of about 10–15 mm. Tapering abruptly, it continues as the loop of Henle, a thin-walled straight tubule which at first descends into the medulla (descending limb) and then reverses course, ascending back into the cortex (thin and thick ascending limbs). The loop of Henle may be 20 mm long. It continues as the distal tubule that is about 5 mm long. The distal tubule continues as the collecting duct which receives many tributaries from other nephrons. It enters the papillary region and, after traversing the renal papilla, opens into the adjoining calyx discharging the filtrate (now urine) into the renal pelvis.

The epithelial cells that line the proximal tubule are highly differentiated with a well-developed brush border, except for the S2 segment. Large numbers of mitochondria and a highly developed lysosomal system are present proximally and decrease distally. The proximal tubule selectively reabsorbs most of the water, salts, sugars and amino acids, with approximately 70% of water and solutes reabsorbed from the glomerular filtrate. This includes sodium, calcium, potassium, chloride, phosphate, magnesium, bicarbonate ions, amino acids, glucose and other physiologically critical materials. Virtually all of the filtered low molecular proteins are reabsorbed here by specialised endocytotic processes. Active secretion by the

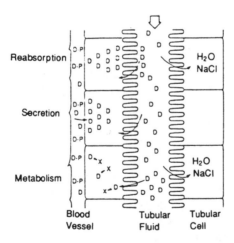

Figure 7.2 The proximal tubule. Chemicals or drugs (D) can reach the tubular epithelial cell either via the tubular fluid (urine) or blood. Once the chemical enters the cell it can remain unchanged, undergo metabolism to a metabolite (X), or be bound to protein (P). (From Hewitt *et al.* 1991, with permission.)

tubular epithelium from the adjacent capillary network into the urine also occurs and is primarily responsible for excretion of certain organic compounds and for elimination of hydrogen and potassium compounds. The specialised transport systems utilised for these physiologic processes are also able to extract and/or concentrate toxic chemicals from the basolateral side of the tubule. Biotransformation is a function of the proximal tubule, especially the S3 segment, due to high concentrations of P450 enzymes, glutathione, as well as cysteine beta-lyase which can bioactivate glutathione conjugates (Schnellmann 2001).

The proximal tubule is the most common site of toxic injury (Goldstein 1994) due to concentration of transport systems in this region allowing accumulation of toxicants (for example, HgCl, cadmium, cisplatinum, aminoglycosides); sensitivity to hypoxia and ischaemia (lack of blood supply) because of high energy and substrate requirements; and presence of high concentrations of biotransforming enzymes (for example, chloroform, trichloroethylene, haloalkene-S-conjugates) (Schnellmann 2001; Their et al. 2002).

Epithelial cells that line the loop of Henle do not have microvilli and are thinner than other tubular epithelial cells throughout the nephron. Reabsorption of water as well as chloride, potassium and sodium occurs in this segment. While cells in the descending limb are able to absorb water, cells in the ascending limb are relatively impermeable to water and urea.

The cells of the distal tubule contain numerous mitochondria, but do not have a well-developed brush border or lysosomal system. Specialised cells in the macula densa, which is part of the juxtaglomerular apparatus and found adjacent to the glomerulus, provide a feedback loop triggered by increased solute concentration. This signal results in decreased blood flow into the glomerulus, thereby decreasing GFR in the nephron (tubuloglomerular feedback). The distal tubule is relatively impermeable to water. The distal tubule and collecting duct reabsorb sodium chloride and water, and acidify urine.

The cells of the collecting ducts have fewer mitochondria than those of the distal tubule. The collecting ducts are involved with final regulation of urinary volume and composition. Reabsorption of water takes place which concentrates the filtrate including urea, one of the more important waste products. Overall, reabsorption is a very efficient process with 100 litres or so of filtrate a day being converted to excretion of 1–1.5 litres of urine per day in humans.

Few toxicants target the more distal sites of the tubular system. Those that do target the concentrating ability and/or processes controlling acidification and result in an antidiuretic hormone (ADH)-resistant polyuria (for example, methoxyfluorane, amphotericin B, cisplatinum) (Schnellmann 2001). The loop of Henle, which allows the kidney to concentrate urine by the countercurrent mechanism, can also result in concentration of chemicals, such as analgesics, in the deep medulla or papilla. Underlying renal pathophysiology compromising blood flow as well as prostaglandin metabolism are implicated in analgesic injury. Papillary necrosis is a well-known sequela of analgesic abuse.

Additional renal functions

While filtration of the blood is probably the most important function of the kidney, it is also involved in other functions. The principal functions of the kidney are summarised in Table 7.1.

The kidney is an endocrine organ. In addition to its capacity to metabolise and excrete certain hormones, the kidney is the site of production of renin, a blood pressure regulating hormone; erythropoietin, a hormone that regulates red blood cell production; prostaglandins, long chain fatty acids involved in a wide variety of physiological functions; 1,25-dihydroxycholecalciferol, a vitamin involved in calcium metabolism; and kinins, peptides which cause changes in blood vessel function. It is the target organ for hormones such as the parathyroid hormone, atrial naturetic peptide, antidiuretic hormone, angiotensin and aldosterone.

The kidney is also involved in the breakdown of small molecular weight proteins and in metabolic interconversions that regulate the composition of body fluids. The ability of the kidney to convert certain organic acids (lactic, α-ketoglutamic) to glucose (a neutral substance) is an example of a metabolic interconversion that minimises potential changes in plasma pH.

Kidney disease and occupation

The primary cause of occupational kidney damage is from chemical exposures (Wedeen 1992). It is estimated that nearly four million workers in the USA are exposed to chemicals at concentrations that may damage the kidney (nephrotoxicity) (Landrigan et al. 1984). Although the best documented toxic causes of renal failure are drugs (de Broe et al. 1998), occupational exposure to chemicals, such as heavy metals and solvents, also plays a role in the development of chronic renal failure (Fels 1999; Loghman-Adham 1997). Chemicals can induce renal injury due to a primary effect on the kidney, or secondarily through systemic effects such as cardiac failure or intravascular haemolysis (de Broe et al. 1996).

The two common manifestations of chemically induced renal disease are acute and chronic renal failure (Table 7.2). Acute renal failure (ARF) is characterised by a rapid decline in GFR with resultant azotemia. Tubular injury and increased intrarenal vascular resistance are the most common causes of acute failure. In

Table 7.1 Principal functions of the kidneys

Maintenance and endocrine regulation of extracellular fluid volume of the body
Maintenance of acid–base balance
Excretion of metabolic waste products (urea, uric acid, creatinine)
Biotransformation and elimination of exogenous toxicants (drugs, chemicals)
Endocrine regulation of blood pressure (renin)
Endocrine control of red blood cell mass (erythropoietin)
Endocrine control of mineral metabolism, notably calcium and phosphate (vitamin D_3)
Breakdown of low molecular weight proteins (such as insulin, calcitonin, β_2-microglobulins)
Metabolic interconversions

Table 7.2 Manifestation of toxicant injury to the kidney

Type of injury	Mechanism	Effect
Acute renal failure	Renal vasoconstriction	Hypoperfusion/hypofiltration
	Glomerular injury (often immunological)	Glomerulonephritis
	Direct tubular injury	Tubular necrosis
	Intertubular obstruction	Obstruction
	Immunological	Tubulointerstitial nephritis
Chronic renal failure	Immunological, chronic tubular injury	Chronic tubulointerstitial nephritis
	Cell injury, ischaemia	Papillary necrosis

addition to other types of intrarenal injury, such as glomerular damage, pre- and postrenal injury can also result in acute failure. Chronic renal failure (CRF) can occur following prolonged exposure to renal toxicants or following severe injury. The compensatory processes that occur in these situations are often maladaptive and contribute significantly to chronic failure and end-stage renal disease. Substantial exposure to specific chemicals (for example, heavy metals, solvents) may also be a risk factor for renal cancer.

Detection and evaluation of kidney damage

There are a number of clinical signs in patients with renal injury (Schrier 1999), as follows:

- anuria (no urine output), oliguria (decreased urine output) or less commonly polyuria (increased urine output)
- oedema from severe hypoalbuminaemia due to glomerular dysfunction (nephrotic syndrome)
- hypertension
- anaemia
- abnormalities of the urine.

Kidney damage is identified by evaluating indicators of kidney function in blood and urine. Since the kidney is able to compensate extremely well for loss of renal functional mass, changes in renal function may not be identified until there is significant (more than 70%) nephron damage or loss. Decreased GRF can cause azotemia characterised by elevation of the blood urea nitrogen (BUN) and creatinine. Changes in electrolytes and acid–base balance may also be present in the blood of patients with renal disease. Urine volume can be determined and urine can be examined (urinalysis) for the presence of cells, proteins, casts, and crystals which are not normally present (Kosnett 1990). The terminology for reporting this includes the cell or protein of interest, for example, albumin, and its presence in urine, albuminuria.

Indicators of renal damage include:

- albuminuria or proteinuria (albumin, protein)
- aminoaciduria (amino acids)
- globulinuria (globulins), either myoglobin or haemoglobin
- glycosuria (glucose)
- haematuria (blood)
- inflammatory cells, either free in the urine or as cellular casts
- tubular cells, either free in the urine or as cellular casts.

Presence of these substances in the urine may indicate injury to the glomerulus resulting in increased permeability and/or lack of reabsorption in the tubules. Proteinuria is often an early and sensitive indicator of renal injury. High molecular weight proteins in the urine indicate glomerular dysfunction, while low molecular weight proteins indicate tubular dysfunction. If large-scale loss of function has occurred (above 67–75% of renal mass) serum concentrations of small molecules such as urea, creatinine or low molecular weight proteins such as β_2-microglobulin may be elevated. Additionally, enzymuria, the urinary excretion of enzymes, may also be of value although it is usually transient. Enzymes present on the brush border (for example, γ-glutamyl transferase, alkaline phosphatase) or more general indicators of cell damage (for example, lactate dehydrogenase) may be used.

Other methods that can be used in the detection and evaluation of renal injury are listed in Table 7.3.

Because of the marked improvement of workplace hygiene in recent years, most patients with chemically induced renal injury due to workplace exposure present with mild abnormalities, such as proteinuria. A history of occupational exposure, identification of exposure to known or suspected nephrotoxicants and the

Table 7.3 Methods for detecting and evaluating chemical-induced renal injury

Clinical examination
Urinalysis and microscopy
Estimation of glomerular filtration rate
- Plasma creatinine
- Blood urea nitrogen
- Clearance of creatinine, inulin or labelled compounds

Assessment of tubular function
- Urine acidification and concentration ability
- Phosphate clearance

Radiographic and radioisotope investigations
Examination of renal biopsy
- Light microscopy
- Electron microscopy
- Immunofluorescence

confirmation of exposure through detection of the toxicant, its adduct, or other specific effect in blood, urine or tissues, are most important for diagnosis. It is important to exclude other possible causes of kidney injury, including nonoccupational exposure to chemicals and drugs, infection or metabolic abnormalities.

Screening for occupational renal diseases remains controversial (Rocskay and Robins 1994). Noninvasive testing (Tolkoff-Rubin *et al.* 1988) and biomarkers (Roels *et al.* 1999; Wedeen *et al.* 1999) are important areas of investigation in occupational toxicology.

Classification of damage to the kidney

As noted above, the kidneys receive a high blood flow, and about 25% of the cardiac output, with approximately 85% of the total renal blood flow distributed to the cortical region and the remainder to the medulla. Thus, blood-borne chemicals in the body will reach the kidney and predominantly enter the cortex. Chemical-induced renal injury can be classified according to functional and structural characteristics of chemicals, mechanisms of injury (see below), or target sites within the kidney (Weening 1989).

By target site

The most useful classification for consideration of the functional consequences and therefore for clinical evaluation is classification according to target site.

The primary target sites for chemicals found in the workplace are the proximal tubule and, less commonly, the glomerulus. However, in clinical disease it is common to see involvement of both sites. The medulla and papilla are target sites of injury due to analgesic abuse and the use of nonsteroidal antiinflammatory drugs (NSAIDs) (Vanherweghem 1999).

The proximal tubule is the most common site of chemically induced injury, including injury from heavy metals, antibiotics, hydrocarbons, herbicides, and organic solvents. The type of injury is dependent on the conditions of exposure (dose and duration) and the properties of the toxicant. Acute injury generally results in tubular epithelial necrosis that may be reversible by epithelial regeneration, depending on the severity of the insult and persistence of the compound. Chronic injury presents as tubulointerstitial nephritis that is often irreversible and can progress to renal failure.

Injury to the glomerulus characterised by inflammation (glomerulonephritis) can occur due to immune-mediated mechanisms that result in the deposition of circulating or in-situ formed immune complexes in the glomerulus. This results in abnormal loss of protein in the urine, predominantly albumin. Gold, mercury, and D-penicillamine have been incriminated in this type of injury.

Renal cell carcinomas make up approximately 2% of all cancers in humans excluding those of the skin (Khan and Alden 2002). Major risk factors identified in an international study include smoking, obesity, hypertension, unopposed oestrogen use, asbestos, petroleum products and heavy metals (Moyad 2001; Savage 1999). While numerous occupational exposures (for example, asbestos, coke oven emissions, heavy metals, gasoline, and solvents) have been identified epidemiologically as risk factors in the development of renal cancer (IARC 2000; McLaughlin and Lipworth 2000), few chemicals have been demonstrated unequivocally to be human carcinogens. However, with the advent of molecular epidemiology and the recognition of genetic susceptibility to some toxicants, more refined evaluations can be made, as has been recently documented for occupational exposure to trichloroethylene (Goldstein 1994). In humans, trichloroethylene is metabolised to its active metabolite through the glutathione transferase pathway with damage to the proximal tubular epithelium. The trichloroethylene exposed population was shown to have different susceptibility to renal cancer based on a genetically determined enzyme supply (Goldstein 1994).

Mechanisms of injury

Renal injury may occur as a result of biochemical, haemodynamic or immunological mechanisms, as listed in Table 7.4 (Foulkes 1994).

Biochemical mechanisms are important in chemically induced tubular injury (Campistol and Sacks 2000). The toxicant generally enters the tubular epithelial cell by physiological transport processes. Toxicity usually develops in a dose-dependent fashion once a critical concentration of the toxicant has accumulated in the cell. There the toxicant may be metabolically activated and then bind to cellular macromolecules. Direct perturbation of subcellular organelle function can then occur. For example, aminoglycoside antibiotics result in lysosomal overload and dysfunction, while heavy metals, such as lead, cadmium and mercury cause a more general perturbation of cell function. Bioactivation of most organic chemicals, such as chloroform, is required for toxicity. Chloroform is biotransformed by the cytochrome P450 enzyme system to phosgene, a reactive compound that depletes glutathione and covalently binds to tissue macromolecules. Perturbation of renal haemodynamics can lead to ischaemia or hypoxia that may rarely initiate, but more frequently exacerbates, chemically induced renal injury. As stated above, immune-mediated injury is a primary mechanism for chemically induced glomerular injury.

The few isolated studies of combined metal exposures indicate that renal injury caused by such exposures may be altered due to unknown interactions of these metals within the kidney (Madden and Fowler 2000).

Table 7.4 Mechanisms of kidney injury

Biochemical
Direct disturbance of cellular or subcellular organelle function, e.g. heavy metals, aminoglycoside antibiotics
Reactive intermediate formation and/or oxidative stress, e.g. organic solvents, halogenated alkenes
Disturbance of endogenous or nutritive substrate, e.g. iron, zinc, ethylene glycol

Haemodynamic
Disturbance of renal haemodynamics, e.g. in hypertension

Immunological
Immune-mediated injury, e.g. mercury

Examples of damage to the kidney from workplace exposure

Many chemicals have been reported to cause renal toxicity following workplace (or environmental or therapeutic) exposure. Examples are given in Table 7.5.

Table 7.5 Examples of workplace exposures that have resulted in kidney damage

Glycols
Diethylene glycol
Ethylene glycol

Halogenated solvents
Bromobenzene
Bromodichloromethane
Carbon tetrachloride
Chloroform
Dibromethane
1,2-Dibromochloropropane
1,2-Dichloroethane
Hexachlorobutadiene
Pentachloroethane
Tetrachloroethylene
Tetrafluoroethylene
Trichloroethylene

Metals
Arsenic (as arsine)
Beryllium
Bismuth
Cadmium
Chromium
Gallium
Gold
Indium
Lead
Mercury
Nickel
Platinum
Uranium

Miscellaneous
Benzidine
Cisplatin
Enflurane
Industrial chemicals in the manufacture of plastics and resins
JP Jet fuels
Lithium
D-Lysine
Nonsteroidal antiinflammatory drugs
Oxalic acid
D-Penicillamine
D-Serine
Silica
Sulfonamides
Tartrates
Tris(2,3-dibromopropyl)-phosphate
Unleaded petrol

Mycotoxins/botanicals
Aflatoxin B
Citrinin
Monocrotaline
Ochratoxin A
Pyrrolizidine alkaloids

Organic solvents and hydrocarbons
Acrylonitrile
Hexachloro-1,3-butadiene
d-Limonene
Toluene
2,2,4-Trimethylpentane
Styrene

Pesticides
Diquat
Lindane
Paraquat
Succinimides
2,4,5-Trichlorophenoxyacetate

Classification of these chemicals according to structural or functional characteristics of the chemical can be useful since members of such classes often act through similar mechanisms. The most important nephrotoxic chemicals in the workplace are metals (Madden and Fowler 2000), hydrocarbons and organic solvents (Nelson *et al.* 1990; Pai *et al.* 1998; Ravnskov 2000; Roy *et al.* 1992), with silica of somewhat lesser importance (Goldsmith and Goldsmith 1993). Additional chemicals that are risk factors for occupational renal disease continue to be identified (Nuyts *et al.* 1995). Further, nephrotoxicants which are unlikely to be encountered in occupational environments may be associated with rare cases of occupational kidney disease, for example during manufacture or in specialised occupational situations. These include:

- non-narcotic analgesics/antiinflammatory agents (such as acetaminophen and nonsteroidal antiinflammatory drugs)
- antibiotics (such as aminoglycosides and sulfonamides)
- antifungal agents (such as amphotericin B)
- immunosuppressive agents (such as cyclosporin A)
- mycotoxins (such as ochratoxin and citrinin)
- radiocontrast agents (such as diatrizoate derivatives).

Rehabilitation of the individual with kidney damage

The problems of rehabilitation of the individual differ according to whether the individual suffers from acute or chronic disease. The possible responses to a nephrotoxic insult are shown in Figure 7.3 below. If a 'critical mass' of the kidney is not affected then repair, through cell proliferation, and compensatory mechanisms are initiated and renal function can return to normal. However, if injury is severe or exposure continues then a return to normal is not possible.

The first consideration in worker rehabilitation is to ensure that a worker is not returned to the same work that precipitated the injury. This requires an assessment of

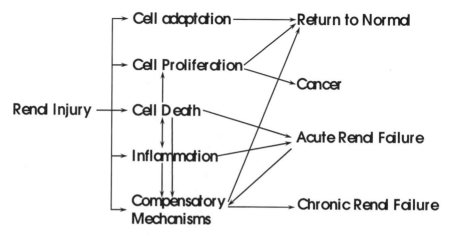

Figure 7.3 Possible responses to a nephrotoxic insult.

the workplace for nephrotoxicants, and elimination or control of any found. Further, it involves identification of other risk factors, not necessarily of an occupational nature, that may interact with occupational nephrotoxic factors (Porter 1989).

In the case of severe renal injury, which is usually chronic, uraemia can occur. This condition was thought to be so serious in the past that it was considered terminal. Kidney transplants and/or periodic dialysis offer some hope for rehabilitation, although not for occupations which require anything more than moderate effort.

In acute renal disease (whether of occupational or other causes), the prospects for rehabilitation depend on the extent of recovery. Resumption of normal work is more difficult for individuals who still have residual kidney damage. However, subject to continuing medical evaluation, a gradual return to work of a 'light duties' type may be permitted (Rom 1998). Return to any job where manual labour is required is unlikely to be permitted.

Summary

The functional integrity of the kidney is vital. The kidney plays a number of roles and has a number of functions, including maintenance of fluid volume, excretion of soluble wastes and metabolism. While the kidney has considerable functional reserve and regenerative capacities, the nature and severity of kidney damage has the potential for significant health problems.

There are a range of workplace risks that can affect kidney function. Most, but not all of these can be characterised as chemical; others, such as traumatic shock, heat stroke and crush syndrome also exist.

The action of nephrotoxic substances may be direct or indirect. In the former, a chemical interacts directly with renal cells, and has an effect on the structure or functioning of the kidney; the effect is normally dose-related. Indirect action means that the nephrotoxic effect is produced by reactive metabolites or by immune mechanisms; these tend to be delayed in onset and less likely to be dose-related. In addition, toxicants can have a secondary effect on the kidney through effects elsewhere (for example, systemic hypertension). Underlying disease, be it renal or systemic, can also affect the manifestation of nephrotoxicity.

Mild kidney damage can be identified easily, and full rehabilitation is possible. More severe cases may be fully or partially rehabilitated. However, very severe cases where kidney failure has occurred, often means the end of employment, requiring long-term dialysis treatment and/or kidney transplantation.

As ever, the risk management approach of identifying, assessing and controlling offers the best approach for preventing occupational kidney disease. Often, identification of even mild kidney damage indicates significant, and entirely preventive, exposure to workplace risks.

References

Bennett, W.M. and Porter, G.A. (1993) Overview of clinical nephrotoxicity. In: Hook, J.B. and Goldstein, R.S. (eds) *Toxicology of the Kidney*, 2nd edn. New York: Raven Press, pp. 61–97.

Brenner, B.M, (ed.) (2000) *Brenner and Rector's The Kidney*, volume 1, 6th edn. Philadelphia: Saunders.

Campistol, J.M. and Sacks, S.H. (2000) Mechanisms of nephrotoxicity. *Transplantation* 69 (Suppl. SS): 5–10.

De Broe, M.E., d'Haese, P.C., Nuyts, G.D. and Elseviers, M.M. (1996) Occupational renal diseases. *Curr. Opin. Nephrol. Hyper.* 5: 114–121.

De Broe, M.E., Porter, G.A., Bennett, W.M. and Verpooten, G.A. (1998) *Clinical Nephrotoxins: Renal Injury from Drugs and Chemicals*. New York: Kluwer Academic.

Fels, L.M. (1999) Risk assessment of nephrotoxicity of cadmium. *Renal Failure* 21: 275–281.

Fenwick, S. and Main, J. (1994) Increased prevalence of renal disease in silica-exposed workers. *Lancet* 356: 913–914.

Foulkes, E.C. (1993) Functional assessment of the kidney. In: Hook, J.B. and Goldstein, R.S. (eds) *Toxicology of the Kidney*, 2nd edn. New York: Raven Press, pp. 37–60.

Goldsmith, J.R. and Goldsmith, D.F. (1993) Fiberglass or silica exposure and increased nephritis or ESRD (end-stage renal disease). *Am. J. Ind. Med.* 23: 873–881.

Goldstein, R.S. (ed.) (1994) *Mechanisms of Injury in Renal Disease and Toxicity*. Boca Raton: CRC Press.

IARC (2000) *Some Chemicals that Cause Tumours of the Kidney or Urinary Tract and some other Substances*. International Agency for Research on Cancer, Lyon.

Khan, K.N.M. and Alden, C.L. (2002) Kidney. In: Haschek, W.M., Rousseaux, C.G. and Wallig, M.A. (eds) *Handbook of Toxicologic Pathology*, 2nd edn, volume II. San Diego: Academic Press Inc, pp. 255–336.

Kosnett, M.J. (1990) Medical surveillance for renal endpoints. *Occup. Med.* 5: 531–546.

Landrigan, P.J., Goyer, R.A., Clarkson, T.W. *et al.* (1984) The work-relatedness of renal disease. *Arch. Environ. Health* 39: 225–230.

Loghman-Adham, M. (1997) Renal effects of environmental and occupational lead exposure. *Environ. Health Persp.* 105: 928–939.

Madden, E.F. and Fowler, B.A. (2000) Mechanisms of nephrotoxicity from metal combinations: a review. *Drug Chem. Toxicol.* 23: 1–12.

McLaughlin, J.K. and Lipworth, L. (2000) Epidemiologic aspects of renal cell cancer. *Sem. Oncol.* 27: 115–123.

Moyad, M.A. (2001) Review of potential risk factors for kidney (renal cell) cancer. *Sem. Urol. Oncol.* 19: 280–293.

Nelson, N.A., Robins, T.G. and Port, F.K. (1990) Solvent nephrotoxicity in humans and experimental animals. *Am. J. Nephrol.* 10: 10–20.

Nuyts, G.D., Vanvlem, E., Thys, J. *et al.* (1995) New occupational risk factors for chronic renal failure. *Lancet* 346: 7–11.

O'Callaghan, C. and Brenner, B. (2000) *The Kidney at a Glance*. New York: Blackwell Science.

Pai, P., Stevenson, A., Mason, H. and Bell, G.M. (1998) Occupational hydrocarbon exposure and nephrotoxicity: a cohort study and literature review. *Postgrad. Med. J.* 74: 225–228.

Porter, G.A. (1989) Risk factors for toxic nephrologies. *Toxicol. Lett.* 46: 269–279.

Ravnskov, U. (2000) Hydrocarbon exposure may cause glomerulonephritis and worsen renal function: evidence based on Hill's criteria for causality. *Quart. J. Med.* 93: 551–556.

Rocskay, A.Z. and Robins, T.G. (1994) Assessment of a screening protocol for occupational renal disease. *J. Occup. Med.* 36: 1100–1109.

Roels, H.A., Hoet, P. and Lison, D. (1999) Usefulness of biomarkers of exposure to inorganic mercury, lead, or cadmium in controlling occupational and environmental risks of nephrotoxicity. *Renal Failure* 21: 251–262.

Rom, W.N. (ed.) (1998) *Environmental and Occupational Medicine*, 3rd edn. New York: Lippincott Williams and Wilkins.

Roy, A.T., Brautbar, N. and Lee, D.B. (1992) Hydrocarbons and renal failure. *Nephron* 58: 385–392.

Savage, P.D. (1996) Renal cell carcinoma. *Curr. Opin. Oncol.* 8: 247–251. Cited in Cotran, R.S., Kumar, V. and Collins, T. (eds) *Robbin's Pathologic Basis of Disease*, 6th edn. London: WB Saunders Company, 1999.

Schnellmann, R.G. (2001) Toxic responses of the kidney. In: Klaassen, C.D. (ed.) *Casarett and Doull's Toxicology. The Basic Science of Poisons*, 6th edn. New York: McGraw-Hill, pp. 491–514.

Schrier, R.W. (series ed.) (1999) *Atlas of Diseases of the Kidney*, five volumes. New York: Blackwell Science.

Stratta, P., Canavese, C., Messuerotti, A., Fenoglio, I. and Fubini, R. (2001) Silica and renal diseases: no longer a problem in the 21st century? *J. Nephrol.* 14: 228–247.

Their, R., Golka, K., Bruning, T., Ko, Y. and Bolt, H.M. (2002) Genetic susceptibility to environmental toxicants: the interface between human and experimental studies in the development of new toxicological concepts. *Toxicol. Lett.* 127: 321–327.

Tolkoff-Rubin, N.E., Rubin, R.H. and Bonventre, J.V. (1988) Noninvasive renal diagnostic studies. *Clin. Lab. Med.* 8: 507–526.

Vanherweghem, J.L. (1999) Toxic nephropathies. In: Glassock, R.J., Cohen, A.H. and Grünfeld, J.P. (eds); Schrier, R.W. (series ed.) *The Schrier Atlas of Diseases of the Kidney*, volume 2. New York: Blackwell Science.

Wedeen, R.P. (1992) Renal diseases of occupational origin. *Occup. Med.* 7: 449–463.

Wedeen, R.P., Udasin, I., Fiedler, N. *et al.* (1999) Urinary biomarkers as indicators of renal disease. *Renal Failure* 21: 241–249.

Weening, J.J. (1989) Mechanisms leading to toxin-induced impairment of renal function, with a focus on immunopathology. *Toxicol. Lett.* 46: 205–211.

Chapter 8

Occupational toxicology of the nervous system

C. Winder

Introduction

In occupational and environmental health, diseases of the nervous system are seldom attributed with their true importance (WHO 1986). Classical neurological conditions, such as milliner's madness, lead wrist drop and foot drop, and carbon disulfide psychosis are all but extinct, although some, such as *n*-hexane neuropathy or chronic manganism, still exist (Rosenstock and Cullen 1986). Fortunately, improvements in hazard control have produced a decrease in the incidence of nervous lesions of toxic or traumatic origin, although borderline or 'subclinical' effects may occur and often go undetected.

The involvement of the nervous system in systemic intoxication offers diagnostic and investigative challenges to the occupational toxicologist. Often early evidence of neurotoxicity is vague; it can be of a highly subjective nature and may be complicated by the presence of neurotoxic signs and symptoms due to other causes (O'Donahue *et al.* 1977). However, over the past 20 years or so, a number of epidemiological, clinical and laboratory studies have shown that the nervous system is a target for many toxic substances (Feldman 1999; MacPhail 1992).

Organisation of the neurological system

The nervous system consists of the central and peripheral nervous systems.

The central nervous system (CNS) comprises the brain and spinal cord, and the brain comprises:

- the cerebral cortex which regulates the higher brain functions including body movement and long-term memory
- the hippocampus and limbic systems which control strong emotions and short-term memory
- the cerebellum which coordinates movement

- lower brain structures such as the thalamus (which coordinates sensory information), the hypothalamus (which regulates heart rate, body temperature, sleep functions and sexual development) and the pons (which integrates information distribution onwards into the central nervous system)
- the spinal cord which carries messages between the brain and the rest of the body.

The peripheral nervous system (PNS) comprises nerves and ganglia. The nerves are further categorised according to whether they stimulate skeletal muscle (somatic) or smooth/cardiac muscle (autonomic system).

Operationally, the brain can be considered as a number of integrated structures, all doing different things at different levels. The spinal cord carries information to and from the body to the brain stem, which passes the information to the hypothalamus. This structure is involved with integrating signals and does many of the automatic functions that keep the body going.

Signals pass from the hypothalamus to the limbic system, which in turn has connections through to the frontal cortex (the part of the upper parts of the brain) and there to the rest of the brain. The cortex of the brain in humans has a wide number of functions, but they include memory, speech, coordination of behaviour and higher forms of behaviour. There are also connections back from cortex, through limbic system to hypothalamus to brain stem to spinal cord.

In species where the cortex is not as well developed, the limbic system serves to process information from the olfactory system (the sense of smell). It was initially considered that the limbic system processed olfactory information in humans, but the sense of smell is less important in humans and the limbic system almost certainly is involved in processing other sources of information as well as olfactory inputs. In humans, the limbic system is entirely covered by the more recently evolved cortex.

Whereas in humans the hypothalamus is involved with automatic body functions and the cortex is involved with higher mental performance, one function of the limbic system is to control expression of the emotions, especially some of the stronger emotions (such as avoidance behaviour, parenting behaviour, rage and sexual behaviour). From a biological viewpoint, the emotions act by directing behaviour into specific patterns. For example, rage behaviour produces a different set of behavioural responses than, say, fear behaviour or sexual behaviour.

Further, while many of the patterns of such behaviours are 'hard wired' through evolutionary processes and are now part of the repertoire of human behaviour, others can be altered through learning. This happens with extraordinary sophistication in the cortex, and to a lesser extent in other parts of the nervous system. For example, the brain can be functioning quite normally, until it detects a strong threat. The cortex analyses the situation, recognises the threat, and then tells the limbic system to set the rest of the brain in readiness for a suitable response. The limbic system can also do this without advice from the cortex, presumably through its own information inputs. This may be the basis of phobias and phobic behaviours.

Types of cells found in the nervous system

The structure of nervous systems can be quite complex, but at the cellular level (see Figure 8.1) it normally consists of neurones, support cells and other structures.

The neurone

The neurone or nerve cell consists of:

■   a cell body containing the nucleus and most of the organelles responsible for maintaining the cell

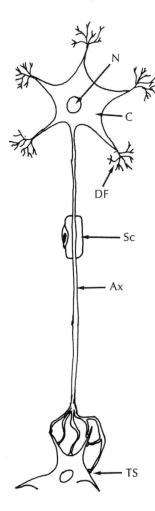

A single neurone or nerve cell.

Each cell has a **cell body**, containing:

○   a **nucleus (N)**,

○   **cytoplasm (C)**, and

○   several branch-like **dendrites** that cover a dendritic field **(DF)**.

A single process, called the **axon (Ax)** extends from the cell body and is wrapped by support cells **(Sc)** over the length of the axon (only one shown).

Terminal synapses at the end of the axon link it to other nerve cells.

Figure 8.1 The structure of a nerve cell.

- several short processes called dendrites which increase the surface area available for connecting with axons of other neurones (dendrites typically conduct electrical currents towards the cell body)
- a long process or axon stretching out of the cell body, sometimes over a very long distance which is responsible for transmitting signals from the neurone to other cells (axons conduct impulses away from the cell body)
- specialised cell junctions called synapses between its axon and other cells which allow for direct communication from one cell to another. Figure 8.2 shows a drawing of a neurone cell body containing a substantial number of synapses.

The axon is often a very long cell process, in some cells being 5000–10,000 times as long as the cell body is wide. This presents some difficulties in resource management for the cell, since the genetic material and the bulk of the synthetic material is located in or close to the nucleus. Nerve cells have evolved a specialised axonal transport process to facilitate the physical movement of cell products to the terminal (anterograde transport). Similarly the terminal may absorb substances which are important as metabolic precursors or as regulators of cell processes which are transported back to the cell body (retrograde transport).

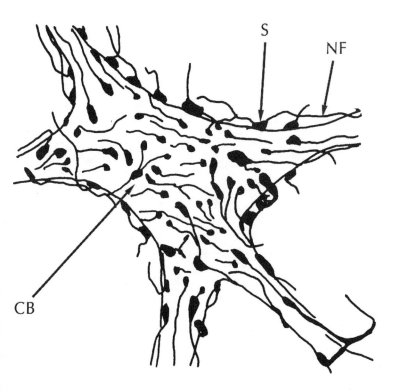

Figure 8.2 Synaptic connections to the cell body of a nerve cell. A drawing of a motor cell body (CB) in the spinal cord, showing connection with a large number of synaptic connections (S) from many nerve fibres (NF).

Neurones display an extremely high level of metabolic activity. They have a massive surface area compared to other cells in the body. This requires a huge energy input. In addition, they need to produce electrochemical gradients that are used in generating resting state and nerve impulses – these also require a high level of metabolic activity. This high level of activity is reflected in the cells' cytological appearance. The nucleus is typically large, rounded and with a prominent nucleolus which represents a high degree of cellular activity. There is also abundant rough endoplasmic reticulum that secretes the proteins required and is visible in the cytoplasm (this is often called the Nissl substance). There is a well-developed Golgi apparatus to provide secretory products, especially neurotransmitters. In addition there are large numbers of mitochondria needed to produce the large amount of energy required by the neurone.

Other cells

These comprise the neuroglial support cells and other structures. Supporting neuroglial cells in the central nervous system are called oligodendrocytes and in the peripheral nervous system they are called Schwann cells. Both types of cell can associate with axons in two different ways to produce either a myelinated or a nonmyelinated neurone. These cells are mainly involved in the process of myelination, by which neuronal cells become insulated from the surrounding environment.

Other cell types include endothelial cells, blood cells, phagocytosing cells and connective tissue cells.

Support cells and myelination

The axon is a protoplasmic process that extends from the cell body. If the cell body is in the spinal cord, and the axon extends out of the spine to innervate a muscle, the axon can be quite long. It will not be able to carry an electrical impulse any distance in the ionic environment of the body unless it is insulated in some way.

This insulation is supplied by the supporting neuroglial cells, which produce a lipid (fatty) insulant called myelin. Myelin is a substance that is pale in colour (it is this colour which leads to the terms white matter and grey matter within the nervous system – the white matter mainly contains myelinated axons and the grey matter mainly nonmyelinated neurone cell bodies). The main functions of myelin are to increase the electrical capacitance of neurones and to insulate against any leakage of the bioelectrical nerve impulse. The higher the capacitance and the better the insulation, the faster the nerve impulse will travel along the neurone.

The process of myelination begins when the neuroglial cell wraps its cell membrane around the axon. The main difference between the Schwann cell and the oligodendrocyte is that the former will myelinate only one axon whereas the latter may myelinate a number of axons.

The process of myelination involves the surrounding of a section of the axon with a neuroglial support cell, which then slowly wraps, rotates and winds itself around the fibre, eventually encircling the cell many times. The long finger (or in three dimensions, a sheet) of support cell that winds itself around the nerve fibre is called the neurilemma (see Figure 8.3).

An axon will have many neuroglial cells along its length, forming a segmented sheath around the axon known as the myelin sheath. Along the length of the axon, each neurological support cell is separated from the next by a short length (approximately 1 mm) of nerve fibre. These tiny gaps are known as the nodes of Ranvier.

Nervous system function

The function of the nervous system as a whole depends very much upon the complexity of the network of connections between the various neurones rather than the specific features of any single neurone.

In general, the purpose of a neurone is to generate and/or propagate electrical excitation in the form of an action potential. Neurones are therefore cells in the nervous system which are responsible for sending messages. They have three major purposes:

- to gather and send information from the senses such as touch, smell, sight and balance
- to send appropriate signals to effector cells such as muscles and glands
- to process all information gathered and provide a memory and cognitive ability allowing organisms with such integrating processes to take voluntary actions on information received.

The transmission of nerve impulses around the body depends on two properties which are unique to neurones. These are excitability and conductivity.

Excitability refers to the fact that nerve cells are able to respond to a stimulus, which may be internal or external. Muscle fibres are also able to demonstrate excitability, giving them the ability to contract if stimulated. **Conductivity** refers

Cross section of an axon onto which a support cell is enveloping

The envelopment continues until the neurilemma is beginning to encircle the axon

A fully myelinated axon, where encirclement has occurred many times

Figure 8.3 The myelination of a nerve axon by a support cell.

to the property that neurones alone have of transferring excitability along their length and on to other neurones or even muscle tissue.

It is these two properties together which allow neurones to deliver messages to appropriate parts of the body as and when required.

Neurotransmission in nerves

The main function of the nerve is the transmission of nerve impulses.

- Nervous impulses are electrochemical events initiated by various stimuli.
- Alterations in membrane permeability cause sodium ions to enter the cell.
- The resting charge on the cell changes (depolarisation).
- A wave of depolarisation spreads over the surface of the cell, and down the axon.
- Resting state cell charge is restored by the diffusion of potassium ions out of the cell (repolarisation).
- The wave of neurotransmission will produce a range of effects depending on the effector system at the end of the axon.

The way in which neurones transmit signals along their length is controlled by an electrical (ionic) gradient across their cell membranes. Along with muscle fibres, neurones are said to be excitable cells.

Resting state

At rest, the neurone cell membrane has a negative membrane potential of -70 millivolts (mV). Neuronal electrical activity is associated with membrane depolarisation, which creates a positive potential of around $+40$ mV. This is possible because, unlike other cells in the body, nerve cells and muscle cells are able to alter reversibly their transmembrane potential. The significant feature which makes this possible is the difference in the ion concentrations inside the cell and outside. This difference is partly as a result of the imbalance between sodium and potassium ions on either side of the neurone cell membrane which is selectively permeable.

In a resting neurone, the concentration of potassium ions inside the cell is around 30 times greater than it is outside, whereas the concentration of sodium ions is around 14 times lower inside the cell than outside. The maintenance of the neurone at rest seems to be due to the action of the sodium–potassium ion pump. The precise action of this mechanism is not as yet fully understood but appears to utilise adenosine triphosphate (ATP) as an energy source.

Even when neuronal cells are not conducting nerve impulses they are continually sending ions across the membrane through the action of this pump. Sodium ions are actively transported out and potassium ions are actively transported in.

The action potential

Information flow within neurones is based on electrical signals from the connecting cell being converted into a chemical signal by the release of a neurotransmitter, which makes contact on to the dendrite that the synapse is connected to, where the signal is passed to the receiving cell. The dendritic fields of neurones can have many scores or hundreds of synaptic inputs from many different cells, and thus each different synapse can release a different neurotransmitter. There are many different neurotransmitters that may excite or inhibit or otherwise affect the synaptic receptors they connect to. Some, like acetylcholine, the catecholamines, or γ-amino butyric acid (GABA) are critical to normal nervous system function.

Therefore, the dendritic fields receive a complex input of information, and collect, process and convey information from synapses to the cell body. Neuronal cells can have many synapses, some releasing different neurotransmitters that may have different functions, such as inhibition, stimulation and neurotransmission.

At defined neurocellular criteria, depending on this input and unique for each neurone, the cell will generate a self-sustaining electrical impulse called an action potential, which will be sent through a single neuronal process (the axon).

Axons can be long or short depending on cell type and function, and innervate either other nerves or organs and tissues (such as muscle cells) in the body. At the end of the axon, the single process branches into a number of processes that ultimately end in synaptic processes to other neurones or specialised structures, such as the motor end plate in muscle cells, where a neurotransmitter specific to the nerve cell is released. This occurs if an action potential is received at the end of the axon. The released neurotransmitter then contacts the receptors on the receiving cell, and the neuronal signal or message is transferred and actioned. The process of synaptic transmission is shown in the Figure 8.4.

Neurotransmission at synapses

Nerve-to-nerve connections are not hardwired. These are called synapses. The onward flow of neurotransmission relies on a process of chemotransmission, whereby:

- the electrical signal arrives at the end of the nerve (the presynaptic process); it is converted to the release of a neurotransmitter chemical, which is discharged into the gap between the two nerve processes (the synaptic cleft)
- the neurotransmitter crosses from the pre-synaptic process to the other side (the postsynaptic process) and binds to specific receptors on that cell membrane to cause a change in the cell's electrical field
- if enough neurotransmitter causes enough change in the electrical field, the wave of neurotransmission is re-initiated.

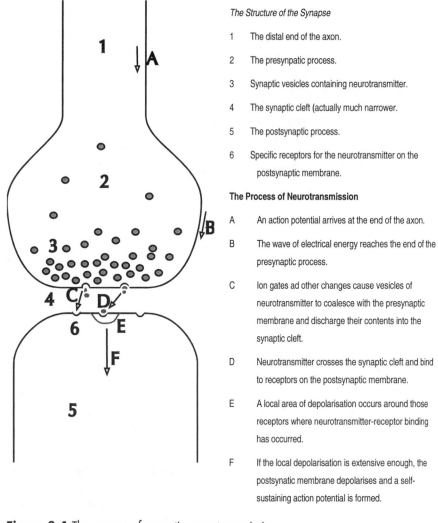

The Structure of the Synapse

1   The distal end of the axon.

2   The presynpatic process.

3   Synaptic vesicles containing neurotransmitter.

4   The synaptic cleft (actually much narrower.

5   The postsynaptic process.

6   Specific receptors for the neurotransmitter on the postsynaptic membrane.

The Process of Neurotransmission

A   An action potential arrives at the end of the axon.

B   The wave of electrical energy reaches the end of the presynaptic process.

C   Ion gates ad other changes cause vesicles of neurotransmitter to coalesce with the presynaptic membrane and discharge their contents into the synaptic cleft.

D   Neurotransmitter crosses the synaptic cleft and bind to receptors on the postsynaptic membrane.

E   A local area of depolarisation occurs around those receptors where neurotransmitter-receptor binding has occurred.

F   If the local depolarisation is extensive enough, the postsynatic membrane depolarises and a self-sustaining action potential is formed.

Figure 8.4 The process of synaptic neurotransmission.

Nervous system function products

There are two main products of nervous system function:

- coordination and control of body processes
- behaviour.

The first of these functions is shown in Figure 8.5. In each case, the arrows represent bundles of nerves carrying information to and from the brain.

This simple model of sensory input, central integration and effector output, is of course much more sophisticated in reality. Axonal processes are found throughout the nervous system and the body. Where they collect together they are called

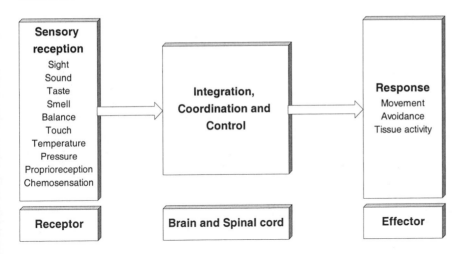

Figure 8.5 Function of the nervous system.

nerves (the term nerve can be used for a single nerve process or a 'bundle' of nerves). Conventionally, the body's nerves are anatomically divided into the central nervous system (the brain and spinal cord) and the peripheral nervous system. Nerves are also divided into those that send messages to the central nervous system (sensory nerves) and those that transmit messages from the nervous system (motor nerves). Sensory nerves are also called afferent nerves; and motor nerves are also called efferent.

Other anatomic and physiological classifications are also made, such as the autonomic nervous system, sympathetic and parasympathetic systems. Nerves generally make connections between a part of the central nervous system and some other body region. Depending on their function, nerves are known as sensory, motor, or mixed.

On the output side, there are those that are automatic (such as control of respiration and heartbeat) and voluntary (such as movement of muscles). The second of these is organised into the somatic or voluntary nervous system, and the first into the autonomic system, itself divided into the sympathetic and parasympathetic systems (see Figure 8.6).

The somatic, motor or voluntary nervous system includes all nerves that control voluntary muscle movement. The areas of brain that control motor function are

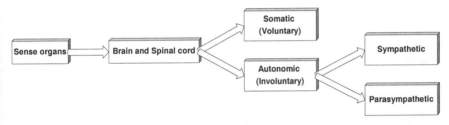

Figure 8.6 Output of central nervous system function.

proportional to the size of the muscle, with some areas such as thumb, lips or tongue, having disproportionally larger areas. The 'motor homunculus' demonstrates this phenomenon.

The parasympathetic nerves include four cranial nerves that connect directly to the brain (the olfactory, optic, oculomotor and trochlear nerves, eight cranial nerves from various parts of the brainstem and spine, and other parasympathetic nerves from the lower lumbar and sacral parts of the spine). While some cranial nerves connect to one organ only, others can affect a number of other organs. For example, the tenth (X) cranial nerve is called the vagus nerve, which innervates the heart, lung, stomach, spleen, adrenals and small intestines.

All sympathetic system nerves arise from the spinal cord, and all arise from 15 segments that make up the thoracic and lumbar parts of the spine.

The second main product of nervous system function is behaviour.

Injury to nerves and neuronal function

Basic mechanisms

Apart from damage by trauma, there are three basic pathophysiological processes of damage to nerves and nerve fibres

- Damage to basic biochemical or neurochemical processes in nerve or support cells, such as damage to energy-producing systems. Agents such as carbon disulfide and acrylamide act by interfering with enzymes of glycolysis, and methyl *n*-butyl ketone reduces ATP levels in nerve tissue. The result of a lack of cellular energy is disruption to high-energy demand systems like axonal transport and nerve fibre polarisation, which are essential for neurotransmission.
- Damage to the cell-to-cell connections, which are crucial for nervous system integrity. Particularly the connection of the nerve fibre to the motor neurone (connection is essential for its survival), and the connection of the fibre to the myelin-producing neuroglial cell. Disruption to this connection will lead to breakdown of the myelin sheath and loss of nerve insulation (Wallerian degeneration). If the sheath degenerates, the nerve fibre degenerates as well, which may explain why damage occurs in the more distal parts of the limbs and the longest fibres ('dying back' neuropathies).
- Specific neurotoxic mechanisms. The well-established polyneuropathy of organophosphorus compounds is due to phosphorylation and eventual modification of an enzyme in nerve cells, the 'neurotoxic esterase'. Also, paediatric lead encephalopathy is due to damage to the immature endothelial cells of the cerebellum (and eventually cerebrum).

Measuring neurotoxic effects

Neurotoxicity can be defined as the capacity of chemical, biological, or physical agents to cause an adverse functional or structural change in the peripheral or central nervous system (Walker 2000). Chemical-induced changes in the structure or persistent changes in behaviour, neurochemistry, or neurophysiology of the nervous system are regarded as neurotoxic effects. Reversible effects occurring at doses that could endanger performance in the workplace, are associated with a known neurotoxicological mechanism of action, co-vary with a known neurotoxicological effect, or are latent effects uncovered by pharmacological or environmental challenge, could be considered to be neurotoxic effects (Tilson 1996).

There are a variety of conventional neurological, neurophysiological, neuropsychological, and biochemical diagnostic tests that are sensitive to neurotoxicological exposures, but that must be applied and interpreted appropriately (Feldman et al. 1999) as listed below.

- Methodological approaches have been used to develop methods to identify neurotoxicants (Simonsen et al. 1994).
- Biochemical markers have been used, though not with much success (Maroni and Barbieri 1988).
- Electrophysiological techniques can help in evaluating the hazards of occupational exposure to toxic substances and in detecting or characterising neurotoxicological disorders (Aminoff 1985).
- Evoked potentials (EPs) have also been used as an index of toxic insult to the nervous system. They are obtained by averaging successive samples of electroencephalogram time-locked to the presentation of stimuli. Components of the resulting waveform can be measured for amplitude, latency, and distribution. Some evidence suggests that these techniques can measure subtle toxicant-induced change (Arezzo et al. 1985).
- Various neurophysiological methods, including electroencephalography, electromyography, nerve conduction velocities, and evoked potential techniques, have been used to detect early signs of neurotoxicity in humans (Seppäläinen 1988).
- Neurobehavioural testing may be used as a means of evaluating the health effects of exposure to toxic chemicals (Baker and Letz 1986).
- Standardised neurobehavioural test batteries for studies of adverse health outcomes in communities are being developed (Johnson et al. 1994).
- Lastly, the interactions between the nervous and immune systems have been recognised in the development of neurodegenerative disease. This can be exploited through detection of the immune response to autoantigens in assessing the neurotoxicity of neurotoxic chemicals (El-Fawal et al. 1999).
- Behavioural endpoints are being used with greater frequency in the hazard identification phase of neurotoxicology risk assessment for workplace and environmental toxicants.
- Animal studies also have some utility. One reason behavioural procedures are used in animal neurotoxicology studies is that they evaluate neurobiological functions known to be affected in humans exposed to neurotoxicological agents, including alterations in sensory, motor, autonomic, and cognitive function. In hazard identification, behavioural tests are used in a tiered-testing context: tests in the first tier are designed to determine the presence of neurotoxicity (for example, functional observational batteries and motor activity); second tier tests are used to characterise neurotoxicant-induced effects on sensory, motor, and cognitive function (Kyrklund 1992; Tilson 1993).

The central nervous system

The central nervous system is normally considered to consist of the brain and the nerves of the spinal cord and autonomic nervous system. Further subdivisions of the brain include:

- the cerebral cortex (which regulates many higher brain functions, including body movement and long-term memory)
- the hippocampus and limbic systems (which control 'strong emotions' and short-term memory)
- the cerebellum (which coordinates movement).

The main product of the central nervous system is behaviour, which has been shown to be sensitive to the effects of a number of chemicals. The predominant central nervous system disorders are shown in Table 8.1.

Neurological disorders

- Spinal cord syndromes: The spinal cord is considered fairly resistant to occupational hazards, except for polyneuritis caused by poisoning by thallium and triorthocresyl phosphate), and decompression effects.

Table 8.1 Occupational diseases of the central nervous system

Neurological effects	Agents
Acute toxic encephalopathy	Organochlorine pesticides, arsenic (and arsine)
Damage to basal ganglia	Carbon monoxide, manganese, carbon disulfide
Damage to cerebellum	Organic mercury
Damage to occipital cortex	Organic mercury
Intracranial pressure	Lead, organotin compounds
Chronic toxic encephalopathy	Carbon disulfide, organic solvents, lead, inorganic mercury, arsenic, manganese
Encephalic syndromes	
Parkinsonian movement disorders	Carbon monoxide, manganese, carbon disulfide, methyl phenyl tetrahydropyridine
Dyskinesias and tremors	Methyl bromide
Toxic tremors	Mercury, organophosphorus compounds, tetrachloroethane, hexacyclohexane
Isolated convulsive crises	Organochlorine pesticides, carbon tetrachloride, nitrophenols and nitrocresols, lead
Bulbar (brain stem) syndromes	Hydrocyanic acid
Spinal cord syndromes	
Polyneuritis	Thallium, triorthocresyl phosphate
Cancer	Acrylonitrile, vinyl chloride (see Tables 8.2 and 8.3)

Table 8.2 Occupations associated with an excess of brain cancer

Aluminium workers	Oil refinery workers
Chemists	Petrochemical workers
Dentists	Pharmaceutical workers
Lead smelter workers	Rubber workers
Machinists	Veterinarians
Medical personnel	Vinyl chloride workers

Table 8.3 Agents capable of producing brain cancers in experimental animals

2-Acetylaminofluorene	Elaiomycin
Acrylonitrile	Ethyl methanesulfonate
1-Aryl-3,3-dialkyltriazenes	Ethylynitrosobiuret
Azoxyethane	Ethylnitrosourea
Azoxymethane	1-Methyl-2-benzyl-hydrazine
Butylnitrosourea	Methylnitrosourea
Cycasin	Methylnitrosourethane
1,2-Diethylhydrazine	Procarbazine
Diethylnitrosoamine	Propane sulfone
Diethyl sulfate	Propylene imine
Dimethylbenzanthracene	Pyrrolizidine alkaloids
Dimethyl sulfate	Vinyl chloride

- Bulbar syndromes are encountered with exposure to hydrocyanic acid which suspends bulbar function.
- Encephalic syndromes: These include **Parkinsonism** which results from acute carbon monoxide poisoning and is caused by haemorrhage and oedema induced by central anoxia. Chronic manganese poisoning is initially manifested by disorders of mental function (subjective syndrome and character disorders). Carbon disulfide can also give rise to paleostriatal (Parkinsonism) and neostriatal (chorea and athetosis) disorders. In rare instances, Parkinsonian symptoms may be observed in methyl chloride, lead or mercury poisoning. **Dyskineas and tremors** are common symptoms of methyl bromide poisoning. Of the **toxic tremors**, mercurial tremor is the best known of these conditions, although tremors may be observed following exposure to organophosphorus compounds, tetrachloroethane and hexacyclohexane. **Isolated convulsive crises** can result from the central action of organochlorine pesticides, carbon tetrachloride, nitrophenols, nitrocresols and lead.
- Cancer: This can arise as a result of occupational exposures. However, epidemiological associations of primary brain tumours to specific agents are undetermined, with the possible exception of ionising radiation. However, an increase of brain cancer in a number of occupations has been reported (see Table 8.2). The results of animal studies also suggest a number of agents cause cancer of the brain (see Table 8.3).

Autonomic (vegetative) nervous system disorders

The autonomic nervous system is divided into the sympathetic and parasympathetic systems. These can be subject to damage or dysfunction by a number of occupational hazards.

The specific inhibitory action of organophosphorus compounds on cholinesterases, leading to postsynaptic excitation of cholinergic systems, explains their very homogeneous parasympathetic effects, including muscular fibrillation, severe muscular clonus, tonic contractions of limb and trunk muscles, and motor incoordination.

Acroparaesthesia, Raynaud's phenomenon, and intense sweating of the hands are caused by vibration (operation of vibrating tools).

Tachypnoea, tachycardia, hypertension, vasoconstriction of the extremities, and diminution of intestinal peristalsis are autonomic reactions caused by noise. Ultrasound may provoke hyper-reactions of the sympathetic nervous system and attacks of sweating.

Colic cause by a number of metals (notably lead) is thought to be a spasm of smooth muscles of the intestine, sometimes of the bladder, but also as a generalised arterial spasm, notably in the mesentery.

Some chlorinated solvents, such as trichloroethylene, dichloroethane, calcium cyanamide dust and dithiocarbamates produce effects on alcohol oxidation and result in a reddening of the face (degreaser's flush) and possibly arterial hypotension with a risk of collapse.

The peripheral nervous system

The peripheral nerves are composed mainly of the afferent (sensory) and efferent (motor) nerves. These perform the functions of carrying **sensory information** to the spinal cord, through the brain stem sensory tracts to the somoesthesic area in the cerebral cortex, including:

- mechanoreception (touch, pressure, kinaesthesia, proprioreception, posture)
- thermoreception (heat and cold)
- pain reception.

The peripheral nerves also carry **motor impulses** from the motor cortex of the brain through the spinal cord to their effectors, which are the spinal muscles.

Effects on the peripheral nervous system are outlined in Table 8.4.

Polyneuropathy

Uncontrolled long- or medium-term exposure to neurotoxic substances may produce nerve fibre impairment with the clinical development of a polyneuropathy (a lesion of or damage to several nerves). Agents such as n-hexane, methyl n-butyl ketone, acrylamide and carbon disulfide have been reported to cause polyneuropathy. However, this is a condition now quite uncommon in modern industrial exposures.

Table 8.4 Occupational diseases of the peripheral nervous system

Neurological effect	Agents
Toxic neuropathy (and polyneuropathy)	See Table 8.3
Traumatic injury	
Interstitial (vascular)	Vibration
Parenchymal	Compressive or entrapment neuropathy
Neuromuscular junction blockade	Organophosphate compounds

The features of polyneuropathy are segmental demyelination and axonal degeneration. Initially, the myelin sheath around the nerve (the axon) unravels and becomes 'leaky'. Degeneration of axon is a more severe neuropathy occurring at the distal segments of the nerve fibre, evolving to the more proximal segments (also called distal axonopathy).

Clinically, many nerves are involved simultaneously. Longstanding paraesthesias (tingling, pins and needles, coldness and cramping pain in the calves) may precede for years the appearance of clear-cut clinical findings. This leads to flaccid muscular weakness in the distal muscles and of stocking-glove sensory loss; the sense of position is occasionally altered with consequent ataxia (failure of muscular coordination; irregularity of muscular action). Motor impairment is generally prominent in the distal muscles of the limbs and in severe cases foot drop and wrist drop (lead poisoning). Proximal involvement is rare, but has been described in some neuropathies caused by glue-sniffing. The cranial nerves are occasionally affected, mainly the optic nerve (carbon disulfide, methanol and other solvents) and the cochleovestibular nerve (trichloroethylene).

Sometimes purely motor neuropathies (lead, tri-orthocresol phosphate) or sometimes purely sensory neuropathies occur. However, generally both sensory and motor nerves are affected in most cases.

Occupational neurotoxicants

Table 8.5 below is a list of substances for which peripheral nerve fibre damage in exposed workers has been recognised or suspected on clinical grounds. However, depending on dose and mode of exposure a far greater number of chemicals can probably induce toxic lesions of both the central and peripheral nervous systems. In general, compounds that can damage the peripheral nervous system can also damage the central nervous system.

Acknowledged neurotoxic therapeutic agents

The list of neurotoxic therapeutic substances is continuously increasing, including: antibiotics (sulfonamides, streptomycin, nitrofurantoin); antineoplastic (vincristine); psychotropics (metaqualone, imipramine); anticonvulsants (phenytoin); antirheumatics (gold salts, chloroquine, indometacin) and cardiovascular agents (hydralazine, perhexilline, clofibrate).

Table 8.5 Agents causing occupational neuropathy

Category	Recognised neurotoxicants	Suspected neurotoxicants
Metals	Arsenic, lead, mercury (organic)	
Organic solvents	Carbon disulfide *n*-hexane, methanol, methyl *n*-butyl ketone, trichloroethylene	Toluene
Pesticides	Lead arsenate, organophosphorus compounds, organotin compounds, thallium, chlorinated phenol derivatives (2,4-D), chlorodecane	Cyclopentadiene compounds (such as aldrin, dieldrin), DDT, hexachlorobenzene
Gases	Methyl bromide, carbon dioxide	
Other organic compounds	Acrylamide, *ortho*-arylphosphates (such as TOCP), dimethylaminopropionitrile, styrene, TCDD	Polybrominated biphenyls

Neurophysiological effects

There has been growing concern about less severe cases known as subclinical neuropathies. These essentially consist of sensory symptoms and electromyographical anomalies which, if exposure continues, can eventually result in development of clinical signs of polyneuropathy. The main investigations of these effects are electromyography (EMG) and nerve conduction tests, also termed electroneurography (ENG).

Electromyography

This technique measures electrical potentials in muscle in resting or contraction states, and can furnish evidence of denervation of muscle fibres resulting from axonal or Wallerian degeneration. This can result from long-term use of vibrating tools acting either by percussion (pneumatic hammers and chisels), rotation (electric cutters and grinders) or mixed movements (drills). Clear EMG signs can occur after 2 years in workers using chisels all day, whereas clinical neuropathies were not present before (but after) 3 years of exposure. The mechanism of peripheral neuropathy from vibrating tools is not certain, though direct action of vibrations, ischaemic nerve disease or compression of nerve trunks from damaged connective tissue have all been suspected.

Electroneurography

This measures conduction rates of the impulses of both motor and sensory nerves. The use of this technique has some problems. A drop of skin temperature of 1°C can cause a reduction of nerve conduction velocity of 2–2.4 m/s. Age, alcohol consumption, therapeutic drug use and biochemical diabetes can all decrease nerve conduction rate. However, slowing of motor and sensory nerve conduction velocities is considered to be a sign of incipient neuropathy, or of incomplete recovery if it persists after clinical signs have regressed. The technique has been widely used in the screening of neuropathies in exposed groups, and it is believed that if interfering factors are taken into account, it can be useful in the prevention of industrial neuropathies. The technique has sufficient reliability that the results of electroneurographical studies can also contribute to the exposure standard-setting process.

Mental disorders

A range of mental disorders is shown in Table 8.6.

Toxic psychoencephalopathies may be caused by exposure to a wide range of materials, including most solvent vapours, alkyl lead compounds, poisons that cause methaemoglobinuria, carbon monoxide and cyanides. This intoxication can be resolved by removal from exposure, though a state of confusion may be a temporary effect.

Confusional states result from a deeper and more prolonged acute action. Pantophobia (methyl bromide), hallucinations (carbon disulfide), schizoid reactions, Korsakoff's syndrome and delirium tremens. The role of alcohol in the production of these signs and symptoms varies, and may even interact with other exposures.

Subjective and behavioural toxic syndromes include a complex of nonspecific disorders such as headache, mental fatigue, neuropsychological or psychotoxic dysfunction, personality changes, and so on.

Psychotic effects from carbon disulfide is now well known (and well controlled), and may be induced by organophosphorus compounds, though the mechanism of action remains inhibition of nervous system cholinesterases.

Table 8.6 Occupational mental disorders

Neurological effect	Agent(s)
Toxic psychoencephalopathies	Organic solvent vapours, alkyl lead compounds, poisons that cause methaemoglobinuria, carbon monoxide, cyanides
Confusional states	Organic solvents, asphyxiant gases, insecticides, manganese, organotins
Pantophobia	Methyl bromide
Hallucinations	Carbon disulfide
Subjective and behavioural toxic syndromes	A wide range of organic chemicals
Psychotic effects	Organophosphorus compounds

Hazards and the nervous system

Groups of hazardous agents that have the nervous system as a target organ of toxicity include solvents, metal and pesticides.

Solvents and the nervous system

General neurotoxicity of most solvents

Of the volatile solvents, most have been shown to cause effects on behaviour and nervous system function, typically producing signs of central nervous system disturbance. The short-term effects of solvents on the nervous system are headaches, tiredness, a sense of intoxication, drowsiness, dizziness and eventually, unconsciousness (NOHSC 1999). Exceptionally high exposures can lead to convulsions and death. Generally, most of these symptoms follow moderate to high short-term exposure, and will wear off following cessation of exposure (EPA 1990).

The effects of prolonged exposure to low concentrations of solvents can be far more serious. Often, the first symptoms are nonspecific or subjective, and difficult to critically evaluate (WHO 1985). The study of the long-term effects of solvents on nerve function uses psychological evaluation. These specialised tests measure components of behaviour such as language comprehension, logical and spatial thinking, power of observation, coordination and memory. Results of these tests indicate that some functions are more affected by some solvents than others.

In the last few years, the term 'solvent neurotoxic syndrome' has been coined to describe the major symptoms of nervous system effects caused by exposure to solvents. These symptoms include difficulties in concentration, forgetfulness, headaches, irritability, insensitivity, personality disorders, mental disabilities and even suicidal tendencies.

Carbon disulfide

Carbon disulfide produces nervous system toxicity beyond that usually seen with many organic solvents. In addition to the narcotic effects common to most solvents, carbon disulfide can induce psychiatric disturbances at prolonged low doses, and at higher exposures, a toxic encephalopathy that may include a toxic psychosis with hallucinations, delirium and manic-depressive derangements.

In addition to CNS effects, repeated exposures can cause a mixed sensorimotor peripheral neuropathy. Optic neuritis and visual disturbances have also been described.

The peripheral neurotoxicity of n-Hexane

n-Hexane produces an acute polyneuropathy, the effects of which are related to the toxicity of its more toxic metabolites. n-Hexane is biotransformed to 2-hexanol, then further to 2,5-hexanediol, and then to the major metabolite 2,5-hexadione. These three substances, solvents in their own right, all produce similar polyneuropathies on exposure, although 2,5-hexadione is many times more potent than n-hexane or 2-hexanone in producing neuropathy. The neurotoxicity of 2,5-hexadione is fairly specific to its γ-diketone structure, as 2,3-, 2,4-hexadione and 2,6-heptanedione are not neurotoxic, while 2,5-heptanedione, 3,6-octadione and other γ-diketones are neurotoxic.

One further important interaction is that of n-hexane and methyl n-butyl ketone (2-butanone). This interaction is synergistic in nature, with exposed animals showing earlier onset and greater severity of neurotoxic signs, and produces greater concentrations of 2,5-hexadione than animals exposed to 2-hexanone alone.

Metals and the nervous system

A number of metals produce neurotoxic effects, some following occupational exposures (Friberg *et al.* 1979). The most important of these are lead, manganese and mercury.

Aluminium

Neurological effects have been reported from aluminium exposure, including encephalopathy, tremor, incoordination and cognitive defects. Elevated aluminium levels have also been found in the autopsy material taken from the brains of individuals with Alzheimer's disease, though its role as a neurotoxic factor is yet to be clarified.

Arsenic

Arsenic toxicity is well studied, owing to its frequent use as a poisoning agent. Many organs and systems are affected, including the nervous system, with degeneration of myelin, encephalopathy, and paraesthesias.

Lead

Lead has been shown to induce effects in a number of body systems including both the central (neuraesthenia, slowed conduction velocity, neurobehavioural disturbances, encephalopathy) and peripheral (peripheral neuropathy seen as wrist and foot drop) nervous systems.

Lithium

Most lithium poisoning occurs through its therapeutic use as an antidepressant.

Manganese

A classic occupational condition known as 'manganism' results from chronic exposure to dust from manganese ores or fumes from manganese steels. The features of this condition are neurological and psychological, including apathy, confusion, bizarre behaviour, increased muscle tone, difficulty with speech and fine motor movement, and loss of balance. Onset is progressive and insidious. One feature of chronic manganism is parkinsonian symptoms, due to manganese-induced damage to the CNS extrapyrimidal system (notable the highly dopaminergic red nucleus). This lesion is similar to that in Parkinson's disease.

Mercury

The main neurotoxic effects are observed in the central nervous system (psychological changes, erethism, tremor abolished by sleep). Insomnia, loss of appetite, loss of memory, shyness and timidity characterise mercury-induced psychosis. Polyneuropathy has also been reported.

Thallium

The onset of thallium poisoning is insidious, but the nonspecific effects eventually progress to nervous system effects (disorientation, lethargy, insomnia, optical neuritis, psychosis, neuropathy, convulsions, cerebral oedema and coma). Some central and peripheral nervous system effects of thallium may persist (ataxia, tremor, memory loss, neuropathy).

Pesticides and the nervous system

The main neurotoxic pesticides are the organochlorine and organophosphorus pesticides (Hayes 1982).

Organochlorine pesticides

The main mechanism of action of organochlorine pesticides, at adequate dosage, is that they interfere with axonic transmission of nerve impulses and therefore disrupt the function of the nervous system, principally that of the brain.

Apprehension and excitability usually occurs after a few hours, with muscle twitching leading to tremors and convulsions (often eliptiform). Other symptoms include dizziness, headache, disorientation, weakness, paraesthesia, pallor and cyanosis.

Respiration is initially accelerated but respiratory depression and then apnoea may supervene. Effects may be delayed for up to 48 hours following exposure. Maintenance of pulmonary ventilation may require intubation and mechanical measures.

Once the acute stage is passed, there is substantial likelihood of complete recovery if vital functions are sustained and convulsions can be controlled.

Long-term sequelae are not generally reported, though liver damage (including hepatocellular carcinoma) has been reported in experimental studies in animals and would not be unexpected in humans.

Organophosphorus pesticides toxicology

Human toxicity to organophosphorus compounds has been known at least since 1899, when neurotoxicity to phosphocreosole (then used in the treatment of tuberculosis) was reported (Echobion 1993). The study of the toxicity is extensive, yielding two very well-established mechanisms for esterases and neurotoxic esterases.

Poisoning with organophosphates

The organophosphorus compounds are generally characterised by a toxicity of inhibition of the esterase enzymes, most particularly cholinesterases (Aldridge 1954) and neurotoxic esterases (Johnson 1975). The mechanism of effect is phosphorylation (Earl and Thompson 1952). The effect is a specific mechanism of organophosphate toxicity.

An organophosphorus molecule can be represented by the general structure:

$$R_3 - \overset{\displaystyle \overset{O}{\|}}{\underset{\displaystyle R_2}{P}} - R_1$$

where P is the phosphorus atom, O is an oxygen atom and R_1-R_3 represents organic structures that can give the molecule a wide range of properties.

Because cholinesterases break down endogenous choline esters, inhibition of these enzymes produces an accumulation of levels of choline esters. Most critical of these esters is acetylcholine, a neurotransmitter molecule released throughout the cholinergic nervous system. Any organ or tissue that receives a cholinergic input will become more active or excited if cholinesterases are not available to catalyse the breakdown of acetylcholine. Indeed, cholinergic overstimulation produces most, if not all, of the symptoms of poisoning from single and short-term exposure to organophosphates.

Signs of low-level intoxication include headache, vertigo, general weakness, drowsiness, lethargy, difficulty in concentration, slurred speech, confusion, emotional lability and hypothermia (Eyer 1995). The reversibility of such effects has been questioned (Kilburn 1999).

Signs of poisoning are usually foreshadowed by the development of early symptoms related to acetylcholine overflow and include salivation, lacrimation, conjunctivitis, visual impairment, nausea and vomiting, abdominal pains and cramps, diarrhoea, parasympathomimetic effects on heart and circulation, fasciculations and muscle twitches (Minton and Murray 1988).

This is the basic site of inhibition for all OP molecules (Johnson 1975; Metcalf 1982).

Organophosphate-induced delayed neuropathy

There is a second reaction that leads to further neurotoxic and neuropathological changes.

Inhibition of neurotoxic esterases can lead to a neuropathological condition of progressive neuronal damage, called organophosphorus-induced delayed neuropathy (OPIDN) (Baron 1981; Metcalf 1982). The mechanism of toxicity is now fairly well understood, as indeed are the organophosphorus structures which are predicted to cause OPIDN (Johnson 1990). Basically, all organophosphorus molecules react with any OH (hydroxyl) groups on the active site of the enzyme:

$$\text{Enzyme—OH} + R_3 - \overset{\displaystyle\overset{O}{\|}}{\underset{\displaystyle\underset{R_2}{|}}{P}} - R_1 = \text{Enzyme—O} - \overset{\displaystyle\overset{O}{\|}}{\underset{\displaystyle\underset{R_2}{|}}{P}} - R_1$$

The basic process is the initial phosphorylation of a group of esterases called the neurotoxic esterases. This is followed by a second reaction of enzyme 'ageing', where the enzyme structure (or its microenvironment) was modified so that it can no longer function properly. The basic mechanism is a break in the P—O—R bond, resulting in a negatively charged P—O group, and a free R group. A determinant of toxicity is the extent of inhibition of these enzymes, in that marked toxicity occurs after inhibition of over 50% (Lotti and Johnson 1980).

Several theories about the significance of these events in the development of OPIDN (Lotti 1992), and a pathway of events have been proposed (Abou-Donia and Lapadula 1990).

The likelihood of this reaction occurring is dependent on the molecular structure of the OP molecule. Where either or both of the R_1 or R_2 groups are linked to the phosphorus with a P—O—R bond (instead of a P—R bond), OPIDN can develop. These OP structures are:

$$\text{Enzyme—O} - \overset{\displaystyle\overset{O}{\|}}{\underset{\displaystyle\underset{R_2}{|}}{P}} - O - R_1 \qquad \text{Enzyme—O} - \overset{\displaystyle\overset{O}{\|}}{\underset{\displaystyle\underset{\underset{\textstyle R_2}{|}}{O}}{P}} - R_1 \qquad \text{Enzyme—O} - \overset{\displaystyle\overset{O}{\|}}{\underset{\displaystyle\underset{\underset{\textstyle R_2}{|}}{O}}{P}} - O - R_1$$

The main classes of organophosphorus molecules that have the potential to cause OPIDN are phosphates (two P—O—R bonds) and phosphonates (one P—O—R bond). A further group known to cause OPIDN is the phosphoroamidates, where the oxygen in the P—O—R bond is replaced by nitrogen (R—N—R).

Where the OP molecule only contains P—R bonds, ageing (and therefore delayed neuropathy) will not occur. The main classes of organophosphorus molecules that have these structures are the phosphinates (Johnson et al. 1982).

Not all animal species are susceptible to developing OPIDN: for example, rodents are not particularly sensitive (Barnes and Denz 1953), although neurological damage can be produced in the rat (Padilla and Veronesi 1985). However, along with the cat (Abou-Donia et al. 1986) and chicken (Cavanagh 1954; Ehrich et al. 1993), humans are considered to be among the most sensitive species (Abou-Donia 1981).

OPIDN is caused when the organophosphate molecule binds with neurotoxic esterases in the long processes of the nerves (the axons). The enzymes have functions related to transport of nutrients and energy molecules from the cell body to the end of the nerves. Phosphorylation of such proteins results in localised disruption of axoplasmic transport. If prolonged, these effects are followed by swelling of the axon, followed by degeneration from the site of the damage to the end of the axon. If exposure continues, this process can continue up the axon by the phosphorylation of more proteins. Lesions are characterised by degeneration of axons followed by degeneration of the cells that surround (and contribute to the insulation of the fibres) the myelin containing support cells (Abou-Donia 1981). This effect can occur in sensory or motor nerves in either the central or peripheral nervous systems (Marrs 1993). Initially, the condition arises as a distal symmetrical sensorimotor mixed peripheral neuropathy mainly affecting the lower limbs with tingling sensations, burning sensations, numbness and weakness. In severe cases paralysis may develop (Lotti *et al.* 1984). Longer nerves are affected more, probably because or their requirements for active nutrient supply (shorter nerves may continue to get supplied through passive mechanisms, such as diffusion). Regeneration is possible if exposure ceases and damage is not too extensive (IPCS 1986; Lotti 1992).

The intermediate syndrome

OPIDN is severe. It is quite likely that such a severe condition would be presaged by a range of clinical and preclinical signs and symptoms. These have been reported extensively, and an 'intermediate syndrome' was defined in 1987 (Senanayake and Karalliede 1987). Symptoms of the intermediate syndrome include: proximal limb paralysis, weakness of neck muscles, inhibition of respiratory muscles and cranial nerve involvement. The mechanism of effect is different from poisoning or OPIDN effects, and is considered to be due to the effect of the organophosphate at the level of the neuromuscular synapses (Sedgwick and Senanayake 1997).

Chronic organophosphate neuropsychological disorder

More recently, chronic exposure to organophosphates has been associated with a range of neurological and neuropsychological effects (Dille and Smith 1964; Rosenstock *et al.* 1991; Savage *et al.* 1988; Steenland 1996; UK DoH 1999). Such symptoms (mainly neurological and neurobehavioural symptoms) may also be seen in individuals who have been sufficiently fortunate in not having exposures that were excessive enough in intensity or duration to lead to clinical disease. A distinct condition, chronic organophosphate neuropsychological disorder (COPIND) has been described with neurological and neuropsychological symptoms (Jamal 1997). These include:

- diffuse neuropsychological symptoms (headaches, mental fatigue, depression, anxiety, irritability)
- reduced concentration and impaired vigilance
- reduced information processing and psychomotor speed
- memory deficit and linguistic disturbances.

COPIND may be seen in exposed individuals either following single or short-term exposures leading to signs of toxicity (Savage *et al.* 1988), or long-term low level repeated exposure with (often) no apparent signs of exposure (Steenland 1996). The basic mechanism of effect is not known, although it is not believed to be related to the esterase inhibition properties of organophosphorus compounds. It is also not known if these symptoms are permanent.

Carbamate pesticides

The intensity, onset and duration of acute toxic effects vary from one carbamate to another. However, all carbamates have the same mechanism, which is reversible carbamylation and hence inhibition of the enzyme cholinesterase. This enzyme is critical for normal nervous activity. At a sufficient exposure, inhibition of this enzyme causes accumulation of acetylcholine at cholinergic nerve terminals (these include terminals in muscle, brain and nervous system, and some glandular systems).

While the toxic effects are similar to those following organophosphorus pesticide exposures, onset and recovery are more rapid. This is due to the more rapid association and disassociation of the carbamate–enzyme complex.

The symptomatology of poisoning by carbamates is that for anticholinesterase agents, and is described in the section on organophosphorus compounds.

Summary

The nervous system is a unique system with unique structures and functions. It is one of the more sensitive systems to toxic insult, and its two main functions (integration of body functions and behaviour) may be affected by chemical exposures at lower levels than other systems.

Neurotoxicological exposures continue to be identified, and not in any particular order, these include: acrylamide (Smith and Oehme 1991), aluminium (van der Voet *et al.* 1991), carbon disulfide (Beauchamp *et al.* 1983; Graham *et al.* 1995), *n*-hexane (Spencer *et al.* 1980; Tahti *et al.* 1997), lead (Landrigen *et al.* 1990; Goodman *et al.* 2002), manganese (Mergler and Baldwin 1997) methyl *n*-butyl ketone (Bos *et al.* 1991), mercury (Cranmer *et al.* 1996), organophosphorus pesticides (Cherniack 1988), solvents (White and Proctor 1997), styrene (Pahwa and Kalra 1993), trichloroethylene (Juntunen 1986), tri-orthocresyl phosphate (Craig and Barth 1999). Associated with Parkinsonian symptoms: manganese, carbon disulfide, organic solvents, carbon monoxide, and the piperidine derivative, MTPT (Tanner 1992).

As the situations in which neurotoxic agents have been recognised in exposed workers have grown, so has the importance of occupational neurotoxicology as a speciality (Costa and Manzo 1998).

There is one last critical point to note: primary prevention of neurotoxicological disorders is essential, because once significant symptoms have been produced, treatment options are extremely limited.

References

Abou-Donia, M.B. (1981) Organophosphorus ester-induced delayed neurotoxicity. *Ann. Rev. Pharmacol. Toxicol.* 21: 511–548.

Abou-Donia, M.B. and Lapadula, D.M. (1990) Mechanisms of organophosphorus ester-induced delayed neurotoxicity: Type I and II. *Ann. Rev. Pharmacol. Toxicol.* 30: 405–440.

Abou-Donia, M.B., Trofatter, L.P., Graham, D.M. and Lapadula, D.M. (1986) Electromyographic, neuropathologic and functional correlates in the cat as the result of tri-*o*-cresylphosphate delayed neurotoxicity. *Toxicol. Appl. Pharmacol.* 83: 126–141.

Aldridge, W.N. (1954) Tricresyl phosphate and cholinesterase. *Biochem. J.* 56: 185–189.

Aminoff, M.J. (1985) Electrophysiologic recognition of certain occupation-related neurotoxic disorders. *Neurol. Clin.* 3: 687–697.

Arezzo, J.C., Simson, R. and Brennan, N.E. (1985) Evoked potentials in the assessment of neurotoxicity in humans. *Neurobehav. Toxicol. Teratol.* 7: 299–304.

Baker, E.L. and Letz, R. (1986) Neurobehavioral testing in monitoring hazardous workplace exposures. *J. Occup. Med.* 28: 987–990.

Barnes, J.M. and Denz, F.A. (1953) Experimental demyelination with organophosphorus compounds. *J. Pathol. Bacteriol.* 65: 597–605.

Baron, R.L. (1981) Delayed neurotoxicity and other consequences of organophosphate esters. *Ann. Rev. Entolomol.* 26: 29–48.

Beauchamp, R.O., Bus, J.S., Popp, J.A., Boreiko, C.J. and Goldberg, L. (1983) A critical review of the literature on carbon disulfide toxicity. *Crit. Rev. Toxicol.* 11: 169–278.

Bos, P.M., de Mik, G. and Bragt, P.C. (1991) Critical review of the toxicity of methyl *n*-butyl ketone: risk from occupational exposure. *Am. J. Ind. Med.* 20: 175–194.

Cavanagh, J.B. (1954) The toxic effects of tri-*ortho*-cresyl phosphate in the nervous system: An experimental study in hens. *J. Neurol. Neurosurg. Psych.* 17: 163–172.

Cherniack, M.G. (1988) Toxicological screening for organophosphorus-induced delayed neurotoxicity: complications in toxicity testing. *Neurotoxicology* 9: 249–271.

Costa, L.G. and Manzo, L. (eds) (1998) *Occupational Neurotoxicology.* Boca Raton: CRC Press.

Craig, P.H. and Barth, M.L. (1999) Evaluation of the hazards of industrial exposure to tricresyl phosphate: a review and interpretation of the literature. *J. Toxicol. Environ. Health* 2: 281–300.

Cranmer, M., Gilbert, S. and Cranmer, J. (1996) Neurotoxicity of mercury: Indicators and effects of low-level exposure: overview. *Neurotoxicology* 17: 9–14.

Dille, J.R. and Smith, P.W. (1964) Central nervous system effects of chronic exposure to organophosphate insecticides. *Aerospace Med.* 35: 475–478.

Earl, C.J. and Thompson, R.H.S. (1952) Cholinesterase levels in the nervous system in tri-ortho-cresyl phosphate poisoning. *Br. J. Pharmacol.* 7: 685–694.

Echobion, D.J. (1993) Organophosphorus ester insecticides. In: Echobion, D.J. and Joy, R.M. (eds) (1993) *Pesticides and Neurological Diseases*, 2nd edn. Boca Raton: CRC Press.

Ehrich, M., Jortner, B.S. and Padilla, S. (1993) Relationship of neuropathy target esterase inhibition to neuropathy and ataxia in hens given organophosphate esters. *Chem. Biol. Interact.* 87: 431–437.

El-Fawal, H.A., Waterman, S.J., De Feo, A. and Shamy, M.Y. (1999) Neuroimmunotoxicology: humoral assessment of neurotoxicity and autoimmune mechanisms. *Environ. Health Persp.* 107 (Suppl. 5): 767–775.

EPA (1990) *Indoor Air Health: Neurotoxic Effects of a Controlled Exposure to a Complex*

Mixture of Volatile Organic Compounds. US Environmental Protection Agency, Research Triangle Park.

Eyer, P. (1995) Neuropsychopathological changes by organophosphorus compounds – a review. *Human Exp. Toxicol.* 14: 857–864.

Feldman, R.G. (2000) *Occupational and Environmental Neurotoxicology.* New York: Lippincott-Williams and Wilkins.

Feldman, R.G., Ratner, M.H. and Feldman, E.S. (1999) Approach to neurotoxicity tort cases. *Neurol. Clin.* 17: 267–281.

Friberg, L., Nordberg, G.F. and Vouk, V.B. (eds) (1979) *Handbook on the Toxicology of Metals.* Amsterdam: Elsevier/North Holland.

Goodman, M., LaVerda, N., Clarke, C., Foster, E.D., Iannuzzi, J. and Mandel, J. (2002) Neurobehavioural testing in workers occupationally exposed to lead: Systematic review and meta-analysis of publications. *Occupat. Environ. Med.* 59: 217–223.

Graham, D.G., Amarnath, V., Valentine, W.M., Pyle, S.J. and Anthony, D.C. (1995) Pathogenetic studies of hexane and carbon disulfide neurotoxicity. *Crit. Rev. Toxicol.* 25: 91–112.

Hayes, W.J. (1982) *Toxicology of Pesticides.* Baltimore: Waverly Press.

IPCS (1986) Organophosphorus pesticides: A general introduction. *Environmental Health Criteria,* 63: WHO International Program in Chemical Safety, Geneva.

Jamal, G.A. (1997) Neurological syndromes of organophosphorus compounds. *Adv. Drug React. Toxicol. Rev.* 16: 133–170.

Johnson, B.L., Grandjean, P. and Amler, R.W. (1994) Neurobehavioral testing and hazardous chemical sites. *Neurotoxicol. Teratol.* 16: 485–487.

Johnson, M.K. (1975a) Organophosphorus esters causing delayed neurotoxic effects. *Arch. Toxicol.* 34: 259–288.

Johnson, M.K. (1975b) Structure activity relationship for substrates and inhibitors of hen brain neurotoxic esterase. *Biochem. Pharmacol.* 24: 797–805.

Johnson, M.K. (1990) Organophosphates and delayed neuropathy – Is NTE alive and well? *Toxicol. Appl. Pharmacol.* 102: 385–399.

Johnson, M.K., Hodgson, E., Bend, J.R. and Philpot, R.M. (1982) The target for initiation of delayed neurotoxicity by organophosphorus esters: Biochemical studies and toxicological applications. *Rev. Biochem. Toxicol.* 4: 141–212.

Juntunen, J. (1986) Occupational toxicology of trichloroethylene with special reference to neurotoxicity. *Develop. Toxicol. Environ. Sci.* 12: 189–200.

Kilburn, K.H. (1999) Evidence for chronic neurobehavioral impairment from chlorpyrifos and organophosphate insecticide (Dursban) used indoors. *Environ. Epidemiol. Toxicol.* 1: 153–162.

Kyrklund, T. (1992) The use of experimental studies to reveal suspected neurotoxic chemicals as occupational hazards: acute and chronic exposures to organic solvents. *Am. J. Ind. Med.* 21: 15–24.

Landrigan, P.J., Silbergeld, E.K., Froines, J.R. and Pfeffer, R.M. (1990) Lead in the modern workplace. *Am. J. Publ. Health* 80: 907–908.

Lotti, M. (1992) The pathogenesis of organophosphate polyneuropathy. *Crit. Rev. Toxicol.* 21: 465–483.

Lotti, M. and Johnson, M.K. (1980) Repeated small doses of a neurotoxic organophosphate: monitoring of neurotoxic esterase in brain and spinal cord. *Arch. Toxicol.* 45: 263–271.

Lotti, M., Becker, C.E. and Aminoff, J. (1984) Organophosphate polyneuropathy: pathogenesis and prevention. *Neurology* 34: 658–662.

MacPhail, R.C. (1992) Principles of identifying and characterizing neurotoxicity. *Toxicol. Lett.* 64–65: 209–215.

Maroni, M. and Barbieri, F. (1988) Biological indicators of neurotoxicity in central and peripheral toxic neuropathies. *Neurotox. Teratol.* 10: 479–484.

Marrs, T.C. (1993) Organophosphate poisoning. *Pharmacol. Therapeut.* 58: 51–66.

Mergler, D. and Baldwin, M. (1997) Early manifestations of manganese neurotoxicity in humans: an update. *Environ. Res.* 73: 92–100.

Metcalf, R.L. (1982) Historical perspective of organophosphorus ester-induced delayed neurotoxicity. *Neurotoxicology* 3: 269–284.

Minton, N.A. and Murray, V.S.G. (1988) A review of organophosphate poisoning. *Med. Toxicol.* 3: 350–375.

NOHSC (1990) *Industrial Organic Solvents*. National Occupational Health and Safety Commission/AGPS, Canberra.

O'Donoghue, J.L., Nasr, A.N. and Raleigh, R.L. (1977) Toxic neuropathy: An overview. *J. Occup. Med.* 19: 379–382.

Padilla, S. and Veronesi, B. (1985) The relationship between neurological damage and neurotoxic esterase inhibition in rats acutely exposed to tri-*ortho*-cresyl phosphate. *Toxicol. Appl. Pharmacol.* 78: 78–87.

Pahwa, R. and Kalra, J. (1993) A critical review of the neurotoxicity of styrene in humans. *Vet. Human Toxicol.* 35: 516–520.

Rosenstock, L. and Cullen, M.R. (1986) *Clinical Occupational Medicine*. Philadelphia: Saunders.

Rosenstock, L., Keifer, M., Daniell, W.E., McConnell, R. and Claypoole, K. (1991) Chronic central nervous system effects of acute organophosphate pesticide intoxication. *Lancet* 338: 225–227.

Savage, E.P., Keefe, T.F., Mounce, L.M., Heaton, J.A. and Burcar, P.J. (1988) Chronic neurological sequelae of acute organophosphorus pesticide intoxication. *Arch. Environ. Health* 43: 38–45.

Sedgwick, E.M. and Senanayake, N. (1997) Pathophysiology of the intermediate syndrome organophosphate poisoning. *J. Neurol. Neurosurg. Psych.* 62: 210–202.

Senanayake, N. and Karalliede, L. (1987) Neurotoxic effects of organophosphorus esters. An intermediate syndrome. *N. Engl. J. Med.* 316: 761–763.

Seppäläinen, A.M. (1988) Neurophysiological approaches to the detection of early neurotoxicity in humans. *Crit. Rev. Toxicol.* 18: 245–298.

Simonsen, L., Johnsen, H., Lund, S.P., Matikainen, E., Midtgård, U. and Wennberg, A. (1994) Methodological approach to the evaluation of neurotoxicity data and the classification of neurotoxic chemicals. *Scand. J. Work Environ. Health* 20: 1–12.

Smith, E.A. and Oehme, F.W. (1991) Acrylamide and polyacrylamide: a review of production, use, environmental fate and neurotoxicity. *Rev. Environ. Health* 9: 215–228.

Spencer, P.S., Schaumburg, H.H., Sabri, M.I. and Veronesi, B. (1980) The enlarging view of hexacarbon neurotoxicity. *Crit. Rev. Toxicol.* 7: 279–356.

Steenland, M. (1996) Chronic neurological effects of organophosphate pesticides. *Br. Med. J.* 312: 1311–1312.

Tahti, H., Engelke, M. and Vaalavirta, L. (1997) Mechanisms and models of neurotoxicity of n-hexane and related solvents. *Arch. Toxicol. Suppl.* 19: 337–345.

Tanner, C.M. (1992) Occupational and environmental causes of parkinsonism. *Occup. Med.* 7: 503–513.

Tilson, H.A. (1993) Neurobehavioral methods used in neurotoxicological research. *Toxicol. Lett.* 68: 231–240.

Tilson, H.A. (1996) Evolution and current status of neurotoxicity risk assessment. *Drug Metab. Rev.* 28: 121–139.

UK DoH (1999) Toxicology of OPs and the mechanisms involved. In: *Organophosphates.*

Committee on Toxicity of Chemicals in Food, Consumer Products and the Environment. London: UK Department of Health, HMSO, pp. 49–58.

van der Voet, G.B., Marani, E., Tio, S. and de Wolff, F.A. (1991) Aluminium neurotoxicity. *Progr. Histochem. Cytochem.* 23: 235–242.

Walker, B. (2000) Neurotoxicity in human beings. *J. Lab. Clin. Med.* 136: 168–180.

White, R.F. and Proctor, S.P. (1997) Solvents and neurotoxicity. *Lancet* 349: 1239–1243.

WHO (1985) *Organic Solvents and the Central Nervous System.* Environmental Health 5. World Health Organisation, Copenhagen.

WHO (1986) *Early Detection of Occupational Diseases.* Geneva: World Health Organisation.

Chapter 9

Reproduction, development and work

C. Winder

Introduction

Work hazards produce a range of health effects, some of which are easy to observe, such as dermatitis or asthma. Other associations are harder because of background or latency factors, such as cancer or chronic back pain.

Effects on reproduction and development are one such health consequence where it is difficult, in the main, to show that a workplace hazard has produced an adverse effect (Winder 2000). There are exceptions (see below), and two workplace exposures – dibromochloropropane and lead – are regulated on the basis of their potential to cause problems in developmental or reproductive processes (Babich *et al.* 1981; Landrigan *et al.* 2000). However, one major end result of reproductive or developmental problems is infertility which, at best, is a poor outcome for the identification of causal agents.

Nevertheless, significant attention has been focused on reproduction and work. This has been partially due to the increase in women joining the workforce and entering traditionally hazardous occupations, but in turn, this has focused attention on problems in male reproduction (Morello-Frosch 1997).

The reproductive system

Reproduction is the process of species continuance through production of offspring (Pinon 2002). While bacteria and other simpler life forms are able to reproduce asexually (that is, by a process where a new organism is produced from one parent, for example by budding or fission), most complex organisms reproduce through a process of sexual reproduction, where genetic material from two parents (male and female) are fused to make the offspring.

Cellular events

All cells in the body contain genetic material, which regulates growth, structure and function. The genetic material is organised into chromosomes, and most organisms have more than one chromosome. For example, normal body **somatic** cells from humans have 46 chromosomes, organised into 23 pairs (see Figure 9.1).

At the cellular level, somatic cells reproduce asexually, replicating the 23 pairs of chromosomes and then allocating a set of each chromosomes to the offspring cells (thereby ensuring that they still have 23 pairs). This process of somatic cell division is called mitosis (Gwatkin 1992).

However, in the reproductive process specialised cells that contain half the normal complement of genetic material are produced (that is one set of 23 chromosomes, not 23 pairs). These specialised cells are called gametes, are unique to the reproductive organs (gonads), and comprise spermatozoa in the male and ova in the female. The process of producing these cells (that is producing gametes with half the genetic material) is different from mitosis, and is called meiosis.

Sexual reproduction in mammals then consists of bringing the male and female gametes together, fusing them into one cell, called the zygote. The process of fusing the genetic material from both parents is called fertilisation. As the zygote has one set of 23 chromosomes from each parent, the normal cell content of 23 pairs of chromosomes are restored.

If the fertilisation process is successful, the zygote will begin dividing. This process, called cleavage in the early stages, produces similar cells, with no apparent structural complexity. At this stage, the zygote gets its nutritional requirements from the fluid in which it is suspended. When the zygote has developed into a

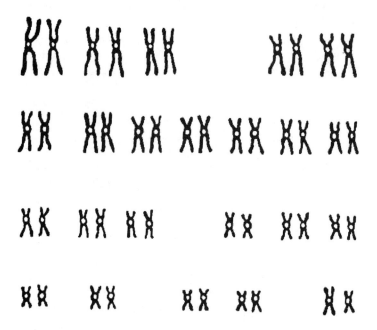

Figure 9.1 The human chromosome.

sphere-like structure of about 100 cells, it begins to form a hollow ball-like structure, called a blastocyst. It is the blastocyst that attaches itself to the lining of the uterus in a process called implantation.

The structure of the blastocyst is now changing rapidly, with formation of the primary cell layers that give rise to all body organs and tissues. These are the ectoderm (skin and nervous system), mesoderm (many body structures) and the endoderm (endocrine glands, mucosal tissue). As these structures develop, the outside of the blastocyst becomes convoluted, increasing the surface area to allow optimal uptake of nutrients. Later this structure becomes the placenta, which supplies nutritional needs to the developing organism.

Continuing development of the blastocyst sees the formation of discrete structures. The embryonic body soon becomes the embryo, where the processes of differentiation of the major organ systems occur (embryogenesis). At some stage of embryonic development, all the major organ systems have been laid down (albeit in different stages of development). From this point, with most of the developmental activities complete, the embryo has to grow and increase the complexity of its structures. At this stage, the embryo is called the foetus. The process of foetogenesis is characterised by dramatic growth (Dhindsa and Bahl 1986).

A critical timetable of development

The timetable of reproductive events is critical for each mammalian species, and is shown for humans in Table 9.1 (Hardy and Garbers 2001; Pryor *et al.* 2000).

Impairment of the developmental process

While the range of things that can affect the reproductive processes is large (see below), there are only four principal manifestations of impairment of the developmental process:

- death of the conceptus (foetal death)
- structural abnormality
- altered growth
- functional deficit.

In some cases, they can be correlated to the different developmental stages:

- pre-implantation: embryolethality
- organogenesis; embryolethality, birth defects
- foetal growth retardation
- functional deficits: foetal death, transplacental carcinogenesis
- neonatal: growth retardation, functional alterations of nervous, immune and endocrine systems, childhood cancer.

Table 9.1 Events of reproduction in humans

Day	Event
Before day 0	Formation of gametes
Day 0	Ovulation; release of ova, sexual intercourse; presence of sperm in uterine tube
By day 1	Fertilisation, formation of zygote (one cell), cleavage begins
Day 3	Zygote reaches uterus (16 cells)
Days 5–6	Formation of blastocyst (about 100 cells)
Day 7	Implantation of blastocyst into uterine wall
Week 2	Implantation complete, placenta-like structure forming
Week 3	Placenta functioning
Week 4	Heart formed and pumping blood
Week 8	Embryogenesis; all organ systems present, at least in rudimentary form; all major brain regions, liver present; bone formation initiated; length 3 cm
Weeks 9–12	Foetogenesis; brain continues to enlarge; facial features present; gender tissues present; blood cell formation in bone marrow; bone growth accelerating; muscle of visceral organs forming; length 9 cm
Weeks 13–16	Eyes, ear and other sensory organs present; kidneys formed; most bones present; length 14 cm
Weeks 17–20	Limbs in final shape; mother senses spontaneous muscular activity (quickening); length 19 cm
Weeks 21–28	Large increase in weight; eyes open; spinal cord myelination begins; bone marrow sole site of blood cell formation; fingernails and toenails present; testes enter scrotum (in males); length 28 cm
Weeks 27–28	Foetus may survive if born prematurely
Weeks 28–40	Continued increase in weight; fat deposits laid down under skin; length 35–40 cm
Weeks 37–42	Labour and birth (parturition)
Neonatal period	Lactation by mother; continued growth of all body systems, including nervous system, immune system, gastrointestinal system

Embryonic loss

Embryonic loss is a normal part of the reproductive process. Only about one-quarter to one-third of all embryos that are conceived (zygotes) develop to become live born infants. Embryonic loss is higher at the earlier stages of pregnancy, often before a diagnosis of pregnancy is made.

Foetal malformations

The period of development basically coincides with the processes of organ formation or embryogenesis. While some further organ development occurs later, most consequent activity is involved in growth. This is the period of foetogenesis. The distinction between development and growth is important, as a change introduced (such as a toxic insult) during foetogenesis will affect growth; however 'catch-up' is possible in such circumstances, because basic systems are still present. A change introduced during embryogenesis leading to loss of tissue is much more critical. No catch-up is possible if tissue containing the only source of information on final structure and function is lost. If this happens, the developing body tries to patch the missing parts and, if such changes do not lead to embryonic or foetal death, will produce an offspring with a structural or functional malformation. Such changes are called terata (from the Greek *teras*, meaning monster) and the study of developmental changes is called teratology.

Foetal malformations (more generally called birth defects) are usually thought of as structural or anatomical changes in neonatal and postnatal organisms. In recent years, the term has also come to include physiological, functional or behavioural defects as well. Therefore, the current meaning of the word teratogen includes any agent that induces structural malformations, metabolic or physiological dysfunction, or psychological or behavioural deficits in offspring, either at birth or in a defined postnatal period.

The background developmental abnormality rate in the community is between 2% and 3% (Clayson 1985). This means that of every 100 babies born, two or three will be malformed for no apparent reason. Figures for the incidence of congenital defects are often underestimated as a number of congenital disorders (mental reproduction, defects of the sense organs, enzyme disorders, and abnormal sexual development) are not detectable during early postnatal life. Approximately half of the total numbers of birth defects are apparent at birth; the remainder become apparent within the first few years of life. The most common malformations involve the cardiovascular system and the male urogenital system.

Indeed, the detection rate of overall incidence of congenital defects may not be useful for identifying specific causal factors, as slight increases in some malformations may be masked by background incidence.

The causes of the majority of birth defects are invariably unknown, as is the aetiology of the majority of malformations (Meyers 1988). The best estimates of causal effects of birth defects are shown in Table 9.2.

Table 9.2 Causes of malformations in humans

Cause of malformation	Percentage
Known genetic transmission	20
Chromosomal aberration	3–5
Environmental causes	
Radiation	Less than 1
Infections	2–3
Maternal metabolic imbalance	1–2
Drugs and environmental chemicals	4–6
Unknown	65–70
Potentiating interactions	?

Since drugs and chemicals account for about 5% of the abnormalities found in humans, it could be argued that the therapeutic use of drugs (largely intentional) and occupational or environmental exposure to chemicals (usually unintentional) appear not to represent a significant source of teratogenic potential. However, since 5% or so of birth defects could be simply avoiding exposure to known teratogens, the prevention of one or two anomalies in every 1000 births is easily achievable and would be a reasonable accomplishment.

The archetypical reprotoxicant in the female: thalidomide

Thalidomide was a sedative drug introduced in the early 1960s as a drug that prevented the sickness of early pregnancy (morning sickness) (London Sunday Times 1979). Thalidomide was therefore prescribed during the period of embryogenesis. One of the more critical organ systems developing at this stage were the upper (and then lower) limb buds, which grow into the arms and legs. Thalidomide caused thinning of the upper surfaces of the limb buds, and in some cases, breakdown and loss of such surfaces.

At this stage in development, a rudimentary limb bud contains cells that have all the necessary information to form the fully developed limb: no other cells contain such information. The loss of these cells leads to loss of information, and therefore loss of the fully developed limb (Stephens and Fillmore 2000).

This is what happened in pregnant women taking thalidomide (McBride 1977). Tissue containing critical information on limb development was lost, it was not possible for the limb to develop normally, and therefore children were born missing limbs (lower limbs if thalidomide was taken later, upper limbs if thalidomide was prescribed earlier).

Reproductive processes in the female

The functions of the female reproductive system are:

- production of ova (female gametes or germ cells)
- production of hormones to support reproductive function
- nurture and protection of the developing foetus.

The components of the female reproductive system are:

- ovaries
- uterine tubules (fallopian tubes) including the fimbria
- uterus
- vagina
- external genitalia.

The structure of the female reproductive system is shown in Figure 9.2 (Manassiev and Whitehead 2002; Robboy and Anderson 2001).

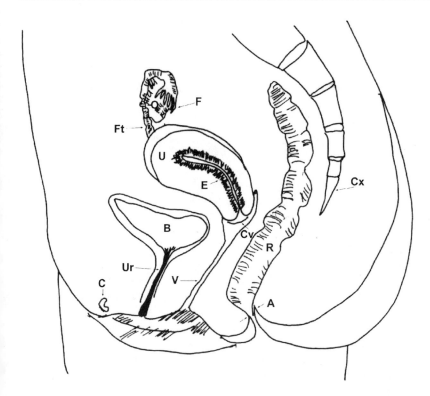

Figure 9.2 The female reproductive system: A, anus; B, bladder; C, clitoris; Cv, cervix; Cx, coccyx; E, endometrium; F, fimbria; Ft, fallopian tube; R, rectum; O, ovary; U, uterus; Ur, urethra; V, vagina.

The ovaries

These are paired structures the size and shape of almonds located against the lateral wall of the pelvis. The functions of the ovaries include:

- production of hormones
- production of eggs (ova)
- regulation of the menstrual cycle.

Hormone production

The ovaries produce the reproductive hormones oestrogen and progesterone. This release starts at puberty when the ovaries become active. Oestrogen is a steroid hormone with a number of activities, including:

- stimulation of the development of secondary sex characteristics
- regulation of the menstrual cycle, including growth of the egg (ovum) follicle
- maintenance of pregnancy
- stimulation of breast milk production.

Oestrogen is released by the developing follicle (see below) and, after ovulation, the corpus luteum.

Progesterone is another steroid hormone that also has a number of activities:

- regulates the menstrual cycle, including thickening of the endometrium in the uterus
- makes implantation of the embryo possible
- controls reproductive activity during pregnancy, including inhibition of ovulation and induction of mammary gland (breast) development during pregnancy.

Progesterone is released by the corpus luteum after ovulation.

Hormone production ceases during the period of permanent cessation of the menstrual cycle, known as the menopause.

Production of eggs (ova)

Egg (ovum) production in the female is a process known as oogenesis.

At birth females already have the total supply of eggs in their ovaries they will use in their lifetime. At this stage, the ova are present as oocytes, each surrounded by follicle cells: this structure is called a primary follicle. Normally, there are 400,000–700,000 primary follicles containing immature ova.

From puberty onwards, the anterior pituitary gland secretes a trophic hormone called follicle stimulating hormone (FSH). This stimulates the release of oestrogen, and a small number of follicles (and the oocytes they contain) to mature each month.

At this stage the stimulated oocytes undergo meiosis to form the ovum, and the follicle cells proliferate, growing in number and forming a central chamber. A mature ovarian follicle is called a Graafian follicle, and prior to bursting, may be 2 cm in diameter (in contrast to the microscopic oocyte prior to stimulation).

The process of maturation of the ova takes approximately 14 days. At the end of this time only one mature follicle, now an outswelling of the ovary, bursts and discharges its contents (including the ovum) into the abdominal cavity. This is known as ovulation. Ovulation is also stimulated by another trophic hormone secreted by the anterior pituitary, known as luteinising hormone (LH).

After ovulation, the remainder of the follicle, now known as the corpus luteum, produces progesterone for about 10–14 days. This stimulates the lining of the uterus (the endometrium) to thicken and develop a rich blood supply. If fertilisation does not occur, the corpus luteum degenerates, and the withdrawal of hormonal support of the endometrium causes the epithelium to be shed in a process known as menstruation.

The menstrual cycle

Aspects of the menstrual cycle have been introduced above. The menstrual cycle consists of cyclical changes in the female body, by which the ova is matured and released, and the uterine wall is stimulated to prepare it to receive a fertilised egg (should fertilisation occur).

The complete menstrual cycle involves ovulation processes as outlined above, and the proliferation of the endometrial lining of the uterus during the first half, an intervening receptive period, followed by subsequent breakdown of the endometrial lining if fertilisation does not occur (see Figure 9.3). The endometrium is receptive to fertilised ova for only a short period each month, about 7 days after ovulation.

Several hormones are important in control of the menstrual cycle, including follicle stimulating hormone, which stimulates follicle development in the ovaries; as they mature they produce oestrogens. Luteinising hormone triggers ovulation and causes the ruptured follicle to be converted to the corpus luteum and produce progesterone (see Figure 9.3).

Stages in the menstrual cycle

The menstrual cycle is normally considered to begin at the first day of the menstrual period, even though a more logical starting point would be the first day of the proliferation stage. This is because menstruation onset is readily identified in all women, while the date of the beginning of proliferation is difficult to judge. The stages of the menstrual cycle are shown in Table 9.3 (Ferin and Jewelewicz 1993).

The menstrual cycle is normally 28 days long, although this may be highly variable: women may experience 22–35 day cycles. Menstrual cycle problems are known as dysmenorrhoea.

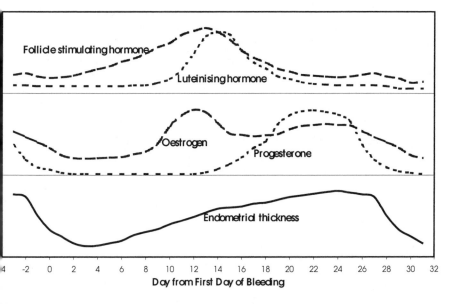

Figure 9.3 Hormonal, ovum development and endometrial thickening during the menstrual cycle.

Table 9.3 Stages of the menstrual cycle

Days	Stage	Events
1–5	Menses	The thick endometrial lining sloughs off with bleeding for 3–5 days.
6–14	Proliferative stage	Under the influence of oestrogens produced by the growing follicles of ovaries, the endometrium increases in thickness, glands are formed and the blood supply is increased.
15–28	Secretary stage	The blood level of progesterone increases due to its production by the corpus luteum of the ovary. The blood supply to the endometrium increases more. Glands secrete nutrients into the uterus to sustain the developing blastocyst (if present) until implantation. If fertilisation occurs then the corpus luteum is maintained and it continues to produce hormones through the early part of pregnancy (and until the placenta takes over). If fertilisation does not occur then the corpus luteum degrades, decreasing the levels of the hormones, so that the endometrial wall begins to die. This sets the stage for menses to begin on day 28.

The uterine tubules (fallopian tubes)

The uterine tubules are also known as the fallopian tubes. These are paired tubular structures about 10 cm long, extending between the upper part of the uterus and the ovaries. The tube is not continuous with the ovaries. The end closest to the ovaries is expanded and has finger-like projections called fimbriae that partly enclose the ovaries.

At ovulation, the ovum (or sometimes ova) is released into the peritoneal cavity.

The ciliated cells on the fimbriae wave rhythmically to direct the ova into the uterine tube, from where it moves towards the uterus due to peristaltic movement and the beating of cilia. This process can be inefficient, and some ova are not collected and are lost in the abdominal cavity. This lack of a complete connection between the ovary and the fallopian tube places women at risk from a range of health problems, including pelvic inflammatory disease and endometriosis.

The journey of the ovum from the ovary to the uterus takes 3 to 4 days. However, the ovum is viable only for about 24 hours. This means that the ovum will probably not be capable of fertilisation by the time it reaches the uterus. Therefore, if present, sperm cells must ascend against the beating of the ciliated epithelium lining of the uterus into the fallopian tube, and if timing is suitable, must meet the ovum there for fertilisation to take place.

Uterus

The uterus (or womb) is located between the bladder and rectum, above and connected to the vagina. It is a hollow muscular organ that can be divided into three regions: the body, the fundus and the cervix.

The thick walls of the uterus are comprised of three layers:

- endometrium, where the blastocyst will implant to produce an embryo (it is this layer which is sloughed off during menses)
- myometrium, a thick layer of smooth muscle which contracts during childbirth (parturition) to expel the foetus
- epimetrium, a serous outer membrane similar in structure to the peritoneum.

Cancer of the cervix is very common in women in the 30–50 year age group. Risk factors include increased sexual activity with many sexual partners, multiple pregnancies, sexually transmitted diseases and frequent cervical inflammation. The pap smear (Papanicolaou stain test) is a diagnostic test by which cells of the cervix are removed and examined for the presence of premalignant or malignant cells. The risk of cervical cancer can be reduced by periodic (every 1–2 years) pap smears in women in the age risk group.

Vagina

The vagina (also called the birth canal) is a thin-walled muscular structure, which extends from the outside of the body to the cervix of the uterus.

A membrane called the hymen partially covers the outside of the vagina in young females. This is ruptured in a range of activities, such as vigorous sports, insertion of a tampon, during pelvic examination, or during the first episode of sexual intercourse.

Other reproductive structures

The breasts

The breasts (or mammary glands) are present in both genders, but only function biologically in females. The main function of the breast is to produce milk for the nourishment of a newborn infant. The breasts develop in females at puberty through stimulation by hormones, especially oestrogens. Anatomically, each breast consists of glandular tissue organised into 15–25 lobes, separated from each other by connective tissue and fat. Slightly below the centre of each breast is a pigmented area, the areola, which surrounds a central protruding nipple. During the process of lactation, milk is secreted from the glandular tissue, and passes along the lactiferous ducts from each lobule, which connects to the nipple.

Milk is an emulsion of water and fatty components, the water being similar in content to plasma. Milk production varies during the lactation period. Colostrum is a thin watery milky fluid secreted around the childbirth period which contains more minerals and less fat and carbohydrate than milk, but also 20% protein, including antibodies found in maternal blood, therefore transferring some of the mother's natural immunity to the child. Milk may be a source of toxic contaminants if they are present in the mother's circulation (particularly the plasma).

Cancer of the breast is a leading cause of death in women in the developed countries.

The external genitalia

The external genitalia include a range of structures, such as the mons pubis (a fat pad overlying the pubic symphysis), the labia major and labia minora (folds on either side of the vaginal entrance), the greater vestibular glands (a pair of mucous-secreting glands located on either side of the vagina, which lubricate the distal end of the vagina during sexual intercourse), and the clitoris (a small structure comprised of erectile tissue which becomes engorged with blood and enlarged with excitation).

Pregnancy

Changes to the mother during pregnancy

The process of developing a human baby is an immense physiological assault on the mother. There are a range of changes that can affect the mother. Nausea and vomiting occur in the early stages as the mother adapts to the changing hormone levels. The metabolic rate of the mother's body increases. Calcium metabolism changes, including some loss of calcium from body stores to the developing foetus. The maternal blood volume increases, which may sometimes lead to high blood pressure. Enlargement of the uterus as the foetus grows may result in difficulty in breathing due to upward pressure on the diaphragm; frequent urination due to increased pressure on the bladder; backache due to increased muscle stress from the shift in the centre of gravity; and varicose veins due to compression of the inferior vena cava and pelvic veins. There will be alterations in kidney function as the mother must filter the baby's wastes as well as her own. The mother may experience fluid retention, leading to swollen feet and legs. The glandular tissue of the breasts increases as the glands prepare to produce milk.

The placenta

This is a large glandular structure which develops during pregnancy, to provide a selectively permeable barrier between mother and foetus, allowing important nutrition to reach the foetus and removing wastes from the foetus. It also produces hormones – oestrogen and progesterone.

The placenta is connected to the foetus through a tough, fibrous tube called the umbilicus or umbilical cord. This contains two tubes, an artery and a vein. While the foetal heart pumps blood around the developing body, it also pumps blood through the placenta.

The placenta consists of foetal and maternal components separated by the placental membrane, which allows transport of substances between the mother and the foetus. Large molecules, such as proteins, cannot transfer across the membrane; smaller molecules can. The membrane allows the transport of oxygen, nutrients, minerals and other essential materials to the foetus; some toxic substances such as alcohol, nicotine, sedatives, antidepressants, some antibiotics, toxic gases and other absorbed toxicants can also pass through the membrane and gain access to the foetus.

Parturition

This is the process of giving birth. It normally occurs about 40 weeks after fertilisation. Three stages are recognised:

- the dilation stage begins with the onset of regular contractions less than 10 minutes apart, and ends with full dilation of the cervix; it lasts on average 7–12 hours
- the expulsion stage ends with the delivery of the baby and usually lasts 20–25 minutes
- the placental stage occurs after delivery of the baby when the placenta and membranes should detach from the uterine wall and also be delivered.

Reproductive processes in the male

The functions of the male reproductive system are the production of male sex hormones (mainly testosterone) and the production of sperm (Wang 1999). The primary reproductive organs are the testes, accessory glands (such as the prostate) and external genitalia (penis and scrotum). The structure of the male reproductive system is shown in Figure 9.4.

The testes

The testes are olive-sized glands suspended in the scrotal sac outside the abdominal cavity. Location of the testes outside the body is necessary because sperm production requires a temperature slightly lower than body temperature.

The wall of the scrotum contains smooth muscle (dartos muscle) which contracts in response to cold. The internal aspect of the scrotum is lined by a sac known as the tunica vaginalis testis. This has parietal and visceral layers. The parietal is attached to the scrotum, while the visceral layer adheres to the testes.

The testes are covered by a protective dense connective tissue, the tunica albuginea. Extensions of this tissue form septa inside the testis dividing it into a number of lobes. Each lobe contains 1–4 tightly coiled seminiferous tubules. The tubules are lined with germinal epithelium, which consists of two basic cell types: the sustentacular or Sertoli cells which have a supportive function; and the seminiferous tubules where sperm are produced. In between the tubules are the interstitial cells which produce testosterone. The seminiferous tubules empty into another set of tubules, the rete testis. From the rete the sperm travel into the epididymis located outside the testes.

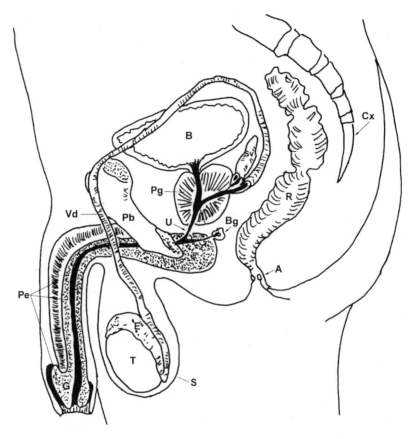

Figure 9.4 The male reproductive system: A, anus; B, bladder; Bg, bulbourethral gland; Cx, coccyx; E, epididymus; G, glans; Pb, pubic bone; Pe, erectile tissue of the penis; Pg, prostate gland; R, rectum; S, scrotum; Sv, seminal vesicle; T, testes; U, urethra; Vd, vas deferens.

The main functions of the testis are:

- production of the male hormone, an endocrine function of interstitial cells that is controlled by interstitial cell stimulating hormone (ICSH; basically the same as luteinising hormone in the female) secreted from the anterior pituitary
- production of sperm (spermatogenesis), exocrine function of spermatogonia cells controlled by follicle stimulating hormone (it takes about 2 months for mature sperm to be formed from the spermatagonia)

Epididymis

This is a coiled tubular organ arising from the upper back part of the testes. Its function is to provide storage for immature sperm. While in the epididymis sperm mature to become fully functional. With sexual stimulation the walls of the epididymis contract to expel sperm into the ductus deferens.

Ductus deferens

This long tube extends from the epididymis into the pelvic cavity, passing over the upper part of the bladder. It is enclosed by a connective tissue sheath, the spermatic cord, along with nerves, lymphatic and blood vessels and the cremaster muscle. The cremaster muscle elevates or depresses the testis in response to the ambient temperature, to maintain the optimum temperature in the testis for spermatogenesis.

The ductus deferens is a highly muscular tube containing smooth muscle arranged in inner longitudinal, middle circular and outer longitudinal layers. It is lined by mucosa. The ductus empties into the ejaculatory duct, which carries sperm through the prostate gland to empty into the prostatic urethra.

Seminal vesicles

The seminal vesicles produce seminal fluid that constitutes 60% of the volume of semen. It is very rich in fructose (a sugar) which nourishes the sperm after it has been deposited in the female reproductive tract. The duct from the seminal vesicles and the ductus deferens join to form the ejaculatory duct, so that sperm and seminal fluid enter the urethra together during ejaculation.

Spermatogenesis

Sperm production (spermatogenesis) begins during puberty and continues throughout life. Sperm formation occurs in the spermatogonia cells located next to the basement membrane of the seminiferous tubules of the testes. Prior to puberty, these cells grow through mitosis to build up a stem-cell line (that is, when they divide, they produce two more stem cells). During and after puberty, follicle stimulating hormone is secreted in ever increasing amounts from the anterior pituitary. Stem cell division changes, in that the two offspring cells are not two stem cells (as they were before puberty) but one stem cell and one spermatocyte-type cell.

Under this new type of cell division, the stem cells stay at the periphery of the seminiferous tubule, and the spermatocyte cell begins moving towards the centre of the tubule. The spermatocyte undergoes meiosis to form secondary spermatocytes, and then spermatids. These are still basically cellular structures, and must undergo further structural changes to become sperm. This includes losing excess cytoplasm, packing the cell DNA into the head, and formation of a three-piece tail.

The mature sperm is a very specialised cell equipped with a high rate of metabolism and a means of propulsion. Once formed, the sperm is released into the lumen of the seminiferous tubule. At this stage, it is incapable of self-directed movement. It is moved through the tubules to the epididymus, where it is stored until needed.

From formation of a spermatocyte to release of immature sperm into the lumen of the seminiferous tubule, the process of spermatogenesis takes over 2 months. Millions of sperm are made every day.

Sperm production is affected by a variety of environmental threats. Drugs such as antibiotics, social drugs such as alcohol, marihuana and tobacco, and environmental or occupational exposures (see below) are examples of exposures that can affect sperm production.

The accessory glands

The prostate gland is a chestnut-sized gland located below the bladder around the upper urethra. It is well supplied with fibromuscular tissue that helps to expel its secretions during ejaculation. It secretes an alkaline fluid that activates sperm and protects it against the acid conditions of the female reproductive system. The fluid is rich in the enzyme acid phosphatase, which helps activate sperm.

Bulbourethral glands are pea-sized glands located below the prostate. They secrete a mucous-like lubricating substance into the penile urethra prior to intercourse.

External genitalia

The **penis** is an organ that delivers sperm into the female reproductive tract. It consists of two main parts, the root, which is attached to the urogenital area and the body, which hangs free from the lower part of the abdomen. The body is further subdivided into the shaft, glans penis and prepuce or foreskin.

The shaft contains spongy erectile tissue called spongiosum. This surrounds the penile urethra (one corpus spongiosum) and form much of the glans (two corpora cavernosa).

During sexual excitation, these erectile bodies become engorged with blood, venous return from them is blocked, and an erection occurs. The corpus spongiosum does not become as hard as the corpora cavernosa, allowing the free passage of sperm during ejaculation.

General aspects of reproduction and development

Background considerations

The complexity of the continuum called reproduction is masked by a tendency to focus on discrete components of the process, such as sperm or egg cell or the embryo. However, reproductive capacity also encompasses pregnancy, embryonic and foetal development, lactation, child health, development and growth, puberty, behavioural development, reproductive senescence and the integration of reproductive functions with the overall health of the individual.

Fertility

Approximately 7–10% of couples in which the female partner is of childbearing age (15–44 years) are unintentionally infertile. This may be temporary. The causes of infertility are multifactorial:

- heritable, such as genetic malformations
- nutritional
- ethnic
- age-related
- iatrogenic (medication-induced)
- pathological, including infectious diseases

- environmental, including pollution
- sociobehavioural, including alcohol, tobacco, exercise, stress
- work related.

Analysis of these factors reveals large gaps in scientific knowledge, and even sparse information on possible interactions with these and other (for example, occupational) factors.

Other reproductive effects

Reproduction is a complex and dynamic sequence of events, which are critically timed and under strict hormonal control. This process can be impaired at many points through a variety of mechanisms (Mattinson 1983). The range of possible reproductive effects is outlined in Table 9.4.

Hazards and reproduction

A number of hazardous agents have been associated in varying degrees with impairment of reproductive function and the health of the developing embryo (Conover 1994; Greaves 1994; O'Rahilly and Muller 2001; OTA 1985; Paul 1993). These agents include:

- aspects of lifestyle (for example, tobacco, alcohol, drug use)
- ionising radiation
- nonionising radiation

Table 9.4 Range of possible effects on reproduction

Prior to conception	Conception/pregnancy	After delivery
- Altered libido	- Effects on implantation	- Effects on labour
- Effects on genetic material (mutation)	- Effects on the mother	(abnormal, prolonged) or parturition
- Effects on gamete production (ovulation and spermatogenesis)	- Spontaneous abortion	- Low birth weight
	- Congenital malformation	- Low APGAR score
	- Effects on embryogenesis	- Effects on the mother
- Abnormal sperm production/transport	- Effects on pregnancy	- Effects on lactation
- Ejaculatory disorders and impotency	- Altered/prolonged gestation	- Postnatal abnormalities leading to developmental disabilities
- Ovulatory disorders	- Hypertension of pregnancy	
- Abnormal menses	- Pre-eclampsia	- Effects on postnatal development
- Infertility	- Foetal death	
- Reversible or irreversible sterility	- Transplacental carcinogenesis	- Neonatal or childhood death
		- Altered reproductive capacity (duration of menses, impairment of spermatogenesis)

- hot or cold environments
- hyper/hypobaric environments
- noise
- vibration
- infectious agents
- overexertion; stress
- some pharmaceuticals
- some chemicals, including occupational exposure to such chemicals.

This is a wide field, so this chapter will deal mainly with the reproductive effects of hazardous chemicals, and primarily, but not exclusively, with chemicals encountered in the work environment.

Chemicals are regulated for a range of health effects, which do not often include those on reproduction. To date, only lead is regulated in Australia because of its potential to produce impairment on reproductive or procreative capacity, though the regulatory restrictions placed on a number of chemicals by, for example, stringent poison scheduling, may be due in part to reported effects on reproduction or development.

Occupational health and safety issues

The reproductive and developmental effects of industrial chemicals are a relatively new concern in occupational health and safety (Frazer and Hage 1997; Hemminki *et al.* 1985; Williams 1990). This has been due in part to the emphasis of research and study in other areas, such as occupational injury or carcinogenicity. It has also been due in part to the increasing number of women present in the workforce (Chamberlain 1984). As a spin off to this phenomenon, interest in male reproductive capacity has also increased.

There is little information available and considerable uncertainty regarding the potential for reproductive and developmental toxicity of industrial chemicals (Barlow and Sullivan 1982; Kirsch-Volders 1981; Shepard 2001).

Estimating reproductive function

Reproductive function in adults can be assessed by relatively simple means, including a detailed patient history, physical examination and analysis of blood and urine samples (Scialli 1991).

The sort of information required in a reproductive history should be from:

- identification (name, gender, age)
- familial health (medical conditions, diseases and conditions in relatives)
- lifestyle characteristics (smoking, alcohol, social drugs, domestic exposures)
- medical history (injuries, medical conditions/diseases, surgical procedures, medication)
- reproductive history (previous and present exposures)
- reproductive history (past reproductive outcomes, reproductive difficulties or disorders).

In men, it is also possible to collect a semen sample without too much difficulty. This can be subjected to analysis:

- ejaculate appearance, pH and volume
- sperm density, counts, motility, vitality and morphology.

The correlation of changes in these parameters with exposure is not easy, especially as the significance of a number of functional changes is not clear. However, there is some evidence to suggest that sperm counts and morphology may be sensitive to exogenous factors.

Demonstrating adverse reproductive function

The demonstration of occupational causes of infertility can be made through the following steps:

- nonoccupational causes must be excluded
- prior ovarian, uterine, cervical or testicular injury
- surgery on the organs of reproduction, such as hysterectomy, tubal ligation, vasectomy, and so on
- mumps
- primary endocrinopathy
- gonadotoxic exposures, such as cytotoxic drugs, dibromochloropropane
- urological abnormalities, such as ductile obstruction, retrograde ejaculation, variocele
- a gonadal dysfunction should be demonstrated, through reproductive history, clinical and biochemical measures or by analysis of semen
- demonstration of exposure to an established or suspect agent sufficient to cause reproductive dysfunction
- in many cases, deficits in reproductive performance (with the possible exception of azoospermia) are likely to be reversible. Therefore, improvement on removal from exposure can be supportive in providing evidence of diagnosis.

Occupational reproductive toxicology

The study of occupational reproductive toxicology involves the experimental, clinical and epidemiological sciences. The evolution of developmental and reproductive toxicity studies in experimental animals has produced nonstandardised protocols, and the strategies employed in such testing tend to be selected on the basis of regulatory requirements, rather than design (Manson *et al.* 1982). The study of human reproductive capacity and function involves case study, prospective and retrospective epidemiology studies, but they generally lack statistical power because of unsure endpoints and small sample sizes (Hemminki *et al.* 1983). Further, as in many epidemiological approaches, estimating exposure is a problem (Lemasters and Selevan 1984).

Experimental studies

Reproductive as opposed to developmental toxicity

It is important to make the distinction between reproduction toxicity and developmental toxicity, as they represent distinct concepts.

Reproductive toxicity is the effect of a toxic agent on the process of reproduction (Dixon 1985). Examples of toxic effects include those on specific reproductive parameters, such as:

- gametogenesis
- fertility
- lactation
- delayed or impaired growth.

Other biological systems that can indirectly effect reproductive capacity may also be affected. These include systems such as impaired metabolic capabilities, a damaged or reduced immune system, or compromised liver function. If there is likely to be an effect on reproductive capacity, it tends to be produced by long-term exposure to low levels of chemicals and often it is difficult to detect casual relationships between a nonclinical reproductive effect and exposure to a chemical (for example, a nonspecific infertility).

Experimentally, reproductive toxicology is tested in multigeneration (usually two or three) reproduction studies (sometimes incorporating a teratology component) using long-term exposure to the chemical over two or more generations. Such studies are required for all new food additives, pesticides and therapeutic substances.

Developmental toxicity is the effect of a toxic agent on the process of development, that is, within the maternal/foetal unit. The main effect is teratogenesis, the production of physical defects on the developing embryo, including death from lethal defects. Exposure to the teratogenic agent must occur during critical developmental periods and will generally occur at doses that do not cause observable maternal toxicity (Christian *et al.* 1989).

Signs of maternal toxicity are not normally considered to be developmental effects (Khera 1987). Note that foetal death can be from either reproductive (for example, impairment in nutrient transfer to the foetus) or developmental (a lethal abnormality) effects.

Scope of the problem

It has been estimated that there are approximately 70,000 chemicals in commercial use (in the USA). Of these, approximately 2800 chemicals have been tested for teratogenicity (Schardien 2000). Of these:

- 180 are clearly positive in two or more species, representing about 6% of the total tested

- just over 600 or 21% are probably positive (based on limited testing or positive in the majority of species tested)
- 291 or 10% are possibly teratogenic (equivocal or variable reaction, and/or less than obvious response)
- the remaining 62% of tested chemicals are probably not teratogenic.

It is also possible to establish that about 50 of these are human teratogens.

A sample size of 2800 of a total sample of 70,000 is a very small proportion (4%). The 50 or so human teratogens represent about 1.8% of the 2800 sample size. Extrapolation to a total of 70,000–100,000 produces an estimate of 12–1700 human teratogens that have not yet been identified (see Figure 9.5).

Of course, the selection of sample chemicals for testing is often based on potential human exposure, and a significant proportion of the 2800 chemicals that have been tested would be therapeutic substances and agricultural chemicals. Under such circumstances, it is quite likely that industrial chemicals, the single largest group of chemical substances, would be considerably under-represented in the 2800 chemicals that have been tested for teratogenicity. It is this category of substances most often encountered in the workplace (Kolb 1993).

Validity of teratology testing

A teratology test should be able to answer three questions:

- Can the agent induce developmental defects (teratogenic potential)?
- What are the effective doses (teratogenic potency)?
- Are the effective doses below adult toxic doses (teratogenic hazard)?

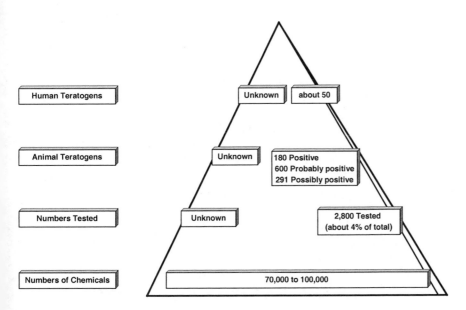

Figure 9.5 Developmental toxicity testing.

The principles of teratology

There are three main principles, which can be best illustrated by the axiom: a teratogenic response depends on the administration of a specific treatment of a particular dose to a genetically **susceptible species** when the embryo(s) are in a critically **susceptible stage of development**.

Susceptible species

Not all species are equally susceptible or sensitive to teratogenic influences. The susceptibility of developing processes to insult is likely to be species specific. There are a number of factors that contribute to this susceptibility, including metabolic capacity of the maternal or embryonic organism, placenta type, the length of gestation, species and strain differences. Animal data should be applied to humans with caution. In general, humans appear more sensitive to effects than do other animals at comparable doses. In the case of thalidomide (almost a prototype teratogen) the human is probably the most susceptible species, with the rabbit and monkey next in sensitivity. Some species (notably the hamster) do not appear to be sensitive at all.

Susceptible stages of development

The timing of exposure is an important determinant of effect. During the predifferentiation period, the conceptus is generally resistant to the induction of birth defects, though embryonic death or resorption (in some species) may occur. During embryonic differentiation (organogenesis), the embryo is highly susceptible to teratogenic insult, and because organ development is continual, a greater specificity of effects is observed as the embryo develops. Following differentiation, the foetus becomes progressively less susceptible to teratogenic stimuli. Therefore, **for an exposure of an agent to have a teratogenic effect, it must occur during the period of organogenesis**.

The importance of this principal can again be illustrated with thalidomide, where time of treatment and not dosage was shown to be the decisive teratological factor.

Figure 9.6 shows the susceptibility of particular organs to a hypothetical developmental toxicant administered at a constant dose on different days of gestation in the pregnant rat (the gestation period for the rat is 20–22 days). Note that organ systems are affected differently at different days, as they are developing differently (and therefore, they show maximum susceptibility to disruption to toxic insult).

Having said all that, effects on reproductive and developmental processes may arise even if exposure occurred before conception. Reasons for this include:

■   hazardous agents or metabolites may be retained or persist in the body for long periods of time (for example polychlorinated biphenyls in fat or lead in bone)

■   prior damage to germ cells that is irreversible or has not been repaired

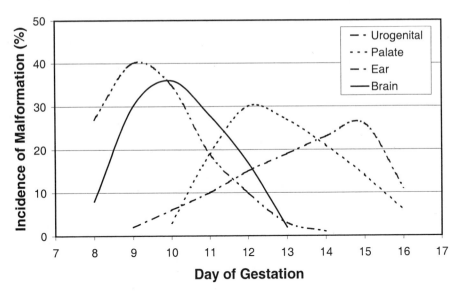

Figure 9.6 Susceptibility and day of gestation of dosing.

- sperm toxicity is usually repaired by cell turnover in several months
- ovarian damage may persist for years.

However, the key point remains: damage arises when delivery of an insult to the critical process occurs at the critical time.

Dose dependency

Teratogenicity is governed by dose–effect relationships, though the dose–response curve is generally quite steep. Moreover, every teratogen that has been realistically tested has been shown to have a no observable effect level. This can generally be interpreted to mean that the doses or exposures required to produce no discernible effect, induce birth defects or cause foetal death generally lie within a narrow range. It is probable that developmental abnormality and death are simply degrees of reaction to the same stimuli.

While the duration of exposure or treatment is an important factor, single or acute exposures are of more teratogenic insult to the developing embryo.

In the maternal animal

Exposures that cause severe effects on the embryo or neonate may have no effect on the mother. Also, exposures that cause maternal illness are unlikely to affect development.

One further issue is that a safe level of exposure (the no observable effects level for developmental effects) can be set in most if not all developmental toxicity studies (see Figure 9.7).

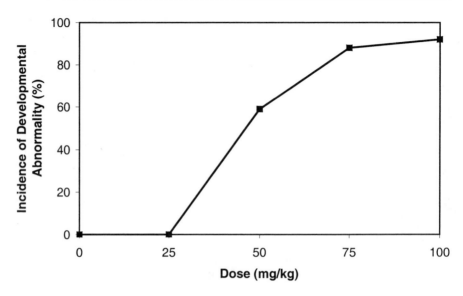

Figure 9.7 Developmental effects and the dose–response level.

If a no effect level can be set, then the potential exists to set a safe level of exposure to an agent with known effects on development. This may be important in the therapeutic or occupational setting.

Extrapolation from animals to humans

Experimentally, developmental toxicology is tested through classic teratology animal studies following subacute exposure to the agent to pregnant animals (usually rabbits or small rodents).

A quantitative evaluation of the predictive value of teratogenicity tests is lacking, though arguments have been presented on qualitative grounds. The accuracy (and therefore the validity) of such testing is impaired by:

- different observed lesions in humans and animals
- human exposure data is often poorly estimated
- dose–response data in humans are largely inaccurate, unsatisfactory or absent
- animal studies are often inconsistent, even between studies using the same species and doses.

The predictive ability of animal results to known human teratogens varies on the species used, and there is no *a priori* basis for selecting a certain species as a suitable model for predicting teratogenic hazard in humans. Indeed, no one species has been demonstrated to be the one of choice (though the rabbit and rat are the species most often used) (Hemminki and Vineis 1985). The ideal animal for teratology testing is one that:

- absorbs, metabolises and eliminates test substances like humans

- transmits test substances and their metabolites across the placenta like humans and has embryos and foetuses with developmental and metabolic patterns like humans
- is inbred to remove unwanted genetic traits
- has a low rate of spontaneous abnormalities
- is easily bred, has a short gestation and produces many offspring
- can be maintained cheaply under laboratory conditions
- does not make objectionable noises or smells, and does not bite, scratch or kick.

Epidemiological and clinical studies

Epidemiology is the study of relationships between the frequency and distribution, and the factors that may influence frequencies and distribution of diseases and injuries in human populations. Epidemiological studies can be categorised into three broad classes:

- descriptive (case studies or surveillance studies)
- analytical (cohort or case–control studies)
- experimental.

The first two of these are often used for studying reproductive or developmental impairment (Thiel and Klug 1997).

Epidemiological studies often suffer methodological problems because sample sizes of worker populations may be too small to significantly demonstrate effects on reproductive or developmental endpoints whose frequency is low in the general population (such as congenital malformation). Many reproductive endpoints (for example, spontaneous abortion, depressed libido) are difficult to measure.

Some study designs do not adequately control for the possibility of paternally, rather than maternally, mediated effects.

Exposure is nearly always difficult to estimate and characteristics of lifestyle (such as alcohol, tobacco, drugs) can confound study results. Worker cooperation or data measurement may be difficult to obtain if the employer/management is unsympathetic to the study.

All these factors tend to contribute to the problems inherent in interpreting the results and conclusions of epidemiological studies. However, criteria exist for the recognition of a new teratogenic agent in humans. These have been used to identify the teratogenic agent in two teratological catastrophes, namely those caused by rubella (German measles) and thalidomide. These criteria are:

- an abrupt increase in the frequency of a particular defect or association of defects (syndrome)
- coincidence of this increase with a known change in exposure conditions, such as sudden exposure to a chemical or widespread use of a drug
- known exposure to a known change in exposure conditions in pregnancies yielding characteristically defective infants
- absence of factors common to all pregnancies yielding infants with characteristic defects.

For the criteria to work, they need the presence of a distinctive defect or pattern of defects, and sufficient numbers of cases. In the case where the defect is rare, fewer cases are needed.

Using such criteria, some 18 therapeutic drugs, one social drug (alcohol) and one chemical (methylmercury) have been definitely identified as probable teratogens to humans.

The study of male reproduction

The study of adverse reproductive consequences in the male has long been a topic of historical neglect in research and testing for a number or reasons (Messing 1999):

- research priorities were generally on the study of female reproduction
- there is a lack of appropriate toxicity testing protocols for male reproductive effects (this lack continues)
- there is a predominance of acute exposure protocols, when many human exposures are of a repeated dose nature.

Effects of chemicals

The structures and biological activities of chemicals with confirmed or suspected adverse effects on reproduction range broadly.

Relatively unreliable laboratory tests, indeterminate clinical endpoints, inaccurate monitors of exposure (often to more than one chemical), nonspecific indicators of biological effects, and poorly documented epidemiological or case-study investigations have each contributed to the difficulty of defining the role of occupational (and environmental) exposure to chemicals in the mechanisms and genesis or reproductive dysfunction (Levine 1984; Schrag and Dixon 1985; Tas et al. 1966).

For some chemicals, there is scientific evidence regarding a cause and effect association with exposure and dysfunction in humans. This is based on:

- the number of studies that have reported toxicity
- insufficient occupational evidence is supported by the results of animal studies
- the effect of the chemical is predictable based on known biological activity.

Exposures to classes of chemicals, such as pesticides, oral contraceptives, or organic solvents and industries or processes where several chemicals are used, such as the rubber industry, or laboratory work, should also be included as producing evidence or reproductive or developmental dysfunction in occupationally exposed individuals.

Data are also available from animal studies alone (Richardson 1993). While such information should always be evaluated carefully with particular regard to mechanisms of toxicity and likely occupational exposures, it does often serve as a useful pointer to potential problems.

Reported health effects

By occupation

There are many difficulties in establishing causal relationships between reproductive toxicity and infertility. There are three main reasons for this:

■   limited parameters evaluated
■   sperm analysis and its relationship with fertility
■   availability of other measures of fertility, especially in women
■   significance of foetal outcome
■   unavailability of histopathological assessment (except in animal studies)
■   limited data resources in predominantly non-English or non-peer-reviewed studies
■   absence of systematic structure–activity studies.

However, a small number of occupations have been identified as having an increased risk, or an association with effects on reproductive function (Hemminki *et al.* 1993; MacDonald and MacDonald 1988; Moore 1993; Roeleveld *et al.* 1990; Wilson 1973). These are shown in Table 9.5.

In female workers

For many years, the study of the interaction between occupational health and reproduction has focused on the female, especially with the recent increase in numbers of women joining the workforce. In many industries, this has caused a change in the way that hazards are controlled and workers are exposed. For example, policies such as exclusion of women and foetal protection are being re-examined and, in many cases, discarded.

There are a number of older and now emerging hazards on reproduction and development in the female (particularly female workers) and these are shown in Table 9.6.

In male workers

One important spin-off of the increased study of reproductive hazards in women has been an increase in attention on reproductive effects in men (Bonde *et al.* 1996). Effects of other agents on reproduction and development in the male (particularly male workers) are shown in Table 9.7 (Alcser *et al.* 1986; Figa-Talamanca *et al.* 2001; Lamb and Foster 1988; Whorton 1982; Winder 1989).

The archetypical reprotoxicant in the male: DBCP

The archetypal male reproductive hazard is dibromochloropropane (DBCP), a nematocide used on pineapples and other crops. Initial indications of long-term (some argue permanent) azoospermia in a small cohort of production workers only came to light in the most trivial circumstances (Babich *et al.* 1981).

Table 9.5 Occupations and reported effects on reproduction

Occupation	Reported effect(s)
Construction workers	CNS and musculoskeletal malformations
Dental technicians	Increased spontaneous abortion
Factory workers	Decreased viability at birth; low birth weight; CNS and musculoskeletal malformations
Food industry workers	Musculoskeletal malformations
Laboratory technicians	Increased abortion and birth defects; chromosomal damage; increased perinatal death rate
Lead manufacturing (males)	Reduced libido, lowered sperm counts, increased rate of miscarriage in partners
Operating theatre personnel	Decreased fertility; increased spontaneous abortion; low birth rate; possible birth defects
Plastics industry workers	CNS defects
Printers	Oemphacele/gastroschisis
Pulp and paper industry workers	Pregnancy complications
Registered nurses	Increased birth defects, especially cleft lip
Teachers	Oral clefts
Telephone operators	Oral clefts
Transportation/communications workers	Oral clefts

Animal data

Data is also available from animal studies alone. While such information should always be evaluated carefully with particular regard to mechanisms of toxicity and likely occupational exposures, it does often serve as a useful pointer to potential problems (see Table 9.8).

The list below has been compiled to indicate the numbers of chemicals or exposures that have been implicated in varying degrees of causing reproductive and/or developmental effects in humans or animals. It is not meant to be exhaustive, though it has been extracted from a large number of references.

Not all data reports strong associations. This may be due to several reasons, including:

- the effect has not been observed in all studies evaluated
- doses of chemicals used in animal studies were high enough to cause toxicity in other systems, suggesting a nonspecific effect

Table 9.6 Effects reported in female workers

Adverse outcome	Agent
Menstrual and other gynaecological disorders	Aniline, benzene, chloroprene, formaldehyde, inorganic mercury, polychlorinated biphenyls, styrene, toluene
Abortion or infertility	Anaesthetic gases, aniline, arsenic, benzene, cytotoxic drugs, ethylene oxide, formaldehyde, lead, 2,4-trichlorophenol
Decreased foetal growth, low birth weight or poor survival	Carbon monoxide, formaldehyde, polychlorinated biphenyls, toluene, vinyl chloride
Prematurity	Lead, thermal stress
Teratogenic effects	Hexachloroprene, radiation, organic mercury, vinyl chloride
Maternal death related to pregnancy	Beryllium, benzene
Cancer	Diethylstilboestrol, hepatitis B

Table 9.7 Effects reported in male workers

Adverse outcome	Agent
Decreased libido and impotence	Chloroprene, manganese, inorganic and organic lead, inorganic mercury, toluene diisocyanate, vinyl chloride
Testicular damage or infertility	Chloroprene, kepone, dibromochloropropane, inorganic and organic lead
Spermatoxicity	Carbaryl, carbon disulfide, cytotoxic drugs, dibromochloropropane, lead, radiation, thermal stress, triglycidylisocyanurate (TGIC), toluene diamine/dinitrotoluene mixture

- epidemiological studies reporting effects suffer from uncertain or inappropriate design.

With regard to occupational exposures, very few of these reported effects were from studies in which the amount of chemical was in the order of that found at the workplace. Chemicals that are likely to cause effects following workplace exposures above currently recommended exposure standards (such as threshold limit values (TLVs) or permissible exposure limits (PELs)) are marked with an asterisk (*).

Table 9.8 Effects reported in animals (but not humans)

Adverse outcome	Agent
Testicular damage or reduced male fertility	Benzene, benzo(a)pyrene, boron, cadmium, epichlorohydrin, ethylene dibromide, polybrominated biphenyl (PBB)
Spermatoxicity	Arsenic, chloroprene, ethylene glycol ethers, ethylene oxide, halothane, kepone, mercury, nitrous oxide, trichloroethylene, triethyleneamine
Embryolethal or foetotoxic effects	Chloroform, dichloromethane, ethylene dichloride, ethylene dioxide, inorganic mercury, nitrogen dioxide, PBB, selenium, tetrachloroethylene, thallium, trichloroethylene, vinylidine chloride
Teratogenic effects	Arsenic, benzo(a)pyrene, chlorodifluoromethane, chloroprene, monomethyl formamide, acrylonitrile, methyl ethyl ketone, tellurium, vinyl chloride
Transplacental carcinogenesis	Arsenic, benzo(a)pyrene, vinyl chloride

Of these agents, it is noteworthy that a number are in fairly widespread use in Australian industry. These include:

- Acrylonitrile
- 1,3-Butadiene
- Cadmium
- Carbon monoxide*
- Chromium trioxide
- Dichloromethane
- Dimethyl sulfoxide
- Epichlorhydrin
- Ethanol (acetaldehyde)
- 2-Ethoxyethanol*
- Ethylene dibromide*

- Ethylene dichloride
- Ethylene oxide
- Manganese
- Butyl methacrylate
- Methyl methacrylate
- 2-Methoxyethanol*
- Pentachlorophenol
- Phthalates
- Tetrachloroethylene
- 1,1,1-Trichloroethane
- 1,1,2-Trichloroethane

Action for reproductive hazards in the workplace

Controlling reproductive hazards

There is no particular reason why hazards that cause reproductive or developmental effects should be considered any different from hazards that cause any other health effect. There is probably a slightly greater emphasis on counselling and medical intervention, but such hazards should still be identified, evaluated, controlled and measured using all the options available to the health professional (Taskinen 1992; Thomas and Ballantyne 1990). While screening for susceptible individuals may be an option with its own problems (Jancovic and Drake 1996), in the main the available options for the control of occupational reproductive hazards are modelled on the hierarchy of controls (Geiser 1993), and are shown in Table 9.9:

In general, the choice of option needs consideration of a number of factors, including:

- the magnitude of the risk (to the extent that it is known)
- the intensity of likely exposures, and their significance to the risk
- the significance of possible health consequences
- any legal rights that apply.

The possible risks and recommended options for control of workplace reproductive hazards are shown in Table 9.10.

It may also be possible to implement some of these options at different stages of pregnancy. For example, transfer or job modification in the third trimester might be possible where work includes physiological stresses such as heat, standing for long periods, regular lifting or vibration.

Controlling new reproductive hazards

When a new association between exposure and adverse reproductive effects or infertility has been demonstrated in workplaces where reproductive hazards are not well controlled, there are a number of steps that should be taken (see Table 9.11).

Table 9.9 Options for control

Control options	Comments
1 Mandatory job transfer	Where it may be required by law (for example in the lead industry)
2 Hazard substitution	For example, use of hexachlorophene for hand washing
3 Rigorous control	Exposure should be controlled through engineering controls
4 Modification of work	The main aim of modification of work is to minimise exposure
5 Job transfer	Where the risk is high (for example with preparation of cytotoxic drugs, anaesthetic gases, organic mercurial compounds)
6 Personnel protection	Stringent use of personnel protection should only be used as an interim or short-term measure while better controls are developed, or where it is the only viable alternative (for example, the wearing of lead-containing shields in X-ray rooms)

Table 9.10 Options and risks for workplace action	
Risk	*Recommendation*
High risk for humans has been demonstrated at comparable levels	Eliminate or substitute hazard Transfer away from hazardous exposures
High risk is suspected at exposure levels encountered at the workplace	Eliminate or substitute hazard Transfer away from exposure Rigorous control of exposure
Minimal exposure to established or suspect agents	Eliminate or substitute hazard Transfer away is desirable but not obligatory Simple modification of work to eliminate or reduce contact

Table 9.11 Controlling reproductive hazards

Action to affected workers:
   advise on hazard and exposure
   remove from exposure
   if necessary, workers should undergo medical evaluation and treatment

The workplace should be investigated to identify sources of hazards and conditions of exposure

If appropriate, the workplace should be monitored to establish levels of exposure and efficiency of existing control measures

Workplace conditions should be improved by abolition or reduction of hazardous exposures using the hierarchy of controls

If possible, workers at risk should be monitored using specific measures of exposure or effect

As infertility is not a compensable condition, removal from exposure may cause economic hardship on the affected individual, unless relocation is negotiated with the employer.

Providing advice when occupational risks impact on reproductive processes

Providing advice on reproductive hazards and exposures is one of the most problematic questions confronting the occupational health and safety professional. Counselling may be required for a number of reasons, including:

- providing information about hazards that cause reproductive effects
- providing advice to workers who are considering having children

- providing advice to infertile workers
- providing advice to newly pregnant workers
- providing advice to newly pregnant workers who have had exposure to reproductive hazards
- providing advice to working mothers about breast feeding or other potential work-related risks to their infants
- counselling parents who have suffered an adverse reproductive outcome.

The most important principle in counselling is that **advice must only be provided by trained counsellors or appropriately trained medical practitioners**.
Following this, there are a number of general guiding rules:

- the nature of the hazard should be determined (known or suspected reproductive agent, animal or human evidence)
- the conditions of exposure established (intensity of exposure, duration of exposure, and whether exposure is one off, short-term or chronic)
- the interactions between hazard, exposure, effect and likelihood of effect
- the patient must be provided with all the facts, and given information that allows an informed judgement to be made.

Ultimately, the type of advice or counselling that can be given depends on:

- the nature of the agent
- the nature of the exposure.

Advice following a potentially foetotoxic exposure in the pregnant worker

There are really only two options – *reassurance* and *therapeutic abortion*. In line with the general guiding rules above:

- Determine whether significant exposure to a known human reproductive or developmental agent has occurred. If not, the patient should be reassured.
- If a potentially harmful exposure has occurred, the degree of exposure should be established with the employer, relevant agency or expert body, or by measurement of the agent under representative conditions of exposure. If not, the patient should be reassured.
- Even if data confirms some risk, reassurance may still be the correct choice. With many toxic exposures, the worst known risk in humans is spontaneous abortion, prematurity or small-for-date babies. With other exposures an increased risk of say a malformation (for example, cleft palate) still implies a low absolute risk.

A simplified approach to choosing options is shown in Table 9.12.
In general, therapeutic abortion should only be recommended in exceptionally rare circumstances (Sever 1994).

Table 9.12 Options following a potentially foetotoxic exposure in a newly pregnant worker

Type of agent	Type of exposure		
	Low to mild	Moderate	Significant
Suspect	Reassurance	Reassurance	Reassurance
Possible	Reassurance	Reassurance	Reassurance
Probable	Reassurance	Reassurance	Reassurance
Established	Reassurance	Reassurance	Reassurance/therapeutic abortion

Summary

There are many possible adverse health effects that may result from exposure to hazardous chemicals. Exposure of both males and females to such chemicals during the reproductive cycle can also have adverse outcomes on development of the embryo, foetus and infant (Kipen and Zuber 1994). Such effects manifest themselves through foetal death, malformation, retarded growth and organ dysfunction. Knowledge concerning these health effects, including the interaction of a host of factors (familial, lifestyle, medical, and so on) is extremely limited.

These unknowns indicate that the central issue in the protection of reproductive health and procreative capacity of working men and women is the management of uncertainty. Options for control of reproductive hazards are presently limited in scope and application (Paul 1997). Nevertheless, while policymakers and employers may never have complete information regarding the full extent of reproductive dysfunction and its causes, the provisions of occupational health and safety legislation are that employers must provide as safe a working environment as is reasonably practicable. It becomes axiomatic that a cautious preventive approach should be taken to limit worker exposure to chemicals to a level as low as reasonably practicable.

Further research efforts should be intensified (Polakoff 1990). These should concentrate on better and more efficient methods of identification (both experimental and epidemiological), increasing the precision of risk assessment processes (including its assumptions, methodologies and research needs), and optimising the usefulness of preventive strategies for exposed populations.

References and further reading

Alcser, K.H., Brix, K.A., Fine, L.J., Kallenbach, L.R. and Wolfe, R.A. (1986) Occupational mercury exposure and reproductive ability in male workers. *Am. Ass. Occup. Health Nurs. J.* 34: 277–279.

Babich, H., Davis, D.L. and Stottzky, G. (1981) Dibromochloropropane (DBCP): A Review. *Sci. Total Environ.* 17: 207–221.

Barlow, S.M. and Sullivan, F.M. (1982) *Reproductive Hazards of Industrial Chemicals.* London: Academic Press.

Bonde, J.P., Giwercman, A. and Ernst, E. (1996) Identifying environmental risk to male reproductive function by occupational sperm studies: logistics and design options. *Occup. Environ. Med.* 53: 511–519.

Chamberlain, G. (ed.) (1984) *Pregnant Women at Work*. London: Royal Society of Medicine/Macmillan.

Christian, M.S., Galbraith, W.M., Voytek, P. and Mehlman, M.A. (eds) (1989) *Assessment of Reproductive and Teratogenic Hazards*. Advances in Modern Environmental Toxicology III. Princeton: Princeton Scientific Publishers.

Clayson, D.B. (1985) Methods for predicting human birth defects. In: *Prevention of Physical and Mental Congenital Defects, Part C: Basic and Medical Science, Education and Future Strategies*. New York: Alan R. Liss, pp. 197–203.

Conover, E. (1994) Hazardous exposures during pregnancy. *J. Obs. Gynecol. Neonat. Nurs.* 23: 524–532.

Dhindsa, D.S. and Bahl, O.P. (1986) *Molecular and Cellular Aspects of Reproduction*. Kluwer Academic.

Dixon, R.L. (ed.) (1985) *Reproductive Toxicology*. Target Organ Toxicology Series. New York: Raven Press.

Ferin, M. and Jewelewicz, R. (1993) *Menstrual Cycle: Physiology, Reproductive Disorders, and Infertility*. Oxford: Oxford University Press.

Figa-Talamanca, I., Traina, M.E. and Urbani, E. (2001) Occupational exposures to metals, solvents and pesticides: recent evidence on male reproductive effects and biological markers. *Occup. Med.* 51: 174–188.

Frazer, L. and Hage, M.L. (1997) *Reproductive Hazards of the Workplace*. New York: John Wiley and Sons.

Geiser, K. (1993) Protecting reproductive health and the environment: Toxics use reduction. *Environ. Health Persp.* 101 (Suppl. 2): 221–225.

Greaves, W.W. (1994) Reproductive hazards. In: McCunney, R.J. (ed.) *A Practical Approach to Occupational and Environmental Medicine*, 2nd edn. Boston: Little, Brown and Com, pp. 447–464.

Gwatkin, R.B.L. (ed.) (1992) *Genes in Mammalian Reproduction*. New York: John Wiley and Sons.

Hardy, D.M. and Garbers, D.L. (2001) *Fertilization*. New York: Elsevier Science and Technology.

Hemminki, K. and Vineis, P. (1985) Extrapolation of the evidence on teratogenicity of chemicals between humans and experimental animals: chemicals other than drugs. *Teratogen. Carcinogen. Mutagen.* 5: 251–318.

Hemminki, K., Axelson, O., Niemi, M.L. and Ahlborg, G. (1983) Assessment of methods and results of reproductive occupational epidemiology: spontaneous abortions and malformations in the offspring of working women. *Am. J. Ind. Med.* 4: 293–307.

Hemminki, K., Kyyrönen, P. and Lindohm, M.-L. (1993) Spontaneous abortions and malformations in the offspring of nurses exposed to anaesthetic gases, cytostatic drugs and other potential hazards, based on registered information of outcome. *J. Epidemiol. Comm. Health* 39: 141–147.

Hemminki, K., Vanio, H. and Sorsa, M. (eds) (1985) *Occupational Hazards and Reproduction*. London: Taylor and Francis.

Jankovic, J. and Drake, F. (1996) A screening method for occupational reproductive health risk. *Am. Ind. Hygiene Ass. J.* 57: 641–649.

Khera, K.S. (1987) Maternal toxicity in humans and animals: effects on foetal development and criteria for detection. *Teratogen. Carcinogen. Mutagen.* 7: 287–295.

Kipen, H.M. and Zuber, C. (1994) Occupational and environmental impacts on reproductive health. *Ann. N.Y. Acad. Sci.* 736: 58–73.

Kirsch-Volders, M. (1981) *Mutagenicity, Carcinogenicity and Teratogenicity of Industrial Pollutants*. New York: Plenum Press.

Kolb, V.M. (1993) *Teratogens: Chemicals Which Cause Birth Defects*. Elsevier.

Lamb, J.C. and Foster, P.M. (1988) *Physiology and Toxicology of Male Reproduction*. Amsterdam: Elsevier Science and Technology.

Landrigan, P.J., Boffetta, P. and Apostoli, P. (2000) The reproductive toxicity and carcinogenicity of lead: A critical review. *Am. J. Ind. Med.* 38: 231–243.

Lemasters, G.K. and Selevan, S.G. (1984) Use of exposure data in occupational reproductive studies. *Scand. J. Work Environ. Health* 10: 1–6.

Levine, R.J. (1984) Methods for detecting occupational causes of male infertility: reproductive history vs. semen analysis. *Prog. Clin. Biol. Res.* 160: 363–373.

London Sunday Times (1979) *Suffer the Children: The Story of Thalidomide*. London: Penguin.

MacDonald, A.D. and MacDonald, J.C. (1988) Foetal death and work in pregnancy. *Br. J. Ind. Med.* 45: 148–157.

Manassiev, N. and Whitehead, M.I. (2002) *Female Reproductive Health*. Boca Raton: CRC Press.

Manson, J.M., Zenick, H. and Costlow, R.D. (1982) Teratology test methods for laboratory animals. In: Hayes, A.W. (ed.) *Principles and Methods of Toxicology*. New York: Raven Press, pp. 141–185.

Mattinson, D.R. (1983) The mechanisms of action of reproductive toxins. *Am. J. Ind. Med.* 4: 65–79.

McBride, W.G. (1977) Thalidomide embryopathy. *Teratology* 16: 79–82.

Messing, K. (1999) One-eyed science: Scientists, workplace reproductive risks and the right to work. *Internat. J. Health Serv.* 29: 147–165.

Meyers, V.K. (ed.) (1988) *Teratogens: Chemicals which Cause Birth Defects. Studies in Environmental Science* 31, Amsterdam: Elsevier.

Moore, R.M. (1993) An overview of occupational hazards among veterinarians, with particular reference to pregnant women. *Am. Ind. Hygiene Ass. J.* 54: 113–120.

Morello-Frosch, R.A. (1997) The politics of reproductive hazards in the workplace: class, gender, and the history of occupational lead exposure. *Internat. J. Health Serv.* 27: 501–521.

O'Rahilly, R. and Muller, F. (2001) *Human Embryology and Teratology*, 3rd edn. New York: John Wiley and Sons.

OTA (1985) *Reproductive Hazards in the Workplace*. Washington: Office of Technology Assessment, US Government Printing Office.

Paul, M. (ed.) (1993) *Occupational and Environmental Reproductive Hazards: A Guide for Clinicians*. Baltimore: Williams and Wilkins.

Paul, M. (1997) Occupational reproductive hazards. *Lancet* 349: 1385–1388.

Pinon, R. (2002) *Biology of Human Reproduction*. University Science Inc, Enfield, NH.

Polakoff, P.L. (1990) Prevention of reproductive disorders requires more research, vigilance. *Occup. Health Safety* 59: 37, 50.

Pryor, J.L., Hughes, C., Foster, W., Hales, B.F. and Robaire, B. (2000) Critical windows of exposure for children's health: The reproductive system in animals and humans. *Environ. Health Persp.* 108: 491–503.

Richardson, M. (ed.) (1993) *Reproductive Toxicology*. New York: John Wiley and Sons.

Robboy, S.J. and Anderson, M.C. (2001) *Pathology of the Female Reproductive Tract*, 4th edn. New York: Churchill Livingstone.

Roeleveld, N., Zielhuis, G.A. and Gabreels, F. (1990) Occupational exposure and defects of the central nervous system in offspring: review. *Br. J. Ind. Med.* 47: 580–688.

Schardien, J.L. (2000) *Chemically Induced Birth Defects*, 3rd edn. New York: Marcel Dekker.

Schrag, S.D. and Dixon, R.L. (1985) Occupational exposures associated with male reproductive dysfunction. *Ann. Rev. Pharmacol. Toxicol.* 25: 567–592.

Scialli, A.R. (1991) *Clinical Guide to Reproductive and Developmental Toxicology.* Boca Raton: CRC Press.

Sever, L.E. (1994) Congenital malformations related to occupational reproductive hazards. *Occup. Med.* 9: 471–494.

Shepard, T.H. (ed.) (2001) *Catalog of Teratogenic Agents,* 10th edn. Baltimore: Johns Hopkins University Press.

Stephens, T.D. and Fillmore, B.J. (2000) Hypothesis: Thalidomide embryopathy-proposed mechanism of action. *Teratology* 61: 189–195.

Tas, S., Lauwerys, R. and Lison, D. (1966) Occupational hazards for the male reproductive system. *Crit. Rev. Toxicol.* 26: 261–307.

Taskinen, H. (1992) Prevention of reproductive hazard at work. *Scand. J. Work Environ. Health* 18 (Suppl. 2): 27–29.

Thiel, R. and Klug, S. (1997) *Methods in Developmental Toxicology and Biology.* Blackwell Publishers.

Thomas, J.A. and Ballantyne, B. (1990) Occupational reproductive risks: sources, surveillance, and testing. *J. Occup. Med.* 32: 547–554.

Wang, C. (1999) *Male Reproductive Function.* Kluwer Academic.

Whorton, M.D. (1982) Male occupational reproductive hazards. *West. J. Med.* 137: 521–524.

Williams, L. (1990) *Reproductive Health Hazards in the Workplace.* Baltimore: Lippincott, Williams and Wilkins.

Wilson, J.G. (1973) *Environment and Birth Defects.* New York: Academic Press.

Winder, C. (1989) Reproductive and chromosomal effects of occupational exposure to lead in the male. *Reprod. Toxicol.* 3: 221–233.

Winder, C. (2000) Reproduction, development and work. In: Bohle, P. and Quinlan, M. (eds) *Managing Occupational Health and Safety: A Multidisciplinary Approach,* 2nd edn. Melbourne: Macmillan, pp. 410–427.

Chapter 10

Genetic toxicology

A.D. Mitchell[1]

Introduction

Genetic toxicology involves the identification of agents that interact with nucleic acids to alter the hereditary material of living organisms; agents that produce such alterations are termed genotoxic. Genetic toxicology is both an investigative science and an applied science which incorporates principles from the fields of genetics, microbiology, cell biology, molecular biology, biochemistry, and toxicology to develop and utilise tests that are predictive of mutagenesis and carcinogenesis (Moutschen 1985).

Genotoxicology is the study of physical and chemical agents interacting with genetic material and with the short- and long-term consequences of such interactions. Researchers in genotoxicology work on the 'three Rs' of DNA: replication, recombination and repair. The ability of cells to maintain genomic integrity is vital for cell survival and proliferation. Lack of fidelity in DNA replication and maintenance can result in deleterious mutations leading to cell death or, in multicellular organisms, cancer (Arelett and Cole 1988; Cairns 1981; Holbrook and Fornace 1991; Shackelfod *et al.* 1999). Alterations, if they occur in the proto-oncogenes or tumour suppressor genes that are involved in controlling cell growth or differentiation, may lead to the development of cancer (Venitt 1987; Weisberger and Wynder 1984). Genetic alterations in germ cells may also lead to reproductive failure or genetic disorders in subsequent generations (Keshaya and Ong 1999; Kirsch-Volders 1984).

Thus, an understanding of these disciplines, and especially the basic genetic processes that may be influenced by genotoxicants, is necessary not only for the development and appropriate use of assay systems that measure the effects of

---

[1] Dr Ann Michell died unexpectedly in July 1999, and additional updates of this text are provided by the editors. The editors dedicate this revision to her memory.

potentially genotoxic agents, but also to interpret the significance of results that are obtained.

During the relatively short time that the field of genetic toxicology has been recognised, extensive efforts have been devoted to defining and evaluating the utility of numerous testing approaches for predicting heritable genetic effects and cancer (Venitt and Parry 1984). Today, because genetic toxicology tests are well defined, assess effects directly related to those observed in humans, and are relatively efficient and economical, they are mandated as an initial step for regulatory assessments (Gatehouse and Kirkland 1990). To provide an introduction to genetic toxicology, the rapid development of the field will be summarised, some of the more extensively used tests, as well as the current application of genetic toxicology approaches for human monitoring and for product registration, will be described, and references to suggested reading will be provided for those who may wish to pursue this topic further.

Background to DNA, genes and genetics

Structure of DNA

Deoxyribonucleic acid (DNA) and the processes involved in cell activity and replication are, like respiration, fundamental to all life on earth. DNA is found only in the cell nucleus where, combined with groups of proteins such as polymerases and histones, it forms the fundamental genetic molecule. DNA consists of nucleotides connected to two extremely long phosphate-deoxyribose chains.

The phosphate-deoxyribose chains are on the outside of the DNA structure, with the nucleotides facing inwards, and with one nucleotide linked across the centre to another by hydrogen bonding (this is called a 'base pair'). These hydrogen bonds are specific to each nucleotide, with adenine in one strand linking to thymine in the other, and, in a similar fashion, between guanine in one strand linking to cytosine in the other (as shown in Figure 10.1).

Hydrogen bonds are weaker bonds than the conventional interatomic ionic or covalent bonds, and are prone to enzymic separation and access by toxicants such as alkylating agents (or other methods of separation in the laboratory, such as heat or nucleolytic chemicals). Breaking these bonds separates the two chains (geneticists talk about 'unzipping' the two chains).

The term 'base pairs' is used to describe this unit of DNA structure. Adenine, thymine, guanine and cytosine (A, T, G, and C) molecules are the main (in some cases only) nucleic acid bases found in DNA. The sequences of the four bases, A, T, G, and C, determine how an individual differs from other individuals and from other living things. The bases are paired, A with T, and G with C. The bases can be arranged in the nucleotide chain in any sequence, but the hydrogen bonding arrangements are such that the nucleotide sequence in one chain automatically determines the sequence in the complementary chain.

The sequence of nucleotide bases on the DNA structure contains coded information that is required to construct a living organism and to direct the way it functions. This also has consequences of very great importance in the process of replication of the DNA molecule.

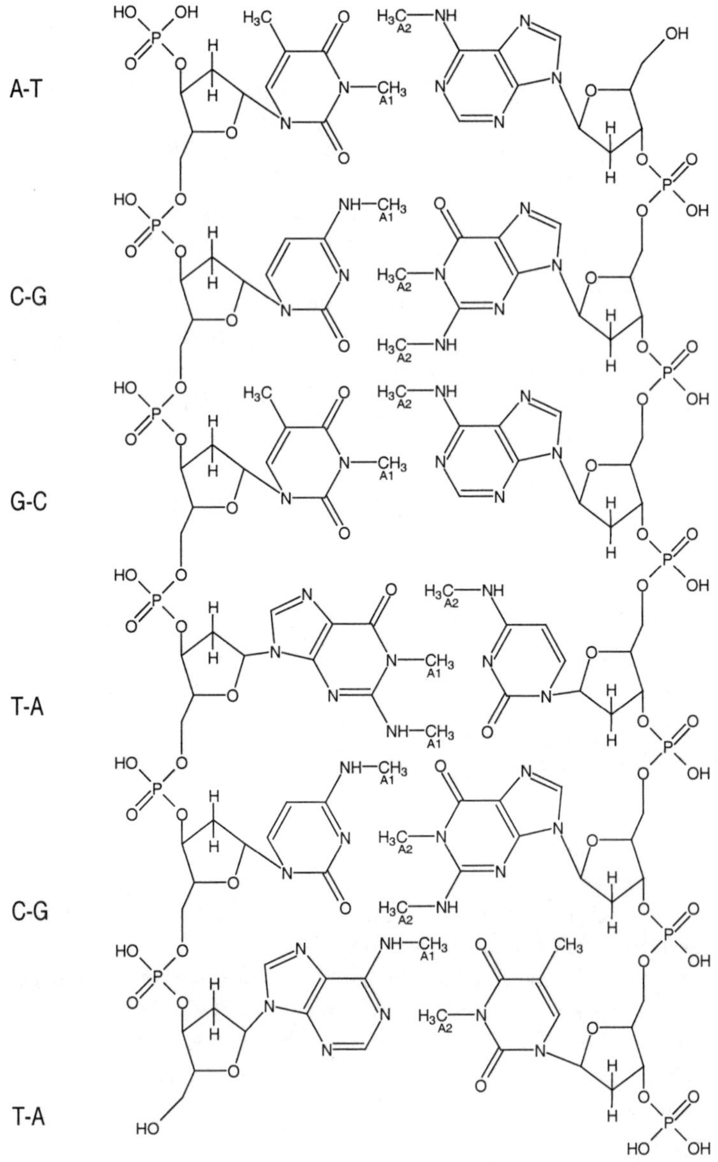

Part of a polynucleotide chain in DNA showing two opposing strands connected by hydrogen bonds, designated by the subscripts A1 (NH-CH₃----O or N-CH₃----O), and

A2 (N□----H₃C or CH₃□----N).

A: Adenine; T: Thymine; G: Guanine; C: Cytosine

Figure 10.1 Structure of DNA.

Evidence that the two-stranded polynucleotide exists in the form of a double helix was first put forward in 1953 (Watson and Crick 1953), and this has since been confirmed by many different workers (see Figure 10.2).

Organisational structures for DNA

Therefore, DNA is a huge molecule. It is a twin-stranded double helix chain with backbone of deoxyribose sugar units and containing twin nucleic acids linking across the two chains (Watson 1969). The DNA in each human cell contains three billion (3,000,000,000) of these base pairs. Such a structure cannot be loosely disorganised, floating in the fluid of the cell nucleus. It has ancillary support structures and a systematic molecular organisation. This hierarchy of structures begins with the double chain of nucleotides, organised into genes, structured into DNA coils and packaged into chromosomes.

The sequence of base pairs is the nuclear code. Discrete parts of the DNA contain the code for each specific protein in the body. A specific sequence of base pairs that codes for a protein is called a gene.

There are about 100,000 genes in the DNA of human nucleated cells. However, most of the DNA (about 98%) does not appear to have any function.

In normal cellular life, the genetic material is dispersed throughout the nucleus in an undissociated state called chromatin, where it directs the activities in the cell.

At other times, for example, during cellular division, the genetic material is organised into coiled structures which then condenses itself into rod-like structures called chromosomes.

DNA is therefore a complex and extremely long molecule, organised into genetic material containing genes and other sequences. The 100,000 or so genes

Diagrammatic representaion of the DNA structure

The two ribbons symbolise the two phosphate sugar chains, and the horizontal rods, the pairs of nuleotide bases holding the chains together.

The vertical line marks the axis of the helix

Figure 10.2 Diagrammatic representation of the structure of DNA.

organised along the DNA are distributed among the distinctive structures known as chromosomes. A complete copy of an organism's DNA is termed the 'genome'.

Chromosomes contain two copies of the DNA molecule, in structures called chromatids. These are generally two tightly wound structures of DNA with their associated structures linked together at a centromere (usually in the shape of an χ).

All living organisms have a number of chromosomes each with different properties and functions. Each individual human chromosome is made up largely of a tightly coiled double strand of DNA. With a few exceptions, each somatic cell in the human body contains 22 pairs of chromosomes, plus either an XY or XX combination, making a total of 23 pairs of chromosomes. The chromosomal constitution of a cell is called the karyotype.

The cell cycle

The nucleus contains the genetic material of the cell (including its DNA). The genetic material holds the 'blueprint' and instruction manual of the cell, directing its development, growth and function. This blueprint provides all the instructions for the cell to construct proteins for structural or other purposes. The genetic material in all cells is based on a specific biochemical structure called deoxyribonucleic acid (or DNA). Specific regions of each DNA chain, referred to as genes, contain the information for making one protein. The DNA is associated with other structures such as proteins like histones and enzymes such as polymerases.

As well as directing cellular activities, the DNA is also involved in replicating itself.

The process of development and growth of the cell is (obviously) a significant part of cell life. Some, but not all, cells can have one further biological function to replicate themselves (by cell division). For example, cells such as neurones have little or no capacity to divide, being in their final differentiated state; while others, such as liver, skin and bone marrow cells, have the property of continuous cell division. The rate of cell division varies considerably in such cells. The normal activities of the cell are directed and controlled by the biochemical and genetic mechanisms that convert the genetic code of the nuclear DNA into instructions and actions for the cell to carry out. These mechanisms also include the process of cell division.

Cell division is an important mechanism for both normal and abnormal cells (for example, cancer cells). While replication of DNA was outlined above, the accompanying cellular processes in cell replication are more complex, and beyond this chapter. However, a short outline of the cell cycle is shown below. The cell cycle is that process by which cells grow, divide, proliferate and differentiate (Johnson and Walker 1999). Initially it was thought that the cell cycle was composed of only two stages, mitosis or cell division, and interphase (the interval between mitoses).

However, interphase is now known to be a period of intense biochemical activity, divided into a number of distinct stages (see Figure 10.3).

A cell is initially in the G_0 or G_1 phase. In the G_0 phase the cell may be dormant or in normal activity, but not considering the process of cell division,

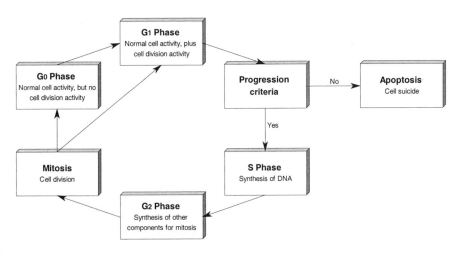

Figure 10.3 The cell cycle.

although a cell in the G_0 phase may at a later time enter the G_1 phase. If in the G_1 phase, then as well as its normal activities, the cell is preparing itself for cell division, the chromosomes are greatly extended, and the cell undergoes enzymatic synthesis of proteins and nucleotides, the precursors of the DNA molecule. However, before the cell can enter into cell division, it must comply with in-built progression criteria, such as levels of growth factors or oncogene activity or lack of DNA damage. If criteria are not met, apoptosis (programmed cell death or loosely, cell suicide) may occur.

Once cell division criteria are met, the cell proceeds to DNA synthesis (S phase). During S phase, the stage generally most sensitive to mutagens, nucleotides are incorporated during DNA synthesis and replication of the whole genome occurs. After the completion of DNA synthesis, each chromosome consists of two daughter chromatids, and the cell enters G_2 phase, preparatory to cell division, where synthesis of the other components of cell division occurs. Cell division (mitosis) then follows. Offspring cells may then be in G_0 or G_1 phases. If in G_0, the cells will no longer divide (some normal cells, after a certain number of divisions, may leave the cycle and become nondividing, terminally differentiated cells) or the cells may still be able to divide, in which case they are in a G_1 phase.

For most *in vitro* mammalian cell mutagenesis and cytogenetics tests, continuous cell cultures, without a G_0 stage, are used. And for tests using human lymphocytes, mitogenic stimulation is used to induce G_0 cells to re-enter the cell cycle and, after a lengthy G_1 stage, to resume the process of cell replication.

Cell division (mitosis)

At the end of the cell cycle and just prior to cell division (between G_2 and G_0/G_1), the genetic material is condensed into chromosomes containing two copies of the genetic material (that is, two chromosomes containing two chromatids each).

The total number of pairs of chromosomes is different from species to species.

However, each somatic cell contains two sets of chromosomes. This is called the diploid number and as there are two sets, this is usually denoted as *2n*.

When a cell reproduces, all of its DNA molecules are duplicated, so that when the nucleus divides, each of its offspring cells receives the same genetic information as the original or mother cell, that is, one chromosome containing two chromatids. This is known as *mitosis* or mitotic division (see Figure 10.4).

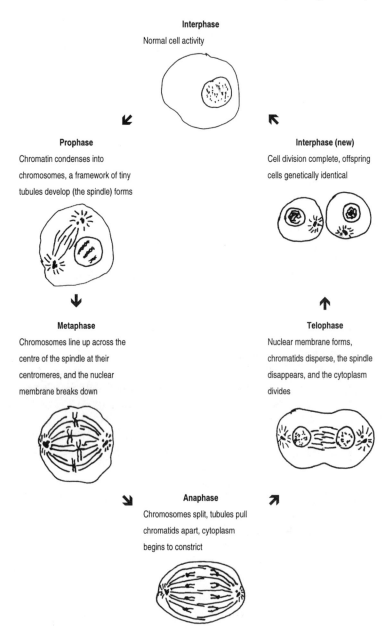

Figure 10.4 The stages of mitosis.

Cell division (meiosis)

Mitosis is the process whereby offspring cells that divide receive the same number of chromosomes from the parent cell. That is at some point in the cell cycle, the 23 pairs of chromosomes are replicated, and each offspring cells gets a set on cell division. This happens for all body (somatic) cells.

The only cells that contain only one set of chromosomes are the germ cells (ova or sperm cells). In such cases cell division continues until the germ cells only contain one set of chromosomes; this is usually denoted as the haploid number (or n). The process of producing n haploid cells is called meiosis. The $2n$ diploid number of chromosomes is reformed in the fertilised egg, when one chromosome of every pair is inherited from each parent.

The study of adverse effects on genetic material

The study of the molecular structure and function of genetic material is called cytogenetics and the study of the toxicology of genetic material is called genotoxicology (Preston and Hoffmann 2001).

While direct heritable damage to genetic material is termed a mutation (and the process of forming mutations is called mutagenesis), it is also possible that heritable genetic damage can be caused indirectly, for example, by damage to proteins in and around the DNA, impairment of repair mechanisms, hormonal changes. Agents causing such effects are called clastogens, and the process clastogenicity. The terminology of cytogenetics and genotoxicology is shown in Figure 10.5 (Heflich *et al.* 1991).

Changing the DNA sequence of bases (either by inserting or deleting a base pair) can change its function. If inherited in an offspring cell, these changes are called mutations. Such changes are the basis for a variety of heritable traits including eye colour and body type. Unfortunately, small changes in the DNA sequence of a gene, such as substitution of one base for another or omission or repetition of a small segment of the sequence, can change the gene's function and result in a developmental or metabolic disorder.

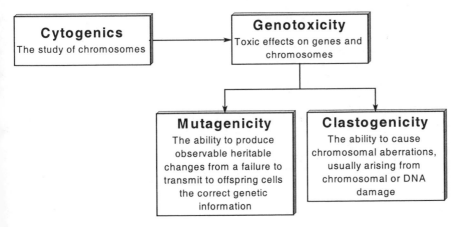

Figure 10.5 The terminology of cytogenetics and genotoxicology.

Mutagenesis

A mutation is a spontaneous or induced change in genetic material that produces observable heritable changes from a failure to transmit to offspring cells the correct genetic information. The basis of these changes may be at the molecular level (Josephy 1996), with alteration in the sequence of the genetic code, or it may be at the structural level with deletion of alterations in chromosomal structures (see Figure 10.6).

Essentially all basic genetic toxicology test systems measure the induction of DNA damage and/or resulting mutations in DNA: with the exception of some RNA viruses this is the molecule containing hereditary information for all living organisms. Damage to DNA may arise from a range of lesions, including single and double strand breaks in the phosphate-deoxyribose backbone, cross links between DNA bases and proteins, and adduct addition to the DNA bases (see Table 10.1).

Formation of damage to DNA may not lead to mutation. Endogenous DNA repair mechanisms must also be considered. Again, there are a range of DNA repair mechanisms (see Table 10.2).

If DNA is repaired, it may:

- be correctly repaired with no genetic consequences
- lead to cell death, again with no genetic consequences
- be incorrectly repaired at a site that is not of genetic consequence (for example, not part of a gene sequence)
- be incorrectly repaired at a site of genetic consequence and be replicated with the damage misrepaired.

Only the last of these leads to mutations, defined as DNA alterations that are propagated through subsequent generations of cells or individuals (Douglas 1980). Mutations do occur spontaneously and any increase in the rate of mutation

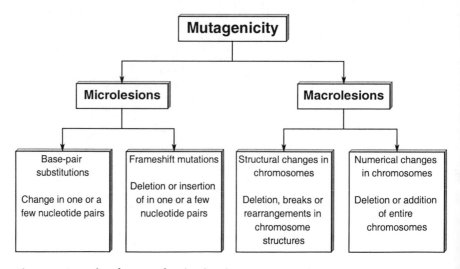

Figure 10.6 Classification of molecular changes in DNA that result from mutation.

Table 10.1 Types of DNA damage

Type of DNA damage	Type of agent
Alkylation of nucleic acid bases	Alkylating agents
Damage to the nucleic acid bases	X-rays
Radical formation	Electromagnetic radiation
Formation of apurininc or apyrimidinic sites	Alkylating agents
Single strand breaks	Ionising radiation
Damage to the phosphate-deoxyribose backbone	Alkylating agents
Double strand breaks	Ionising radiation
Intercalations	Acridines
Interstrand cross links	Alkylating agents
DNA–protein cross links	X-rays and alkylating agents

Source: Modified from Heflich *et al.* 1991; Kirsch-Volders 1984; Preston and Hoffmann 2001; Venitt and Parry 1984.

Table 10.2 Types of DNA repair

Direct reversal of damage to DNA
Base excision repair
Nucleotide excision repair
Double strand break repair
Mismatch repair

induction is the consequence of unusual circumstances. While biological diversity may be considered to be a result of the accumulation of beneficial mutations over time, it is generally accepted that many mutations are deleterious, and it is the concern of many genetic toxicologists that the human genome received from our ancestors be none the worse when it is passed on to future generations.

Mutations are now generally classified into three types.

- Point or gene mutations, which are changes in nucleotide sequence at one or a few coding segments within a gene.
- Chromosomal mutations, which are structural aberrations or morphological alterations in the gross structure of chromosomes.
- Genomic mutations, which are changes to the numbers of chromosomes.

While point or gene mutations are identified in any organism (including microorganisms), both chromosomal mutations and genomic mutations can only be identified in eukaryotic structures.

Carcinogenesis

Carcinogenesis is known to proceed through a number of steps, and the first step, initiation, is generally believed to involve mutation in somatic cells (Morelly

2000). A somatic cell mutation would be of no consequence if it resulted in death of the cell or if the mutated cell failed to divide, but it would be of consequence if it occurred in a dividing cell, or if a second mutation (for example, in an onco-gene) resulted in resumption of cell division of an otherwise nondividing mutated cell. Support of the role of mutagenesis in carcinogenesis includes the clonal origin of tumours, the association of specific chromosomal abnormalities with certain cancers, information about genetically determined cancer-prone conditions, and the number of carcinogens that have been shown to be mutagens (Lawley 1994). However, some carcinogens, including hormones and inert materials, have not been shown to be mutagenic, and it is thought that they may act through nongenotoxic, 'epigenetic' mechanisms such as alterations in cell regulatory mechanisms, including the induction of proliferation in otherwise nondividing cells.

A significant number of carcinogens are mutagens, but correlations between carcinogenicity and mutagenicity are not complete, as was predicted by Malling and Chu (1974). The degree of correlation observed between these two endpoints is highly dependent upon the test system used, the precision of testing, the class of chemicals tested, and the methods used to evaluate the results. For example, a number of rodent carcinogens that have yielded 'negative' results in in vitro genetic toxicology tests are insoluble in vitro (Tennant et al. 1987).

However, because resources are insufficient for testing more than a small per-centage of commercially important chemicals in long-term in vivo cancer bio-assays, and because a reduction in the requirements for testing in animals is a generally accepted goal, over 200 short-term relatively inexpensive genetic toxi-cology tests have now been developed to assess the genetic hazards of chemicals and to assist in predicting the results of long-term testing (Preston and Hoffmann 2001).

Development of genetic toxicology

In comparison to toxicology, the science of poisons, which can trace its origins to antiquity (Gallo 2001), the origin of genetics dates only to the rediscovery of the works of Gregor Mendel, shortly after the turn of the last century (Henig 2000). Genetic toxicology is an even newer discipline; the first published study on radia-tion-induced mutations, in Drosophila, were those of H.J. Muller, in the late 1920s (Muller 1927); not until after World War II could the first work on chemically induced mutations (by mustard gas) be published by Charlotte Auerbach (Wassom 1989).

At least another decade elapsed before geneticists recommended that tests for mutagenicity be developed as a routine component of toxicology testing because of concerns that some chemical products of benefit to industrialised society may adversely affect human health and the environment. During the 1960s, several lab-oratories focused on developing methods suitable for testing chemicals for muta-genicity, and by the end of the decade, the first in vivo tests to assess heritable effects had been defined (Butterworth 1979; Styles 1977).

The cost of a long-term, repeated-dose carcinogenicity bioassay is expensive. Also, the tests are lengthy, often taking over 3 years to complete. It is impossible to envisage spending such time and money testing all the chemicals in current use. Short-term tests offer one solution.

Substantial evidence indicates that one step in carcinogenesis is damage to or mutation of DNA. The first short-term test was the Ames test, introduced by Bruce Ames and colleagues in the early 1970s (Ames *et al.* 1973). The test relies on the fact that many carcinogens initiate cancer through damage to DNA (that is, a mutagenic event), and the Ames test sought to measure mutagenicity. The test relies on mutated strains of the bacteria *Salmonella typhimurium* to back-mutate to a form that will grow on special media. Over the last 30 years, the Ames test has become more sophisticated, with the number of strains being standardised, and with refinement of test conditions.

Since 1973 a number of short-term genotoxicity tests have been developed. The short-term tests currently available can be broadly classified into three major groups:

- those that detect gene mutations and chromosome aberrations
- those that are capable of detecting transformation of mammalian cell *in vitro*
- those that measure interactions of chemicals with critical molecules (primarily DNA).

The identification of the field as 'genetic toxicology' and its exponential growth occurred during the 1970s when numerous methods to evaluate the genotoxic effects of chemicals were developed and published. It was discovered that many chemicals underwent metabolic changes before they were capable of inducing the processes leading to cancer (Williams 1983). This led to the development of *in vitro* techniques for bringing about such metabolic transformations and, consequently, enormous progress was made in the fields of mutagenesis and carcinogenesis (Brusick 1994). It soon emerged that there was a reasonably strong correlation between the carcinogenic activity of a chemical, particularly in rodents, and its mutagenic properties, as demonstrated in a wide variety of *in vitro* and *in vivo* experimental systems involving bacteria, yeasts, insects, rodents, and mammalian cells in tissue culture (Ames *et al.* 1975; Hofnung and Quillardet 1986).

Of great importance to many of these approaches was the development of *in vitro* metabolic activation systems that permitted *in vitro* tests to reflect more closely *in vivo* metabolism (Malling 1971). Also during this time there were substantial efforts to enhance awareness of the public and governmental agencies to the potential threat of chemicals to the human gene pool, which resulted in the promulgation of regulations mandating that industry evaluate the potential genotoxic hazards of their products.

Objectives of genetic toxicology

Genetic toxicology testing has two objectives:

- The identification of agents capable of inducing heritable effects – the initial goal.

- The prediction of carcinogenicity – a goal that gained momentum in the mid-1970s following the publication of test results which suggested that carcinogens are mutagens (Ames 1979; Bridges 1976; McCann *et al.* 1975a, 1975b; Sugimura *et al.* 1976).

And, indeed, when most of the currently available genetic toxicology approaches were developed, their efficacy was first demonstrated by testing a group of classical mutagens that are also carcinogens.

A further aim, of reducing the numbers of animals required for conventional toxicity testing, is also one objective of genotoxicity.

The spectrum of genetic toxicology tests

The virtual universality of DNA as the genetic material and the observation that chemicals that induce genetic effects in one species or test system frequently induce similar effects in other species or test systems provides the basis for a diversity of test systems for bacteria, yeast and fungi; plants; submammalian animal species, for example, *Drosophila*; mammalian cells in culture; and rodents (for assessing the mutagenic activity of chemicals). Mutations can range from a single nucleotide change to changes in chromosome number. Thus, they may be classified according to complexity or the extent of the genetic alteration, as gene mutations, chromosomal mutations, or genomic mutations.

A wealth of literature is available for those who wish to pursue further the development and use of specific test systems, including texts on genetic toxicology (Brusick 1980; Li and Heflich 1991), the series *Chemical Mutagens* (Hollaender and de Serres 1971–1983), and a number of established journals in the field, including *Environmental and Molecular Mutagenesis*, *Mutagenesis*, and *Mutation Research*. In the latter may be found the series of US Environmental Protection Agency (EPA) Gene-Tox reviews of specific test systems.

The main genotoxicity assays used are shown in Table 10.3.

To illustrate the diversity of test systems available, representative examples will be briefly described in this chapter. Descriptions of other test systems that have been developed but that are not widely used may be found in the texts by Vennit and Parry (1984), Gatehouse and Kirkland (1990), Heflich *et al.* (1991), Douglas (1980), Brusick (1980, 1984) and Hollaender and de Serres (1983). They include tests in yeast, fungi (for example, *Neurospora*), and plants, measurements of DNA adducts (DNA binding), and the use of host-mediated assays. The tests that are summarised in this chapter are also detailed in these references.

Tests for gene mutations

A gene is the simplest functional unit in a DNA molecule, and gene (or point) mutations are changes in the nucleotide sequence at one or a few coding segments. Base pair or base substitution mutations occur when one nucleotide base is replaced with another; frameshift mutations occur through the addition or deletion of one or more bases, which alters the sequences of bases in DNA, and, hence, the reading frame in RNA. These changes may occur at the site of the original muta-

Table 10.3 Utility and application of assays

Assays that may be used for mutagen and carcinogen screening
Salmonella typhimurium reverse mutation assay
Escherichia coli reverse mutation assay
Gene mutation in mammalian cells in culture
Gene mutation in *Saccharomyces cerevisiae*
in vitro cytogenetics assay
Unscheduled DNA synthesis *in vitro*
in vitro sister chromatid exchange assay
Mitotic recombination in *Saccharomyces cerevisiae*
In vivo cytogenetics assay
Micronucleus test
Drosophila sex-linked recessive lethal test

Assays that confirm in vitro *activity*
In vivo cytogenetics assay
Micronucleus test
Mouse spot test
Drosophila sex-linked recessive lethal test

Assays that assess effects on germ cells and that are applicable for estimating genetic risk
Dominant lethal assay
Heritable translocation assay
Mammalian germ cell cytogenetic assay

Source: Reproduced with permission from OECD 1986.

tion or at a second site on the chromosome. Missense mutations are gene mutations that specify the insertion of an 'incorrect' amino acid into a polypeptide and usually result in impaired function of the affected cell or organism. Nonsense mutations introduce terminator codons into the DNA, resulting in the premature cessation of polypeptide synthesis, which is usually associated with complete loss of function of the altered gene. Forward mutations lead to loss of normal function of a gene product; reverse, or back, mutations restore the normal function of the gene.

Using molecular biology techniques, an increasing number of gene mutations are being identified in the human population, and molecular genetic techniques have now become sufficiently powerful to allow the routine identification of the specific DNA sequence changes responsible for mutant phenotypes. But, as yet, these techniques are, for the most part, used to enhance understanding of molecular genetic mechanisms rather than as genetic toxicology test systems. Tests for gene mutations use phenotypic changes in nutritional requirements or drug resistance to indicate that genotypic changes have occurred (at the nucleotide level) in large numbers of cells or organisms in order to demonstrate statistically and/or biologically significant increases over spontaneous mutation frequencies.

Gene mutations are tested most frequently in bacteria, insects, and mammalian cells. Although extensive literature exists on mutagenesis testing in yeast (*Saccharomyces cerevisiae*) and mould (*Neurospora crassa*), the use of these organisms has

declined because of problems associated with chemical permeability and the presence of endogenous target cell metabolism (Brusick 1989). Currently, there are no
well-established *in vivo* gene mutation assays. However, various laboratories have
used host-mediated assays when bacteria or mammalian cells are implanted in the
peritoneal cavity of animals, the animals are exposed to the test material, and then
the cells are removed for continuation of the test *in vitro*. Also, a number of laboratories are currently evaluating gene mutation assays in transgenic mice, for
example, mice carrying a bacterial gene that, following *in vivo* exposure, is 'rescued'
and reinserted into bacteria, which are then cloned to assess mutagenesis. Because
the transgenic mouse mutagenesis assays can assess mutations in the DNA from
essentially any tissue, which cannot be done with the host-mediated assay, it is
anticipated that one or more of the transgenic mouse gene mutation assays, when
sufficiently developed and evaluated, will meet the critical need for an *in vivo* point
mutation assay.

Tests for gene mutations in bacteria

The most extensive testing for gene mutations is in bacteria, particularly using
reverse mutation assays in *Salmonella typhimurium* and *Escherichia coli*. Bacterial
tests present the advantages of relative ease of performance, economy and efficiency, and the ability to .identify specific DNA damage that is induced (for
example, frameshift or base pair mutations). Bacterial tests can also provide
information on the mode of action of the test chemical, as the bacterial strains
used vary in their responsiveness to different chemical classes.

The *Salmonella typhimurium* microsome reverse mutation assay (the Ames test)
and the *Escherichia coli* reverse mutation assay are among the most economical and
rapid genetic toxicology assays. The Ames test, which is the most widely used and
considered by many to be the cornerstone of genetic toxicology testing, features
unique strains of histidine-dependent *Salmonella typhimurium* designed for sensitivity in detecting gene mutations, specifically frameshift mutations and base pair sub-
stitutions, which revert the bacteria to histidine independence. The *Salmonella*
strains are histidine auxotrophs by virtue of mutations in the histidine operon:
when these histidine-dependent cells are grown on minimal medium agar plates
containing a trace of histidine, only those cells that revert to histidine independence are able to form colonies. The small amount of histidine allows all the plated
bacteria to undergo a few divisions; in many cases this growth is essential for mutagenesis to occur, and the histidine revertants are easily visible as colonies against a
slight lawn of background growth. The spontaneous reversion rate of each strain is
relatively constant (McCann *et al.* 1975a; Maron and Ames 1983), although these
rates may vary with different solvents, and when the bacteria are exposed to mutagens, the mutation rates are increased, usually in a dose-related manner.

In addition to the histidine operon, most of the indicator strains carry a deletion that covers genes involved in the synthesis of the vitamin biotin (*bio*), and all
carry the *rfa* mutation that leads to a defective lipopolysaccharide coat and makes
the strains more permeable to many large molecules. All strains also carry the *uvrB*
mutation which results in impaired repair of ultraviolet (UV)-induced DNA
damage and renders the bacteria unable to use accurate excision repair to remove

certain chemically or physically damaged DNA, thereby enhancing the strains' sensitivity to some mutagenic agents. Additional strains containing the resistance transfer factor, plasmid pKM101, are believed to cause an additional increase in error-prone DNA repair, leading to even greater sensitivity to most mutagens (McCann *et al.* 1975a, 1975b). Plasmid pKM101 also confers resistance to the antibiotic ampicillin, which is a convenient marker for detecting its presence in the strains.

In Ames testing, one strain is used for a preliminary concentration range-finding assay, then mutagenesis assays are conducted with four or five strains. For example, some protocols use five well-established strains: TA1535, TA1537, TA1538, TA98, and TA100; other protocols may use the four strains recommended by Maron and Ames (1983): TA97a (formerly TA97), TA98, TA100, and TA102. Strains TA1535 and TA100 are reverted to histidine-independence by base pair mutagens; strains TA1537, TA1538, TA97a, and TA98 are reverted by frameshift mutagens, and strain TA102 detects a variety of oxidants and cross-linking agents that are not detected by the other strains (Levin *et al.* 1982a, 1982b, 1984).

E. coli strain WP2 *uvr*A (pKM101) is a tryptophan auxotroph by virtue of a base pair substitution mutation in the tryptophan operon. The *uvr*A mutation renders it deficient in the repair of some physically or chemically induced DNA damage, and the presence of plasmid pKM101 enhances sensitivity to most mutagens, as was described above. Prior to the development of the *S. typhimurium* strains TA100 and TA98, the *E. coli* WP2 assay was strongly recommended to complement the Ames test; its use has continued, particularly in conjunction with the *Salmonella* strains, to address international regulatory requirements based on the finding that some chemicals that are negative in *Salmonella* are detected by *E. coli*.

Basic microbial mutagenesis protocols use the plate incorporation approach or a preincubation modification of this approach (Maron and Ames 1983). They are outlined in Figure 10.7.

In the standard plate incorporation protocol, the test material, bacteria, and either a metabolic activation mixture (S9) or buffer are added to liquid top agar which is then mixed, and the mixture is immediately poured on a plate of bottom agar. After the agar gels, the bacteria are incubated, at 37°C, for 48–72 hours; then, the resulting colonies are counted. The preincubation modification is used for materials, including volatile and water-soluble compounds, which may be poorly detected in the plate incorporation assay. In this protocol, the test material, bacteria, and S9 mixture (if used) are incubated for 20–30 min at 37°C before top agar is added, mixed, and the mixture poured on a plate of bottom agar. Increased activity with preincubation in comparison to plate incorporation is attributed to the fact that the test compound, bacteria, and S9 are incubated at higher concentrations (without agar present) than in the standard plate incorporation test (Prival *et al.* 1979). Other modifications of the assays provide for exposing the bacteria to measured concentrations of gases in closed containers and/or the use of metabolic activation systems from a variety of species.

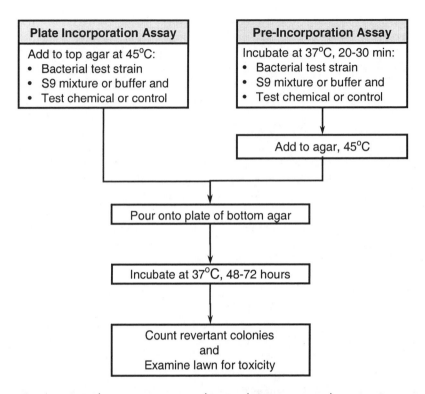

Figure 10.7 Plate incorporation and preincubation assays in bacteria. Comparison of representative plate incorporation and preincubation assays for reverse mutation in bacteria. For each strain and dilution of the test material or positive control, three sterile 13×100 mm test tubes are placed in a heating block; six test tubes are used for each negative (solvent) control. In the plate incorporation assay, the heating block is maintained at 45°C and the following are added for each sample: (a) 2.0 ml of molten top agar containing histidine for *Salmonella* or tryptophan for *Escherichia coli*, NaCl, Vogel–Bonner medium E, and glucose; (b) 0.5 ml of the metabolic activation mixture or buffer; (c) a dilution of the test material or the positive or negative control; and (d) 0.1 ml of the tester strain (about 10^8 bacteria) from an overnight culture. This mixture is then mixed and poured on a previously prepared plate containing about 25 or 30 ml of bottom agar (minimal agar with glucose). After the top agar sets (about 15 min), the plates are inverted and incubated for 48–72 hours for growth of mutant colonies. Preincubation assays differ in that top agar (at 45°C) is not added until the bacteria have been incubated, at 37°C, with the chemical and the metabolic activation mixture or buffer for 20–30 min. Because the top agar contains a trace of histidine or tryptophan, which is soon depleted, the bacteria form a background lawn, but only those bacteria which have reverted to histidine- or tryptophan-auxotrophy can continue to grow and form colonies. A reduction in the background lawn indicates toxicity.

Tests for gene and chromosomal mutations in mammalian cells

Although measurements of mutagenesis in mammalian cells are more time-consuming and expensive than bacterial assays, the test results obtained reflect the greater complexity of mammalian cells and chromosomes in comparison to those of prokaryotes, and, thus, they more closely approximate the genetic effects of chemicals in rodent species and humans. In contrast to bacteria, mammalian cells are essentially diploid ($2n$, with two copies of each chromosome), and their chromosomes are located within the nucleus of the cell and contain nonfunctional and noncoding, as well as functional, coding sequences. And whereas in bacterial cells all genes are usually expressed, in diploid cells there are usually two copies of each gene (one on each chromosome), and one dominant form of the gene may be expressed while the other recessive gene remains unexpressed, unless both copies are recessive. If both copies of the gene are the same the cell or organism is homozygous for that trait; a heterozygous condition exists if the copies are different; and if only one chromosome is present to carry the trait the condition is hemizygous. Hemizygous traits are found on the X chromosome in mammals because males have only one X chromosome and, in female cells, only one X chromosome is expressed.

In mammalian cell mutagenesis assays, the chemical exposure step must be followed by an expression period, during which mutant (and nonmutant) cells increase in number and the nonmutant protein (enzyme) present in the mutated cells and the RNA coding for that protein are depleted. Then a selective agent is added which permits only the mutated cells to grow and form colonies. The most extensively used mammalian cell mutagenesis tests employ established cell lines from mice and Chinese hamsters. Three such well-characterised protocols are shown in Table 10.4.

The L5178Y/$tk^{+/-}$ and AS52/$xprt$ gene mutation assays

The L5178Y/$tk^{+/-}$ and AS52/$xprt$ assays measure mutations at heterozygous, rather than hemizygous, loci; therefore, cells with chromosomal mutations, as well as cells with less extensive damage (gene mutations), survive and are enumerated. These protocols are used to determine whether a test material will induce chromosomal effects as well as gene mutations *in vitro*. In contrast, the Chinese hamster ovary (CHO) and V79/$hgprt$ assays measure mutagenesis at a hemizygous locus; thus, cells with more extensive (chromosomal) damage do not survive and cannot be enumerated. For a number of years these *hprt* mutation assays were used at least as extensively as the L5178Y mouse lymphoma cell mutagenesis assay; then, in many

---

Table 10.4 Examples of tests for mammalian cell mutagenesis

Chinese hamster ovary (CHO) *hgprt* gene mutation assay
V79 (Chinese hamster lung fibroblast) *hgprt* gene mutation assay
L5178Y/$tk^{+/-}$ mouse lymphoma cell mutation assay (with colony sizing)
AS52/*xprt* mutation assay in Chinese hamster cells

laboratories, the *hprt* assays were favoured because there was a suggestion that the L5178Y assay was too sensitive, that is, yielded too many 'false-positive' results. However, it has been found that the L5178Y assay is not overly sensitive when the appropriate protocols and evaluation criteria are used, and its use in comparison to that of the *hprt* assays is now increasing because the latter necessitates the parallel use of an *in vitro* chromosomal aberration assay.

The CHO/*hprt* and V79/*hprt* gene mutation assays

The *hprt* gene is located on the X chromosome; therefore, mutations at this locus are measurable in virtually all diploid or near-diploid mammalian cell lines. The CHO and V79 target cells have a rapid growth rate (a 12–14 hour cell doubling time), may be grown as monolayer cultures or in suspension, have a high cloning efficiency, a relatively low spontaneous mutation frequency, and may be used for cytogenetic studies (described below). CHO and V79 cells have two X chromosomes, but only one of them is actively transcribed and used in cellular processes. Thus, there are two possible genotypes: (1) $hprt^+$, in which the enzyme is produced and provides a 'salvage' pathway to use hypoxanthine and guanine for DNA synthesis; and (2) $hprt^-$, in which the enzyme is either not produced or is defective, and the cell cannot use the salvage pathway for DNA synthesis. For these assays, after a 6–8 day expression period, the cells are plated in thioguanine (TG)-containing medium, and *hprt*-deficient mutants are detected as surviving colonies. An outline of the procedure is shown in Figure 10.8.

The L5178Y/$tk^{+/-}$ gene and chromosomal mutation assay

The L5178Y mouse lymphoma assay measures gene and chromosomal forward mutations at the thymidine kinase locus, $tk^{+/-} \rightarrow tk^{-/-}$, and it yields high concordances with carcinogenicity test results. When trifluorothymidine (TFT) is used as the selective agent, many mutagens yield a biphasic curve of colony sizes which is resolved into small (σ) colonies of slowly growing mutant cells and large (λ) colonies of more rapidly growing mutants. It has been shown that most σ colony mutants have chromosomal mutations, as well as gene mutations, and that most λ colony mutants have only gene mutations (Clive and Moore-Brown 1979).

L5178Y mouse lymphoma cells, specifically clone 3.7.2C, grow in suspension culture with a relatively short cell doubling time of 9–10 hours. After the expression period, the cells are cloned in culture dishes in medium containing sufficient agar to immobilise the cells to measure mutagenesis (with trifluorothymidine present in the cloning medium) and survival. An approximately 2-week incubation period is required before colonies can be counted, as, with shorter times, the small colonies may not grow to a sufficient size to be visualised. Colony counting and sizing are conducted using colony counters that can discriminate between size ranges of the objects (colonies) counted. Both σ and λ colony mutation frequencies, as well as total mutation frequencies, should be reported for positive and negative controls and for treated cultures that yield a positive response.

At least two modifications of the L5178Y mouse lymphoma cell mutagenesis assay have been used for testing chemicals. In the first, the cells are cloned in

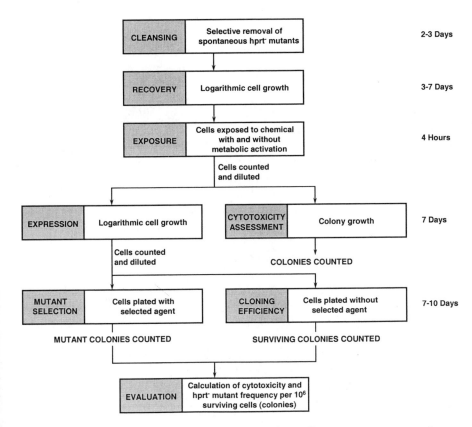

Figure 10.8 Representative CHO or V79 mammalian cell mutagenesis assay. Before each assay, spontaneous *hprt* mutants are removed by maintaining a monolayer culture of cells for 2–3 days in medium containing thymidine, hypoxanthine, methotrexate and glycine (THMG). The cells are then allowed to recover from cleansing for 1 day in medium containing THG (no methotrexate); then they are grown for another 2–6 days, with subculturing each 2–3 days, before establishing the cultures to be exposed. Each culture of cells is exposed to a dilution of the test material, the negative (solvent) control, or the positive control, in the absence and presence of metabolic activation, for 4h at 37°C. Following exposure, the cells are rinsed, detached with trypsin, centrifuged, resuspended, and counted. Then 200 cells are plated in each of three dishes and grown without subculturing for 7 days before straining and counting colonies to assess cytotoxicity, and two cultures of 1.5×10^6 cells are grown, with subculturing each 2–3 days, for expression. After expression, the two cultures are pooled and counted; 2×10^5 cells are plated in medium containing 6-thioguanine (the selective agent) in each of 12 dishes for growth of mutant colonies, and three dishes of 200 cells are plated (without the selective agent) to determine cloning efficiency of each culture. After growth of colonies (without subculturing) for 7–10 days, the colonies are stained and counted to determine the number of *hprt* mutants per 10^6 cells that were cloned.

microtitre plates in the absence of agar. In the second, an *in situ* variant, the cells are cloned in soft agar cloning medium after the exposure step, and TFT in liquid medium is added 2 days later, following the expression period. Because the presence of agar can retard colony growth, the microtitre approach yields more accurate measures of mutation frequencies and survival than when cells are cloned in agar; however, it is more time-consuming than cloning in soft agar, and, to date, colony sizing must be accomplished subjectively with visual (microscopic) observation of the colonies. The *in situ* approach also yields a more accurate measure of mutation frequencies, as the more slowly growing mutants are not lost by dilution during the expression period. With this approach, spontaneous and induced mutation frequencies (particularly σ colony mutation frequencies) are several times higher than with the standard approach, but to date, no chemical has been found positive with the *in situ* approach that was not also positive in the standard assay. Therefore, its future use may be for research purposes rather than for testing.

The AS52/*xprt* gene and chromosomal mutation assay

The AS52/*xprt* cell mutation assay provides many features of the L5178Y cell mutagenesis assay, but CHO cells are used, with essentially the same procedures as for the CHO/*hprt* assay (Tindall *et al.* 1986). The AS52 cell line carries a single functional copy of the bacterial *gpt* gene integrated into the CHO genome at a site that allows the recovery of chromosomal mutations. Although, as yet, the AS52/*xprt* assay has been successfully used in only a few laboratories, it is anticipated that, with greater availability and more extensive evaluation of its performance, this system may become more widely used in the future.

The *Drosophila* sex-linked recessive lethal test

The sex-linked recessive lethal (SLRL) test in the fruit fly, *Drosophila melanogaster*, is a forward mutation assay that detects the occurrence of both point mutations and small deletions in germ cells, and it is capable of screening for mutations at about 800 loci on the X chromosome (hence, sex-linked mutations). The *Drosophila* SLRL test is of historical importance because the first demonstration of mutagenesis was in *Drosophila* (Muller 1927). In the many decades that have elapsed since then, *Drosophila* has been extensively characterised genetically, hundreds of chemicals have been tested in this species, and the SLRL has been found to be responsive to mutagens from different chemical classes.

In this assay, wild-type males (3–5 days old) are treated with the test material and mated to an excess of virgin females from the Basc or Muller-5 stock, with specially marked and arranged chromosomes. The females are replaced with fresh virgins every 2–3 days to cover the entire germ cell cycle. Heterozygous female offspring (F_1) are mated individually to their brothers (also F_1), and the progeny (F_2) from each of these crosses are scored for phenotypically wild-type males. Absence of these males indicates that a sex-linked recessive lethal mutation has occurred in a germ cell of the parental male.

Advantages of this assay include the presence of a xenobiotic metabolic system in insects and significant economies of time and resources over similar tests in mammals. However, the SLRL is not used as an initial test for mutagenesis because of concerns that insect metabolism may be too far removed from human metabolic responses. Even though the test is more economical and rapid than tests in mammals, the assay is still labour intensive in comparison to many testing approaches because of the time required to examine thousands of progeny. At this time the SLRL is used primarily in research laboratories.

Tests for chromosomal mutations

Chromosomal abnormalities are associated with:

- spontaneous abortion
- congenital malformation
- neoplasia
- infertility.

They occur in approximately 0.6% of live births in humans. It has been estimated that up to 40% of spontaneous abortuses have chromosomal defects. The rationale for measuring chromosomal effects includes: the increasing evidence that specific chromosomal defects in somatic cells are associated with most neoplasias and evidence that chromosomal rearrangement may play a role in oncogene activation; the relevance of chromosomal mutation to inheritance of genetic defects in germ cells; and the observation that substances which produce chromosomal structural changes also produce point mutations and that few gene mutagens produce no aberrations.

An understanding of mitosis and the stages of the somatic cell cycle, and their derangements, is important in the design and interpretation of results for several genetic toxicology test systems including, most particularly, cytogenetic assays. The cell cycle and mitosis were introduced and described earlier in this chapter.

To visualise chromosomes, cells are arrested in metaphase, treated with a hypotonic solution to swell the chromosomes, fixed, transferred to microscope slides, and stained. The first metaphase (M) following chemical exposure, M_1, is the time when the greatest number of chromosomally damaged cells may be observed because many cells with extensively damaged chromosomes are incapable of progressing through additional cell cycles. The progression of severely damaged cells through the cell cycle to mitosis is retarded in direct relation to the extent of damage, a phenomenon that must be considered in the design of cytogenetic tests.

Chromosome breakage, which is necessary for chromosomal rearrangements (Hsu 1979), is the classical endpoint in chromosomal aberration assays designed to assess the potential health hazards of chemicals. Agents that break chromosomes are termed clastogens. Following exposure to a clastogen, the number of cells with chromosome breakage and rearrangements increases in M_1 cells in a dose-related manner but declines rapidly after M_1 because of the greatly extended cell cycle times of some cytogenetically damaged cells. A chromosomal exchange results when the broken ends of the same or different chromosomes rejoin in an aberrant manner. An inversion results from two breaks in one chromosome with the broken piece turned 180° before rejoining and cannot be visualised without special chromosome staining techniques to reveal chromosome bands. A translocation results from an interchange between different nonhomologous chromosomes.

Chromosomal mutations (structural aberrations) are morphological alterations in the structure of eukaryotic chromosomes and, hence, they may affect the expression of numerous genes, with gross effects, or be lethal to affected cells. To assess chromosomal damage, test systems are used that provide for exposures *in vitro* and *in vivo*. While the L5178Y/*tk*$^{+/-}$ and AS52/*xprt* cell mutation assays indicate chromosomal, as well as gene, mutations, they do not provide for direct observations of broken and aberrant chromosomes or indicate possible cell-cycle specificity of test materials. Tests for chromosomal mutagenesis include *in vitro* and *in vivo* tests for chromosome aberrations, the micronucleus test, tests for dominant lethal mutations, the heritable translocation test, and mammalian germ cell cytogenetics.

In vitro *chromosome aberrations*

Two types of cells are routinely used for *in vitro* chromosome aberration assays, which are conducted in the absence and presence of exogenous metabolic activation, CHO cells and human lymphocytes stimulated to divide *in vitro*. Mitotic cells in the first metaphase (M$_1$) following chemical exposure are evaluated for the presence of chromosome and chromatid breakage and aberrations. Preliminary concentration range-finding assays are conducted with mitotic index determinations and/or assessments of cell growth (generation times) in culture to select concentrations for testing in which there will be sufficient numbers of mitotic cells for analyses.

Because some chemicals induce a delay in cellular progression to mitosis, some protocols specify use of a second, delayed, harvest time, and, if delayed harvest times are used, both the preliminary assay and the cytogenetic assay should include an analysis of whether the mitotic cells are in the first division, M$_1$, following exposure. This is accomplished by growing the cells in the presence of bromodeoxyuridine (BrdUrd) and then staining the chromosomes of the harvested cells using the fluorescence-plus-Giemsa (FPG) technique, as will be described for sister chromatid exchanges. When this is done, only M$_1$ cells will contain uniformly darkly stained chromosomes; the chromosomes of M$_2$ cells will have one darkly stained and one lightly stained chromatid, and M$_3$ cells will have a mixture of chromosomes with differentially stained and lightly stained chromatids.

The *in vitro* chromosome aberration assay in human lymphocytes is used to enhance the relevance of the effects studied to humans. To visualise chromosomes, freshly drawn blood from healthy human donors is diluted with cell culture medium containing phytohaemaglutinin (PHA), which stimulates the lymphocytes to divide *in vitro*. After a prolonged G$_1$ stage, the cells enter S phase, which should be the time of addition of the test chemical to the cells. For a typical protocol, chemical exposure is initiated at 48 hours, colcemid, a mitotic spindle inhibitor is added at 70.5 hours, and metaphases are harvested at 72 hours. The cells are then treated with a hypotonic solution to swell the chromosomes, fixed, and dropped onto slides. The chromosomes are stained and coverslips are attached before microscopic analysis.

in vitro chromosome aberration assays in CHO cells present several advantages: Chinese hamster cells have a low chromosome number ($2n = 22$), a short generation time, and no donor-to-donor variability, as could potentially influence the results of studies with human lymphocytes. Also, because the CHO cells grow as attached monolayers, metaphase-rich populations can be obtained using mitotic shake-off procedures, which reduces the time required for microscopic analysis (although mitotic index determinations are not informative with this approach). In spite of the suggestion that the analysis of aberrations in CHO cells lacks sensitivity (Tennant *et al.* 1987), a number of laboratories have found the assessment of aberrations in CHO cells, but not in human lymphocytes, to be overly sensitive (yielding 'false positive' results), particularly when a delayed harvest time is used. As with other assays, it is probable that such difficulties will be resolved by identifying insoluble chemicals as 'not testable', and with greater attention to procedural details, such as ensuring that the population of cells analysed is in M_1.

In vivo *chromosomal aberrations*

Measurements of chromosome and chromatid breakage and aberrations in the rapidly dividing population of cells in rodent bone marrow, following *in vivo* exposures, are reasonably well predictive of carcinogenesis in rodents, an outcome that may be due to the fact that the target cells are exposed to chemical metabolites under physiological conditions. Preliminary dose range-finding studies are necessary, using an evaluation of mitotic indices to select a high dose in which a sufficient number of M_1 cells will be present for analysis. Experience has shown that this dose may be lower than the LD_{50} defined in acute toxicology testing.

One testing approach utilises a single exposure (of each dose) followed by three killing times to allow for variations in metabolic rates and cell-cycle delay. More recently, it has been shown that this variability is negated by repeated dosing to achieve a steady state, followed by a single killing time.

The latter approach conserves animals and significantly reduces the time required for microscopic analysis of the cells. For either approach, it is recommended that BrdUrd incorporation (accomplished by implanting BrdUrd in the test animals) followed by FPG staining of the cells be used to insure that the chromosomes analysed are in M_1.

Micronucleus tests

Because of the need for careful microscopic analysis of a sufficient number of cells to determine the statistical significance of results, chromosomal aberration assays have been, historically, time and labour intensive. Furthermore, the analysis of chromosomal aberrations requires a high level of training and expertise to preclude the introduction of subjective bias. However, most regulatory agencies consider the micronucleus test to meet the requirements for an assay to assess cytogenetic damage *in vivo*, and micronucleus testing has virtually replaced the use of *in vivo* chromosome aberration tests, because the micronucleus test is more rapid, considerably less expensive, and requires minimal training.

Micronuclei result when nuclear membranes form around broken pieces of chromosomes; therefore, micronucleus tests in rodents measure chromosome breakage, the process necessary for the induction of essentially all cytogenetic aberrations, but not rearrangements. Following exposure, micronuclei may be readily observed microscopically in stained preparations of (otherwise anucleate) erythrocytes from the bone marrow of rats or mice or in the peripheral blood of mice as, in mice, the spleen does not remove micronucleated cells from the blood. Bone marrow cells are a more heterogeneous population and give a more informative index of toxicity, but peripheral blood erythrocytes are evaluated more efficiently. Further, because mice are less expensive than rats, most micronucleus testing is now conducted with mice. As was described, above, for *in vivo* chromosomal aberration tests, either single dosing followed by three killing times or multiple dosing to achieve a steady state of exposure, followed by one killing time can be used, with the latter providing economies of time and animal usage.

The dominant lethal test in rodents

The dominant lethal test in rodents has been used for over 50 years to assess the effects of mutagenic agents. A dominant lethal mutation is one that occurs in a germ cell (gamete) without causing dysfunction of the cell, but that is lethal to the fertilised egg or developing embryo. Male rats or mice are exposed to the test material then mated to untreated virgin females, and susceptibility of the various germ cell stages can be tested by the use of sequential mating intervals. After an appropriate interval, the pregnant females are killed, and the contents of the uteri are examined to determine the numbers of implants and live and dead embryos. A positive response is indicated by an increase in number of dead implants in the treated group in relation to the number observed in the untreated group, and indicates that the test substance has affected germinal tissue of the test species. Dominant lethal effects are generally considered to be the result of structural and numerical chromosomal aberrations (that is, genomic mutations), but gene mutations and toxic effects cannot be excluded.

Because the correlation between positive results in this assay and animal carcinogenicity is poor, the dominant lethal test is not generally recommended as a predictive test for carcinogenicity. However, it has been recommended as a second-tier test for chromosomal mutations, for example, to indicate whether chromosomal effects observed in bone marrow cells could also be observed in germinal tissues. In contrast with the tests that will be described for genomic mutations, the dominant lethal test is relatively economical and can be performed in most toxicology laboratories.

The heritable translocation test

The mouse heritable translocation test detects structural and numerical chromosome changes, specifically reciprocal translocations (between chromosomes) and, if female progeny are included, X-chromosome loss in germ cells as recovered in the F_1 generation. The heritable translocation test is an important third-tier test for assessing the potential of a test material to induce heritable chromosomal mutations. However, because of the large numbers of mice that are required and the time and expense associated with this test, its current usage is limited to research, rather than testing, laboratories.

To test for heritable translocations, sexually mature male mice are administered the test material (using either single dosing or repeated dosing over several weeks), then mated with two or three virgin females, each; then the females are caged individually. When the females give birth, the date, litter size and sex of the progeny are recorded; all F_1 males are weaned, and all F_1 females are discarded unless the experimental design includes assessment of X-chromosome loss. F_1 males (300–500 per initial dose level) are then raised to maturity, and each is mated to approximately three virgin females to test for fertility. The presence of heritable translocations is detected by reduced fertility (litter size) of the F_2 generation, and is confirmed by cytogenetic analysis of meiotic cells from the F_1 males. XO-females are recognised by a change in the sex ratio among their progeny from 1:1 males versus females to 1:2, and are confirmed cytogenetically by the presence of 39 (instead of 40) chromosomes in bone marrow mitoses.

Tests for genomic mutations

In germinal tissues, eukaryotic chromosomes undergo meiosis, when the diploid ($2n$) chromosome number is reduced to a haploid (n) number, ensuring that each gamete has one copy of each pair of genes. During meiosis, homologous chromosomes (one of paternal origin and one of maternal origin) are paired then assorted independently, and a high level of chromosome recombination may occur; therefore, each gamete carries a discrete complement of genetic traits. Without these processes, genetic variability would be limited to rare mutational events that survive natural selection. Interference with normal meiosis may result in genomic mutations which are changes in the number of chromosomes, or aneuploidy, and are also referred to as numerical aberrations.

Genomic mutations include: monosomy, when one of a pair of chromosomes is lost; trisomy, the addition of a chromosome; and polyploidy, when the complete set of chromosomes is increased in number. Virtually all genomic mutations arise in germ cells of the previous generation and are detrimental to the health of the individual. In humans, well-known genomic mutations include several trisomies such as Down's syndrome (trisomy 21) and Kleinefelter's syndrome (XXY). At present, there is no direct evidence that specific chemicals induce inherited defects in humans and no way to measure the induction of specific mutations in human germ cells. However, the induction of mutations by physical and chemical agents in a wide variety of species suggests that humans are also susceptible to mutations by the same or similar agents.

There is some redundancy in the classification of genomic mutation tests. For example, genomic mutation tests include the *Drosophila* sex-linked recessive test (which is also a test for gene mutations), the heritable translocation test (which is also a test for chromosomal mutations), and the specific locus test in *Neurospora* (de Serres 1992). In the design of dosing times and interpretation of test results for heritable damage, it is necessary to be aware of the time involved in cells traversing each stage of meiosis to the formation of gametes. But, apart from research on stage-specific effects, testing protocols usually incorporate dosing over the entire meiotic sequence.

The mouse visible specific locus test, initially developed by W.L. Russell in

1951 to study the effects of radiation and later used to study chemical mutagens, and the more recently developed biochemical specific locus test in mice (Malling and Valcovic 1978), are generally considered to be definitive tests for heritable germ cell mutations. The mouse visible specific locus test involves the use of developed strains that differ in selected gene loci so that mutations in these loci can be visualised as a phenotypic change due to the expression of recessive genes in the F_1 generation. The biochemical specific locus test is similar, using mice from defined inbred strains, but presumptive mutants are identified by variations in the electrophoretic mobility of proteins. For each mutagenic agent that is evaluated in each specific locus test, thousands of mice are examined to assess the induced mutation frequency. For this reason, and because the number of laboratories in which they can be performed competently is limited, the specific locus tests are infrequently used as third, or final, tier tests. Rather, their importance stems from the information they provide on basic processes of mutagenesis in mammals.

Tests for the repair of primary DNA damage

As it is thought that the incorrect repair of damage to DNA may be an early step in mutagenesis and/or carcinogenesis, two short-term tests, in particular, have been used to measure the repair of primary DNA damage. These are tests for sister chromatid exchanges (SCEs) and tests for unscheduled DNA synthesis (UDS).

Sister chromatid exchange

The induction of DNA lesions by genotoxicants leads to the formation of sister chromatid exchanges (SCEs), which may be related to recombinational or postreplicative repair of DNA damage at the chromosomal level. SCEs are revealed by a 'harlequin' pattern of differentially stained chromatid segments in chromosomes from cells grown in the presence of BrdUrd for two rounds of DNA replication, then stained with the FPG technique so that one chromatid is more lightly stained than the other. Because SCEs can be enumerated more rapidly than chromosomal aberrations, SCE assays are more efficient and economical than aberration assays. SCEs can also be measured *in vitro*, for example, in CHO cells or human lymphocytes, or *in vivo*, primarily in mice. For the latter, BrdUrd is implanted in mice exposed to the test material.

SCE assays have been used as supplementary tests rather than as alternative assays for measuring cytogenetic damage. However, the apparently high sensitivity of SCE assays, uncertainty about the significance of SCE induction, and a lack of a close correlation with results from chromosome aberration tests, has led to a virtual discontinuation of SCE testing for regulatory submissions. Today, UDS is used for submissions to those agencies still requiring assessments of primary DNA damage and its repair.

Unscheduled DNA synthesis

Unscheduled DNA synthesis (UDS) is the incorporation of DNA precursors at times other than the scheduled, S phase, synthesis of DNA in the cell cycle. It indicates that an active form of a chemical has damaged cellular DNA and that the cell was capable of (correctly or incorrectly) repairing the damage. UDS is evaluated *in vitro* by exposing cells to the test chemical in the presence of tritiated thymidine ([^{3}H]-TdR). Then, autoradiographic techniques are used to reveal the extent of incorporation of [^{3}H]-TdR in the nuclei of non-S-phase cells that have undergone UDS, that is, the [^{3}H]-TdR-exposed silver grains in an autoradiographic emulsion over the cell nuclei are counted, a process that may be semiautomated. When tested with and without addition of a metabolic activation system, UDS may be examined *in vitro* in any cells blocked from entering S phase, or *in vivo* in various tissues. Rat hepatocytes are routinely used because they are metabolically competent and normally exhibit little scheduled DNA synthesis; therefore, no S phase inhibition or exogenous metabolic activation is required.

As UDS testing is an indirect measure of mutagenesis, and because UDS testing in hepatocytes has failed to detect a number of hepatocarcinogens, regulatory agencies are discontinuing requirements for UDS testing in initial batteries of tests. Instead, UDS tests are often used in a second tier to evaluate whether the test material has interacted with DNA.

Mammalian cell transformation

Morphological transformation is sometimes used as a test for *in vitro* carcinogenesis, rather than mutagenesis, and is the process whereby mammalian cells with a limited number of generations in culture, and/or cells that proliferate with an oriented pattern on a plastic surface, are converted into cells with unlimited capacities to divide and random patterns of growth. The relevance of tests for transformation stems from their measurement of other than strictly mutagenic events leading to carcinogenesis. Hence, they extend the range of short-term test batteries.

Examples of mammalian cell transformation protocols include tests in BALB/c-3T3 cells and C3H-10T1/2 cells. The BALB/c-3T3 cell line has been studied extensively for postconfluent inhibition of cell division, when the cultures have a cobblestone-like morphology. This inhibition is reversed by various treatments including exposure to carcinogens, with and without metabolic activation, to yield transformed foci of cells. The C3H mouse prostate cell line C3H-10T1/2 is acutely sensitive to postconfluent inhibition of cell division and has been transformed with a variety of carcinogens to produce foci of cells that yield malignant fibrosarcomas when administered to irradiated recipient mice. The C3H-10T1/2 cell line has a lower rate of spontaneous transformation than BALB/c-3T3 cells, but the C3H-10T1/2 cell line is customarily used for testing only in the absence of metabolic activation. Mammalian cell transformation assays are currently used more extensively for screening antineoplastic agents for efficacy rather than as genetic toxicology tests.

Genetic toxicology testing for product registration

Genetic toxicology tests that have been proposed for product registration will be described below. With increasing use and evaluation of the test systems, it has been found that only a few of the tests are appropriate for routine use. Many have been inadequately evaluated, and some have been found to be insensitive to important classes of chemicals, while others appear to yield too many 'false positive' results, and still others are too time-consuming and/or costly. Nowadays, relatively few genetic toxicology tests are routinely used for initial assessments of genetic effects, with the selection of protocols often based on published regulatory guidelines and the experience of industry. To assist organisations wishing to market chemicals and pharmaceuticals in diverse geographical areas, concerted efforts are being made by regulatory agencies to reach agreement on which genetic toxicology testing protocols should be universally required for regulatory submissions.

Many testing schemes utilise a battery of tests which usually fall into three tiers as shown in Table 10.5. When currently proposed changes in regulations are implemented, it is probable that the basic, initial tier of tests recommended by various agencies will be limited to tests for:

- gene mutations in bacteria
- gene mutations in mammalian cells
- chromosomal mutations *in vitro*
- chromosomal mutations *in vivo*.

There is less agreement on which tests should be used after the initial tier. Tests that have been recommended and used in a second tier include unscheduled DNA synthesis (UDS), the dominant lethal test, the sex-linked recessive lethal test in *Drosophila*, *in vivo* chromosomal aberrations, and mammalian cell transformation. Assessments of genomic mutations predominate in the third tier of tests, but, in general, tests for genomic mutations use large numbers of animals, are labour-intensive and, hence, expensive, and the number of laboratories proficient in these tests is limited. Further, as discussed by Prival and Dellarco (1989), the development of pharmaceuticals and industrial chemicals is often discontinued when initial test results indicate mutagenicity. Although regulatory agencies have used

---

Table 10.5 Sample tiered battery of genetic toxicology tests

Tier 1: Simple and rapid tests for
  Gene mutations in bacteria
  Gene and chromosomal mutations in mammalian cells
  Chromosomal mutations *in vivo* (possibly)

Tier 2: Intermediate tests for
  Mutations in *Drosophila*
  Dominant lethal in mice
  Repair of DNA damage

Tier 3: Long-term tests in animals
  Visible or biochemical specific locus test in mice
  Heritable translocation test in mice

---

positive results in initial mutagenicity tests to justify requests for additional information on the mutagenicity and/or carcinogenicity of chemicals, because of the time and expense involved in this additional testing, a request for additional information may preclude a new product from being marketed.

To establish that a test chemical is negative, *in vitro* tests must be conducted in the absence and presence of exogenous metabolic activation, and positive and negative controls must be within historical ranges. The latter is also applicable to *in vivo* tests, and, in addition, animals must be maintained under conditions that minimise the influence of environmental variables, and bioavailability must be considered when selecting the route of administration of a test material. Some current testing guidelines specify that the results obtained should be demonstrably reproducible in independent experiments. However, many scientists believe that it is not necessary to repeat appropriately conducted assays that are clearly positive or negative, particularly for *in vivo* tests, as this is contrary to the objective of minimising the use of animals. Also, replication for its own sake (that is, an exact repeat of the first assay) may not be informative; instead, repeat assays should be designed to correct any deficiencies of the first assay. If one assay results in a negative response and the other a positive response, and if there is no reason to give greater weight to either one of them, the overall evaluation must be equivocal, and another repeat of the assay should be conducted.

Applicability of genetic toxicology for human population monitoring

As described by Albertini and Robinson (1991), human population monitoring involves quantitation and characterisation of genotoxicant exposures and genetic effects *in vivo* in humans for human health risk assessments. Many of the laboratory assays used for monitoring are similar or identical to those that will be described for traditional genetic toxicology. However, traditional genetic toxicology deals with the identification of potential genetic hazards and is important for interpreting the results of population monitoring while human population monitoring deals with another level of risk, the identification of effects that may have occurred in humans. Population monitoring is also related to two other activities concerned with human populations: human genetic screening, which defines genetic susceptibilities within populations; and epidemiology, which is the topic of another chapter in this text.

Measures of human genotoxicant exposure related to the tests that will be described include: the use of *Salmonella* mutagenesis to detect mutagens or their metabolites in urine concentrates; assessing UDS in human tissues (for example, blood) shortly after chemical exposure to measure genotoxicant interactions with DNA; evaluating chromosome damage in peripheral blood lymphocytes as an index of irreversible genotoxic effects; assessing *hprt* mutations in human blood cells as a measure of human somatic cell gene mutations; and using markers of both chromosomal- and gene-level genetic effects which are detectable in human sperm. These and other approaches are described in Albertini and Robinson (1991), and it is anticipated that with the current rapid developments in the field of molecular biology, additional measures of human exposure will be used in the future.

The future of genetic toxicology

This chapter has outlined the origins and development of genetic toxicology and many of its current practices. The use of genetic toxicology tests for product registration gained acceptance from 1973 and has matured with the more precise definition and more extensive evaluation of many of the tests. International consensus has been sought on those tests and protocols that are most predictive and useful (Ashby et al. 1983).

Thus, it is anticipated that the emphasis of applied genetic toxicology in the immediate future may be on the more extensive use of specific tests to reduce the backlog of chemicals that are as yet untested. The emphasis of investigative genetic toxicology is to draw upon developments in related disciplines to understand the basis of tests for assessing the potential risk of exposing humans to various agents. The use of molecular biology techniques has led to the definition of many human genetic diseases. Thus, the development of polymerase chain reaction (PCR) in combination with molecular techniques to identify changes in DNA, and the use of fluorescence in situ hybridisation (FISH), pulsed-field gel electrophoresis, restriction enzyme mapping and associated techniques for probing the larger genetic changes should also lead to the definition of new and improved approaches for probing genotoxic mechanisms. These, in turn, should bring the field closer to the goal of accurately assessing effects directly related to those observed or with the potential to occur in humans.

Further work is still required to elucidate aspects of in genotoxicity important to workers and the workplace:

- Methods of genotoxicity assays and their application. For example, the use of metabolising systems to increase the efficiency of genotoxicity assays (Paolini and Cantelli-Forti 1997), or the relevance of concentrations in assay media and its relevance to human exposure (Baan et al. 1989)
- Some multispecies, multisite nongenotoxic carcinogenesis models may not be relevant for humans. Perhaps in the near future, it may be possible to use data-based decisions in risk assessment as opposed to hypothetical low-dose extrapolations and mathematical modelling, building on the wealth of current understanding in human and rodent cancer processes in arenas beyond those that pharmaceutics have produced (Alden 2000)
- The methods used for genetic monitoring are either substance specific, for example, the quantitation of identified DNA adducts, or substance unspecific as in the measurement of DNA repair (Knudsen and Sorsa 1993). The usefulness of such methods in the occupational environment remains unclear (Ong et al. 1990).
- The variety of background or spontaneous rates of genetic damage (Lutz 1990). The influence of metabolic genotypes on biological indicators of genotoxic risk in environmental or occupational exposure has also been studied; and there is sufficient evidence that genetically based metabolic polymorphisms, as determined using urinary metabolites for monitoring exposure to genotoxic substances, must be given due consideration (Pavanello and Clonfero 2000).

- The role of antagonism of mutagenic or clastogenic activity by such antimutagens as vitamins, fatty acids, thiols, tannins and other phenolics has also been studied (Waters *et al.* 1998), although the application of such agents in workers remains unclear.
- Exposure to nonchemical genotoxicants (McCann *et al.* 1993) (for example, the use of genotoxicity assays for assessing the genotoxicology of fibres (Jaurand 1996)) is another area of investigation, as is co-exposure to environmental or lifestyle genotoxicants (Forbes 1998; Phillips and Venitt 1999).
- The use of genotoxicity testing in health risk assessments (Lauwerys *et al.* 1995).
- The use of quantitative structure–activity relationships (Q)SARs of nongenotoxic carcinogens (Combes 2000; Stouch and Jurs 1985).
- The role of genotoxic effects of exposure to mixtures (Waters *et al.* 1990), including interactions of carcinogens with DNA, and other targets, and their relevance to both molecular dosimetry of exposure and development of cancer (de Matteis and Smith 1995).

Other issues, such as the use of toxicology in cancer risk assessment, are still under development (Choy 2001).

The use of a 'no-threshold' regulatory policy for nongenotoxic carcinogens has been suggested (Perera 1984; Roe 1989) although the scientific basis for the absence of such an effect remains far from clear (Ashby 1992; Lutz and Maier 1988; Shaw and Jones 1994; Sofuni 2000).

Summary

Millions of workers worldwide are potentially exposed each year in occupational settings to hazardous chemicals, particulates or fibres. Some of these agents are genotoxic and may cause genetic alterations in the somatic or germ cells of exposed workers. Such changes have the potential to cause health problems such as genetic diseases, cancer and developmental problems.

Introduction of the Ames test in 1973 revolutionised the way in which methods of genotoxicity and carcinogenicity were studied. The subsequent development of many, many short-term tests, and the refinement of the most suitable of these for research and regulatory attention gave a huge impetus to the knowledge about genotoxicological principles and processes.

However, the promise of obtaining information about the genotoxicity of chemicals in a far shorter time than that required by epidemiology and animal toxicity bioassays at a fraction of the cost has not yet been realised. There were many discrepancies in the findings, so-called false positives and false negatives, and there was much debate concerning the true predictive nature of assay systems, singly and in combination, for rodent carcinogens and the relevance of the findings for human disease. However, the evolution of *in vitro* genotoxicity test systems has produced a better understanding of the molecular effects of chemicals in interaction with genetic materials. Although there are many potential contributing mechanisms to carcinogenesis, mutagenesis remains the dominant driving force behind the process (Grant 2001). Also, the development of a battery or tier of assays in and major international collaboration in codifying and validating the assays has assisted in better screening processes for chemicals under commercial development (Rueff *et al.* 1996).

Chemical carcinogenesis is a complex process dependent on uptake, toxicokinetics and involving at least steps of initiation, promotion and progression (Butterworth 1990). Basic genotoxicity techniques and applications of such techniques in human risk characterisation, including in workplace chemical exposures, remain a challenge for the future.

References

Albertini, R.J. and Robinson, S.H. (1991) Human population monitoring. In: Li, A.P. and Heflich, R.H. (eds) *Genetic Toxicology*. Boca Raton: CRC Press, pp. 375–420.

Alden, C.L. (2000) Safety assessment for non-genotoxic rodent carcinogens: curves, low-dose extrapolations, and mechanisms in carcinogenesis. *Human Exp. Toxicol.* 19: 557–560; 571–572.

Ames, B.N. (1979) Identifying environmental chemicals causing mutations and cancer. *Science* 204: 578–593.

Ames, B.N., Lee, F.D. and Durston, W.E. (1973) An improved bacterial test system for the detection of mutagens and carcinogens. *Proc. Nat. Acad. Sci. USA* 70: 782–786.

Ames, B.N., McCann, J. and Yamasaki, E. (1975) Methods for detecting carcinogens and mutagens with the *Salmonella*/mammalian-microsome mutagenicity test. *Mutat. Res.* 31: 347–364.

Arelett, C.F. and Cole, J. (1988) The role of mammalian cell mutation assays in mutagenicity and carcinogenicity testing. *Mutagenesis* 3: 455–458.

Ashby, J. (1992) Prediction of non-genotoxic carcinogenesis. *Toxicol. Lett.* Spec No: 64–65, 605–612.

Ashby, J., de Serres, F.J., Draper, M.H., Ishidate, Jr, M., Matter, B.E. and Shelby, M. (1983) The two IPCS collaborative studies on short-term tests for genotoxicity and carcinogenicity. *Mutat. Res.* 109: 123–126.

Baan, R.A., Fichtinger-Schepman, A.M., Roza, L. and van der Schans, G.P. (1989) Molecular dosimetry of genotoxic damage. *Arch. Toxicol. Suppl.* 13: 66–82.

Bridges, B. (1976) Evaluation of mutagenicity and carcinogenicity using a three-tier system. *Mutat. Res.* 204: 17–115.

Brusick, D. (1980) *Principles of Genetic Toxicology*. New York: Plenum Press.

Brusick, D. (1989) Genetic toxicology. In: Hayes, A.W. (ed.) *Principles and Methods of Toxicology*, 2nd edn. New York: Raven Press, pp. 407–434.

Brusick, D.J. (1994) *Methods for Genetic Risk Assessment*. Boca Raton: CRC Press.

Butterworth, B.E. (1977) *Strategies for Short Term Mutagens/Carcinogens*. New York: Franklin Book Co.

Butterworth, B.E. (1990) Consideration of both genotoxic and nongenotoxic mechanisms in predicting carcinogenic potential. *Mutat. Res.* 239: 117–132.

Cairns, J. (1981) The origin of human cancers. *Nature* 289: 353–357.

Choy, W.N. (2001) *Genetic Toxicology and Cancer Risk Assessment*. New York: Marcel Dekker.

Clive, D. and Moore-Brown, M.M. (1979) The L5178Y/TL$^{+/-}$ mutagen assay system: Mutant analysis. In: Hsie, A.W., O'Neill, J.P. and McElheny, V.K. (eds) *Banbury Report 2: Mammalian Cell Mutagenesis: The Maturation of Test Systems*. New York: Cold Spring Harbor Laboratory, pp. 421–429.

Combes, R.D. (2000) The use of structure–activity relationships and markers of cell toxicity to detect non-genotoxic carcinogens. *Toxicol. in vitro* 14: 387–399.

de Matteis, F. and Smith, L.L. (eds) (2001) *Molecular and Cellular Mechanisms of Toxicity.* Boca Raton: CRC Press.

de Serres, F.J. (1992) Development of a specific-locus assay in the ad-3 region of two-compartment heterokaryons of *Neurospora*: A review. *Environ. Molec. Mutagen.* 20: 225–245.

Douglas, J.F. (ed.) (1980) *Carcinogenesis and Mutagenesis Testing.* Humana Press.

Forbes, V.E. (ed.) (1998) *Genetics and Ecotoxicology.* New York: Hemisphere Publishing.

Gallo, M.A. (2001) History and scope of toxicology. In: Klaassen, C.D. (ed.) *Casarett and Doull's Toxicology, The Basic Science of Poisons,* 6th edn. New York: McGraw-Hill, pp. 3–10.

Gatehouse, D.G. and Kirkland, D.J. (eds) (1990) *Basic Mutagenicity Tests: UKEMS Recommended Procedures.* Cambridge: Cambridge University Press.

Grant, S.G. (2001) Molecular epidemiology of human cancer: biomarkers of genotoxic exposure and susceptibility. *J. Environ. Pathol., Toxicol. Oncol.* 20: 245–261.

Heflich, R.H., Albert, P. and Li, A.P. (eds) (1991) *Genetic Toxicology.* Boca Raton: CRC Press.

Henig, R.M. (2000) *The Monk in the Garden: The Lost and Found Genius of Gregor Mendel, The Father of Genetics.* New York: Houghton Mifflin.

Hofnung, M. and Ouillardet, P. (1986) Recent developments in bacterial short-term tests for the detection of genotoxic agents. *Mutagenesis* 1: 319–330.

Holbrook, N.J. and Fornace, A.J. (1991) Response to adversity: molecular control of gene activation following genotoxic stress. *New Biol.* 3: 825–833.

Hollaender, A. and de Serres, F.J. (1971–1983) *Chemical Mutagens. Principles and Methods for their Detection,* Volumes 1–8. New York: Plenum Press.

Hsu, T.C. (1979) Forward. *Mammalian Chromosomes Newsletter* 30: 59–70.

Jaurand, M.C. (1996) Use of in-vitro genotoxicity and cell transformation assays to evaluate the potential carcinogenicity of fibres. *IARC Scientific Publications* 55–72.

Johnson, D.G. and Walker, C.L. (1999) Cyclins and cell cycle checkpoints. *Ann. Rev. Pharmacol. Toxicol.* 39: 295–312.

Josephy, D.P. (1996) *Molecular Toxicology.* Oxford: Oxford University Press.

Keshaya, N. and Ong, T.M. (1999) Occupational exposure to genotoxic agents. *Mutat. Res.* 437: 175–194.

Kirsch-Volders, M. (ed.) (1984) *Mutagenicity, Carcinogenicity and Teratogenicity of Industrial Pollutants.* New York: Plenum Press.

Knudsen, L.E. and Sorsa, M. (1993) Human biological monitoring of occupational genotoxic exposures. *Pharmacol. Toxicol.* 72 (Suppl. 1): 86–92.

Lauwerys, R.R., Bernard, A., Roels, H. and Buchet, J.P. (1995) Health risk assessment of long-term exposure to non-genotoxic chemicals: application of biological indices. *Toxicol. Lett.* 77: 39–44.

Lawley, P.D. (1994) Historical origins of current concepts of carcinogenesis. *Adv. Canc. Res.* 65: 17–111.

Levin, D.E., Hollstein, M., Christman, M.F. and Ames, B.N. (1984) Detection of oxidative mutagens with a new *Salmonella* tester strain (TA102). *Meth. Enzymol.* 105: 249–254.

Levin, D.E., Hollstein, M., Christman, M.F., Schwiers, E.A. and Ames, B.N. (1982a) A new *Salmonella* tester strain (TA102) with A–T base pairs at the site of mutation detects oxidative mutagens. *Proc. Nat. Acad. Sci. USA* 79: 7445–7449.

Levin, D.E., Yamasaki, E. and Ames, B.N. (1982b) A new *Salmonella* tester strain, TA97, for the detection of frameshift mutagens. A run of cytosines as a mutational hot-spot. *Mutat. Res.* 94: 315–330.

Li, A.P. and Heflich, R.H. (1991) *Genetic Toxicology.* Boca Raton: CRC Press.

Lieber, M.R. (1998) Pathological and physiological double strand breaks. *Am. J. Pathol.* 153: 1323–1332.

Lutz, W.K. (1990) Endogenous genotoxic agents and processes as a basis of spontaneous carcinogenesis. *Mutat. Res.* 238: 287–295.

Lutz, W.K. and Maier, P. (1988) Genotoxic and epigenetic chemical carcinogenesis: one process, different mechanisms. *Trends Pharmacol. Sci.* 9: 322–326.

Malling, H.V. (1971) Dimethylnitrosamine: Formation of mutagenic compounds by interaction with mouse liver microsomes. *Mutat. Res.* 13: 425–429.

Malling, H.V. and Chu, E.H.Y. (1974) Development of mutational model systems for the study of carcinogenicity. In: Ts'o, P.O. and Dipalo, J.A. (eds) *Chemical Carcinogenesis*, Part B. New York: Marcel Dekker.

Malling, H.V. and Valcovic, L.R. (1978) New approaches to gene detection in mammals. In: Flamm, G. and Mehlman, M. (eds) *Advances in Modern Toxicology*, Volume 4. New York: Hemisphere Press, pp. 149–171.

Maron, D.M. and Ames, B.N. (1983) Revised methods for the *Salmonella* mutagenicity test. *Mutat. Res.* 113: 173–215.

McCann, J., Choi, E., Yamasaki, E. and Ames, B.N. (1975a) Detection of carcinogens as mutagens in the *Salmonella*/microsome test: Assay of 300 chemicals. *Proc. Nat. Acad. Sci. USA* 72: 5135–5139.

McCann, J., Dietrich, F., Rafferty, C. and Marton, A.O. (1993) A critical review of the genotoxic potential of electric and magnetic fields. *Mutat. Res.* 297: 61–95.

McCann, J., Spingarn, N.E., Kobori, J. and Ames, B.N. (1975b) Detection of carcinogens as mutagens: Bacterial tester strains with R factor plasmids. *Proc. Nat. Acad. Sci. USA* 72: 979–983.

Morelly, M.A. (2000) Industry viewpoint on thresholds for genotoxic carcinogens. *Toxicol. Pathol.* 28: 396–404.

Moutschen, J. (1975) *Introduction to Genetic Toxicology*. New York: John Wiley and Sons.

Muller, H.J. (1927) Artificial transmutation of the gene. *Science* 64: 84–87.

OECD (1986) *Guidelines for the Testing of Chemicals*. Paris: Organisation for Economic and Cooperative Development, Paris.

Ong, T., Stewart, J.D. and Whong, W.Z. (1990) Use of bacterial assay system for monitoring genotoxic complex mixtures in the occupational setting. *IARC Scientific Publications* 101–106.

Paolini, M. and Cantelli-Forti, G. (1997) On the metabolizing systems for short-term genotoxicity assays: a review. *Mutat. Res.* 387: 17–34.

Pavanello, S. and Clonfero, E. (2000) Biological indicators of genotoxic risk and metabolic polymorphisms. *Mutat. Res.* 463: 285–308.

Perera, F.P. (1984) The genotoxic/epigenetic distinction: relevance to cancer policy. *Environ. Res.* 34: 175–191.

Phillips, D.H. and Venitt, S. (eds) (1999) *Environmental Mutagenesis*. New York: Academic Press.

Pitot, H.C. and Dragan, Y.P. (2001) Chemical carcinogenesis. In: Klaassen, C.D. (ed.) *Casarett and Doull's Toxicology, The Basic Science of Poisons*, 6th edn. New York: McGraw-Hill, pp. 241–319.

Preston, R.J. and Hoffmann, G.R. (2001) Genetic toxicology. In: Klaassen, C.D. (ed.) *Casarett and Doull's Toxicology, The Basic Science of Poisons*, 6th edn. New York: McGraw-Hill, pp. 321–350.

Prival, M.J. and Dellarco, V.L. (1989) Evolution of social concerns and environmental policies for chemical mutagens. *Environ. Molec. Mutagen.* 14(S16): 46–50.

Prival, M.J., King, V.D. and Sheldon, Jr, A.T. (1979) The mutagenicity of dialkyl nitrosamines in the *Salmonella* plate assay. *Environ. Mutagen.* 1: 95–104.

Roe, F.J. (1989) Non-genotoxic carcinogenesis: implications for testing and extrapolation to man. *Mutagenesis* 4: 407–411.

Rueff, J., Chiapella, C., Chipman, J.K. *et al.* (1996) Development and validation of altern-ative metabolic systems for mutagenicity testing in short-term assays. *Mutat. Res.* 353: 151–176.

Shackelfod, R.E., Kaufmann, W.K. and Paules, R.S. (1999) Cell cycle control, checkpoint mechanisms, and genotoxic stress. *Environ. Health Persp.* 107 (Suppl. 1): 5–24.

Shaw, I.C. and Jones, H.B. (1994) Mechanisms of non-genotoxic carcinogenesis. *Trends Pharmacol. Sci.* 15: 89–93.

Sofuni, T., Hayashi, M., Nohmi, T., Matsuoka, A., Yamada, M.N. and Kamata, E. (2000) Semi-quantitative evaluation of genotoxic activity of chemical substances and evidence for a biological threshold of genotoxic activity. *Mutat. Res.* 464: 97–104.

Stouch, T.R. and Jurs, P.C. (1985) Computer-assisted studies of molecular structure and genotoxic activity by pattern recognition techniques. *Environ. Health Persp.* 61: 329–343.

Styles, J.A. (1977) A method for detecting carcinogenic organic chemicals using mam-malian cells in culture. *Br. J. Canc.* 36: 558–563.

Sugimura, T., Sato, S., Nagao, M. *et al.* (1976) Overlapping of carcinogens and mutagens. In: Magee, P.N., Takayama, S., Sugimura, T. and Matsushima, T. (eds) *Fundamentals in Cancer Prevention*. Baltimore, MD: University Park Press, pp. 191–215.

Tennant, R.W., Margolin, B.H., Shelby, M.D. *et al.* (1987) Prediction of chemical carcino-genicity in rodents from *in vitro* genetic toxicity assays. *Science* 236: 933–941.

Tindall, K.R., Stankowski, Jr, L.F., Machanoff, R. and Hsie, A.W. (1986) Analyses of muta-tion in pSV2gpt-transformed CHO cells. *Mutat. Res.* 160: 121–131.

Venitt, S. (1987) Use of mutagenicity assays in studies of human cancer. *Anticanc. Res.* 7:5B 949–954.

Venitt, S. and Parry, J.M. (eds) (1984) *Mutagenicity Testing*. London: Oxford University Press.

Wassom, J.S. (1989) Origins of genetic toxicology and the environmental mutagen. *Soc. Environ. Molec. Mutagen.* 14(S16): 1–6.

Waters, M.D., Lewtas, J., Nesnow, S., Moore, M.M. and Daniel, F.B. (1990) *Genetic Toxi-cology of Complex Mixtures*. Amsterdam: Kluwer Academic Publishers.

Waters, M.D., Stack, H.F. and Jackson, M.A. (1998) Inhibition of genotoxic effects of mammalian germ cell mutagens. *Mutat. Res.* 402: 129–138.

Watson, J.D. (1969) *The Double Helix*. New York: Mentor.

Watson, J.D. and Crick, F.H.C. (1953) Molecular structure of nucleic acids. *Nature* 171: 737–738.

Weisberger, J.H. and Wynder, E.L. (1984) The role of genotoxic carcinogens and of pro-moters in carcinogenesis and in human cancer causation. *Acta Pharmacol. Toxicol.* 55 (Suppl. 2): 53–68.

Williams, G.M. (1983) Genotoxic and epigenetic carcinogens: their identification and significance. *Ann. N.Y. Acad Sci.* 407: 328–333.

Chapter 11

Carcinogens and cancer

H. Vainio and C. Winder

Introduction

The history of occupational and environmental carcinogenesis bears the strong imprint of the clinical observations that led to successful identification of aetiological factors. The story of the first identified carcinogen is shown in Table 11.1, and begins with the classic description of cancers of the scrotum among chimney sweeps in 1775 (Pott 1775), then moves to the accounts of similar cancers in oil, tar and paraffin workers by the mid-19th century, and the experimental production in rats of skin cancer after exposure to coal tar in 1915 (Yamagiwa and Ichikawa 1915), culminating with the identification of polycyclic aromatic hydrocarbons such as benzo-a-pyrene in coal tar in 1933 (Cook *et al.* 1933).

Thus, over a 160-year period, one class of chemicals in which carcinogens occur (polynuclear aromatic hydrocarbons) was discovered, and a specific carcinogenic agent (benzo-a-pyrene) was isolated. The relative roles of epidemiology and of experiments in gathering the information relevant to the identification and

Table 11.1 The history of cancer and the polyaromatic hydrocarbons

Year	Author	Results
1775	Pott	Cancer of the scrotum in chimney sweeps
1850s		Cancer of the scrotum, skin, vulva from shale oils, tars and paraffin
1915	Yamagiwa and Ichikawa	Animal model established (rabbit ear carcinoma from tar and derivatives)
1933	Cook	Isolation of benzo-a-pyrene

prevention of the environmental causes of cancer have since been the object of much debate.

Many of the early epidemiological successes were based on observations made either in the occupational setting or in other situations of high-level exposure (for example, to drugs and cigarette smoking). In contrast, much of today's cancer epidemiology deals with general environmental exposures (for example, to pollutants in ambient air and in drinking water) and with personal behaviour patterns (for example, dietary habits, use of contraception, sexual behaviour, reproduction). Such exposure situations are more difficult to study epidemiologically than some of the earlier ones.

Toxicological approaches in nonhuman test systems can be used to avoid many of the problems of epidemiological studies. The capacity of animal model systems to serve as the only means for evaluating potential hazards is limited, however; therefore, the two approaches have complementary roles in the identification and prevention of environmental cancer risks.

In this chapter, aspects of environmental carcinogenesis are discussed and methods of identifying carcinogens as well as means of prevention are given.

What are the main causes of cancer?

Various types of agents cause various types of cancers. These include physical (such as sunlight), chemical (such as tobacco smoke) and biological (such as viruses) agents. An agent that causes cancer is called a **carcinogen**.

The ability of a carcinogen to cause cancer is really no different from any other toxic agent causing a toxic effect. Nevertheless, carcinogens have a number of characteristics that make them different from other types of toxicants. They have:

- long induction periods
- additive or synergistic effects
- nonapplicability of thresholds (that is a safe level of exposure), though many carcinogens show dose–response effects
- exposure to some carcinogens is unavoidable as they occur naturally in the environment
- for some carcinogens there appear to be levels below which the effects cannot be distinguished from background levels
- for some potent carcinogens any level of exposure is an unacceptable risk.

In an attempt to estimate the proportion of cancer deaths (in the USA) which could be attributed to various factors, Doll and Peto (1981) grouped the factors and classes of factors that cause cancer as shown in Table 11.2.

From the preventive point of view it is noteworthy that lifestyle factors (such as smoking and diet) feature very prominently in the development of cancer.

It is also noteworthy that public concerns about the things that cause cancer tend to be non-lifestyle factors such as pollution and industrial products, and these actually count for a relatively small proportion of the causes of cancer.

Table 11.2 The causes of cancer

	Percentage of all cancer deaths	
Factor or class of factors	Best estimate	Range of acceptable estimates
Tobacco	30	25–40
Alcohol	3	2–4
Diet	35	10–70
Food additives	Less than 1	−5*–2
Reproductive and sexual behaviour	7	1–13
Occupation	4	2–8
Pollution	2	Less than 1–5
Industrial products	Less than 1	Less than 1–2
Medicines and medical procedures	1	0.5–3
Geophysical factors	3	2–4
Infection	10?	1–?
Unknown	?	?

*Some food additives, such as antioxidants, may protect against cancer.

Cancer as a public health burden

About 20% of the people in the developed part of the world will contract cancer.

The total number of new cancer cases (excluding nonmelanoma skin cancer) in the year 2000 was 10.1 million, just over half of which occur in the developing countries. Of the estimated 6.2 million deaths in the year 2000 worldwide from cancer (excluding nonmelanoma skin cancer), over half occurred in developing countries. Also, 22 million people are living with cancer (within 5 years of diagnosis) (Parkin 2001).

The most common cancer in the world today, in terms of incidence, is lung cancer, accounting for 12.3% of cancers of the world total and 1.1 million deaths a year. The estimated number of cases worldwide has increased by 20% since 1990 (17% in men and 27% in women).

Breast cancer is the second most frequent cancer in the world (1.05 million new cases), and is by far the most common cancer among women (22% of all new cases). Stomach cancer is the fourth in frequency (almost 9% of all new cancers and 647,000 deaths) and colorectal cancer ranks the third (9.4% of the world total). Liver cancers accounting for 5.6% of new cancer cases worldwide. There are large differences in the relative frequency of different cancers by world area. The major cancers of developed countries (other than those already named) are cancers of the colon–rectum and prostate gland, and in developing countries, cancers of the cervix uteri and oesophagus (Parkin *et al.* 1999).

In men, the risk of dying from cancer is highest in Eastern Europe, with an age-standardised rate for all sites of 205 deaths per 100,000 population. Mortality rates in all other developed regions are around 180. The only developing area with an overall rate of the same magnitude as that in developed countries is southern Africa. All of eastern Asia, including China, has mortality rates above the world average, as do all developed countries. The region of highest risk among women is northern Europe (age-standardised rate = 125.4), followed by North America, southern Africa and tropical South America. Only south-central and western Asia (Indian subcontinent, central Asia and the middle-eastern countries) and Northern Africa are well below the world average of 90 deaths per 100,000 population annually (Pisani *et al.* 1999).

Diet and lifestyle are considered to be the most prevalent causes of cancer. Cancers at two of the leading sites in males, lung and mouth/pharynx, are due largely to the effects of tobacco, whether chewed or smoked, and abuse of alcohol. Cancer of the large bowel has been linked to diet, as has that of prostate. Breast and cervix cancers, though confined to the female gender, are high in the rank order for both sexes combined. Female breast cancer, the incidence of which seems to have stabilised in several western populations but is increasing rapidly in populations that have hitherto been at low risk (such as in Japan), is also believed to have a dietary component. Lung cancer, sixth in rank among females, is likely to become much more common because of the current smoking habits of women. By the year 2000, lung cancer was the first ranking cancer in the two sexes combined (Parkin 2001). Tomatis (1990) and Stewart and Kleihues (2003) give further discussion of the cancer burden and of target sites.

The process of carcinogenesis

Carcinogenesis is a complex process involving sequential genetic events which result in altered functions of the genes that control normal cellular growth, with subsequent clonal growth of the resulting 'preneoplastic' or neoplastic cells (Farber 1984; Boyd and Barrett 1990). Chemical carcinogens may act by inducing mutations and/or by altering gene and cellular growth control. Figure 11.1 presents a simplified schematic illustration of the development of cancer.

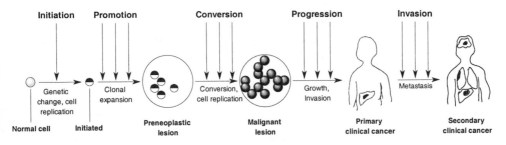

Figure 11.1 Development of cancer – initiation, promotion and progression. Schematic illustration of the development of cancer, depicting the stages of initiation, promotion and progression. Initiation results from the action of a chemical, physical or biological agent to alter irreversibly the information in the genetic material. Initiated cells have the potential to develop into a clone of preneoplastic cells. Promotion is characterised by alteration of genetic expression and the (reversible) clonal growth of the initiated cell population. Progression, which is the final stage of cancer development, is characterised by demonstrable changes in the number and/or arrangement of chromosomes associated with increased growth rate, invasiveness and metastasis.

Therefore, cancer is a multistage process involving multiple genetic and epigenetic events (indirect changes to genetic material made through biotransformation, hormonal, injury response or other mechanisms). The stages in cancer are outlined in Table 11.3.

Cancer growth is usually measured by estimating the number of cell doublings from induction. Figure 11.2 below shows the number of cell doublings required before a large mass of cells is identified by clinical measures.

The incidence of many human cancers rises sharply with age. For cancers of the stomach and lung, for instance, the age-specific incidence rates increase exponentially with age, with exponents of 4–7 for ages between 30 and 65 years. This time incidence curve reflects a multistage process for carcinogenesis that requires 4–7 independent 'events' (Armitage 1985). Because of the heritable nature of cancer, these 'events' are probably genetic changes. Tumour development in the human colon, for instance, appears to require more than one type of genetic change; these include hypomethylation, gene mutation, recombination and chromosomal aberrations (Fearon et al. 1987; Vogelstein et al. 1988). Some of these changes represent DNA damage, while others are constitutive processes that regulate gene expression.

Thus, the accumulated experience in the field of carcinogenesis clearly supports the concept that cancer development is a multistage process and that multiple genetic changes are required before a normal cell becomes fully neoplastic.

Table 11.3 The multistage process of cancer

Stage	Cell activity	Activities leading to cancer	Modifying factors
Normal activity	Basal activity		
Exposure	Absorption and circulation	Delivery to target tissues/cells	Excretion
Disposition	Biotransformation	Activation	Detoxification
Genetic change	Binding to genetic materials	Change in DNA structure	DNA repair mechanisms
Initiation	Change in cell function	Production of an initiated cell	Growth control, resistance to cytotoxicity
Promotion	Selective clonal expansion	Production of a preneoplastic lesion	Inactivation of tumour suppressor genes
Conversion	Further genetic changes	Production of a malignant tumour	Immune function
Growth and invasion	Invasion of tissues/ organs	Clinical cancer	Angiogenesis
Progression	Metastasis	Clinical cancers	

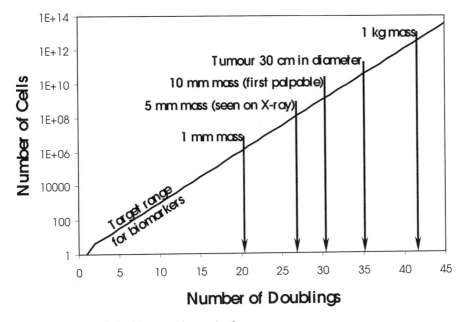

Figure 11.2 Cell doublings and growth of cancers.

Likewise, studies of human tumours suggest that the multistage paradigm, together with similar genetic events, is involved in the development of cancer in humans, and that the carcinogenic process is indistinguishable among mammals, from laboratory rodents to humans. This finding plus the knowledge that all chemicals known to induce cancer in humans that have been studied under adequate experimental protocols also cause cancer in laboratory animals lead most prudent investigators to the speculation that the obverse would hold true: chemicals shown unequivocally to induce cancer in laboratory animals are capable of causing cancer in humans.

As more advances are made in molecular carcinogenesis, knowledge about the mechanism of cancer induction in mammals will shed more light on the usefulness of animals as predictive surrogates for humans. The genes that control cellular growth and differentiation are likely to play a fundamental role in basic cellular function in view of the degree to which they have been conserved throughout eukaryotic evolution; for instance, the protein sequence of the H-ras gene is similar in man and rats (Barbacid 1987).

Proto-oncogenes are cellular genes that are expressed during normal growth and development (Weinberg 1985). The proteins coded by proto-oncogenes are involved in mitogenic regulation, including growth factors, protein kinases and signal transducers, all of which, when behaving aberrantly, may lead to uncontrolled cell proliferation (Torry and Cooper 1991). Proto-oncogenes can be activated to cancer-causing oncogenes by point mutations or by gross DNA rearrangements (chromosomal translocation or gene amplification) (Anderson and Reynolds 1988; Bos 1989; Reynolds and Anderson 1991; Simon *et al.* 1995). These lesions are especially revealing for chemicals that are apparently nonmutagenic in standard genotoxicity tests and yet cause point mutations in exposed animals (Reynolds *et al.* 1987). Loss of specific regulatory functions (for example, tumour suppressor genes) represents an important feature in neoplastic transformation (Weinberg 1989). Inactivation of tumour suppressor genes, like activation of proto-oncogenes, can occur via mutations but can also result from loss of the gene or portions of the chromosome containing it. The absence of specific chromosomes in tumours frequently indicates that loss of tumour suppressor gene(s) function is associated with the etiology of the tumour (Knudson 1985). Hereditary inactivation of a tumour suppressor gene is the strongest risk factor for human cancer known to date – an indication of the importance of this class of genes in human carcinogenesis (Friend *et al.* 1986; Harbour *et al.* 1988).

In certain malignancies, such as retinoblastoma, loss of function of only a single suppressor gene is implicated. However, in other cancers, studies of restriction fragment length polymorphism studies have indicated that loss of multiple tumour suppressor genes may be necessary for progression of the full malignant condition. Examples of such malignancies include small cell lung cancer (Yokota *et al.* 1987) and colorectal cancer (Vogelstein *et al.* 1988).

Critical regulators of genomic integrity, as exemplified by mismatch repair genes, have also been implicated as tumour-suppressor genes. The microsatellite instability genes, *MLH1* and *MSH2*, are important to maintenance of genomic integrity by repairing mismatched base pairs that arise with a stable frequency during DNA replication (Kolodner and Marsischky 1999). Without such genes, repairs are not made and mutations are introduced into newly synthesised DNA. Alternatively, the stress of mismatch structure may fragment the DNA. Both of these possibilities can lead to cell growth or death.

Types of cancer risk factors

The categories of known risk factors for cancers are listed in Table 11.4; these may act individually or in combination.

Genetic factors are inherited at conception and their control is presently not feasible. Individuals have some control over their behaviour, including tobacco use, diet, alcohol use, exposure to sunlight, sexual behaviour patterns and general personal hygiene. Environmental factors include occupational exposures to carcinogens, exposures during medical procedures, as well as factors that occur naturally or are synthetic and contaminate water, air and soil. These factors are beyond the individual's control, and their effective control thus requires broad social action. For example, there is an obligation that employers who employ workers exposed to known or suspected carcinogens will take active steps to eliminate or minimise exposures to levels as low as practicable.

Identifying carcinogens

Many, if not all, human tumours are clonal in origin and result from the expression of a single transformed cell. This argues that a single step (probably mutation) is common to all cancer processes. As noted above, molecular biology studies of normal and tumour cells have demonstrated that proto-oncogenes and tumour sup-

Table 11.4 Risk factors for cancer

Category	Example
Endogenous	
Genetic predisposition	Xeroderma pigmentosum
Hormones	Oestrogens
Exogenous	
Chemicals	Benzene
Industrial processes	Rubber industry
Dusts	Asbestos
Hormones	Oestrogen replacement therapy
Ionising radiation	X-rays; radiotherapy
Nonionising radiation	Ultraviolet light
Viruses	Hepatitis B virus
Cultural habits	Tobacco smoking
Iatrogenic exposures	Cytotoxic drug therapy
Diet	Excessive caloric intake, low fibre diet
Socioeconomic	Less favoured occupational or social class

pressing genes, present in all cells, appear to influence the initiation, expression and expansion of most tumours following their breakage by mutation, chromosome breakage or gene amplification.

For the vast majority of potential carcinogens, the scientific basis for determining whether they will increase the incidence of human cancer is highly uncertain. Because of this uncertainty, labelling a substance as a 'carcinogen' for regulatory purposes almost always involves a hypothetical exercise.

- Does this substance cause cancer in humans?
- Should this substance be treated as if it causes cancer in humans?

Methods for assessing carcinogenicity

There are a range of approaches used in detecting carcinogens:

- Epidemiology of human populations.
- Use of long-term animal toxicity bioassays.
- Use of validated short-term mutagenicity assays (either *in vitro* or *in vivo* tests).
- Use of predictive structure-activity relationship (SAR) modelling.

The role of epidemiology

Direct answers to the question 'Is a particular agent carcinogenic for humans?' can be obtained only from studies of human beings, using an epidemiological approach. Epidemiology plays an important role in identifying carcinogenic chemicals in the industrial environment. It is also useful in identifying exposures that may be associated with an increased incidence of cancer, although the agent or agents responsible for the increased incidences may often not be identified.

Sources of epidemiological evidence

Epidemiological studies are prompted by different sources of evidence (Tomatis 1989), which include results from the:

- laboratory; and
- two kinds of observations in humans: anecdotal clinical evidence, and results of exploratory analyses of cancer mortality and morbidity statistics.

Montesano *et al.* (1980) noted that 'the identification of human carcinogens has occurred under conditions of exposure similar to those used in experimental carcinogenesis'.

Application of epidemiological methods to the detection of cancer

In applying epidemiological methods to the detection of cancer in industrial populations, a positive relationship is based on the statistical significance of the data. However, it is important to remember that a positive statistical association does not necessarily mean that a cause–effect relationship has been established. Negative epidemiological studies are also useful in that they help define the upper limit of human cancer risk.

Nine criteria should be met ideally in order to establish whether an observed relationship is likely to be causal, having ascertained that the exposed persons are at higher risk than the unexposed (Bradford-Hill 1965):

- Consistency and unbiasedness of the findings.
- Strength of the association.
- Temporal sequence.
- Biological gradient (dose–response relationship).
- Specificity.
- Coherence with biological background and previous knowledge.
- Biological plausibility.
- Reasoning by analogy.
- Experimental evidence.

An example of the use of epidemiological evidence to establish cancer causality with chemical exposure is the increase in the incidence of angiosarcoma of the liver of vinyl chloride-exposed workers (Creech and Johnson 1974; Doll 1988).

The International Association for Research on Cancer (IARC), in collaboration with the Commission of the European Communities (CEC) have also compiled a list of ten criteria to help in selecting occupational exposures for epidemiological study. Description of the ten criteria are as follows (IARC 1987):

- Number of workers exposed.
- Level of exposure of workers.
- Quality of exposure data.
- Carcinogenic potential.
- Evidence of human carcinogenicity.
- Ongoing exposure to known carcinogens at permissible levels of exposure.
- Trends in exposure.
- Control of confounding factors.
- Cases potentially attributable to exposure.
- Time since first exposure.

Therefore, epidemiology is indeed an important technique. However, there are major limitations.

Limitations of the epidemiological method

There are a number of major weaknesses to the epidemiological approach.

First of all, epidemiology cannot be proactive. It cannot be used to assess the potential carcinogenicity of chemicals to which human exposure is anticipated.

While epidemiology is reasonably good at identifying intermediate and high levels of risk, it is quite weak at identifying the causes of very low levels of risk. Very small differences in risk between a group exposed to some agent as opposed to a group not exposed could be due to a variety of reasons (for example, chance) or other differences between the exposed and unexposed which are not known or cannot adequately be controlled. Because of this, it becomes next to impossible to say with any confidence that a very low level of risk is caused by a similarly low-level exposure to some substance.

Epidemiological methods are insensitive. Because most human exposures are to such low levels as to preclude the direct measurement of risk, negative results do not necessarily demonstrate the absence of hazard because of the insensitivity of epidemiological methods in detecting cancer incidences.

Epidemiology can be very resource intensive. Large-scale epidemiological studies, particularly cohort studies, are time-consuming and can be very expensive.

One fairly critical limitation is the latent period. This is the interval between exposure to a cause of a disease and the manifestation of the disease itself. Latent periods for cancer causing exposures can be quite long – from 5 years to over 50 years. The long latency period between first exposure to a carcinogenic chemical or physical factor and the chemical detection of the tumour limits the usefulness of epidemiology.

Availability of controls is another critical factor. Suitable unexposed control groups are sometimes difficult to obtain if exposure to the chemical is widespread.

Another weakness of the method is its lack of specificity of exposure. Often, in fact usually, epidemiologists are studying the effect of exposure to mixtures of chemicals or other agents, rather than pure exposures to a single toxic substance. Most particularly, retrospective studies are often reliant on poor indicators of exposure.

Background incidences of disease may swamp any effect. Although an increase in the incidence of a rare human cancer may be detected using epidemiology, an increase in common cancers such as cancer of the lung, breast, or colon will not likely be detected except in the case of high incidence of the disease.

A further limitation is the inability to control for unknown risk factors for the disease in question. In an experiment the randomisation procedure hopefully introduces some control for unknown factors. In observational studies, control for unknown confounding factors is possible, but obviously cannot guarantee control for unknown confounders.

The last weakness of the approach encompasses the practical problems involved in doing epidemiology. In order to effectively function, an epidemiologist needs to have access to appropriate information on large groups of people (Weiss 1996). Sometimes this is possible, and sometimes the information just does not exist. In addition, there is currently a lack of enough competent epidemiologists even to evaluate adequately the information that does exist.

In spite of these limitations, epidemiology is a valuable technique. It is the only method currently available that provides direct evidence of the presence or absence of human cancer risks. In future, great hopes are placed on the development of molecular epidemiology with more possibilities for prevention.

Experimental identification of carcinogens

The epidemiological characterisation of carcinogens relies on:

- the production of cancer in a defined group of the population
- the availability of data relating to exposure to a causal or putative causal agent
- an association to be made between exposure and disease.

This process is reactive in nature, and requires the production of disease as proof of causation (that is, a body count). The long latency of cancer also means that even after the association is established, and relevant controls are introduced, cases of cancer to the exposure will still occur for many years. For example, asbestos-related cancers are still being reported, even though exposures began declining in the late 1960s and were largely removed by 1976. These cancers are still occurring, from workers who were exposed to asbestos in the pre-1966 period.

Such processes can never be proactive or preventive. Therefore, it would be more appropriate to define the carcinogenic potential of a material before humans are exposed to it, rather than wait 20–50 years for the epidemiology to establish any effect.

For ethical and other reasons, the characterisation of carcinogenic potential can only be made using nonhuman species.

There are two main approaches to experimental investigation of carcinogenic potential:

- use of long-term repeated-dose toxicity testing in animal models
- use of short-term tests of genotoxicity.

Use of animal models

Animal models, particularly the long-term repeated-dose bioassays, are generally considered to provide the most definitive carcinogen assessment of a chemical (OECD 1993). Reasons for this are manifold:

- The range of doses employed can study the full toxic range of the test material
- Human exposures are sometimes too low, or in too small a population, or of too short a duration to unequivocally distinguish carcinogenicity
- It is possible to investigate pathological changes (such as tumourigenicity) in test animals through gross necroscopy for tumour masses and by extensive bio-chemical, clinical, morphological and histological examinations in a variety of tissues and organs.

Animal bioassays

The essence of the animal bioassay for carcinogenicity is to observe test animals for a major portion of their lifespan for the appearance of tumours resulting from expo-sure to various doses of a chemical or chemical mixture. There is some debate about the utility of long-term animal toxicity studies for the delineation of other toxicity parameters, although the need for such studies for the investigation of car-cinogenicity remains unquestioned.

Typical bioassays consist of continuous exposure of approximately 400 test animals (usually laboratory rats or mice) in groups of 50 sex/dose group. Choice of doses is usually based on a fraction of the maximum tolerated dose (MTD) from other repeated dose studies (Counts and Goodman 1995). Four dose groups are usually employed (control, low-, mid- and high-dose groups). The main aims of the bioassay are to establish nonspecific toxicity in the high-dose group, minor non-significant effects in the low- or (particularly) the mid-dose group, and a no observ-able effect level (NOEL) in the low- or mid-dose group.

Design and conduct of animal bioassays

A number of constraints apply to the design and conduct of animal bioassays for estimating long-term repeated effects of chemicals (OECD 1993).

- The chemical should be examined in both sexes of at least two species (usually rats and mice, although other species, such as monkeys, may be used where the expense is warranted).
- At least two and preferably three doses should be administered.
- The minimum number of animals of each sex at each dose is usually 50.
- A suitable control group of the same size should be included.
- The exposure should start shortly after weaning and continue for a major part of the animal's life (in the rat and mouse, at least 24 months).

To identify a carcinogen, it is necessary to observe an increase in occurrence of tumours above a baseline (control) level, so a nondosed control group must be maintained for comparison. Further confirmation is available if animals receiving higher doses also show increased incidences of cancer.

Interpretation of the results of animal studies

In evaluating the available data, it has to be realised that no single piece of evidence by itself provides the definitive answer. It is only when all the evidence is evaluated, that implications to human health can be defined. It is particularly important when considering the relevance of ambient exposures to humans that the mechanism of action and the probable shape of the dose–response curve at very low doses are considered.

A number of international agencies have developed schemes for that interpretation of the results of animal bioassays. Of these the most well established are those of IARC and the US National Toxicology Program. The former rely on expert review of published evidence relating to exposure of animals to suspect agents; the latter rely on a formal testing program where a suspect agent is given to experimental animals in carefully controlled bioassays. Both the IARC and NTP criteria rely on a qualitative evaluation of the evidence, not a quantitative evaluation. Classification categories such as 'sufficient evidence' and 'limited evidence' refer only to the strength of the experimental evidence that these chemicals are (or are not) carcinogenic and not to the extent of their carcinogenic activity.

In evaluating experimental animal data, it must be recognised that animal testing is not a perfect tool. For example, inorganic compounds of arsenic, which are highly suspected of producing an increased incidence of cancer in workers, but for which animal experimentation has to date been unsuccessful.

The concept of dose–effect relationship and the existence of a 'no effect' dose level are major outcomes of animal experimentation. The establishment of dose–effect relationships and cancer is mainly based on animal experimental work. The zero-effect dose seems thus to be a phenomenon closely related to the number of animals, the species and the route and type of administration, which may be difficult to extrapolate to other population sizes, or to other species, absorption routes or exposure times.

The adequacy of animal tests in human risk characterisation

One method for judging the adequacy of the animal bioassay to identify accurately potential human carcinogens is to examine those chemicals known to be carcinogenic in humans for their ability to cause cancer in animal models.

Purchase (1982) determined how accurately the mouse can predict for the ability of chemicals to cause cancer in the rat and vice versa. Thus, the rat as a predictor of chemically induced cancer in the mouse has a specificity of about 85%. The mouse as a predictor of cancer in the rat has a specificity of about 82%. It is reasonable to assume that the ability of the rat or mouse to predict for the potential or lack of potential of a chemical to cause cancer in man is no better than their ability to predict for cancer in each other.

Adamson and Sieber (1983) studied the carcinogenicity in nonhuman primates of eight model rodent carcinogens. With the exception of urethane, none of these model compounds has yet induced tumours in Old World Monkeys even after prolonged periods of administration and observation. Thus, the use of monkeys is restricted by the small numbers of animals that can be used.

Limitations of animal testing

There are limitations in the use of animal testing:

Relevancy

Some animal studies will not show cancer using materials known to cause cancer in humans. For example, inhalation studies with asbestos, arsenic and cigarette smoke will not *generally* produce cancer in laboratory animals.

Tumour predisposition

Some animal strains used in toxicological testing are known to produce more tumours than other strains of the same species. Other tumour sites, such as the rat liver hepatocellular carcinoma model, or mammary tumours in the mouse, may or may not be predictive of cancer in humans.

Metabolic differences

Dispositional, detoxification and activation differences in animals may produce false-positive or false-negative results (for example, saccharin, DDT). About 80% of human carcinogens are carcinogens in either mouse or rat.

Study biases

The use of doses based on the MTD can lead to artificially distorted tumour incidences. For example, tumour induction may be due to generalised nonspecific toxicity such as impaired hormonal or immune function, or saturation of normally functioning metabolic pathways.

One major shortcoming with long-term repeated-dose studies is their expense. A detailed 2-year rat study can cost $5 million.

Extrapolation from animals to humans

One assumption of toxicology is that effects observed in nonhuman mammals are likely to be observed in humans. However, debate on extrapolation of animal data to humans continues:

■ Humans are genetically more heterogeneous than test animal strains. It is thus more likely that a highly sensitive or highly resistant individual will be found in any human population and that such an individual will be detected among a limited number of animals.

■ There are reasons to believe that a number of endogenous and/or exogenous factors interfere with the response towards a carcinogen, either strengthening or antagonising the carcinogenic action of a substance.

■ The age of the individual at the start of the exposure may be an important parameter in determining the response. Young animals are more sensitive than old animals. Laboratory animals are selected according to age, weight or both at the start of the experiment and are thus more biologically homogeneous. Human populations are not standardised in any respect when occupational exposure starts.

■ Fractionated exposure, that is, low doses over a long time, is a common feature in human populations which may increase the risk of cancer.

■ Human populations are likely to be exposed to more than one potential carcinogen at one time during life. Additive or synergistic effects may therefore be expected. Moreover, carcinogens are metabolised by different species in a different manner and at a different rate into biologically active or inactive metabolites. There are differences in the metabolic mechanisms of man and animals. Extrapolation of 'safe' doses from animal experiments to man is therefore at present still not possible, and in any case it would not ensure safety when the total exposure panorama is considered.

■ The size of an animal test population as compared to man is a considerable problem when discussing zero-effect levels of carcinogens.

In summary, although providing very useful information, experimental studies on chemical carcinogenesis may be insufficient for establishing a risk estimate for human exposure in the work environment (Wadell 1996). The best way to establish a true risk estimate for human exposure is by means of an integrated evaluation of epidemiological studies and proper animal experiments. In the absence of an adequate dose–response curve for a human population as obtained by epidemiological studies, animal data should essentially be used to establish carcinogenicity as such, and possibly for comparing the risk potential from one substance to another. Experimental animal testing remains indeed necessary in the case of new substances for which epidemiological studies are obviously not applicable. From a practical point of view, exposure to an experimental carcinogen should be kept as close to zero as possible in the occupational environment, irrespective of dose level in the test system, animal species, tumour site, type or frequency.

The IARC first stated its position concerning the evaluation of the results of carcinogenicity studies conducted in experimental animals and the possible association of the results of such carcinogenicity studies to humans in 1979 in a preamble in its early monographs. The latest edition of this preamble is now on the IARC website at http://www.itcilo.it/english/actrav/telearn/osh/kemi/iarc.htm#Preamble.

Among other things, the preamble states:

Although this association cannot establish that all agents and mixtures that cause cancer in experimental animals also cause cancer in humans, nevertheless, in the absence of adequate data on humans, it is biologically plausible and prudent to regard agents and mixtures for which there is sufficient evidence of carcinogenicity in experimental animals as if they presented a carcinogenic risk to humans.

Inherent in this statement is the implication that a correlation between carcinogenicity in animals and possible human risk cannot be made on a scientific basis. Rather, the statement is a pragmatic one, made with the intent of assisting regulatory agencies in making decisions related to the control of potentially carcinogenic substances.

Short-term tests of genotoxicity

The cost of a long-term repeated-dose carcinogenicity bioassay is expensive. Also, the tests are lengthy, often taking over 3 years to complete. It is impossible to envisage spending such time and money testing all the chemicals in current use. Short-term tests offer one solution.

Substantial evidence indicates that one step in carcinogenesis is damage to or mutation of DNA. The first short-term test was the Ames test, introduced by Bruce Ames and colleagues in the early 1970s (Ames 1973). The test relies on the fact that many carcinogens initiate cancer through damage to DNA (that is, a mutagenic event), and the Ames test sought to measure mutagenicity. The test relies on strains of the bacteria *Salmonella typhimurium* to back-mutate to a form that will grow on special media. Over the last 20 years, the Ames test has become more sophisticated, with the number of strains being standardised, and with refinement of test conditions.

It important to note that the aim of most short-term tests is to detect changes to genetic material, that is, genotoxic changes. However, as with *in vivo* experiments, a number of terms has arisen, which have produced confusion (see Figure 11.3).

Since 1973, a number of short-term genotoxicity tests have been developed. The short-term tests currently available can be broadly classified into three major groups:

- those that detect gene mutations and chromosome aberrations
- those that are capable of detecting transformation of mammalian cell *in vitro*
- those that measure interactions of chemicals with critical molecules (primarily DNA).

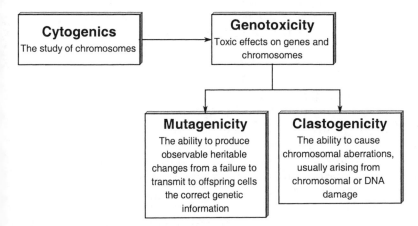

Figure 11.3 The terminology of genotoxicology/cytogenetics.

False negatives and false positives in testing

Some chemicals are active in some short-term tests, but fail to produce cancer in animal models (mutagenic noncarcinogens). These include chemicals such as sodium ascorbate, 1-nitronaphthalene, 3-nitropropionic acid, sodium fluoride and caffeine.

Other chemicals increase tumours in rodent bioassays that are not readily detected in short-term tests (nonmutagenic carcinogens). Examples here include arsenic, chloroform and carbon tetrachloride, ethylene thiourea, progesterone, polychlorinated biphenyls (PCBs), amitrole, phenobarbitol, heptachlor, dieldrin, aldrin, asbestos, dichlorodiphenyltrichloroethane (DDT), 2,3,7,8-tetrachlorodibenzo-*p*-dioxin (TCDD), and diethylstilbestrol (DES).

There are several possible explanations for the activity of nonmutagenic carcinogens:

- Unusual metabolic processes are required for carcinogenic activation not normally available in standard mutagenic assays (amitrole, DES).
- Mutagenic activity is limited to structural or numerical chromosomal changes at the chromosomal, rather than the DNA, level (such as benzene, arsenic, asbestos and DES).
- They inhibit DNA methylation (ethionine and 5-azacytidine).
- They act as tumour promoters (such as phenobarbitol, asbestos, some hormones and TCDD).

Possible explanations for the activity of mutagenic noncarcinogens include:

- Differences in metabolic activation/detoxification or DNA repair.
- A mutagenic change such as aneuploidy might not be relevant to the multi-stage carcinogenic process.
- The initiated cell might not proliferate enough to produce cancer.
- The initiated cell might die before dividing.
- Some mutagenic changes might be so slow in inducing carcinogenesis that the life of the test animal may be too short to detect any pathological change.
- Some *in vivo* mutagenicity tests are insensitive to weak mutagens.

Use of short-term tests

The IARC have evaluated the utility of short-term tests in determining the potential carcinogenicity of chemicals to humans or experimental animals. They conclude that because the mechanisms of carcinogenesis are incompletely understood, the results of the short-term tests *should not be used by themselves to conclude whether or not an agent is carcinogenic*. This also concludes that the results of short-term tests should not be used to predict the carcinogenic potency of a chemical in experimental animals and humans.

Limitations of short-term tests include false positives (a chemical is mutagenic but the test was unable to detect any effect) and false negatives (a chemical is not mutagenic, but the test detected that it was).

Some chemicals are active in some short-term tests, but fail to produce cancer in standard animal models (mutagenic noncarcinogens). These include chemicals such as sodium ascorbate, 1-nitronaphthalene, 3-nitropropionic acid and caffeine.

Other chemicals increase tumours in rodent bioassays that are not readily detected in short-term tests (nonmutagenic carcinogens). Examples here include arsenic, chloroform and carbon tetrachloride, ethylene thiourea, progesterone, PCBs, amitrole, phenobarbitol, heptachlor, dieldrin, aldrin, asbestos, DDT, TCDD and DES.

Structure–activity relationships

An estimate of a chemical to produce cancer in humans or experimental animals by implication may also be obtained by comparing its structure with that of known carcinogenic chemicals. There are three categories of criteria for suspecting chemical compounds of carcinogenic activity:

- structural criteria, considering: formal structural analogies with established chemical carcinogens; molecular size, shape or symmetry; electron distribution; any reactive functional groups; or the possibility of undergoing metabolism to reactive intermediates
- functional criteria, such as mutagenicity, induction of DNA repair, immune suppression and so on
- 'guilt by association' of chemicals which although found inactive under standard bioassays, belong to a chemical class with known carcinogenic properties.

Quantitative structure–activity relationships (QSAR) and molecular modelling are becoming useful tools in cancer research (Kubinyi 1990).

Of the various classes of compounds, those of particular concern are chemically reactive substances, which act as direct alkylating substances such as polyaromatic hydrocarbons, nitrosamines, aromatic nitrocompounds and substituted hydrazines. However, subtle changes in the structures of various members of these and other classes of compounds may markedly alter their carcinogenicity, in relation to other members of the class.

The process of carcinogen identification

The results of epidemiological studies, in combination with the results from animal oncogenicity studies, and studies using various techniques for measuring the ability of the chemicals to alter the structure of DNA in plant and animal cells, allow a more accurate determination of the potential cancer risk to humans than can be arrived at by evaluating the results of each of the various techniques separately.

Although significant progress has been made, several important questions have still not been completely resolved. These include the extent to which human cancers are due to specific causes, such as chemicals, hormones and physical and viral agents, the role of nutritional factors and the interactions of endogenous and environmental factors.

Identification of cancer risk factors

The identification of carcinogens through laboratory experiments relies on two main types of tests:

- long-term carcinogenicity tests in rodents (mice, rats, hamsters)
- short-term tests assessing the effect of an agent on a variety of genetic and related effects.

These tests are valuable to the extent that such effects may reflect underlying events in the carcinogenic process. Indeed, consistent positivity in tests measuring mutations (point mutations or chromosomal mutations) is usually regarded as indicating potential for carcinogenicity of the tested agent. Results from laboratory experiments constitute useful supporting evidence when adequate epidemiological data for the carcinogenicity of an agent exist (for example, vinyl chloride) but become all the more essential when the epidemiological evidence is non-existent or inadequate in quality and/or quantity. In the latter case, although no universally accepted criteria exist to automatically translate data from long-term animal tests or short-term tests in terms of cancer risk in humans, an evaluation of the risk can be made on a judgemental basis using all available scientific evidence.

The IARC has applied this policy for Research on Cancer in a systematic program of 'Evaluation of Carcinogenic Risks to Humans'. The IARC evaluation process is essentially qualitative, aimed at assessing the strength of evidence that an agent is or is not carcinogenic to humans (risk identification) and it does not extend to the subsequent stage of risk quantification nor to risk management.

Although significant progress has been made, several important questions have still not been completely resolved. These include the extent to which human cancers are due to specific causes, such as chemicals, hormones and physical and viral agents, the role of nutritional factors and the interactions of endogenous and environmental factors.

Recognised human carcinogens

For a number of chemical, physical, and biological environmental agents there is direct evidence from epidemiological studies, usually supported by experimental evidence, that they cause malignant neoplasms in humans (Doll and Peto 1981). Nearly 90 agents have been shown to be causally related to human cancer (see Table 11.5).

The majority are chemicals to which humans have been exposed only relatively recently (Tomatis et al. 1989). Most are either chemicals to which people are exposed occupationally, pharmaceutical products or naturally occurring compounds, to which specific groups of people have been exposed, at high concentrations, for long enough so that the increased risk could be detected by the methods of epidemiology.

Among physical agents, UV radiation as part of the solar spectrum, as well as particulate ionising radiations, are clearly carcinogenic in humans.

Table 11.5 Established human carcinogens

Agents and groups of agents (64)
Aflatoxins, naturally occurring
4-Aminobiphenyl
Arsenic and arsenic compounds
Asbestos
Azathioprine
Benzene
Benzidine
Beryllium and beryllium compounds
N,N-Bis(2-chloroethyl)-2-naphthylamine (chlornaphthazine)
Bis(chloromethyl) ether
Chloromethyl methyl ether (technical grade)
1,4-Butanediol dimethanesulfonate (busulfan)
Cadmium and cadmium compounds
Chlorambucil
1-(2-Chloroethyl)-3-(4-methylcyclohexyl)-1-nitrosourea (methyl-CCNU/semustine)
Chromium[VI]compounds
Cyclosporin
Cyclophosphamide
Diethylstilboestrol
Epstein–Barr virus
Erionite
Ethylene oxide
Etoposide in combination with cisplatin and bleomycin
[Gamma radiation: see X- and γ-radiation]
Helicobacter pylori (infection with)
Hepatitis B virus (chronic infection with)
Hepatitis C virus (chronic infection with)
Human immunodeficiency virus type 1 (infection with), human papillomavirus type 16
Human immunodeficiency virus type 1 (infection with), human papillomavirus type 18
Human T-cell lymphotropic virus type I
Melphalan
8-Methoxypsoralen (methoxsalen) plus ultraviolet A radiation
MOPP (methchlorethamine, oncovin, procarbazine, prednisone) and other combined
chemotherapy including alkylating agents
Mustard gas (sulfur mustard)
2-Naphthylamine
Neutrons
Nickel compounds
Oestrogen therapy, postmenopausal
Oestrogens, nonsteroidal

continued

Table 11.5 continued

Oestrogens, steroidal
Opisthorchis viverrini (infection with)
Oral contraceptives, combined
Oral contraceptives, sequential
Phosphorus-32, as phosphate
Plutonium-239 and its decay products (may contain plutonium-240 and other isotopes) as aerosols
Radioiodines, short-lived isotopes, including iodine-131, from atomic reactor accidents and nuclear weapon detonation (exposure during childhood)
Radionuclides, α-particle-emitting, internally deposited
Radionuclides, β-particle-emitting, internally deposited
Radium-224 and its decay products
Radium-226 and its decay products
Radium-228 and its decay products
Radon-222 and its decay products
Schistosoma haematobium (infection with)
Silica, crystalline (inhaled in the form of quartz or cristobalite from occupational sources)
Solar radiation
Talc-containing asbestiform fibres
Tamoxifen
2,3,7,8-Tetrachlorodibenzo-*para*-dioxin
Thiotepa
Thorium-232 and its decay products, administered intravenously as a colloidal dispersion of thorium-232 dioxide
Treosulfan
Vinyl chloride
X- and γ-radiation

Mixtures (12)
Alcoholic beverages
Analgesic mixtures containing phenacetin
Betel quid with tobacco
Coal-tar pitches
Coal tars
Mineral oils, untreated and mildly treated
Salted fish (Chinese-style)
Shale oils
Soots
Tobacco products, smokeless
Tobacco smoke
Wood dust

continued

Table 11.5 continued

Exposure circumstances (14)
Aluminium production
Auramine, manufacture of
Boot and shoe manufacture and repair
Coal gasification
Coke production
Furniture and cabinet making
Haematite mining (underground) with exposure to radon
Iron and steel founding
Isopropanol manufacture (strong-acid process)
Magenta, manufacture of
Painter (occupational exposure as a)
Rubber industry
Strong inorganic acid mists containing sulfuric acid (occupational exposure to)

Source: from the IARC website in February 2002, reproduced with permission.

Among biological agents, hepatitis B virus, hepatitis C virus, human immuno-deficiency virus type and human papilloma virus are established human carcinogens.

Two parasites have been shown to be associated, probably causally, with human cancer (see Tomatis 1990). In some areas of South-east Asia and China, infestation of bile ducts with *Chlonorchis sinensis* or *Opishorchis viverrini*, two liver flukes, is followed by the development of cholangio carcinomas. In some parts of Africa, bladder cancer is one of the most common cancers and there is evidence that bilharziasis due to infestation with *Schistosoma haematobium* plays on aetiological role.

Diet may conceivably influence the development of cancer in a variety of ways. One mechanism is through ingestion of carcinogens formed by fungal or bacterial action during the storage of food (for example, aflatoxin B_1 produced by *Aspergillus flavus*) or through pyrolysis or pyrosynthesis during cooking of meat and fish. A second type of mechanism is by providing substrates for the formation of carcinogens in the body (for example, nitrites and nitrates) or by altering the concentration or duration of contact of carcinogens with cells in the large bowel, as may be derived through varying the quantity of fibres in the diet. Finally, being overweight, which implies overnutrition, whatever the mechanism, is associated with cancer at certain sites (such as endometrium and colon) (NRC 1982).

By far the largest number of proven human carcinogens has been identified from occupational causes (see Table 11.6).

The knowledge derived from epidemiological and laboratory studies is unbalanced, in that most is known about the specific causal factors cancers of the lung, urinary bladder, skin, upper aero–digestive tract and liver, and of leukaemia. There is less certainty about the causes of cancers of the breast, cervix, endometrium, ovary, pancreas, stomach, colon and rectum; and we know very little about the causes of cancers of the brain and prostate and of lymphomas and sarcomas.

Table 11.6 Common occupational cancers

Industry	Type of cancer	Agent	Occupation
Chemical industry	Lung	Chromium	Paints, tanning
		Bis-chloromethyl ether	Solvents, metallurgy
		Benzo-a-pyrene	Asphalt, roofing
		Arsenic	Paper, wood pulp
	Bladder	Benzidine, 2-naphthylamine	Aniline base dyes
		Auramine	Auramine manufacture
		Aromatic amines	Rubber and latex manufacture
	Leukaemia	Benzene	Solvents, petrol additive, degreasing
	Liver angiosarcoma	Vinyl chloride	Plastics
Metallurgy industry	Lung	Chromium	Chrome plating
		Benzo-a-pyrene	Steel production
		Arsenic	Copper smelting
		Asbestos	Ship building
		Cadmium	Cadmium smelting and alloy metals
	Lung and sinuses	Nickel	Nickel refining
	Skin	Arsenic	Copper smelting
	Leukaemia	Benzene	Metal degreasing
Petroleum and gas industry	Skin	Polycyclic tars	Asphalt
	Leukaemia	Benzene	Petrol additive
	Lung	Benzo-a-pyrene	Coke smelters
	Bladder	Naphthylmines	Coal gas production
Construction industry	Lung, and pleural Mesothelioma	Asbestos	Insulation, fibro-cement
	Nasal mucosa	Wood dust	Cabinet making, paper
	Skin	Polycyclic aromatic hydrocarbons, tar	Asphalt, roofing
Mining	Lung	Arsenic	Gold mining
		Asbestos	Asbestos mining
		Radon	Uranium and coal
Agriculture	Lung, skin	Arsenic	Pesticides and fungicides
Medical	Lung, thyroid	Cobalt-90, iodine-131	Medical radioisotopes
	Skin and cutaneous malignant melanoma	Immunosuppressive drugs	Organ transplants
	Vaginal	Stilboestrol	Oestrogens in pregnancy
	Leukaemia	X-rays and γ-rays	Radiation in pregnancy

Potential for cancer prevention

Control of risk factors

Primary prevention of cancer implies avoidance of its occurrence, either by reducing the exposure of individuals to causative agents in the environment or by enhancing resistance to them. Prevention is clearly feasible in relation to tobacco smoking, chemicals, industrial processes, medical drugs and viruses and to a certain extent, radiation. Another measure would be to improve socioeconomic conditions. The variety of cancer risk factors of which we are aware (see Table 11.2) clearly implies that there cannot be a single approach to cancer control and prevention. An effective preventive action must therefore be focused on those factors that cause the greatest numbers of deaths and that are most easily controlled, for example, diet, alcohol drinking, tobacco use and exposure to carcinogens in the workplace.

Diet and physical activity

A high proportion of human cancers could be related directly or indirectly to dietary factors, at least if we consider cancer sites for which such a relationship is conceivable. Dietary epidemiology is notoriously complex owing to the variety of foods and their many contribuents and to intercorrelations and temporal changes in their patterns of use. Being overweight/obese is the best-established diet-related risk factor for cancer. In most populations the proportion of people who are overweight/obese is increasing, and the amount of physical activity is decreasing (IARC 2002). Thus, although obesity/overweight was formerly prevalent only among the most affluent people, it is now becoming an important public health problem in many countries throughout the world. Reversing this trend presents a huge challenge. Primary prevention of cancer through dietary intervention therefore exerts an obvious attraction, and several intervention studies are currently underway in different parts of the world. The possibilities are rendered even more attractive by the facts that not only might we avoid exposure to carcinogens present as such in the diet or formed endogenously from what we eat, but also that certain dietary factors may protect against cancers caused by exposures to other, nondietary factors (Vainio and Hemminki 1989). It would be reasonable to recommend an increased intake of fresh fruit and vegetables, a reduction of animal fat, and the need to control one's weight. These recommendations will result in no harm, and the effects on cancer (and cardiovascular) risk would probably be beneficial. For a full discussion of this topic, with policy recommendations, see IARC (2002).

Alcohol drinking

The abundant epidemiological evidence that consumption of alcoholic beverages increases the risks for certain cancers was reviewed by the IARC in 1988. There is strong evidence that drinking alcoholic beverages increases the risks for developing cancers of the oral cavity, pharynx, larynx, oesophagus and liver. The risks are multiplied in people who also smoke. Pure ethanol has not been shown to be carcinogenic in laboratory animals, but there is no indication in humans that the carcinogenic effect is dependent on the type of beverage consumed. Thus, although the drinking of alcoholic beverages causes certain cancers, we do not know which of their constituents are responsible. The only means of prevention of those cancers, therefore, is avoidance of excessive drinking.

Tobacco use

Tobacco smoke contains at least 40 chemicals that are carcinogenic to animals (Hecht 1999; IARC 1986). Tobacco smoking is the main identified cause of cancer in industrialised countries. Risk increases with the amount smoked and with the duration of smoking.

Tobacco smoking is the major cause of cancers of the lung, oral cavity, larynx, oesophagus (squamous cell carcinoma) and an important cause of cancers of the pancreas, urinary bladder and renal pelvis (IARC 1986). Further, emerging evidence since 1986 indicates that there is now sufficient evidence for a causal association between cigarette smoking and cancers of the nasal cavities and nasal sinuses, oesophagus (adenocarcinoma), stomach, liver, kidney (renal cell carcinoma), uterine cervix and myeloid leukaemia (IARC 2003).

Involuntary (or passive) smoking is exposure to secondhand tobacco smoke and involves exposure to the same numerous carcinogens and toxic substances that are present in tobacco smoke produced by active smoking and involuntary smoking (exposure to secondhand or 'environmental' tobacco smoke) is confirmed to be carcinogenic to humans (IARC 2003).

Many tobacco products increase the risk for cancer: cigarettes, cigars, cigarillos, bidis, other tobacco products smoked in Asia, pipe tobacco and smokeless tobacco, which is chewed or snuffed. In many parts of the world, particularly in the Indian subcontinent, tobacco is chewed in a quid with areca nut and betel leaf. This habit is also carcinogenic, causing cancers of the oral cavity and upper gastrointestinal tract (IARC 1985).

Evidence of synergy between smoking and several occupational causes of lung cancer (arsenic, asbestos and radon), and between smoking and alcohol consumption for cancers of the oral cavity, pharynx, larynx and oesophagus and between smoking and human papillomavirus infection for cancer of the cervix are confirmed (IARC 2003).

Smoking tobacco will be a factor in approximately 3 million deaths per year during the 1990s (Peto and Lopez 1990). Worldwide mortality from tobacco is still rising rapidly, particularly in less developed countries, partly because of the population growth but chiefly because of previous large increases in cigarette smoking by young adults which cause large increases in mortality rates by the time the young adults are middle-aged. On the basis of current smoking habits, the date when worldwide annual mortality from tobacco will exceed 10 million has been estimated to be about 2020.

Strategies for prevention include persuading adult smokers to quit, which would have an immediate impact on incidence and mortality, and discouraging teenagers from taking up the habit. Low-tar and filter cigarettes are known to result in lower, but still substantial risks, and a switch to such brands could also markedly reduce disease.

Occupation

The main result of epidemiological studies of cancer in occupational settings has been to indicate methods for prevention, often long before the specific aetiological agents were identified and their exact relationship to human cancers elucidated. This practice is fundamental in the prevention of exposure to known or potential occupational carcinogens. Examples of carcinogens encountered in the workplace were listed in Table 11.6.

Data on exposures and their association with specific cancers provide a starting point for identifying more specific problems. Firstly, although occupational carcinogens induce a wide variety of cancers, certain organs, tissues and cells have been found to be the primary targets of these agents: lung, urinary bladder, skin, lymphatic system and haematopoietic system. Secondly, there are some occupational exposures for which specific agents have been identified, but many others for which only the occupation or industrial process itself has been shown to be associated with increased cancer risk.

Furthermore, very few of the potential occupational carcinogens have been evaluated by epidemiological research. Since so few of the agents shown by the experimental methods to be potentially carcinogenic have been the subject of epidemiological investigation, it is essential that such studies continue the evaluation and explication of human risk within well-defined occupations and, where possible, in association with specific agents. An impetus for conducting studies of occupational cancer should be the immediacy of the opportunities for prevention that often arise from the findings.

Current scientific knowledge indicates two primary approaches to the prevention of cancer: avoidance of carcinogens or reduction of exposure to them. In the first instance, substances that are known or suspected human carcinogens could be removed from commercial use and replaced with noncarcinogenic materials that serve the same function. However, there is often no adequate substitute and a substitute chemical or material may not have been tested for carcinogenicity. The multiplicative (synergistic) interaction between cigarette smoking and many occupational carcinogens is well documented. Therefore an immediate priority is to develop strategies for smoking prevention and cessation among workers in high-risk occupations. The second approach, reduction of exposure, can be widely applied, since extensive technological advances have resulted in the development of both engineering controls that can be installed at the worksite and protective clothing and equipment that reduce or eliminate worker exposure to hazardous substances. The only limiting factors to this approach are the cost of implementing such interventions and the specificity of available data regarding those substances to which exposure should be limited.

Sunlight

Ultraviolet radiation, especially UV-B, in the 320–290 nm region, appears to be related to skin cancer occurrence. The epidemiological evidence indicates that UV-B is the aetiological agent of the majority of basal cell epithelioma of face and neck, of squamous cell carcinomas of the exposed surfaces, and of the cancers

(Lee 1982; Scotto *et al.* 1982). For malignant melanoma, sunburn in childhood appears to be important, as studies of migrants from Europe to Australia have shown (Muir and Sasco 1990).

Protection against excessive exposure to sun should reduce the burden of skin cancers substantially.

Summary

Over the next 25 years, there will be a dramatic increase in the number of people developing cancer. Globally, 10 million new cancer cases are diagnosed each year and there will be 20 million by the year 2020. The risk for cancer in humans is increased by a variety of factors, ranging from exposure to an identified agent to a culturally determined behaviour, such as smoking, or to socioeconomic conditions. The export of unhealthy western lifestyles – cigarette smoking, excess energy intake, and sedentary occupations – will increase cancer not only in developed countries but also in developing countries, which can least afford the treatment costs. Intervention is possible with regard to some of these carcinogenic factors, while others affect us in as yet undetermined ways. Carcinogenesis is a multi-event process nearly always involving more than one aetiological agent, so complementing reduction of exposures to carcinogens in the occupational and general environment with reductions in exposures that are under the control of the individual will maximise the potential for cancer prevention. Furthermore, exciting developments in epidemiological and experimental research provide leads about factors that may reduce cancer risk. As we learn more about diet, energy balance, and physical activity, for instance, the potential arises for more general approaches to cancer prevention.

References

Adamson, R.H. and Sieber, S.M. (1983) Chemical carcinogenesis studies in nonhuman primates. *Basic Life Sci.* 24: 129–156.

Ames, B.N. (1973) Carcinogens are mutagens: their detection and classification. *Environ. Health Persp.* 6: 115–118.

Anderson, M.W. and Reynolds, S.H. (1988) Activation of oncogenes by chemical carcinogens. In: Sirica, A.E. (ed.) *Pathology of Neoplasia.* New York: Plenum Press, pp. 291–301.

Anonymous (1996) Twelve major cancers. *Sci. Am.* 275: 126–132.

Armitage, P. (1985) Multistage models of carcinogenesis. *Environ. Health Persp.* 63: 195–201.

Barbacid, M. (1987) ras genes. *Ann. Rev. Biochem.* 56: 779–827.

Bos, J.L. (1989) ras oncogenes in human cancer: A review. *Canc. Res.* 49: 4682–4689.

Boyd, J.A. and Barrett, J.C. (1990) Genetic and cellular basis of multistep carcinogenesis. *Pharmacol. Ther.* 46: 469–486.

Bradford-Hill, A. (1965) The environment and disease: Association or causation? President's address. *Proc. Roy. Soc. Med.* 9: 295–300.

Cook, J.W., Hewett, C.L. and Higger, I. (1933) The isolation of a cancer producing hydrocarbon from coal tar. *J. Chem. Soc.* 1: 395–405.

Counts, J.L. and Goodman, J.I. (1995) Principles underlying dose selection for, and extrapolation from, the carcinogen bioassay: dose influences mechanism. *Reg. Toxicol. Pharmacol.* 21: 418–421.

Creech, J.L. and Johnson, M.N. (1974) Angiosarcoma of the liver in the manufacture of polyvinyl chloride. *J. Occup. Med.* 16: 150–151.

Doll, R. (1988) Effects of exposure to vinyl chloride: An assessment of the evidence. *Scand. J. Work. Environ. Health* 14: 61–78.

Doll, R. and Peto, R. (1981) The causes of cancer. *J. Nat. Canc. Inst.* 66: 1197–1308.

Farber, E. (1984) The multistep nature of cancer development. *Canc. Res.* 44: 4217–4223.

Fearon, E.R., Hamilton, S.R. and Vogelstein, B. (1987) Clonal analysis of human colorectal tumors. *Science* 238: 193–197.

Friend, S.H., Bernards, R., Rogdej, S. *et al.* (1986) A human DNA segment with properties of the gene that predisposes to retinoblastoma and osteosarcoma. *Nature* 323: 643–646.

Harbour, J.W., Lai, S.-L., Whang-Peng, J., Gazdar, A.F., Minna, J. and Kay, F.J. (1988) Abnormalities in structure and expression of the human retinoblastoma gene in SCLC. *Science* 241: 353–357.

Hecht, S.S. (1999) Tobacco smoke carcinogens and cancer. *J. Nat. Canc. Inst.* 91: 1194–1210.

IARC (1985a) Tobacco habits other than smoking; betel quid and *Areca* nut chewing; and some related nitrosamines. *IARC Monographs on the Evaluation of the Carcinogenic Risks of Chemicals to Humans.* Lyon: IARC, p. 37.

IARC (1985b) Tobacco smoking. *IARC Monographs on the Evaluation of the Carcinogenic Risks of Chemicals to Humans.* Lyon: IARC, p. 38.

IARC (1987) *An Updating of IARC Monographs Volumes 1 to 42. IARC Monographs on the Evaluation of the Carcinogenic Risks of Chemicals to Humans, Supplement 7.* Lyon: IARC.

IARC (1988) Alcohol drinking. *IARC Monographs on the Evaluation of the Carcinogenic Risks of Chemicals to Humans.* Lyon: IARC, p. 44.

IARC (2002) Weight control and physical activity. *IARC Handbooks of Cancer Prevention.* Lyon: IARC, p. 6 at www.iarc.fr.

IARC (2003) *Tobacco Smoke and Involuntary Smoking. IARC Monographs on the Evaluation of the Carcinogenic Risks of Chemicals to Humans.* Lyon: IARC, p. 83.

Knudson, A.G. (1985) Hereditary cancer, oncogenes and antioncogenes. *Canc. Res.* 45: 1437–1443.

Kolodner, R.D. and Marsischky, G.T. (1999) Eukaryotic DNA mismatch repair. *Curr. Opin. Genet. Develop.* 9: 89–96.

Kubinyi, H. (1990) Quantitative structure–activity relationships (QSAR) and molecular modelling in cancer research. *J. Canc. Res. Clin. Oncol.* 116: 529–537.

Lee, J.A.H. (1982) Melanoma and exposure to sunlight. *Epidemiol. Rev.* 4: 110.

Montesano, R., Bartsch, H. and Tomatis, L. (1980) *Molecular and Cellular Aspects of Carcinogen Screening Tests.* Lyon: IARC.

Muir, C.S. and Sasco, A.J. (1990) Prospects for cancer control in the 1990s. *Ann. Rev. Publ. Health* 11: 143–163.

NRC (1982) *Diet, Nutrition and Cancer. A Report to the Commission of Life Sciences.* Washington: US National Research Council/National Academy Press.

OECD (1993) *OECD Guidelines for the Testing of Chemicals.* Paris: Organisation for Economic Cooperation and Development.

Parkin, D.M. (2001) Global cancer statistics in the year 2000. *Lancet Oncol.* 2: 533–543.

Parkin, D.M., Pisani, P. and Ferlay, J. (1999) Estimates of the worldwide mortality of 25 major cancers in 1990. *Int. J. Canc.* 80: 827–841.

Peto, R. and Lopez, A.D. (1990) The future worldwide health effects of current smoking patterns: 3 million deaths/year in the 1990s, but over 10 million/year eventually. *Seventh World Conference on Tobacco and Health*, Perth, Western Australia, 3 April

Pisani, P., Parkin, D.M., Bray, F. and Ferlay, J. (1999) Estimates of the worldwide incidence of 25 major cancers in 1990. *Int. J. Canc.* 83: 18–29.

Pott, P. (1775) *Chirurgical Observations Relative to Cataract, the Polypus of Nose, the Cancer of Scrotum, the Different Kinds of Ruptures, and the Mortification of the Toes and Feet*. London: Hawes, Clarke and Collins, pp. 63–65.

Purchase, I.F. (1982) International Commission for Protection against Environmental Mutagens and Carcinogens (ICPEMC) Working paper 2/6. An appraisal of predictive tests for carcinogenicity. *Mutat. Res.* 99: 53–17.

Rehn, L. (1985) Blasengeschwülste bei Fuchsin-Arbeitern. *Arch. Klin. Chirurg.* 50: 588–600.

Reynolds, S.H. and Anderson, M.W. (1991) Activation of proto-oncogenes in human and mouse lung tumors. *Environ. Health Persp.* 93: 145–148.

Reynolds, S.H., Stowers, S.J., Patterson, R.M., Maronpot, R.R., Aaronson, S.A. and Anderson, M.W. (1987) Activated oncogenes in B6C3F1 mouse liver tumors: implications for risk assessment. *Science* 237: 1309–1316.

Scotto, J., Fears, T.R. and Fraumeni, J.F. (1982) Solar radiation. In: Schottenfield, D. and Fraumeni, J.F. (eds) *Cancer Epidemiology and Prevention*. Philadelphia: Saunders, p. 254.

Simon, M.M., Sliutz, G. and Luger, T.A. (1995) Proto-oncogenes and oncogenes in epidermal neoplasia. *Exp. Dermatol.* 4: 65–73.

Stewart, B.W. and Kleihues, P. (eds) (2003) World Cancer Report. Lyon: IARC.

Tomatis, L. (ed.) (1989) *Cancer: Causes, Occurrence and Control*. IARC Scientific Publications No. 100, Lyon. Lyon: IARC.

Tomatis, L., Aitio, A., Wilbourn, J. and Shuker, L. (1990) Human carcinogens so far identified. *Jpn. J. Canc. Res.* 80: 795–807.

Torry, D.S. and Cooper, G.M. (1991) Proto-oncogenes in development and cancer. *Am. J. Reprod. Immunol.* 25: 129–132.

Vainio, H. and Hemminki, K. (1989) Multistage process and prevention of cancer. In: *Occupational Cancer in Chemical Industry*. Copenhagen: World Health Organization, Regional Office for Europe, pp. 78–97.

Vogelstein, B., Fearon, E.R., Hamilton, S.R. *et al.* (1988) Genetic alterations during colorectal tumour development. *N. Engl. J. Med.* 319: 525–532.

Waddell, W.J. (1996) Reality versus extrapolation: an academic perspective of cancer risk regulation. *Drug Metab. Rev.* 28: 181–195.

Weinberg, R.A. (1985) The action of oncogenes in the cytoplasm and nucleus. *Science* 230: 770–776.

Weinberg, R.A. (1989) Oncogenes, antioncogenes, and the molecular basis of multistep carcinogenesis. *Canc. Res.* 49: 3713–3721.

Weiss, N.S. (1996) Ambiguities in the IARC criteria for evaluation of carcinogenic risks to humans, and a recommendation. *Epidemiology* 7: 105–106.

Yamagiwa, K. and Ichikawa, K. (1915) Uber die künstliche Erzeugung von Papillom Verk. *Jpn. Pathol. Ges.* 5: 142–148.

Yokota, J., Wada, M., Shimosato, Y., Terada, M. and Sugimura, T. (1987) Loss of heterozygocity on chromosomes 3, 13 and 17 in small cell carcinoma and on chromosome 3 in adenocarcinoma of the lung. *Proc. Nat. Acad. Sci. USA* 84: 9252–9256.

Toxicity by group of chemical

Chapter 12

Toxicity of metals

C. Winder

Introduction

The metals represent a different aspect of toxicology in that they do not undergo breakdown to other metals (to do so would be transmutation of the elements, an old alchemical principle still not possible without huge sources of energy, and a particle accelerator!). The absorption, disposition and excretion of metals are largely dependent on physical factors, such as solubility, ionisation, particle size and chemical species (for metal salts).

About 70–80 elements in the periodic table (see Figure 12.1) are considered metals. Groups Ia and IIa, the 's block' metals, form monovalent and divalent cations respectively. Groups IIIb to VIb constitute the 'p block' elements, which include metals that can have ions of different valencies. These are called the transition elements.

Of the metal elements, about 40 are considered to be 'common' metals. However, less than 30 have compounds that have been reported to produce toxicity. Metals are probably some of the oldest toxicants known to humans. Health effects such as colic were reported following exposure to lead, arsenic and mercury over 2000 years ago. On the other hand, metals such as cadmium, chromium and nickel belong to the modern era. The importance of some of the rarer metals may become apparent with emerging changes in technology, such as microelectronics and superconductors.

The toxicity of a metal is only partially related to its position in the periodic table. Toxicity decreases with the stability of the configuration of electrons in the atomic nuclei. This produces a number of properties that can affect toxicity.

- The very light metals (beryllium and lithium) have a very small ionic radius and therefore a higher charge to mass ratio; these are quite toxic metals.
- Increasing electropositivity will increase toxicity (for example zinc < cadmium < mercury, or aluminium < gallium < indium < thallium). Electropositivity increases to the left and down the periodic table.

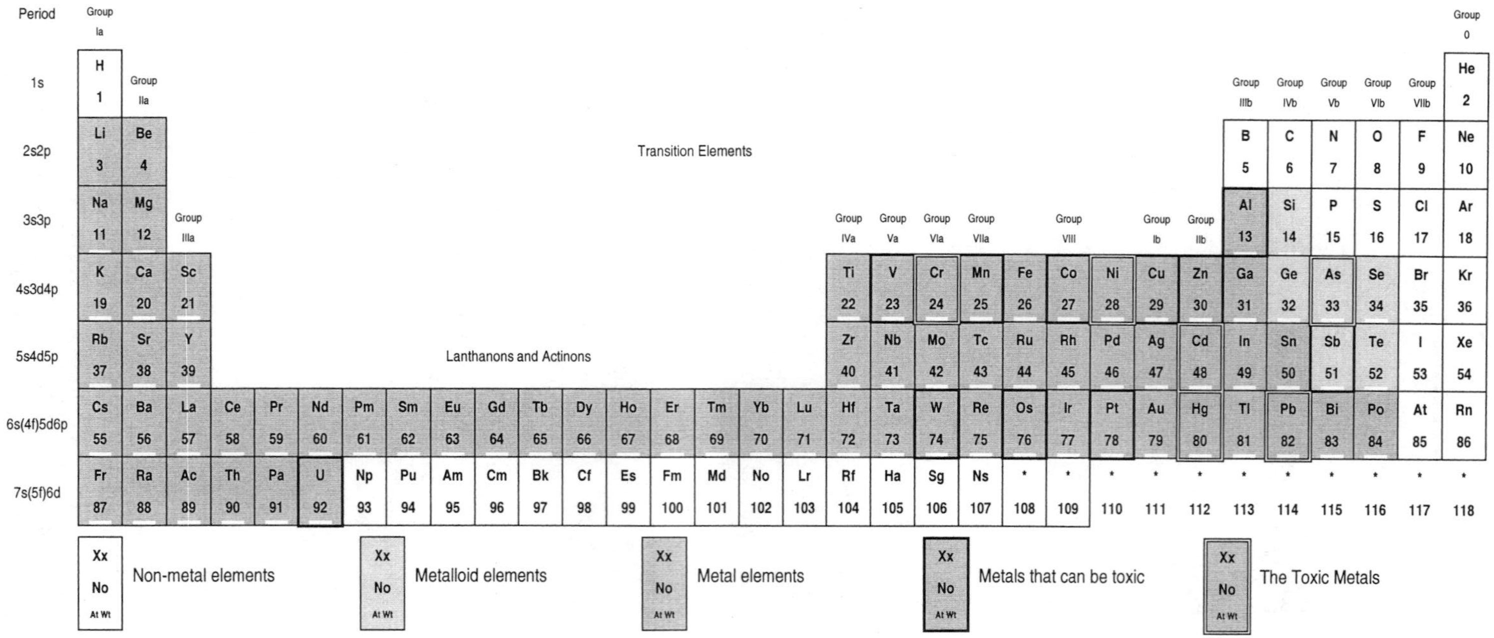

Figure 12.1 The periodic table and toxic metals.

- Highly electropositive elements such as those in the first two columns (the alkali metals and alkali earth metals) appear in the biological environment primarily as free cations (that is, charged ions). This increases toxicity by virtue of increased bioavailabilty.
- Metals lower down the periodic table are potentially the most toxic. However, this toxicity only becomes apparent in those lead, mercury and thallium salts that are relatively soluble.
- Various oxidation states (in the transition elements) are also important. For example, manganese[VII] is more toxic than manganese[II] and arsenic[III] is more toxic than arsenic[V]. Again, electronic configuration is important in expression of toxicity.

One other factor important in expression of the toxicity of metals is the relative amount of the metal produced. The annual production of a metal like chromium is about three million tonnes. The potential for human or environmental exposure is much greater than, say, the platinum group metals, where annual production is in the range of 200 tonnes a year. While it is likely that chromium is generally more toxic than platinum, the huge amount in production and use increases the potential for its toxicity to be expressed.

Sources of metals

Metals can be widely distributed in the environment by geological, biological and anthropogenic activities. For most individuals, the largest source of metal exposure is through metal content in food, with a smaller additional component from air. Other potential exposures to metals in the nonoccupational environment include:

- use as therapeutic materials (barium as an X-ray contrast agent; lithium in treatment of depression; platinum is used as a chemotherapeutic agent; some surgical implants may be metal; aluminium accumulates in dialysis patients; gold colloids in the treatment of arthritis)
- paediatric (such as children swallowing mercury batteries)
- consumer products such as deodorants (zirconium), vitamin and mineral supplements (selenium), hair dyes (silver and lead) and cosmetics (lead, antimony and copper)
- naturally occurring areas of high levels of some minerals, such as selenium
- exposure to industrial wastes and pollution.

For workers, occupational exposure to metals occurs in a wide variety of occupations.

Metal toxicity: general principles

Exposure to metals and metal-containing compounds is common to many industrial, nonindustrial and environmental situations. Absorption of metals can have many effects on the body. Not all of these are adverse, and it must be remembered that some metals are essential for the normal function of the body. Examples include cobalt, copper, iron, magnesium, manganese, selenium and zinc.

General properties which may affect toxicity

Before a discussion of the exposure of metals, it is necessary to consider some general properties that have a bearing on the expression of toxicity:

- Metals seldom interact with biological systems in the elemental form, and are usually active in the ionic form.
- Availability of metal ions to biological processes is often dependent on solubility. Soluble salts of metals readily dissociate in the aqueous environment of biological membranes, making transport into the body easy, whereas insoluble salts are poorly absorbed (for example, reduction of chromium(VI) to the less soluble chromium(III) will decrease its absorption).
- Absorption of soluble salts may be modified by formation of insoluble compounds in biological materials (for example, high dietary levels of phosphate will reduce absorption of lead because of the formation of insoluble lead phosphate).
- Some metals are produced as alkyl compounds. These are often very lipid soluble, and pass readily across the lipid phase of biological membranes (examples include methyl mercury and organotin compounds).
- Strong attractions between metal ions and organic compounds will influence the disposition of metals and their rate of excretion. Most of the toxicologically important metals bind strongly to tissues, are only slowly excreted, and therefore tend to accumulate on continuing exposure. Affinities for different tissues vary – elements like lead are bound in bone, whereas mercury and cadmium localise in the kidney.

The uses of metals

Humanity has been using metals for millennia. The terms 'bronze age' and 'iron age' delineate the early technological development of the human race. Further developments in alchemy, mining, engineering, industry and modern chemistry has enabled human beings to identify, extract, refine and ultimately exploit the mineral resources available on the earth.

Some metals have only minor occupational uses, with an associated low risk of production of adverse effects. These metals may cause adverse effects in other applications (for example, in therapeutic use or paediatric poisoning).

The uses of the metals are enormous, from use of the pure metal, to alloys (mixtures of metals), to inorganic and organic compounds of metals. Table 12.1 shows some applications of some of the commercially useful metals.

The mining, extraction, industrial application, disposal and environmental dispersion of metals is not without its risks. Hazards to workers, the public and the environment are possible in what is essentially a process of increasing the purity of the metal content of minerals to commercially useful concentrations.

Table 12.1 Commercial uses of the metals

Metal	Uses
Aluminium	Packaging
	Building and transport
	Water treatment
	Medical (deodorants)
Antimony	Alloy manufacture
Arsenic and arsine	Pesticide (now banned)
Barium	Radiographic agent
Beryllium	Nuclear industry
Bismuth	Low melting alloys
	Silvering of mirrors
	Dentistry
	Superconductors
	Cosmetics
	Therapeutic agents
Cadmium	Alloys
	Electroplating
Chromium	Electroplating
	Tanning (dichromates)
	Safety match manufacture
	Pigments
Cobalt	Alloy manufacture (jet engines, turbines)
	Radiation source
Copper	Plumbing
	Algicide
	Electrical industry
	Electroplating
Gallium	Integrated circuit boards
	Electronics
	Therapeutic agent
Germanium	Electronics
	Semiconductors
Gold	Jewellery
	Electroplating
Indium	Solder alloys
	Semiconductors
Iron	Iron and steel products
Lanthanides	Steel alloys
	Carbon arc electrodes
	Lens production
	Glass and ceramics pigments
	Cigarette lighter flints

continued

Table 12.1 Commercial uses of the metals

Metal	Uses
Lead	Battery manufacture
	Pigment manufacture
Lithium	Nuclear industry
	Alkaline batteries
	Treatment of depression
Magnesium	Aluminium alloys
	Antacid/laxative
Manganese	Mining and refining
	Alloys (steel industry)
	Dry cell batteries
	MMT petrol additive
Mercury	Chlorine production
	Gold extraction
	Scientific instruments
	Dentistry
	Battery manufacture
	Fungicide
Molybdenum	Steel alloys
	Paint pigments
Nickel	Mining and refining
	Coinage
	Stainless steel production
	Electroplating
Niobium	Steel alloys
	Arc welding rods
Platinum	Jewellery
	Catalyst (chemical manufacturing)
	Cytotoxic drugs
Selenium	Electronics
	Glass and ceramics production
	Pigments
	Topical therapeutics
Silver	Photography
	Electrical equipment
	Coinage
	Jewellery
Tantalum	Medical applications
	Electronics
Tellurium	Copper alloys
	Catalysts

continued

Table 12.1 continued	
Metal	Uses
Thallium	Semiconductors
	Electronics
	Lenses
	Contaminant in other metals
	Rodenticide (now discontinued)
Thorium	Incandescent mantles
	Nuclear industry
Tin	Tin plate
	Food containers
	Solder
	Marine antifouling paint
Titanium	Alloys (aeronautics)
	Paint pigment
	Surgical prosthetics
Tungsten	Tool and drill manufacture
	Electric bulb filaments
	Pigment
Uranium	Nuclear industry
Vanadium	Alloys (steel industry)
	Catalyst (chemical industry)
Yttrium	Television phosphor
	Superconductors
Zinc	Battery manufacture
	Galvanising and electroplating
	Vulcanisation of rubber
	Topical therapeutic
Zirconium	Alloys
	Nuclear industry
	Deodorants

Metal toxicity

The onset of severe gastrointestinal symptoms (for example, bloody diarrhoea, abdominal pain, nausea, vomiting) and upper gastrointestinal haemorrhage together with a metallic taste in the mouth is indicative of metal poisoning. Long-term effects of absorption of metal and metallic compounds are varied, and related to effects in many body systems.

Some metals have only minor occupational uses, with an associated low risk of production of adverse effects. These were shown in Table 12.1. These metals may cause adverse effects in other applications (for example, in therapeutic use or paediatric poisoning).

Other metals have significant occupational uses, but not all are toxic.

Metal toxicity: specific metals

Not all metals are of concern with regard to toxic effects, and not all toxic metals are important from the perspective of occupational and environmental health. Some metals, such as arsenic, lead and mercury, have a long history, mainly from their use as poisoning agents, but also for their now classic causes of disease. Other metals have come to the fore as new technologies and processes require their use (for example, beryllium, chromium or uranium).

The amount of information on the health effects of metals varies from metal to metal, and of course is dependent on a number of factors, including inherent toxicity, availability to workers, the volume of production, and the types of processes in which the metal is employed.

Some of the more important workplace metals are described below.

Aluminium

Aluminium has many uses, from mining, refining and smelting through to use of the metal and its compounds.

A major use of aluminium metal is in the manufacture of aluminium cans – perhaps one-third of all production is consumed in such packages. The metal is also used in the construction industry, in transportation, in the electrical industry and in consumer products. Alloys of aluminium have a range of uses, including powder metallurgical products, coatings and reducing agents. Important inorganic aluminium compounds include aluminium sulfate (for water treatment), potassium alum (for tanning and mordants), synthetic zeolites, in the paper industry and as a concrete accelerator.

Significant recycling of aluminium occurs nowadays.

All soils contain aluminium compounds. However, not all minerals are economically viable. The term bauxite is used for sedimentary rocks that contain commercially extractable aluminium. Generally, bauxite is digested at high temperatures and pressures with caustic soda in the Bayer process. This produces aluminium hydroxide, which is reduced to alumina. Alumina is then reduced to metal aluminium in primary aluminium smelters using the Hall–Heroult method.

However, occupational exposures to aluminium are considered less hazardous than some of the medical exposures discussed below.

Exposure

Exposures to aluminium are shown in Table 12.2.

Toxicity

The study of aluminium toxicity has been related to the study of patients with chronic renal failure on long-term dialysis. These patients develop high aluminium concentrations. Orally, about 12% of aluminium hydroxide is retained, with most body aluminium residing in bone tissue. The main route of excretion appears to be through bile, with renal elimination being more important after high exposures.

Table 12.2 Human exposure to aluminium

Industrial/occupational
Refining and smelting processes
Production of aluminium alloys and compounds
Production of fine aluminium dusts

Environmental
Air: mainly as aluminosilicates associated with dust particles. In rural areas air aluminium levels are normally less than $0.5 \mu g/m^3$, although in urban areas they are higher and can reach levels above $10 \mu g/m^3$ near point sources of pollution, such as cement plants.
Water: some lakes, river, groundwaters and domestic tapwater supplies can contain aluminium in high concentrations, either naturally or because aluminium salts have been added as a flocculant in the purification process. Salinity, pH and biological processes affect the concentration of dissolved aluminium, and concentrations can vary substantially. Soil exists almost exclusively in the form of silicate, hydroxides and oxides. Release is possible from silicate in soil acidification (such as from acid rain). High concentrations of soil aluminium may cause root die back.

Domestic
Food: concentrations of aluminium in food are not high. Cereals, root vegetables, meat and liver contain aluminium in the range of about 5 mg/kg. About 2.5 mg of aluminium is absorbed daily with food; which undergoes minimal absorption through the gastrointestinal tract (0.1%) and total excretion of the absorbed dose through the kidneys. Cooking in aluminium pots, although the contribution to total dietary intake is minimal.
Antacids: a single dose normally represents about 50 times the average daily intake from other sources.
Dialysis patients.
Implicated as a neurotoxic agent in the pathogenesis of Alzheimer's' disease

---

Elevated aluminium levels have also been found in autopsy material taken from the brains of individuals with Alzheimer's disease, though its role as a neurotoxic factor is yet to be clarified.

The lungs and the nervous system appear to be main organs of toxicity following occupational exposure. Pulmonary fibrosis has been reported in workers exposed to fine aluminium dust (aluminosis), and aluminium may exacerbate asthma. Neurological effects have also been reported, including encephalopathy, tremor, incoordination and cognitive defects.

Biological monitoring: measuring human exposure to aluminium

The concentrations of aluminium in blood of nonexposed subjects with normal renal function are extremely low ($1-3 \mu g/L$). Al in plasma values of $5-15 \mu g/L$ have been reported in exposed workers, which is about 10 times lower than the values observed in dialysis patients with encephalopathic symptoms.

Concentration of aluminium in serum and urine reflect current exposure and the amount in the body. Care with respect to contamination is required with sample preparation as the levels are low and the metal is widespread. Urinary levels are regarded as a more sensitive indicator of exposure in people with normal renal function as concentrations in urine may be elevated while blood levels are barely altered. For workers chronically exposed to aluminium, samples collected 1–2 days after exposure finishes probably indicates body burden. Samples collected at the end of a shift are more likely to reflect the very recent exposure.

Antimony

Antimony oxide and other compounds can cause a benign pneumoconiosis, a severe pulmonary oedema and cardiomyopathy following severe exposure. Skin burns, including pruritic papules progressing to pustules known as 'antimony spots' are also observed.

Stibine gas (SbH_3) exposure produces effects similar to that of arsine gas (AsH_3) exposure (see below).

Arsenic

Arsenic has been used in the production of pesticides, though this is declining. Other uses include in glassware, alloy and pigment production, and the use of arsine gas (AsH_3) in the semiconductor industry. Arsenic is a by-product of the extraction of a number of metals, including lead, copper and gold. Intense abdominal pain, bloody diarrhoea and a garlic-like smell on the breath are suggestive of arsenic absorption.

Exposure

Arsenic is widely distributed in nature environment especially in a large number of minerals. It is found most abundantly in sulfide ores as trivalent and pentavalent compounds. The earth's crust and igneous rocks contain about 3 mg/kg arsenic, coal between 0.5 and 93 mg/kg with a mean value of 17.7 mg/kg, and brown coal up to 1500 mg/kg. As a by-product arsenic trioxide is obtained during the production of copper and lead from sulfide ores. It is recovered from the flue dust in a reasonably pure form.

Occupational exposure to arsenic compounds takes place mainly among workers, especially those involved in the processing of copper, gold, and lead ores. In agriculture, one formerly frequent exposure was through application of arsenic as an insecticide, mainly as lead arsenate and less frequently as calcium arsenate and arsenite, sodium arsenite, cupric arsenite and cupric acetoarsenate. Other frequent applications of arsenicals are found as herbicides (weed killers for railroad and telephone posts), desiccants to facilitate mechanical cotton harvesting, fungicides, rodenticides, insecticides, algicides, and wood preservatives. Because of occupational and environmental risks, these applications are all but ended.

A significant source of human exposure to arsenic compounds originates from marine fish and shellfish. There may be more than 1 mg/kg of fish arsenic compounds is normally around 10 μg, but will be much higher if contaminated water or food is consumed. Arsenic (As) in food (both fish arsenic and other compounds) is absorbed effectively from the gastrointestinal tract.

Although arsenic has almost exclusively been associated with criminal poisoning for many centuries, the matter of concern today is its contribution to occupational and environmental pollution through man's use of pesticides, nonferrous smelters and coal-fired and geothermal power plants.

Other exposures to arsenic are shown in Table 12.3.

Table 12.3 Human exposures to arsenic

Industrial/occupational
Alloying constitution (account for about 90% of the production)
Metallic mining
Pest control with arsenical pesticides
To make gallium arsenide for dipoles and other electronic devices
Doping agent in germanium and silicon solid state product
Solders
Cutting and sawing on wood pretreated with arsenic preservation
As catalyst in the manufacture of ethylene oxide
Semiconductor devices
Glass industry (As_2O_3, As_2Se, As_2O_5, metallic arsenic)
Colours for digital watches
Textile and tanning industries
Manufacture of pigments
Antifouling paints
Light filter (thin sheets of As_2O_5)
Ceramic industry (As_2O_5)
Manufacture of fireworks (As_4S_4)

Environmental
Natural occurring sources: volcanoes, sulfide ores
Artificial sources: metallic mining, pesticide application, veterinary medicine, airborne emission from the smelting of metals (mainly nickel–copper smelters)
Air: present mainly in particulate form as arsenic trioxide, with background levels of 1–10 ng/m^3 in rural areas and 20 ng/m^3 in urban areas
Soil: level in soil is about 7 mg/kg, but can be as high as 1000 mg/kg in the vicinity of metal smelters and in the agricultural soil where extensive use was made of pesticides, herbicides and defoliants
Water: ground water normally contaminated by metallic mining, metal smelters, pesticide application and in appropriated industrial waste disposal

Domestic
Criminal poison
In cosmetics as depilatory agents
Eating seafood (fish and shellfish), sprayed fruit, vegetables
Artwork (painting, photography, sculpture)
Ceramics
Tobacco smoking
Drugs (now largely discontinued)

Toxicity

Arsenic compounds present a wide range of toxicities. The most toxic arsenic compound is arsine gas, which is a colourless, nonirritating gas evolving from the mixing of arsenic compounds with acid.

The chemical equation for this reaction is shown by the generic equation:

$$A—AsX + H—Y \rightarrow AsH_3 + A—X + A—Y$$

where As is arsenic, A is the anion of arsenic salt, H is acid, X is the cation of arsenic salt, and Y is the salt of the acid.

The organic arsenic[V] (arsenate, As^{5+}) forms are less toxic than the inorganic arsenic[III] (arsenite, As^{3+}) forms; which is based in part on their lower solubility.

As^{5+} compounds are well absorbed through the gastrointestinal tract, and As^{3+} compounds are more lipid soluble (dermal absorption). Arsenic replaces phosphorus in bone and binds to sulfydryl groups on proteins. It also disrupts metabolic pathways, including inhibition of oxidative phosphorylation and pyruvate metabolism. Almost all of the absorbed arsenic is excreted by the kidney within 4–10 days.

Arsenic toxicity is well studied, owing to its frequent use as a poisoning agent. Many organs and systems are affected:

- cardiovascular (vasodilatation leading to reflex arteriolar constriction, myocardial depression)
- gastrointestinal (submucosal vesicle formation, bleeding)
- kidney (acute tubular necrosis, oliguria, proteinuria and haematuria)
- skin (erythema, brittle fingernails, local oedema, pigmentation, pyoderma and skin cancer)
- nervous system (degeneration of myelin, encephalopathy, paraasthesias)
- liver (hepatic fatty degeneration, cirrhosis).

Biological monitoring: measuring human exposure to arsenic

The primary method used for biological monitoring for arsenic exposure determines inorganic arsenic, monomethylarsonic acid and cacodylic acid in urine. This overcomes the problems of contamination from ingestion of seafood or water with high arsenic levels experienced when total arsenic is measured. The urinary concentration mainly reflects recent exposure. Sampling is best undertaken when the worker is into the normal routine by a day or two because it takes this time for equilibrium to be reached. Arsenic has also been measured in the blood and it also seems to reflect primarily more recent exposures. Hair and nails provide a good indication of inorganic arsenic entering the body during the growth period. External contamination of the hair with arsenic is a contamination problem for measurements involving those occupationally exposed.

Beryllium

Beryllium is the 35th most abundant element in the earth's crust, with an average content of about 6 mg/kg. Apart from the gemstones, emerald (chromium-containing beryl) and aquamarine (iron-containing beryl), only two beryllium minerals are of economic significance. The annual global production of beryllium minerals in the period 1980–1984 was estimated to be about 10,000 tonnes, which corresponds to approximately 400 tonnes of beryllium.

Exposure

In general, beryllium emissions during production and use are of minor importance compared with emissions that occur during the combustion of coal and fuel oil, which have natural average contents of 1.8–2.2 mg beryllium/kg dry weight, and up to 100 μg beryllium/L, respectively. Beryllium emissions from the combustion of fossil fuels amounted to approximately 93% of the total beryllium emission in the USA, one of the main producer countries.

Approximately 72% of the world production of beryllium is used in the form of beryllium–copper and other alloys in the aerospace, electronics, and mechanical industries. About 20% is used as the free metal, mainly in aerospace, weapons, and nuclear industries. The remainder is used as beryllium oxide for ceramic applications, principally in electronics and microelectronics.

In the environment, water contains very little beryllium because the small amount that escapes capture by clay minerals during rock weathering and soil formation is largely adsorbed by the surfaces of mineral grains. In soil, samples collected from about 1300 localities throughout the USA contained from less than 1 to 15 ppm (parts per million) of beryllium, averaging about 1 ppm.

Toxicity

The lung is the major organ affected, with dyspnoea and cough progressing to chronic granulomatous disease similar to sarcoidosis and miliary tuberculosis.

Biological monitoring: measuring human exposure for beryllium

The usefulness of blood and urine beryllium levels as a quantitative indicator of exposure remains to be established. At the moment detection of this metal in blood and/or urine only indicates that exposure has occurred. There is some evidence that urinary beryllium reflects current exposure but it also seems that beryllium can remain detectable in the urine even years after exposure has ceased.

Cadmium

This metal, while being relatively rare, has a wide distribution in the earth's crust and is often found in association with deposits of copper, zinc and lead. Emissions from volcanoes are a major source of natural release of cadmium to the environment. Cadmium has received prominence only in the past 20 years.

Table 12.4 Human exposures to beryllium

Industrial/occupational
  Coal and fossil fuel combustion
  Nuclear industry
  Power plants
  Hardening of copper
  Space optics, missile fuel and space vehicles
  X-ray window
  Alloy component
  Navigational systems, aircraft/satellite and missile parts

Environmental
  Natural occurring sources: abundance in earth's crust, silicate minerals and certain fossil fuels
  Artificial sources: coal and fuel combustion and beryllium-extraction plants, ceramic artists and others
  Air: atmospheric beryllium concentrations at rural sites in the USA ranged from 0.03–0.06 ng/m^3; annual average beryllium concentrations in urban air in the USA were found to range from < 0.1–6.7 ng/m^3
  Soil: soil concentrations generally range from 0.1–40 ppm, with the average around 6 ppm
  Water: in the USA analysis of surface, and rain waters have shown that beryllium concentrations are well below 1.0 μg/L
  Food chain: food is not a significant source of human exposures; so far there is no evidence that beryllium is moving from soils into food or feed plants in detrimental amounts

Domestic
  Ceramic artists
  Dental casting alloy
  Jewellery
  Food

Exposure

Exposures associated with toxic effects in humans have occurred both environmentally and from work-related activities and are mainly related to production, consumption and disposal of cadmium and other nonferrous metals (see Table 12.5).

Primary uses of cadmium which result directly or indirectly in human exposures include production of batteries, as a protective plating for metals and in plastics as pigments and stabilisers. Cadmium appears in the workplace in solder, pigments and some alloys. It may also occur in fumes from the smelting or welding of other metals. It is also used as a component of some metal alloys. Cadmium is also used in other products such as television screen phosphors, photography, lasers and lithography.

Disposal of materials containing cadmium presents a further source of potential

Table 12.5 Human exposures to cadmium

Industrial/occupational
Electroplating of metals
During soldering
Battery production
Alloy production
Pigment production
Plastics production
Smelting of nonferrous metals
Welding

Environmental
Volcano emissions
Black shale deposits
Mining operations involving cadmium
Smelting of nonferrous metals
Production of cadmium products
Disposal of cadmium-containing materials
Phosphate fertilisers
Iron and steel production
Fossil fuel and wood combustion
Garbage and sewage sludge incineration
Cement manufacture

Domestic
Tobacco smoke
Foodstuffs

exposure. Combustion of wood and fossil fuels releases cadmium as well as incineration of sewage sludge and refuse. Its presence in phosphate fertiliser may result in
soil contamination and uptake by plants and subsequently grazing animals. Use of
sewage sludge as a fertiliser can also increase soil levels of cadmium.

Toxicity

Approximately 10% of the available dose is absorbed through lung and gastrointestinal tract. Cadmium is transported in the body bound to metallothionein, a low
molecular weight plasma protein, and about half the body burden is stored in the
kidney and to a much lesser extent, the liver. Elimination of cadmium is a slow
process, with a half-life of 16–33 years. A syndrome of severe arthralgias and osteo-
malacia called *itai-itai byo* (ouch-ouch disease) was observed in postmenopausal
women in Japan, caused by a combination of cadmium contamination of diet and
low calcium and vitamin D intake.

Acute toxic effects are similar to those for metal fume fever (see below) for
inhalational exposure, and gastroenteritis from oral exposure. The kidney is the
main target organ for chronic exposure and proteinuria is the most frequent finding

(glycosuria and aminoaciduria are also common). This is due to proximal tubule damage. This damage may be associated with hypertension. Other effects include emphysema, chronic bronchitis and lung cancer.

Biological monitoring: measuring human exposure to cadmium

Cadmium may be measured in whole blood, this being of greater use than serum because of cadmium binding to red blood cells. Cadmium in blood reflects more recent exposure as distinct from longer term exposure. Cadmium can also be detected in urine, this being more representative of body burden than short-term exposure. This is due in part to the very long half-life of cadmium which is between 10 and 30 years. Great care is required when interpreting biological monitoring data for cadmium because a recent high exposure can increase urinary levels without a corresponding increase in body burden. Furthermore, when kidney damage is considerable (as a result of long-term exposure to cadmium), greater amounts of cadmium are released into the urine due to decreased reabsorption of metallothionein which has cadmium bound to it. This can be in the absence of any recent high exposures. Thus an integrated picture of body levels and likely exposure patterns are needed to best interpret the data.

Detection of low molecular weight proteins, such as β-microglobulin and retinol-binding protein, has been used to indicate early kidney damage as a result of long-term cadmium exposure. However, as they are reflective of an adverse effect they are better regarded as a health surveillance technique. Assay for metal-lothionein in urine has been reported to be a specific and sensitive index of increased levels of cadmium in exposed workers.

Organ concentrations of cadmium have been measured using neutron activation but this is not a technique that can be used routinely. Hair and faeces determinations are also possibilities that have been addressed but they have not found regular application.

Levels of 5 μg cadmium/g creatinine and 0.5 μg cadmium/100 ml have been suggested as appropriate urinary and blood levels that would not be associated with the adverse effects of this metal.

Chromium

Chromium is an important commercial element with many uses. The metal exists is all oxidation states from [II] to [VI], but only the trivalent [III] and hexavalent [VI] compounds, and the metal is of practical importance. Of these, the hexavalent form is 100–1000 times more toxic than most common trivalent compounds.

Biologically active chromium is an essential mineral, and a chromium deficiency has been identified in animals.

Exposure

The most important mineral of chromium is chromite. About 95% of the world's economically viable resources of chromite are located in the southern part of Africa.

Chromium metal is mostly used in the production of special (stainless) steels. It is also used in/to electroplate other metals.

Chromium is present in high concentrations in cement, where it may cause contact dermatitis. Chromium compounds are used as paint pigments and dyes, as a catalyst, to make magnetic tape, in tanning, in wood impregnation as a wood preservative, in safety match production, and so on.

Generally, chromium is found in nonindustrial environments in air at $10\,ng/m^3$, in soil at concentrations of 10–90 ppm, and in fresh water at 1–10 $\mu g/L$. Under such conditions, chromium is found as the chromate. Chromium is found at concentrations in the air of industrial cities up to $70\,ng/m^3$ and in water up to 25 $\mu g/L$.

Toxicity

There are two forms of chromium, the hexavalent chromium[VI] (chromous, Cr^{6+}) and trivalent chromium[III] (chromic, Cr^{3+}). Compounds of the latter form are relatively harmless, owing to their insolubility. Chromium exposures are found in a number of industries, including electroplating, concrete use, tanning, safety match and pigment manufacture. Exposure to chromium is also observed with the use of dichromate compounds. Absorption occurs from gastrointestinal tract and lungs, and systemic effects are possible following skin exposure; about 60% of an absorbed dose is excreted in the urine within 8 hours of absorption.

The toxicity of chromium compounds appears to be related to powerful oxidising properties of the hexavalent compound, which is reduced to the trivalent form once absorbed into the body. Chromium compounds are both skin and mucous membrane irritants, and skin and pulmonary sensitisers. Dermatitis, chrome ulcers

Table 12.6 Human exposures to chromium

Industrial/occupational
Extraction: most of the ore is reduced to ferrochrome, an iron–chromium alloy containing about 60% chromium; this is usually the main source of chromium, as the pure metal is not required for production
Production of ferrochrome
Pigment manufacture
Welding
Electroplating
Timber preservation
Tanning

Environmental
Air: fly ash during/from incineration
Water (for example, from disposal of industrial wastes)

Domestic
Biologically active chromium (high levels are found in liver and cheese)
Cutlery (more than half of dietary chromium comes from sources other than the food itself, such as cutlery and in food preparation)

(a penetrating lesion of skin called chrome holes), corrosion of the nasal septum, conjunctivitis and lacrimation also occur. Hexavalent chromium compounds are carcinogenic (bronchogenic carcinoma). There is some evidence that short-lived pentavalent chromium[V] intermediates are the specific carcinogenic agents as these have been shown to be directly genotoxic.

Biological monitoring: measuring human exposure to chromium

Due to the difference in the toxic effects between hexavalent and trivalent chromium it is important that the contribution of each to levels measured in body fluids can be assessed. It seems that urinary and plasma levels mainly relate to hexavalent chromium exposures, but trivalent exposure may be responsible for some of the chromium measured. Sampling at the end of shift provides an indication of the recent exposure but previous exposure may contribute to this.

With determinations in blood it is relevant to note that exposure to hexavalent chromium results in its uptake by the red blood cells with subsequent reduction to the trivalent species. It has been suggested that differential estimation of plasma and red cell levels of chromium could provide a sensitive internal indication of exposure to the hexavalent form but this requires further development.

There is an indication that chromium levels are raised in hair after exposure.

Cobalt

Cobalt is a nutritionally essential metal, and deficiency results in severe health consequences. It is a relatively rare metal produced primarily as a by-product of the metals, chiefly copper. It is used in high-temperature alloys and in permanent magnets. Its salts are useful in paint driers, as catalysts, and in the production of numerous pigments.

Exposure

Cobalt is found in soil. Its deficiency can cause health problems in humans. Although cobalt deficiency is not yet widespread, it is likely to become a problem in the future, as natural cobalt content of soil is low, and depletion of cobalt is transported through the rivers to the oceans. On the other hand, because of mining activities and its widespread industrial uses, cobalt also belongs to the metals posing potential dangers due to excessive exposures. Primarily at risk are metal workers.

Traces of cobalt are found in all rocks, minerals and soils. The average cobalt content of the earth's crust is 18 ppm Cobalt usually occurs together with nickel and iron. Annually, approximately 21,000 tonnes of cobalt are transported by rivers to the oceans and about the same amount is deposited in deep-sea sediments, whose content is about 74 ppm

Toxicity

Chronic exposure to cobalt produces 'hard metal pneumoconiosis' in industrial workers using cobalt in the manufacture of heat-resistant hard metal alloys. A car-

Table 12.7 Human exposures to cobalt

Industrial/occupational
Metallic mining and metal refining
Cobalt powder handling and grinding
Blue and green ceramic glazes
Coloured glass superalloys (jet engines, gas turbines)
Pigments
Acetic acid makers
Magnet steel workers
As catalyst in the synthesis of heating fuels and alcohol
Nuclear technology

Environmental
Natural occurring sources: natural mineral cobaltite
Artificial sources: production of cobalt-containing metal and cobalt salts
Air: cobalt can be detected during production of hard metals and cobalt salt
Soil: exceedingly low levels of cobalt were observed in soil of some areas of Australia, New Zealand and USA; cobalt sulfate can be used for land treatment
Water: in uncontaminated sample of fresh water, cobalt concentrations are generally low, ranging from 0.1–10 µg/L, however, in waters of polluted rivers, high cobalt levels have been observed, e.g. 4500 µg/L in Mineral Creek, USA
Food chain: only a few plant species accumulate cobalt above the 100 ppm which causes severe phytotoxicity

Domestic
Pigments
Artwork (painting, ceramic art and sculpture)
Ceramics
Alloy as dental materials
Tobacco smoking
Medication: used in γ-ray therapy of cancer, for example, Cobalt-60
Medical uses

diomyopathy has also been observed in workers using cobalt as a defoaming agent (beer drinker's cardiomyopathy).

Biological monitoring: measuring human exposure to cobalt

This metal can be measured in blood, urine, nails and hair. Available data suggest that measurement of pre- and post-shift urinary levels allows estimation of exposure during the shift by calculation of the difference. As the work week progresses levels increase such that sampling at the end of the week represents accumulation over that period. A sample taken at the very beginning of the work week is thought to be more representative of long-term exposure.

Lead

Lead is a metal of antiquity, and has been used for a many purposes for thousands of years. Some of the uses of lead in ancient times (such as lead sheet for lining roofs) remain today. Other uses, such as the use of white lead in paint, have been discontinued for obvious toxicological reasons, and still others have developed for modern applications.

Exposure

Exposure to lead is possible in a variety of occupations: smelting, ship breaking, welding, plumbing, battery, alloy and pigment manufacture, printing. Exposure is also possible from nonoccupational sources, some quite exotic (hair preparations, tonics, cosmetics).

About 40–50% of all lead production is used to make lead–acid batteries (both lead metal and lead oxides are used). A further 20% of production is used as the metal, for example, lead sheet, cable sheathing, solder, ammunition, alloys, weights, ballast, low melting alloys. Lead-based pigments have a long tradition of being used in paints, although they have virtually all been substituted by other pigments in the last 20 years. Lead compounds are used in glassware and ceramics, and as a stabiliser in plastics. Red lead is used to make television tubes. About 10% is converted into alkyl lead compounds and used as antiknock additives in petrol, although this use is in decline as more and more countries move to limit the concentration of such additives in petrol.

The most important minerals of lead are galena (lead sulfide), cerussite (lead carbonate) and anglesite (lead sulfate). All are extracted through processing of the crude ore, roasting, sintering, reduction and refining. The refining process also extracts zinc, copper, gold, silver, antimony and arsenic. In the past, significant domestic contamination has also occurred from workers taking home their overalls. This has decreased with better hygiene and laundry arrangements.

While not strictly a lead exposure, physiological conditions such as pregnancy, infection or menopause may mobilise lead sequestered in body bone stores.

The total daily intake of lead varies considerably from area to area and country to country. Most studies report a daily intake between 20 and 200 μg/day in adults, with a recommendation that lead intake in children be much less, owing to a significantly higher lead intake on a body weight basis.

Absorption and disposition

Approximately 5–10% of an ingested dose is absorbed, though lung uptake can reach 50–70% depending on particle size, retention and solubility. Most recently absorbed lead is stored in red blood cells (half-life of about 20–40 days) and most long-term body lead (90%) is stored in bone tissue (half-life in the order of 20–30 years). Elimination of lead occurs mainly through the kidney.

Table 12.8 Human exposures to lead

Occupational
Extraction process
Lead–acid battery manufacture
Ship breaking
Car radiator repair
Welding
Paint manufacture or application
Plumbing
Petrol manufacture
Plastics manufacture

Environmental
Mining, smelting
Processing
Use
Recycling
Disposal
Air: the major part of lead found in the atmosphere results from the combustion of leaded petrol. (Atmospheric transport of minute particulate airborne lead may range over hundreds and thousands of kilometres. Air lead concentrations vary from below $0.01\,\mu g/m^3$ in remote areas, below $0.02\,\mu g/m^3$ in rural areas, $0.02–2.0\,\mu g/m^3$ in urban areas and $1–2\,\mu g/m^3$ near lead smelters. The WHO recommended limit is $0.5–1.0\,\mu g/m^3$ as a long-term average. The stepwise decremental decreases of lead in air (and lead in blood in the population) from those countries removing lead from petrol establish car exhaust as a major source of environmental lead pollution. Lead smelting and refining (including secondary refining/recycling) is also known to give rise to substantial lead emissions. Decreases in emissions have occurred, due to improvements in air pollution control technology and better air quality regulation.)
Water: the major contributor to lead in water in from fallout of lead in air. Lead in water comprises dissolved lead and suspended matter. Water lead concentrations vary from below $0.02–2.0\,\mu g/L$ in rural areas, $1–40\,\mu g/L$ in urban areas and $10–1000\,\mu g/L$ near lead smelters. However, concentrations in water remote from point sources of pollution can increase by two orders of magnitude (to about $100–1000\,\mu g/L$) where water is plumbosolvent (for example, areas of acidic or 'soft' water).
Soil: lead levels in unpolluted areas range from $10–40\,ppm$, with most values below $20\,ppm$ Except in areas where the underlying rock contains appreciable amounts of lead, the top layers of soil tend to have higher lead levels than deeper levels, due to atmospheric deposition. Significantly increased lead levels have been reported in surface soils in inner city areas, near busy highways and near lead processing industries.
Sediments: in areas where significant discharges have occurred, concentrations of lead in sediments can be substantial. For example, $50\,ppm$ lead has been reported in sediments in the Rhône river.

continued

Table 12.8 continued

Domestic
Plants growing in high lead soils or from surface deposition
Inadvertent addition of lead from food processing
Leaching of lead from cans with soldered seams
Improperly glazed crockery
Lead in cooking water
Alcoholic drinks, notably wine (from bottle sealing) and 'moonshine' spirits (from lead solder in distilling equipment) may contain substantial amounts of lead
Lead in paint (by far the biggest source of domestic lead exposure where lead in paint is found)
Lead plumbing
Hair dyes and colour restorers
Asian cosmetics
Health tonics
Tobacco smoking

Toxicity

Lead toxicity in workers results primarily from inhalational exposure, though oral exposure may be important in cases of poor hygiene. Lead has been shown to induce effects in a number of body systems:

- haematopoietic system (disruption in porphyrin synthesis, anaemia)
- nervous system (neurasthenia, slowed conduction velocity, neurobehavioural disturbances, peripheral neuropathy, encephalopathy)
- gastrointestinal (colic, constipation)
- kidney (lesions of proximal tubule, Fanconi syndrome, chronic interstitial disease)
- reproductive system (stillbirth, spontaneous abortion, decreased sperm activity and counts)
- cardiovascular (hypertension).

Biological monitoring: measuring human exposure to lead

Background levels of lead in human blood are of the order of $5–15\,\mu g/100\,ml$. While blood lead levels continue to fall, recent initiatives recommend that blood lead levels in children, particularly preschool infants, should not increase above $10\,\mu g/100\,ml$.

Since about the late 1960s, biological monitoring of blood lead levels in workers has been employed to assess lead exposures, workers with blood lead levels exceeding a mandated amount being removed. While this practice is not preventive, it establishes the upper limit of exposure in workers. This value was $70–80\,\mu g/100\,ml$ in the early 1970s, but has fallen to $40–60\,\mu g/100\,ml$ in the mid-1990s.

The measurement of exposure of humans to lead can be carried out in virtually any biological media. Lead has been measured in blood, urine, sweat, hair, nails, saliva and milk. The most used methods are lead in capillary or venous blood, and lead in urine (preferably 24-hour collection).

Lead in blood represents recent exposure (over the past 3–8 weeks) and can range up to 10–15 μg/100 ml in non-exposed populations, and up to 40–60 μg /100 ml in workers in lead processes (intervention and medical removal usually occurs in workers with blood lead levels above these values). Occupationally, lead in blood remains the method of choice for biological monitoring.

As well as biological monitoring of absorption, it is also possible to carry out biological measurement of effect. The several enzymes in the haem and porphyrin synthetic pathway are inhibited by lead, and the rate-limiting enzyme in the synthetic pathway, δ-amino-laevulinic acid dehydratase, is one of the most sensitive systems affected by lead, with inhibition beginning at a blood lead level of 10–20 μg/100 ml, and almost total inhibition at 70–80 μg/100 ml. A number of properties of this system have been used in biological monitoring, including: the blood or erythrocyte levels of δ-amino-laevulinic acid dehydratase, the level of its substrate δ-amino-laevulinic acid in blood or urine, levels of intermediates in the porphyrin synthetic pathway, such as uroporphyrin and coproporphyrin intermediates, zinc protoporphyrin (ZPP) and free erythrocyte protoporphyrin (FEP). Of these, ZPP and FEP are closely correlated with blood lead, and are used to measure lead exposure in workers.

Lithium

Lithium is the smallest of the alkali metal ions. It is highly electropositive, and forms compounds with a range of metals.

Exposure to lithium

Lithium is obtained by electrolysis of volcanic brine from lithium-containing materials.

Lithium has found practical applications in medicine, in battery production, is used for cooling in nuclear reactors, is a necessary raw material in the production of tritium, and in the form of its isotope ^6Li, as a thermonuclear fuel. Though lithium is used industrially as a nuclear reactor coolant and in manufacture of alkaline storage batteries and alloys, occupational toxicity is rare.

Toxicity

Most lithium poisoning occurs through its therapeutic use as an antidepressant.

Biological monitoring: measuring human exposure for lithium

Lithium can be measured in urine and plasma but routine biological monitoring for environmental or occupational exposures has not been undertaken.

Table 12.9 Human exposures to lithium

Industrial/occupational
Advanced technologies
Tritium production
Battery manufacture

Environmental
Lithium concentrations in the environment are generally low

Medical
Treatment of depression
Increase tolerance to the side effects of cancer therapy
Topical treatment of genital herpes

Manganese

Manganese is an essential metal and is present in all living organisms. While it is present in urban air and in most water supplies, the principal portion of the intake is derived from food. Human daily intake ranges from 2 to 9 mg.

Exposure

Manganese and its compounds are used in making steel alloys, dry-cell batteries, electrical coils, ceramics, matches, glass dyes; in fertilisers, welding rods; as oxidising agents; and as animal food additives. Heavy exposure may occur in mines and during the production of metals. Workers are mainly exposed to manganese dioxide by inhalation. Manganese is also used in the chemical industry. Organic manganese compounds such as methyl-cyclopentadienyl manganese tricarbonyl (MMT) have been used to increase the octane level of gasoline, although the concept that manganese-containing 'unleaded' petrol is somehow beneficial seems dubious.

Manganese is widely distributed in soil as an abundant element. Its average concentration in soil is probably about 500–900 mg/kg. Manganese pollution of water does not appear to be a problem except possibly in isolated cases of waste disposal. In fresh water both soluble and suspended forms exist. Atmospheric concentration of manganese observed in urban areas can be attributed primarily to man-made sources. A principal source of atmospheric emissions is metallurgical processing.

Toxicity

Acute exposure to manganese fumes produces a metal fume fever (see below), which may develop into pneumonitis. A classic occupational condition known as 'manganism' results from chronic exposure to dust from manganese ores or fumes from manganese steels. The features of this condition are neurological and psychological, including apathy, confusion, bizarre behaviour, increased muscle tone, difficulty with speech and fine motor movement and loss of balance. Onset is progressive and insidious.

Table 12.10 Human exposures to manganese

Industrial/occupational
Mining and raw material transport
Metallurgical processing and reprocessing (90% used in the making of steel)
Boiler making
Shipyard
Welding and cutting torches (the welding rod)
Power plant

Environmental
Metallurgical processing
Power plant
Welding and cutting torches
Waste disposal (waste water and solid waste)

Domestic
Food intake
Dye
Ceramics

One feature of chronic manganism is parkinsonian symptoms, due to manganese-induced damage to the extrapyrimidal system (notable the highly dopaminergic red nucleus) of the central nervous system. This lesion is similar to that in Parkinson's disease.

Biological monitoring: measuring human exposure to manganese

While manganese can be determined in urine, blood, faeces and hair it has been difficult to establish a clear relationship between levels in the body and chronic toxicity. Detection in urine would appear to be satisfactory only to confirm qualitatively that exposure has occurred.

Mercury

Mercury is used in dentistry, battery, medical and scientific equipment manufacture, and in the production of chlorine and caustic soda. The metal is liquid at room temperature, and exists in elemental, inorganic (Hg, or mainly, Hg^{2+}) or organic (Hg^{4+}) forms.

Exposure

Mining, smelting, and industrial discharge of mercury (especially in the paper pulp industry) have been major factors in occupational and environmental contamination in the past. Fossil fuel may contain as much as 1 ppm of mercury, and it is estimated that about 5000 tons of mercury per year may be emitted from burning coal, natural gas, and the refining of petroleum products.

The major source of mercury is the natural degassing of the earth's crust, including land areas, rivers, and oceans, and is estimated to be of an order of 25,000 to 150,000 tonnes per year. Metallic mercury in the atmosphere represents the major pathway of global transport of mercury. As much as one-third of atmospheric mercury may be due to industrial release of organic or inorganic forms. Regardless of source, both organic and inorganic forms of mercury may undergo environmental transformation. Metallic mercury may be oxidised to inorganic divalent mercury, particularly in the presence of organic material formed in the aquatic environment. Methyl mercury, an important form of organic mercury, can be taken up by fish in the food chain, and may eventually be consumed by humans. The intake of mercury from food is generally very low, with a daily intake below 1 μg/day.

Table 12.11 Human exposures to mercury

Industrial/occupational
Mining, smelting, and industrial discharge of mercury
Metal refining
Paper pulp mill
Fossil fuel (coal, natural gas) burning
Production of chlorine and caustic soda
Refining of petroleum products
Catalysts
Pesticide application
Military application (such as detonators)
Production of steel
Cement production
Phosphate and smelting of metals from their sulfide ores
Dental applications
Laboratory usages

Environmental
Metallic industry
Burning of fossil fuel (such as power station)
Production of steel
Cement production
Phosphate and smelting of metals from their sulfide ores
Incinerators
Waste disposal
Dental amalgam filling

Domestic
Dental amalgam
Painting and ceramics
Measurement and control systems (such as thermometers)
Foodstuffs (especially for organic mercury)
Contaminated water and plants
Paints
Battery
Drugs

Toxicity

Toxicological effects and relevant human exposures of mercury have been illustrated over the past few centuries. On the basis of toxicological characteristics, there are three forms of mercury: elemental, inorganic and organic compounds.

Absorption and biological effects vary with these forms, and though the latter is bioaccumulated and bioconcentrated in food chains and can appear in environmental disasters (for instance, the neurological condition Minamata disease), only the first two are important in occupational situations:

- Elemental mercury vapour is well absorbed by the lung (80%) but poorly absorbed by the gut (0.01%). Once absorbed, mercury is oxidised to the Hg^{2+} form. This reaction takes place in a number of tissues, especially the brain.
- Moderate amounts of inorganic mercury are absorbed by the gut (7–15%). These cause most cases of mercury poisoning. Systemic effects are possible from skin and lung exposure. Excretion occurs in urine and faeces, with a half-life of 40–60 days.
- Organic forms of mercury are extremely well absorbed across the gut, though dermal absorption does not appear to be great. Inhalation of methyl mercury compounds results in a classic condition of ataxia, dysarthria and constricted visual fields. The major route of excretion is through the bile.

The main toxic effects are observed in the kidney and gastrointestinal system (inorganic), central nervous system (elemental and organic) and respiratory system (elemental mercury vapour). They include:

- kidney: oliguria, proteinuria, nephrotic syndrome
- central nervous system: psychological changes, erethism, tremor abolished by sleep, polyneuropathy; insomnia, loss of appetite, loss of memory, shyness and timidity characterise mercury-induced psychosis
- gastrointestinal system: gingivitis, salivation, stomatitis, severe mucosal necrosis
- respiratory system: inflammation, ventilation/perfusion defects, hypoxaemia, progressive fibrosis, pulmonary granulomas.

Biological monitoring: measuring human exposure to mercury

Due to the different forms and sources of mercury to which humans are exposed the biological monitoring of the substance and data interpretation are somewhat complicated. Urinary levels relate mainly to exposure to mercury vapour or to inorganic mercury. The number of amalgam fillings in the individual may contribute to the overall levels in the bodily fluids. Coincident exposure to mercury-containing disinfectants could also result in higher levels. Considerable work has been done which indicates that at levels above about $50\,\mu g/g$ creatinine health effects may be experienced. The duration of exposure needs to be appreciated because concentrations in people only recently exposed will not have reached equilibrium.

Attention also seems to be required to consistency of sampling time, correction for specific gravity and ensuring that the person has not been absent from the exposure source for some days. The contribution to body levels of mercury from fish is of lesser importance for urinary levels than it is for blood because it is in the form of methylmercury that is not excreted in the urine. Blood levels primarily reflect more recent exposures and are useful in confirming occasional high exposures, especially in comparison to urinary values. Determination of mercury in saliva has been reported to be consistent with blood and urine levels but it has not developed as a widely used method.

Nickel

Most of the world's nickel is processed from sulfide ores and to a lesser extent, oxides. The largest producers are Canada, the former USSR and New Caledonia.

Exposure

The exposure and toxicity of nickel is based on the various classes of nickel compounds, and the human activities associated with them:

- Divalent nickel compounds have a low toxicity at concentrations found in the environment.
- Exposure to nickel metal (for example, from its use in coinage or stainless steel) is not considered harmful.
- In humans, adverse effects of water-soluble nickel compounds occur after contact to skin (causing contact dermatitis in perhaps 10% of exposed individuals) and after inhalation (which causes respiratory tract irritation and asthma) in workers such as electroplaters.
- Human exposure to inorganic, water-insoluble nickel compounds usually occurs through inhalation of fumes or dusts, which are associated with cancers of the respiratory tract in nickel refineries.
- The organic compound nickel carbonyl ($Ni(CO)_4$) is produced in the Mond refining process, and because of its volatility (boiling point of 43°C) and lipid solubility, is highly toxic and carcinogenic.

Nickel is a constituent of over 3000 metal alloys, and is used for a huge range of purposes, such as coinage (some coins may be 99.8% nickel), stainless steel (which contains about 10% nickel) cooking utensils, corrosion-resistant equipment, aircraft parts, magnetic equipment, jewellery, rechargeable batteries, medical applications, ceramics and so on.

Representative exposure data are difficult to obtain. Time-weighted average concentrations in air in different workplaces where nickel may be found range up to high concentrations (above $1 \, mg/m^3$) in roasting and smelting operations and in electrolytic refining and in foundry operations. Moderate levels ($0.05–1.0 \, mg/m^3$) were found in stainless steel welding, electroplating and in nickel–cadmium battery manufacture.

Table 12.12 Human exposures to nickel

Occupational
Extracting
Refining: either (i) electrolytic refining to yield nickel cathodes; or (ii) the Mond process, whereby the oxide is reduced with hydrogen, reacted with carbon dioxide to produce nickel carbonyl, and thermally decomposed to produce pure nickel; both processes produce exposures that are hazardous to workers, although the Mond process is relatively more toxic
Alloy production
Catalysts in the chemical industry and in petroleum refining
Electroplating processes

Environmental
Weathering
Dissolution of rocks and soil
Atmospheric fallout
Industrial processes
Waste disposal
Air: volcanic emissions and windblown dusts from weathering, from combustion of fossil fuels (both coal and oil), from mining, from industrial processes and from incineration of wastes. Atmospheric levels average $6\,ng/m^3$ in nonurban areas in the USA, and in urban areas ($17\,ng/m^3$ in summer and $25\,ng/m^3$ in winter, indicating a source from energy demand for heating). Close to industrial areas, atmospheric nickel concentrations as high as $170\,ng/m^3$ have been recorded, and close to nickel refineries, air nickel levels can average as much a $1200\,ng/m^3$.
Water: usually contain less than $20\,\mu g/L$, although this can increase owing to pollution of supply and acid rain has a tendency to mobilise nickel from soil and to increase nickel concentrations in groundwaters. This can produce increased uptake in soil microorganisms and plants, and in turn, animals. Nickel leached from dump sites can contribute to contamination of the aquifer, with potential ecotoxicity and risk to humans.
Soil: is dependent on soil type and pH, with high mobility in acid soils. Insoluble nickel may be deposited by precipitation, complexation, adsorption onto clay or silica, and uptake by biota. With variability in microbial activity, ionic strength, particle concentration, these processes may be reversed.
Anthropogenic sources of soil nickel are emissions and wastes from mining and refining, atmospheric deposition from other industrial activities and disposal of sewage sludge.

Domestic
Food: nickel in nuts (5 ppm) and cocoa (10 ppm) are high, compared with a nickel concentration in most foods of below 0.5 ppm Daily human intakes are in the range $100–800\,\mu g/day$, depending on dietary habits.
Jewellery: the incidence of allergic contact dermatitis through exposure to nickel-containing jewellery is relatively high, and possibly in about 10% of exposed individuals. The sensitising exposure is usually from the piercing of ears or other body parts.

Toxicity

The main occupational hazards from soluble nickel compounds are its allergenicity (producing a contact dermatitis called 'nickel itch') and carcinogenicity (lung cancer from inhalation of nickel carbonyl and nickel dusts).

Biological monitoring: measuring human exposure for nickel

Wide variations have been reported in body nickel levels. Further, because of decreased emissions and better analytical methods, exposures are decreasing, suggesting that earlier data are not representative of contemporary conditions.

Background levels of nickel in human blood are less than 0.5 μg/L and about 2–3 μg/L in urine. Both blood and urine are used to monitor workers, although more data are available from urinary measurements. Elevated urine nickel levels are shown in some nickel workers, with the highest in nickel refinery workers (with a mean of over 200 μg/L). All other occupational groups had urinary levels below 20 μg/L. Nickel concentration in bile suggests that biliary excretion may be quantitatively significant.

Nickel can be detected in both urine and plasma. Workers exposed to nickel compounds have elevated levels in both body fluids with plasma being suggested as the more reliable. At the present time there is no clear relationship between health risks and levels in body fluids, except for overexposure to nickel carbonyl. A potentially complicating factor is the exposure to nickel compounds of low solubility. Slow clearance of these compounds may be responsible for elevated levels in workers for some years after cessation of exposure. It is also these low solubility nickel compounds (nickel subsulfide) that are more linked to lung cancer.

Nickel has also been determined in nasal mucosa and in hair, although these methods do not seem to have gained wide use.

Platinum

Use of platinum has followed a number of stepwise increases due to technological developments. Firstly, the metal was jewellery and coinage in the 19th century. Secondly, invention of the process of ammonia oxidation in which platinum served as a catalyst allowed large-scale fertiliser production in the earliest part of the 20th century. The process was also used to make ammunition for two world wars. Thirdly, platinum has been used extensively by the petroleum industry in the catalytic 'cracking' of high-boiling crude oil fractions since the Second World War. Fourthly, with the advent of tighter emission controls on vehicles from the 1970s, platinum (with palladium and rhodium) is now used in exhaust catalysts that oxidise harmful combustion by products, such as nitrogen oxides, partially oxidised hydrocarbons and carbon monoxide.

Platinum is found both as the metal and in a number of minerals; these are mainly associated with ores of nickel and copper. Extraction of platinum is not extensive, at about 100–150 tonnes a year. Total world production of platinum has been estimated at about 1500 tonnes. Naturally occurring levels of platinum in the environment are low.

Exposure

The main use of platinum is in catalytic converters in motor vehicle manufacture, though the metal and its alloys are also used in jewellery and dentistry, as a catalyst in the chemical industry and in fibre glass manufacture. Cisplatinum (*cis*-dichloro-diamine platinum) is a chemotherapeutic agent.

Modern day uses of platinum derive from its exceptional catalytic properties, where the major use (probably over two-thirds of production) is in catalytic converters in vehicles. Further industrial applications also relate to its catalytic properties, and to other characteristics, including resistance to chemical corrosion, high melting point, high mechanical strength and good ductility. Platinum is used as a catalyst in chemical manufacturing, in electrical engineering, in instruments for high temperature measurement, in glass technology, in jewellery, dentistry and for surgical implants.

Specific complexes of platinum (most notably cisplatin – cis-dichlorodiamine platinum) are used therapeutically as chemotherapeutic agents to treat cancer.

Possible emerging uses include cathodic protection of steel and as a catalyst in hydrogen fuel cells.

Toxicity

Workplace toxicity results from the sensitising properties of platinum salt complexes, but not from platinum. These compounds induce immediate hypersensitivity reactions, and the term platinosis has been applied to the spectrum of effects (asthma, rhinorrhoea, dermatitis). The incidence of occupational asthma by these compounds in workers can reach 100%.

Biological monitoring: measuring human exposure to platinum

Background levels in human blood are of the order of 0.1–2.8 µg/L. These concentrations are also reported in workers occupationally exposed to platinum, with the exception of hospital workers exposed to cisplatin where blood levels can range up

Table 12.13 Human exposures to platinum

Industrial/occupational
Mining and processing of platinum from copper and nickel sulfide ores
Soluble platinum salts in the manufacture of metal catalysts and alloys
The cytotoxic drug cisplatin and its analogues

Environmental
Loss of platinum from catalysts in normal use can occur (ammonia oxidation and the catalytic conversion of nitric oxide to nitric acid)
Vehicle exhausts (from catalytic converters)

Domestic
Platinum jewellery

to ten times this concentration. In a study of exposure in hospital personnel, urinary platinum levels ranged from 0.6–23 µg/L, were at the limit of the analytical method and were not different from controls (2.6–15 µg/L). By comparison, an average of 7 mg/L was measured in the urine of cisplatin-treated patients.

Platinum can be measured in both urine and blood although in exposed workers urinary levels have been reported without simultaneously detectable plasma levels.

Selenium

Selenium occurs in nature and biological systems as selenate (Se^{6+}), selenite (Se^{4+}), elemental selenium (Se^0), and selenide (Se^{2-}), and deficiency leads to cardiomyopathy in mammals including humans. In most countries the daily intake of selenium is about 100 µg.

Selenium is an essential mineral, and selenium deficiency can occur.

Exposure

Selenium is widely used in the electronics, glass, ceramics, steel and pigment manufacturing industries. It also has medicinal applications in shampoos, dietary supplements, and so on.

Environmental selenium compounds may originate from metal smelting, coal combustion, or the disposal of waste. The earth's crust contains an average selenium content of 0.05–0.09 ppm Higher concentrations of selenium are found in volcanic rock (up to 120 ppm), sandstone, uranium deposits and carbonaceous rocks. River water levels of selenium vary depending on environmental and geologic factors; 0.02 ppm has been reported as a representative estimate. Selenium has also been detected in urban air, presumably from sulfur-containing materials.

Selenium in foodstuffs provides a daily source of selenium. Seafood (especially shrimp), meat, milk products, and grains provide the largest amounts in the diet. The concentration of selenium in different foodstuffs varies considerably (0.01–1.0 mg/kg), depending on the origin of the food. Some plants accumulate selenium.

Toxicity

The most toxic compound of selenium is Gun blue (2% selenious acid), with a lethal dose of 30–60 ml. Exposure to selenium dusts can cause respiratory tract irritation, nasal discharge, cough, loss of sense of smell and epistaxis.

Chronic selenium toxicity resembles arsenic poisoning.

Biological monitoring: measuring human exposure to selenium

Determination of selenium in blood or urine is often carried out to establish deficiency rather than excess. In selenium deficiency, blood and serum concentrations are below 40 µg/L. Concentrations in bodily fluids vary with dietary intake and geographically and their biological significance remains to be fully established. Concentrations in plasma and urine may reflect recent exposures while chronic exposures are better reflected by levels in red blood cells.

Table 12.14 Human exposures to selenium

Industrial/occupational
Mining, milling, smelting and refining
Burning of fossil fuel
Electrolytic refining of copper
Rectifiers and burned-out rectifiers
Glass manufacture (flat glass, pressed or blown glass and glassware)
Catalyst for oxidation in chemical industry
Electronic usage (photoconductors and photoelectric cells)
Inorganic pigments (in plastics, paints, enamels, inks, rubber and ceramics)
Colour copper and copper alloy
Lubricants

Environmental
Emission from selenium mining, milling, smelting and refining
Coal combustion
Waste disposal
Insecticide
Soil additive

Domestic
Foodstuffs
Food additives
Veterinary pharmaceuticals
Lubricants
Medical usages (drugs and radioactive diagnostic agent)

Thallium

Thallium occurs in small amounts in sulfur-containing ores and potassium minerals. Following decline of its use as a rodenticide, thallium has virtually no economic or technical importance.

Exposure

Thallium has specialised uses in the manufacture of low-temperature thermometers, semiconductors, scintillation counters and optical lenses. The metal has been used as a rodenticide in the past. Minor uses include as: cement additive; deep temperature thermometers; acid-resistant alloys; low-melting glasses; semiconductors; and in scintigraphy of heart and circulatory system.

Some thallium is found in copper and selenium minerals, but generally, thallium minerals are rare.

Toxicity

The average lethal dose for an adult is in the order of 1 g. Thallium is quickly absorbed and distributed widely in the body. Dermal absorption is possible through

Table 12.15 Human exposures to thallium

Industrial/occupational
Extraction from residues from the electrolytic smelting of copper, lead, zinc and iron ores
Rodenticide (now discontinued)
Accidental, deliberate and criminal misuse of thallium led to its prohibition of use in most countries in the 1970s and 1980s

Environmental
Some vegetation damage was reported around a cement plant in Germany, which was associated with thallium emissions; elevated thallium levels were seen in soils, plants and cattle

thallium-containing ointments. Elimination is also rapid, with a half-life of about 2 days.

Thallium has an affinity for sulfydryl groups, and inhibits many enzyme reactions. Thallium also exchanges with potassium and interferes with potassium-dependent reactions. The onset of poisoning is insidious, with gastrointestinal symptoms such as nausea, vomiting and diarrhoea progressing to nervous system effects (disorientation, lethargy, psychosis, insomnia, neuropathy, convulsions, cerebral oedema and coma), cardiorespiratory effects (tachycardia, hypertension, dysrythmias, myocardial necrosis) and ophthalmological effects (optical neuritis, ophthalmoplegias). Blue–grey lines may appear on the gums, and dark pigmentation around hair follicles may also be seen. Alopecia, especially on the face, and a dry scaly skin, with white lines across the nails are late signs. Central and peripheral nervous system effects may persist (ataxia, tremor, memory loss).

Biological monitoring: measuring human exposure to thallium

Determination of urinary thallium is considered to be preferable to blood as an indicator of exposure but not a great deal is known about this relationship.

Tin

Tin is rather unique in the wide variety of its compounds and applications. Ever since the beginning of the bronze age the metal and its alloys have been of importance to humans.

Exposure

Occupational exposure occurs during, for example, tin plating (a major industrial application) and the production of alloys and solders. In general, most of the operations associated with tin extraction and treatment of the tin ore are wet processes. Organic tin compounds also exist and are used as stabilisers in plastics and in some pesticides.

Tin is dispersed in very small amounts in silicate rocks containing 2–50 ppm, the ore casserite (SnO_2) being of major commercial importance. The earth's crust

Table 12.16 Human exposures to tin

Industrial/occupational
Tin extraction from ore
Tin alloy
Tin ore packing (tin dust)
Smelting operation
Tinplate containers (e.g. can) production
Manufacture of collapsible tubes in the pharmaceutical and cosmetic industries
Corrosion-resistant coating
Solder
Brass
Bronze
Catalyst
Mordant in dying of silk
Deposition of SnO_2 film on glass
Organotin: heat stabiliser (such as PVC stabiliser); transformer oil stabiliser; catalysts; biocides

Environmental
Tin extraction industry
Smelting operation
Organotin: agricultural uses: e.g. fungicides, bactericides and slimicide
Disposal of tin-containing materials
Garbage dumping
Organotin residues

Domestic
Food packing
Tinplate containers
Ceramics
Organ pipes
Toothpaste and dental preparations
Ceramic pacifiers and pigments
Die casting
Cooking utensils
Lining of lead pipes for distilled water, beer, carbonated beverages, and some chemicals

contains about 2–3 ppm tin. Tin is rarely detected in air, except near to industrial emission points. The concentration of tin in soil generally ranges from 2 to 200 mg/kg. In fresh water and in most foodstuffs the concentration is low and difficult to measure.

Trace amounts of tin are present in most natural foods. The normal tin intake is still uncertain. It is well known, however, that consumption of canned foods may increase the daily intake of tin considerably. Estimates for the average daily intake vary between 0.2 mg and 1 mg.

Toxicity

Inhalation of dusts or fumes of tin and tin oxide result in a benign pneumoconiosis (stannosis). Inorganic tin compounds have relatively low toxicities. Some tin salts can produce gastrointestinal disturbances at high doses, though systemic toxicity is limited, owing to poor absorption.

Organic tin compounds are potent neurotoxicants and skin irritants. These have wide use as fungicides (now prohibited), antioxidants and plastics stabilisers. Cerebral oedema and hippocampal necrosis has been observed following exposure to dialkyl tin compounds. Severe organic tin exposure has resulted in headache, tinnitis, deafness, memory loss, disorientation, psychosis, respiratory depression and coma. A syndrome of permanent neurological sequelae often occurs, including hyperactivity, inappropriately aggressive behaviour and loss of independence.

Biological monitoring: measuring human exposure to tin

Tin has not been widely measured in biological media and the usefulness of biological monitoring in determining human exposures in relation to health effects are not developed.

Uranium

Uranium is mined almost entirely for energy production in fission reactors, or for military uses in fission or hydrogen bombs. Minor uses include as a negative contrast in electron microscopy, and possibly in superconductors.

Exposure

Low levels of uranium are found in igneous rocks such as granites. The metal only exists as unstable radioactive isotopes that undergo a long chain of radioactive decays to end up finally a stable isotope of lead. Uranium (and indeed all actinides) are hazardous after enrichment.

Table 12.17 Human exposures to uranium

Industrial/occupational
Extraction
Enrichment and transformation into the hexafluoride for the selective enrichment of ^{235}U and ^{239}U
Radioactivity in mine tailings
Recycling of spent nuclear fuel

Environmental
Radiotoxicity of nuclear fallout is marginal
Use of fission bombs
Accidental release of radionuclides (such as during the Chernobyl accident)
Burning of coal

Toxicity

Adverse effects of uranium are related to exposure to gamma radiation (unlikely to exceed time-weighted average exposure standards) and inhalation dust particles containing short-life radon daughters (isotopes of lead, bismuth and polonium). Toxic effects are related to chronic lung disease and lung cancer.

Biological monitoring: measuring human exposure

Urinary uranium detection has been used to determine recent exposures to uranium compounds.

Vanadium

Vanadium is considered a trace element in some species, and presumed an essential mineral in humans, although no deficiency symptoms have been reported. The metal is implicated in disruption of coenzyme A activity, Na^+/K^+ ATPase inhibition and synthesis of cystine and cysteine.

Exposure

The major use of vanadium is in the production of steel. Vanadium steel contains up to 3% and is strong, heat resistant, and can withstand strain and vibration. The metal readily dissolves in iron, and with a boiling point of 3000°C, there is little vanadium in the fumes produced. Vanadium is also an alloying element in high-strength titanium alloys, usually at a concentration of about 4%.

Vanadium pentoxide is used as a catalyst in a variety of reactions, with the most important use the oxidation of sulfur dioxide to sulfur trioxide in the production of sulfuric acid. Other compounds of vanadium are used as catalysts in the production of plastics.

Vanadium is a typical rare element, fairly widely distributed at low concentrations in the earth's crust, and is found in the sulfide and oxidised forms in nature. A few ore deposits exist in Canada, South Africa and Russia which can be mined commercially, but most vanadium is recovered from the mining of other minerals, for example from some ores of iron, aluminium, uranium, magnesium, titanium, lead and zinc.

Vanadium is also found in some crude oils, and the ash from the combustion of such fuels can be a commercial source of vanadium pentoxide.

Toxicity

The vanadate ion (VO^{3-}) is the most common form in biological fluids, and it is one of the most potent inhibitors of the Na^+/K^+ ATPase pump.

Absorption is primarily through the lungs, and absorbed vanadium is largely excreted by the kidneys. Most exposures occur when workers inhale vanadium pentoxide fumes. As vanadium is present in fuel oils, workers have developed vanadium toxicity while cleaning out gas-fired boilers. A green discolouration of the tongue and a metallic taste suggest vanadium exposure.

Table 12.18 Human exposures to vanadium

Occupational
Recovery from ores
Production of vanadium compounds
Metal and catalyst production
Vanadium pentoxide production
Plastics manufacture
Boiler cleaning
Handling of vanadium-containing wastes (vanadium-rich slags are produced in steel production, vanadium-rich sludges are produced in titanium alloy production, disposal of spent vanadium catalysts, handling of vanadium-containing ashes, and vanadium contamination of plastics up to 500 ppm are possible)

Environmental
Fly ash from the combustion of fossil fuels
Leaching from soil or strata containing the element
Levels in water near areas of industrial activity or from deposits where industrial wastes containing vanadium have been dumped

Vanadium compounds produce dermal, mucous membrane and pulmonary irritation (including rhinitis, wheezing, nasal haemorrhage, cough, sore throat and chest pain). The systemic effects result from the ability of vanadium to oxidise, including the inhibition of oxidative phosphorylation. Long-term effects include chronic bronchitis, conjunctivitis and pneumonia.

Biological monitoring: measuring human exposure to vanadium

The background level of vanadium in human urine is below 3 μg/L. Exposed workers (sweeps exposed to vanadium-containing soot) have been reported with urinary levels fluctuating up to 13 μg/L.

The background level of vanadium in human blood is below 2.5 μg/L. Exposed metal workers had blood levels ranging up to 55 μg/L. Urinary levels in these workers were also high, and the cystine content of fingernails was lower than normal.

A rather wide range of vanadium concentrations has been reported in biological fluids which may be due to problems of analysis rather than indicating large differences across individuals or groups. Urinary determinations are preferred and levels can increase markedly from the beginning to the end of a shift where exposure is occurring. Longer term exposure is probably better determined by estimating concentrations 2 days after exposure ceases.

Zinc

Zinc is a nutritionally essential metal, and deficiency results in severe health consequences. On the other hand, excessive levels of zinc are relatively uncommon and require high exposures.

Exposure

Human exposures to zinc are possible in a wide variety of environmental and occupational situations, including soldering, battery manufacture, dentistry, pharmacological manufacture, electroplating, pigment and rubber working. The most common zinc-related condition is metal fume fever, caused by inhalation of zinc-containing fumes by welders. Metal fume fever, as the name implies, is associated with other metals such as aluminium, copper, manganese, nickel and so on.

Wastage results from all stages of production and processing of zinc, leading to emissions into the atmosphere, to water, and to solid wastes. Certain amounts of zinc can be recycled and this is mainly from the melting of alloys.

Zinc is ubiquitous in the environment, so that it is present in most foodstuffs, water, and air. However, zinc does not accumulate with continued exposure, but body content is modulated by homeostatic mechanisms that act principally on absorption and liver levels. Zinc atmospheric levels are increased over industrial areas. The zinc concentration of noncontaminated soils ranging from 10 to 300 mg/kg are comparable with those of their rocky subsoils. On average, levels of 20 mg/kg are found. Zinc applied to soil is taken up by growing vegetables.

Table 12.19 Human exposures to zinc

Industrial/occupational
Metallic mining and metal refining
Alloy production
Soldering
Electroplaters
Battery manufacture
Dentistry
Pharmacological manufacture
Pigments and rubber workers

Environmental
Volcano emission
Mining operation
Waste recycle
Sewage and garbage sludge
Waste dumping ground
Ceramics

Domestic
Artwork (pigments, painting, and sculpture)
Ceramics
Alloy as dental materials
Tobacco smoking
Dermal cream
Foodstuffs
Contaminated tap water

Water exposure may be increased by contact with galvanised copper or plastic pipes.

The average American daily intake is approximately 12 to 15 mg, most from food.

Toxicity

As zinc is an essential element, it is fairly well absorbed from the intestine. Most elimination is through the faeces.

The most toxic salt of zinc is zinc chloride (ZnCl), though powerful emetic properties limit its toxicity. This chemical is used in solder fluxes and can be caustic to the gastrointestinal tract. Inhalation of zinc chloride dust has caused pulmonary toxicity and death.

Biological monitoring: measuring human exposure to zinc

While determination of zinc in blood and urine has been used to detect work-related exposure, the significance of the levels to any health effects has not been established.

Zinc and other metals: metal fume fever

The most common zinc-related condition is metal fume fever, caused by inhalation of welding fumes containing zinc. Welders are the obvious workers this condition affects, although welder's mates and personnel working close by may sometimes be affected.

Metal fume fever is a flu-like illness, occurring 4–6 hours after exposure. Fatigue, chills, myalgias, cough, dyspnoea, thirst, metallic taste in the mouth and salivation are characteristic, though these tend to resolve after 36 hours. Metal fume fever is associated with other metals, including aluminium, copper, manganese, nickel and so on. Other organs affected by zinc include the kidney (tubular necrosis, interstitial nephritis), the pancreas (elevated glucose and amylase and lowered calcium) and lung oedema and pneumonitis.

Treatment of metal poisoning

Chelation

The first 'antidote' specifically designed to antagonise metal poisoning was dimercaprol or BAL (British Anti-Lewisite), which was developed to combat the effects of mustard gas (Lewisite, dichloro-2-chlorovinyl-arsine) in trench warfare. This was the first chelator (from the Greek *chela* for claw). Chelators are chemicals that bind directly with metal ions to form stable, water-soluble complexes which effectively bind the metal and stop it from interacting with biological processes prior to elimination from the body.

Chelators are perceived as having a specific affinity for a particular metal, though this is often not the case, and there is a large amount of nonspecific affinity for other metals. For example, ethylene diamine tetra-acetic acid (EDTA) is a chelator used extensively in metal toxicity (notably lead poisoning), though it will form tight complexes with other metals. When the other metal is calcium (required for physiological control of muscle contraction), hypocalcaemic tetany (muscle spasm) may be produced from chelation of calcium. Nowadays, this problem is removed by administering EDTA as the calcium/sodium salt ($CaNa_2$-EDTA).

Common chelators

The range of chelating compounds, their main uses and the metals that they complex with are shown in Table 12.20.

Summary

The availability of metals has been pivotal in the technological development of the human race. The use of bronze took the human race out of the stone age, and the use of iron produced significant advances in agriculture, warfare (itself not directly beneficial) and technology. Further, the development of the industrial revolution would not have been possible without new processes which relied on the use of an ever-increasing number of metals.

The metals constitute a group of materials that have a significant range of toxicities, from relative innocuousness to significant toxicity. The spectrum of health effects covers most, if not all, body organs and tissues. A separate group of toxic metals have emerged over many years (even centuries), owing to a history of occupational, environmental, domestic, and criminal use.

New metals get added to this group as newer technologies use increasing amounts of metals, which in turn produces workplace conditions which cause adverse effects on health.

Some exposures to metals are universal, for example in mining, extraction, refining and smelting. These activities can produce exposures to workers that cause injury, disease and even death. They can also cause environmental contamination that even today is a sad legacy of exploitation, ignorance and mismanagement.

Other exposures are common to many metals, such as the use of the transition metals in catalyst manufacture and battery production, and the use of virtually all metals in the production of alloys or the production of compounds used as

Table 12.20 Metal chelators

Chelator	Main Use	Metals known to form complexes
$CaNa_2$-EDTA	Lead	Beryllium, cadmium, cobalt, copper, iron, manganese, nickel, silver, zinc
Dimercaprol (BAL)	Arsenic	Cadmium, lead, mercury
Penicillamine	Oral	Copper, lead, mercury, zinc
Deferoxamine	Iron	
DMSA	Lead (paediatric)	

inorganic pigments, pesticides, and ceramic additives. A range of metals is used for the manufacture of electronics, printed circuit boards and semiconductors. Welding is an example of an industrial activity to which workers can be exposed to a range of metals, most with common health consequences.

Other exposures are unique. The use of beryllium in the nuclear industry, the radiotoxicity of uranium enrichment, the use of lithium in the production of tritium, the use of lanthanides for cigarette lighter flints, the use of cisplatin in the treatment of cancer, the use of tin for tin plate, the use of tungsten in electric light filaments and the use of zirconium in deodorants, are just some examples.

Some uses, such as mercurials in millinery, lead arsenate as a pesticide and thallium as a rodenticide are no longer in use. However, new technologies, such as superconductors, have the potential to increase the amounts and numbers of metals in commercial use.

In many cases, the environment has been adversely affected by metals. Natural processes such as volcanism, weathering and sedimentation have assisted in the distribution of metals over the planet. However, anthropogenic activities have greatly added to this. Because of a long use of metals over (in some cases) many centuries, wasteful extraction processes, poorly controlled technologies, and inappropriate waste-disposal methods have seen metals redistributed into every environmental compartment, often with disastrous consequences. Mine tailings, poorly controlled industrial processes, production of hazardous wastes, badly controlled combustion of fossil fuels, domestic activities, such as car use, also contribute to the pollution. Further anthropogenic activities, such as acid rain and nuclear fallout, increase the unpredictability by which environmental metal levels may influence air quality, water quality, biodiversity and human health.

Hopefully, a better understanding of these issues will result in more effective methods of dealing with them.

Further reading

CDC (1991) *Preventing Lead Poisoning in Children*. Washington: US Centers for Diease Control.

Elinder, C.-G., Friberg, L., Kjellstrom, T., Nordberg, G. and Oberdoerster, G. (1994) *Biological Monitoring of Metals*. World Health Organisation IPCS Chemical Safety Monographs. WHO, Geneva.

Ellenhorn, M.J. and Barceloux, D.G. (1988) Metals and related compounds. In: *Medical Toxicology: Diagnosis and Treatment of Human Poisoning*. New York: Elsevier, pp. 1007–1066.

Friberg, L., Nordberg, G.F. and Vouk, V.B. (eds) (1979) *Handbook on the Toxicology of Metals*. Amsterdam: Elsevier/North Holland.

Goyer, R.A. and Clarkson, T.W. (1996) Toxic effects of metals. In: Klaassen, C. (ed) *Cassarett and Doull's Toxicology: The Basic Science of Poisons*, 5th edn. New York: Macmillan, pp. 811–868.

IPCS (1997) *Aluminium. Environmental Health Criteria, 194: International Programme on Chemical Safety*. WHO, Geneva.

IPCS (1981) *Arsenic. Environmental Health Criteria, 18: International Programme on Chemical Safety*. WHO, Geneva.

IPCS (1987a) *Beryllium. Environmental Health Criteria, 61: International Programme on Chemical Safety.* WHO, Geneva.

IPCS (1992) *Cadmium. Environmental Health Criteria, 134: International Programme on Chemical Safety.* WHO, Geneva.

IPCS (1990) *Chromium. Environmental Health Criteria, 106: International Programme on Chemical Safety.* WHO, Geneva.

IPCS (1998) *Copper. Environmental Health Criteria, 200: International Programme on Chemical Safety.* WHO, Geneva.

IPCS (1995) *Lead, Inorganic. Environmental Health Criteria, 165: International Programme on Chemical Safety.* WHO, Geneva.

IPCS (1981) *Manganese. Environmental Health Criteria, 17: International Programme on Chemical Safety.* WHO, Geneva.

IPCS (1991a) *Mercury, Inorganic. Environmental Health Criteria, 118: International Programme on Chemical Safety.* WHO, Geneva.

IPCS (1990) *Methylmercury. Environmental Health Criteria, 101: International Programme on Chemical Safety.* WHO, Geneva.

IPCS (1991b) *Nickel. Environmental Health Criteria, 108: International Programme on Chemical Safety.* WHO, Geneva.

IPCS (1991c) *Platinum. Environmental Health Criteria, 125: International Programme on Chemical Safety.* WHO, Geneva.

IPCS (1987b) *Selenium. Environmental Health Criteria, 58: International Programme on Chemical Safety.* WHO, Geneva.

IPCS (1996b) *Thallium. Environmental Health Criteria, 182: International Programme on Chemical Safety.* WHO, Geneva.

IPCS (1980) *Tin and Organotin Compounds. Environmental Health Criteria, 15: International Programme on Chemical Safety.* WHO, Geneva.

IPCS (1988) *Vanadium Compounds. Environmental Health Criteria, 81: International Programme on Chemical Safety.* WHO, Geneva.

Kang, Y.K. (1997) Part 1: Metals toxicology. In: Massaro, E.J. (ed.) *Handbook of Human Toxicology.* Boca Raton: CRC Press.

Lauwerys, R.R. and Hoet, P. (1993) *Industrial Chemical Exposure – Guideline for Biological Monitoring,* 2nd edn. Boca Raton: Lewis Publishers.

Nriagu, J.O. and Davidson, C.I. (1986) *Toxic Metals in the Atmosphere.* New York: John Wiley.

US (1994) *Environmental Protection Agency's Integrated Risk Information System (IRIS) on Arsenic, Inorganic (CAS 7440-38-2).* National Library of Medicine's TOXNET System.

US EPA (1986) *Air Quality Criteria for Lead, Volumes I-IV.* Washington: USA Environmental Protection Agency.

WHO (1986) *Early Detection of Occupational Diseases.* Geneva: World Health Organisation.

Chapter 13

Toxicity of pesticides

A. Moretto and M. Lotti

Background

Pesticides are used in agriculture and public health to control insects, weeds, animals and vectors of disease. *The Pesticide Manual* lists about 750 active ingredients currently used as pesticides (Tomlin 1997); 75% of the total amount used is made up of about 50 active ingredients (Salem and Olajos 1988). Information on global production and use of pesticides is scarce, while sales data are more readily available. Pesticide sales doubled between 1972 and 1985, and it has been estimated that world production of formulated pesticides increased from about 400,000 tonnes in 1955 to over 3 million tonnes in 1985 (WHO 1990). In 1990, herbicides, insecticides, and fungicides represented 44%, 29% and 21% of all pesticides used, respectively. About 75% of the total amount were used in Western Europe, the USA and Japan. However, 50% of insecticides were used in developing countries (IARC 1991). The fastest growing market is Africa (an increase of 182% between 1980 and 1984) followed by Central and South America (32% increase), Asia (28% increase) and the Eastern Mediterranean region (26% increase) (WHO 1990).

Organophosphorus compounds are likely to continue to be the most used insecticides in developing countries and it has been estimated that demand for these compounds will more than double in the next 10 years (WHO 1990). Moreover, the use of carbamates and pyrethroids will increase, while that of organochlorine compounds will continue to decline. The use of herbicides (mainly carbamates and triazines) will also increase.

New application techniques are developing and include new mixtures or formulations (for example, water-wettable powders) and higher concentrations of the active ingredients. This, together with the use of more selective and potent formulations, will allow a reduction of the dosage per unit of area.

Classification of pesticides

In most countries, pesticides are classified according to their acute mammalian toxicity as suggested by WHO (WHO 1996a) (see Table 13.1). The classification refers to the technical compounds and for each active ingredient it may vary according to the formulation. Different criteria are used when neither the oral nor the dermal is the most relevant route of absorption (for example, fumigants).

This classification is useful for identifying pesticides posing a hazard of acute poisoning, which still represents a major problem in developing countries. However, if the active ingredient produces irreversible damage (such as terata or tumours), is volatile, causes markedly cumulative effects or is found after direct observations to be particularly hazardous or significantly allergenic to man, then adjustments are made by placing the compound in a class indicating a higher hazard.

Pesticides can also be classified according to their chemical structure or according to the target pest, as reported in Table 13.2 (Plestina 1984). Application of pesticides against individual pests implies the use of various equipment and work practices and, therefore, different kinds of exposure.

Epidemiology of acute poisoning

Single exposure poisoning can derive from intentional, occupational or accidental exposure to pesticides but worldwide figures of pesticide poisonings are not

Table 13.1 Classification of pesticides according to hazard

| | | Rat LD_{50} (mg/kg body weight) | | | |
| | | Oral | | Dermal | |
Class		Solids	Liquids*	Solids	Liquids*
1a	Extremely hazardous	≤5	≤20	≤10	≤40
1b	Highly hazardous	5–50	20–200	10–100	40–400
II	Moderately hazardous	50–500	200–2000	100–1000	400–4000
III	Slightly hazardous	>500	>2000	>100	>4000

Source: WHO 1996a.
*Note: The terms 'solids' and 'liquids' refer to the physical state of products and formulations.

Table 13.2 Classification of pesticides according to the target pest or mode of action

Acaricides	Herbicides	Nematocides
Attractants	Insecticides	Plant growth regulators
Defoliants	Larvicides	Repellents
Desiccants	Miticides	Rodenticides
Fungicides	Molluscicides	

Source: Plestina 1984.

available. The WHO (1990) estimates an annual incidence of unintentional acute poisoning of about one million, with an overall mortality rate of about 1% (of which only 1% is in developed countries). The majority of unintentional pesticide poisonings are occupational. Population-based studies in 17 countries gave annual incidence rates of unintentional pesticide poisoning of 0.3–18 per 100,000 (Jeyaratnam 1990).

The estimated annual incidence of intentional single exposure poisoning is about two million with a 5.7% mortality rate (WHO 1990). Such poisoning is more frequent in developing than in developed countries. In Indonesia, Malaysia and Thailand suicide attempts (usually with organophosphorus compounds) represent 60–70% of acute pesticide poisonings (Jeyaratnam 1990) whereas in California all nonoccupational pesticide poisonings represent about 5% of the total (Mehler *et al.* 1990).

Occupational exposures

Application of pesticides has become progressively safer due to technological improvements, to increasing awareness, and to the fact that pesticides are often applied by professional operators. In developing countries, however, pesticides are often applied by farmers who might use poorly maintained equipment, inefficient hand-sprayers, inadequate protection, and have insufficient training and education. Furthermore, in developing countries about 63% of the active population works in agriculture as compared to 11% elsewhere.

Threshold limit values

The American Conference of Governmental Industrial Hygienists (ACGIH) has proposed threshold limit values (TLVs) for about 65 active ingredients (ACGIH 1999). About half of them have the notation 'skin' since dermal exposure entails significant absorption. With regard to biological exposure indexes (BEIs), only limited indications have been given; in most cases the metabolism of the active ingredient (a.i.) is not well understood and the relationship between urinary levels of a metabolite and the toxic effect of the parent compound is not usually known. The ACGIH has established BEIs for parathion (urinary excretion of *p*-nitrophenol) and pentachlorophenol (plasma and urinary levels) and for organophosphorus compounds (red blood cell acetylcholinesterase) (see below).

Re-entry times

Pesticide residues on plants, fruit, soil, and buildings may cause significant exposure in farm and pest-control workers, and inhabitants of the buildings. Cases of poisoning after exposure to residues have been reported (Currie *et al.* 1990; Hodgson *et al.* 1986; Mehler *et al.* 1990; Spear *et al.* 1975). The extent of exposure to residues depends on formulation and amount of pesticide applied, the time elapsed from application and the persistence of the active ingredient (Davis *et al.* 1981). Climatic conditions may influence both degradation of the active ingredient and the exposure to foliar dust particles (Nigg *et al.* 1984). It is also possible that degradation products are more toxic than the parent compound (for example, parathion which might degrade to paraoxon) (Spear *et al.* 1977). General rules cannot therefore be established and mathematical models to calculate re-entry times are difficult to apply (Popendorf and Leffingwell 1982).

Mixtures and impurities

Simultaneous exposures to two or more active ingredients is increasingly frequent for agricultural workers and might influence the toxicity of individual chemicals. For instance, concomitant exposure to two or more anticholinesterase agents (that is, organophosphorus compounds) usually produces an additive effect but potentiation may ensue. Potentiation might be due to inhibition of carboxylesterase which prevents inactivation of organophosphorus compounds containing a carboxylester function. For instance, DEF (*S,S,S*-tri-*n*-butylphosphorotrithioate), EPN (*O*-4-nitrophenyl phenylphosphothioate) and parathion potentiate malathion toxicity because they inhibit its hydrolysis by carboxylesterases (WHO 1986a). Certain impurities (isomalathion and possibly *O,S,S*-trimethylphosphorodithioate) also increased malathion toxicity by similar mechanisms (Baker *et al.* 1978; Talcott *et al.* 1979). Pyrethroids are metabolised by esterases which are probably the same carboxylesterases (Miyamoto 1976) and potentiation of their toxicity might occur with combined exposure to certain organophosphorus esters.

A number of organochlorine compounds are inducers of mixed-function oxidases in mammalian liver and this may explain why, when given in combination, they might change to toxicity of an organophosphorus compound (Keplinger and Deichmann 1967).

Impurities can be present in commercial pesticides and their nature and amount are influenced by a number of factors: the process of synthesis and production, coformulants, condition of storage, and so on. Impurities might either enhance toxicities (higher toxicity of impurities or potentiation) or induce a different kind of toxicity.

It should also be considered that toxicity of some solvents may approach or exceed the toxicity of the active ingredient, for example, DDT (1,1,1-trichloro-2,2-bis(4-chlorophenyl)ethane) dissolved in kerosene or other compounds dissolved in xylene.

Long-term effects

The identification of subjects who have been chronically exposed to pesticides is relatively easy, but biochemical evidence of exposure is seldom available. Moreover, extrapolation from current biomonitoring data to assess past occupational exposures as well as the risk associated with a given pesticide are difficult since most workers use several active ingredients and work practices are different and often changing. Attention has been focused on carcinogenicity of pesticides, but the long time taken for the disease to develop further hampers these studies. Evaluations of the carcinogenic potential of relatively few pesticides have been performed by the International Agency for Research on Cancer (IARC) even though carcinogenicity studies in animals are available for all pesticides. However, most of them have not been published in the open literature. The results obtained so far by IARC are summarised in Table 13.3 (see pp. 350–351). Several criteria are used by the IARC for choosing the compounds to be evaluated and include:

- evidence of human exposure
- some experimental evidence of carcinogenicity and/or some evidence or suspicion of a risk to humans.

Contrary to other national and international committees and agencies, the IARC evaluates data published in the open literature or widely available commercially prepared studies only. In addition to compounds reported in Table 13.3, 'occupational exposure in spraying and application of insecticides' has also been considered by the IARC; the conclusion as based on a few epidemiological studies was that 'spraying and application of insecticides entail exposures that are probably carcinogenic to humans (Group 2A)' (IARC 1991). It is not clear, however, what the basis is for such a generalisation – several chemicals are applied in different ways and settings, using different mixtures and the assessment of exposure was vague or anecdotal. The statement also seems not to be consistent with the evaluations reported in Table 13.3.

Toxicology

The chemical structures of pesticides are extremely variable. The generic structures of the main groups are shown in Figure 13.1.

Organophosphorus pesticides

Most organophosphorus compounds (OPs) are used as insecticides, few are used as fungicides, nematocides or plant regulators. One organophosphorus pesticide (glyphosate) is a widely used herbicide.

The chemical structures of organophosphorus compounds is shown in Figure 13.1. Depending on the atom (oxygen or sulfur) which is double-bonded to the phosphorus, they are called 'phosphates' or 'phosphorothioates', respectively. The latter need to undergo oxidative desulfuration to inhibit esterases. Most commercial organophosphorus insecticides are phosphorothioates since the $P=S$ form is more stable and more lipid soluble; their oxidised forms ($P=O$) are commonly referred to as the 'oxon' (for example, parathion/paraoxon or chlorpyrifos/chlorpyrifos-oxon). Moreover, dimethyl(thio)phosphates are thought to be less toxic than diethyl(thio)phosphates because of the higher rate of reactivation of dimethylphosphorylated acetylcholinesterase (AChE).

Thousands of cases of acute poisoning by organophosphorus compounds have been reported. The majority of cases were suicidal or accidental but several, including some fatal ones, were occupational in origin, mainly due to dermal exposure. Organophosphorus poisoning generally represents a high percentage of total systemic pesticide poisoning, varying from about 30% in California (Mehler et al. 1990) to 77% in China (WHO 1990).

Organophosphorus compounds exert their lethal toxic effects to both insects and

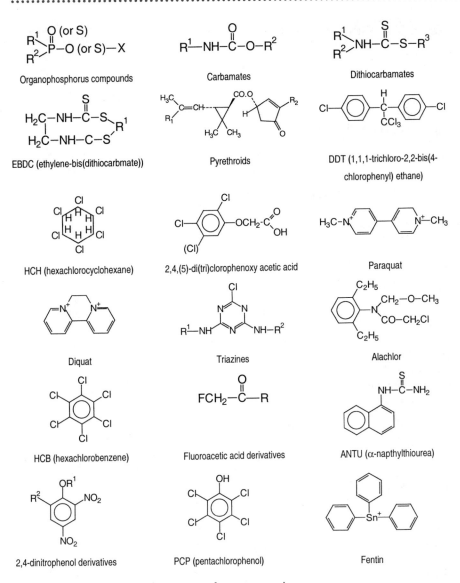

Figure 13.1 The chemical structures of some pesticides.

mammals through inhibition (phosphorylation) of AChE activity at nerve endings (Lotti 1991a). AChE inhibition leads to accumulation of acetylcholine, which is responsible for the cholinergic syndrome (see Table 13.4). Symptoms and signs may occur in various combinations and at different times after exposure. Long-term effects such as reduction in vibrotactile sensitivity and memory deficits on test batteries have been reported in subjects who were poisoned by organophosphorus pesticides years before (Rosenstock *et al.* 1991; Steenland *et al.* 1994). However, the neurological examinations were normal in these patients. Unless the patients had prolonged brain hypoxia and/or seizures, the significance of these findings is unclear.

Table 13.3 IARC evaluation of pesticides

Compound	Year	Degree of evidence for carcinogenicity		Overall evaluation
		Human	Animal	
Insecticides				
Agents and groups of agents				
Aldicarb	1991	ND	I	3
Aldrin	1987	I	L	3
Aramite	1974	ND	S	2B
Arsenic and arsenic compounds	1987	S	L	1*
Carbaryl	1976	ND	I	3
Chlordane/heptachlor	1991	I	S	2B
Chlordimeform	1983	ND	I	3
Chlorobenzilate	1983	ND	L	3
DDT (1,1,1,-trichloro-2, 2-bis(4-chlorophenyl) ethane)	1991	I	S	2B
Deltamethrin	1991	ND	I	3
Dichlorvos	1991	I	S	2B
Dicofol	1983	ND	L	3
Dieldrin	1987	I	L	3
Endrin	1974	ND	I	3
Fenvalerate	1991	ND	I	3
Hexachlorocyclohexanes (HCH)	1987	I		2B
▪ Technical-grade HCH			S	
▪ α-HCH			S	
▪ β-HCH			L	
▪ γ-HCH (lindane)			L	
Malathion	1983	ND	I	3
Methoxychlor	1979	ND	I	3
Methyl parathion	1987	ND	ESL	3
Mirex	1979	ND	S	2B
Parathion	1983	ND	I	3
Permethrin	1991	ND	I	3
Piperonyl butoxide	1983	ND	I	3
Tetrachlorvinphos	1983	ND	L	3
Trichlorfon	1983	ND	I	3
Zectran[†]	1976	ND	I	3
Mixtures				
Terpene polychlorinates (strobane)	1974	ND	L	3
Fungicides				
Captafol	1991	ND	S	2A
Captan	1983	ND	L	3
Chlorophenols	1978	L		2B
▪ Pentachlorophenol	1991	I	L	3
▪ 2,4,5-Trichlorophenol[†]			I	
▪ 2,4,6-Trichlorophenol[†]			S	

continued

Table 13.3 continued

Compound	Year	Degree of evidence for carcinogenicity		Overall evaluation
		Human	Animal	
Chlorothalonil	1983	ND	L	3
Copper 8-hydroxyquinoline	1977	ND	I	3
Ferbam	1976	ND	I	3
Maneb	1976	ND	I	3
o-Phenylphenol	1983	ND	I	3
Quintozene (pentachloronitrobenzene)	1974	ND	L	3
Sodium o-phenylphenate	1987	ND	S	2B
Thiram	1991	I	I	3
Zineb	1976	ND	I	3
Ziram	1991	ND	L	3
Herbicides				
Amitrole	1987	I	S	2B
Atrazine	1991	I	S	2B
Chlorophenoxy herbicides	1987	L		2B
▪ 2,4-Dichlorophenoxyacetic acid			I	
▪ 2,4,5-Trichlorophenoxyacetic acid			I	
▪ 4-Chloro-2-methylphenoxyacetic acid		ND		
Chloropropham	1976	ND	I	3
Diallate	1983	ND	L	3
Fluometuron	1983	ND	I	3
Monuron	1991	ND	L	3
Propham	1976	ND	I	3
Picloram	1991	ND	L	3
Simazine	1991	ND	I	3
Sulfallate	1983	ND	S	2B
Trifluralin				
Other				
1,2-Dibromo-3-chloropropane¶	1987	I	S	2B
Dimethylcarbamoyl chloride‖	1987	I	S	2A
Ethylene dibromide**	1987	I	S	2A
α-Naphthylthiourea (ANTU)††	1987	I	I	3

Notes: Degree of evidence: I, inadequate evidence; S, sufficient evidence; L, limited evidence; ND, no data; ESL, evidence suggesting lack of carcinogenicity. Overall evaluation: 1, the agent is carcinogenic to humans; 2A, the agent is probably carcinogenic to humans; 2B, the agent is possibly carcinogenic to humans; 3, the agent is not classifiable as to its carcinogenicity to humans; 4, the agent is probably not carcinogenic to humans.
*This evaluation applies to the group of chemicals as a whole and not necessarily to all individual chemicals within the group.
†Also a molluscide.
‡Primarily used as chemical intermediates.
¶Soil fumigant/nematocide.
‖Pesticide intermediate.
**Soil fumigant.
††Rodenticide.

Table 13.4 Signs and symptoms of organophosphorus pesticide poisoning

Severity of poisoning	Symptoms and signs		
	Muscarinic	Nicotinic	Central nervous system
Mild (RBC AChE above 40%)	Nausea, vomiting, diarrhoea, salivation/lachrymation, bronchoconstriction, increased bronchial secretions, bradycardia		Headache, dizziness
Moderate (RBC AChE of 20–40%)	Same as above plus meiosis (unreactive to light), urinary/faecal incontinence	Muscle fasciculation (fine muscles)	Same as above plus dysarthia, ataxia
Severe (RBC AChE below 20%)		Same as above plus muscle fasciculation (diaphragm and respiratory muscles)	Same as above plus convulsions, coma

Source: modified from Lotti 1991a.
Note: RBC, red blood cell (erythrocyte); AChE, red blood cell acetylcholinesterase.

An 'intermediate syndrome', apparently noncholinergic, has been described in some patients acutely intoxicated with organophosphorus compounds (Senanayake and Karalleide 1987). After the cholinergic phase, the patients developed paralysis of the proximal limb muscles, neck flexors, motor cranial nerves, and respiratory muscles. This syndrome occurs in patients with high and prolonged AChE inhibition probably due to the long kinetics of the compounds involved (De Bleeker *et al.* 1993) and may lead to death if artificial respiration is not provided.

Some organophosphorus insecticides (that is, chlorpyrifos, dichlorvos, methamidophos, trichlorfon and trichlornat, isofenphos) caused a sensorimotor polyneuropathy known as organophosphate-induced delayed polyneuropathy (OPIDP) (Lotti 1992; Moretto and Lotti 1998; Tracey and Gallagher 1990). Most cases of OPIDP were suicidal, and only a few involved careless occupational exposures to methamidophos. OPIDP is characterised by flaccid paralysis of the lower limbs but the upper limbs might also be affected in severe cases. Sensory peripheral nervous system is affected to a lesser degree (Moretto and Lotti 1998). Histopathology shows degeneration of long and large-diameter axons in peripheral nerves and spinal cord. OPIDP development is unrelated to inhibition of AChE and the putative molecular target is a nervous system protein called neuropathy target esterase (NTE) (Lotti 1992). Since all commercial organophosphorus insecticides display a high potency for acetylcholinesterase, OPIDP always developed after doses causing severe cholinergic syndrome.

Repeated exposures to organophosphorus compounds at doses not causing AChE inhibition do not cause either neuropsychiatric disorders or behavioural disturbances (Lotti 1991b). Persistent electroencephalogram (EEG) changes have been reported in industrial workers who had repeated accidental exposures to sarin (a nerve agent). The exposures caused symptoms and significant inhibition of red blood cell (RBC) AChE. The toxicological significance of these EEG changes has, however, been questioned (Lotti 1991b). Minimal electrophysiological changes have been reported in peripheral nerves of workers repeatedly exposed to organophosphorus compounds (Jager et al. 1970; Roberts 1976; Stokes et al. 1995). However, in other studies where exposure was better characterised, no electrophysiological changes were found (Jusic et al. 1980; Lotti et al. 1983).

Urinary dimethyl(thio)phosphates and diethyl(thio)phosphates (metabolites of dimethoxy and diethoxy organophosphorus compounds, respectively), which represent the majority of organophosphorus insecticides, can be measured. Since organophosphorus compounds with different toxicity may give similar levels of urinary dialkyl(thio)phosphates, the hazard can only be determined if such levels are calibrated against AChE inhibition. The time-course and peak of excretion of metabolites vary according to the compound (different rate of absorption, distribution and metabolism) (WHO 1986a). The leaving group ('X') is also excreted and it can be measured in urine. Since it is typical of one or two compounds only, its determination in the urine is specific. The time-course of urinary excretion of the leaving group can be different from that of the alkylphosphates deriving from the same molecule (Morgan et al. 1977). The ACGIH proposes a BEI only for parathion by means of measuring p-nitrophenol at end of shift, the level of which should be less than 0.5 mg/l (ACGIH 1996).

RBC AChE is biochemically similar to the synaptic enzyme. High RBC AChE inhibition is diagnostic of acute organophosphorus poisoning. Interindividual variability of RBC AChE is high (coefficient of variation 13–16%) and pre-shift measurements are necessary to assess occupational exposure. The minimal differences for statistical recognition of inhibited RBC AChE vary from about 15% with one to about 12% with five pre-exposure values (Gallo and Lawryk 1991). A 20–25% RBC AChE reduction of pre-shift values is diagnostic of exposure but not indicative of hazard; 30–50% indicates overexposure; more than 50% may be accompanied by symptoms of intoxication (see Table 13.4) and the worker(s) must be removed from further exposure until AChE activity returns to within normal values (Plestina 1984).

Plasma also contains pseudocholinesterases (butyrylcholinesterase), which can be inhibited by organophosphorus compounds. No physiological substrate for this enzyme is known and its inhibition has no toxicological significance. Measurements of plasma pseudocholinesterase activity, however, can be used to monitor exposures to organophosphorus, and its sensitivity to organophosphorus inhibition may be different from that of AChE, depending on the organophosphorus involved (Gallo and Lawryk 1991). Interindividual variation of plasma pseudocholinesterase is similar to that of AChE.

Neuropathy target esterase (NTE) is present in peripheral lymphocytes and its inhibition predicts the development of OPIDP. The interindividual variation of lymphocyte NTE activity requires baseline values to evaluate the effects of exposures and situations are known where false-positive and false-negative results might arise (Lotti 1989).

In acute organophosphorus poisoning the muscle response to single stimulation is repetitive while the nerve conduction velocity is usually normal or slightly slowed and the amplitude of the compound action potentials are very mildly reduced (Wadia et al. 1987). However, no such alterations were seen in exposed workers (Jusic et al. 1980; Stålberg et al. 1978; Verberk and Sallé 1977) or in subjects treated with an organophosphorus compound at doses causing about 50% of RBC AChE inhibition (Le Quesne and Maxwell 1981). Thus, it appears that electromyography (EMG) is not a sensitive measurement of exposure to organophosphorus, given also the low reproducibility of these methods in field conditions.

OPIDP patients may display small compound action potentials, prolonged terminal motor latencies, and electromyographical evidence of denervation with relative preservation of maximum motor conduction velocities (Lotti et al. 1984). In severe cases motor action potentials in the lower limbs might not be recorded. Sensory involvement is usually absent or less severe (Moretto and Lotti 1998).

Carbamates

The chemical structure of carbamates and dithiocarbamates are shown in Figure 13.1. Depending on the nature of the R^1 group, carbamates are divided into three groups:

- R^1, methyl: insecticides and nematocides
- R^1, aromatic group: herbicides and sprout inhibitors
- R^1, benzimidazole group: fungicides.

Carbamate insecticides inhibit AChE, while herbicides and fungicides do not. Carbamylation of AChE is short-lasting and regeneration of enzyme activity is rapid (reversible inhibition) when compared with that of phosphorylated AChE. Human exposure to carbamates is therefore less dangerous than that to organophosphorus compounds. Several cases of carbamate poisoning have been described; symptoms usually recover within a few hours after the end of exposure and parallel reappearance of RBC AChE activity (WHO 1986b).

Carbamates with herbicidal activity are not inhibitors of AChE because they have a bulky R^1 group attached to the nitrogen atom (see Figure 13.1). No characteristic toxic effect has been described in mammals for carbamate (phenmedipham, chlorpropham and others), thiocarbamate (molinate, diallate and others) and dithiocarbamate (sulfallate) herbicides. High doses of molinate given to rats have a reversible effect on sperm morphology and reproduction. However, in exposed workers such alterations were not found (WHO 1987a).

Data on urinary metabolites of carbamates in humans are available for carbaryl and propoxur. Measurement of 1-naphthol in workers exposed to carbaryl should be carried out in urine voided 4–8 hours after the end of exposure. Levels of 30 mg/l of urinary 1-naphthol after an 8-hour shift are expected following exposures to $5\,mg/m^3$ carbaryl (Maroni 1988). Following exposure to propoxur, depression of RBC AChE is expected at urinary concentrations of 2-iso-propoxyphenol higher than $5\,mg/l$ (Maroni 1988).

When exposures to carbamates are monitored by measuring RBC AChE, analytical methods must allow for spontaneous reactivation of AChE. Therefore, short time of hydrolysis, minimal dilution of the sample and special care in the storage of blood are needed.

Dithiocarbamates

Dithiocarbamates are mainly used as fungicides, but they have other industrial uses (antioxidants, accelerators and slimicides). Few of them are also used as insecticides or herbicides. The chemical structure is shown in Figure 13.1. Either R^1 or R^2 or both are alkyl groups: the most used compounds are thiram, ziram and ferbam. When diamines are used for the synthesis, two terminal groups are then linked via an ethylene bridge (ethylene-bis-(dithiocarbamates) (EBDC) (see Figure 13.1): the most used compounds are maneb (R=manganese), zineb (R=zinc) and mancozeb (R=manganese polymeric complex with zinc salt). Under certain conditions (presence of oxygen and moisture), dithiocarbamates may decompose to, among other things, ethylenethiourea (ETU).

The acute toxicity of dithiocarbamates is low and equivocal cases of human intoxication have been reported. Dithiocarbamates cause alcohol intolerance by interfering with alcohol metabolism (WHO 1988). High doses of dithiocarbamates cause antithyroid effects in animals, possibly by interfering with iodination of thyroxine precursors. Workers exposed to dithiocarbamates have been studied in this respect, but neither alterations of thyroid function (Moretto and Lotti, unpublished results; WHO 1988) nor an increased incidence of thyroid tumours (IARC 1987) were found.

EBDCs are metabolised to ETU, carbon disulfide (CS_2), hydrogen sulfide and many other compounds. From studies performed in alcoholic subjects given disulfiram it was concluded that measurement of blood CS_2 levels may be useful for biological monitoring of exposure to dithiocarbamates (Maranelli et al. 1991). Mean urinary excretion of ETU in workers exposed to mancozeb or maneb was $2-4\,\mu g$ per 24 hours (Kurttio and Savolainen 1990). The half-life of urinary ETU was found to be about 100 hours. However, such a long half-life may have been due to occasional exposures after completion of application (Kurttio et al. 1990).

Pyrethroids

Pyrethroids are synthetic derivatives of the natural pyrethrins from *Chrysanthemum cinerariaefolium* (see Figure 13.1). Characteristics include high stability in the field (greater than organophosphorus compounds and carbamates), persistence in soil lower than that of organochlorines, greater insecticidal potency and low mammalian toxicity (ratios of rat LD_{50}/insect LD_{50} are generally higher than 1000, whereas they are in the range 1–50 for other pesticides) (Elliot 1976).

Acute intoxication is characterised by dizziness, headache, nausea, muscular fasciculation, convulsive attacks and coma (Chen et al. 1991; He et al. 1989). Two patterns of symptoms are described in rats after acute intoxication, depending on the presence or the absence of an (α)cyano-group substituent: the so-called 'T syndrome' (aggressive sparring, sensitivity to external stimuli and tremor)and the 'CS syndrome' (choreathetosis, salivation and seizures), respectively. Sometimes, however, the two syndromes may combine to give a more complex one (Aldridge 1990). Cases of acute pyrethroid poisoning in China have recently been reviewed but it was not possible to differentiate the two syndromes in humans (He et al. 1988).

Occupational exposures often result in abnormal skin sensations mainly of the face described as burning and tingling (Chen et al. 1991; He et al. 1989; Moretto 1991; Zhang et al. 1991). Symptoms appear shortly after beginning work and disappear within 24 hours. Most pyrethroids caused these sensations with the following order of decreasing potency: deltamethrin > flucythrinate > cypermethrin = fenvalerate > permethrin (Aldridge 1990). This neurotoxicity is due to a local effect since only unprotected parts of the skin are affected. Pyrethroids are known to act on sodium channels, thereby causing repetitive firing of sensory nerve endings of the skin (Aldridge 1990).

Electrophysiological studies performed in the arms and legs of exposed subjects complaining of cutaneous sensations were negative (Le Quesne et al. 1980).

Measurement of urinary levels of the parent compound and/or of metabolites after occupational exposure have been performed, but no correlation with the degree of cutaneous sensations was found (He et al. 1988; Zhang et al. 1991). This is consistent with the local nature of this type of neurotoxicity.

Organochlorines

Organochlorine (OC) insecticides include:

- ethane derivatives (DDT and its analogues)
- cyclodienes
- hexachlorocyclohexane (HCH) (or benzene hexachloride, BHC) and lindane
- toxaphene (camphechlor) and related chemicals
- mirex and chlordecone
- chlordecone, mirex and toxaphene are no longer produced in significant amounts.

Different acute toxicities and toxicokinetics are displayed by individual OCs. In general, signs of acute poisoning are related to neuronal hyperexcitability. In the case of DDT and its analogues, for instance, signs of hyperexcitability are due to slowed repolarisation of neuronal membrane and appear in both the peripheral and the central nervous system. They are due to effects on sodium channels thereby interfering with flow of sodium and potassium ions across the membrane (Coats 1982) with a mechanism similar to that of pyrethroids. The onset of DDT poisoning is characterised by mild effects that progress gradually to convulsions, whereas other OCs may induce convulsions without preceding signs (Smith 1991).

DDT and its derivatives

DDT (1,1,1-trichloro-2,2-bis(4-chlorophenyl)ethane) (see Figure 13.1) has been extensively used, even by direct application to humans. Because of its accumulation along the food chain, DDT was banned in many countries, but it is still used for vector control in many tropical countries. As extrapolated from human exposures, 10 mg/kg of DDT produces illness in some subjects (Smith 1991). However, heavy exposure to concentrated DDT dust, workers being covered with DDT dust while working did not cause adverse effects. Induction of liver microsomal enzymes was demonstrated in some heavily exposed workers (WHO 1979). 2,2-bis(4-chlorophenol)acetic acid (DDA) is the main urinary metabolite of DDT. DDA excretion peaks a few hours after intake and continues for several days (Roan *et al.* 1971). Among workers whose DDT intake was of the order of 35 mg/day, urinary DDA was found to be 1.7 ppm on average. 1,1-Dichloro-2,2-bis(4-chlorophenyl)ethylene (DDE) is the metabolite which is preferentially stored in fat. Low DDE as compared to total stored DDT residues indicates high and recent intake of DDT. Occupational exposure can be monitored either by measuring DDT and/or DDE in plasma or by measuring DDA urinary excretion (WHO 1979).

TDE (tetrachlorodiphenylethane, sometimes called DDD (dichlorodiphenyl-dichloroethane), when a metabolite of DDT), ethylan, methoxychlor, chlorobenzilate and dicofol are less toxic than DDT, and cases of human poisoning have only been reported for chlorobenzilate (one case) and dicofol (one case) (Smith 1991).

Cyclodienes

The cyclodienes used as insecticides include chlordane, heptachlor, and endosulfan. Aldrin, isodrin, dieldrin, endrin and isobenzan are now banned or severely restricted in most countries. All cyclodienes are easily absorbed through the skin. These compounds differ in their toxicity, endrin being the more potent (rat oral LD_{50}: 10–40 mg/kg). Serious poisoning by these compounds usually causes convulsions that may appear without other preceding signs of illness (Rowley *et al.* 1987).

Hexachlorocyclohexane and lindane

Both lindane (which is the γ-isomer of hexachlorocyclohexane (HCH; see Figure 13.1) and HCH are absorbed through the skin. Lindane is rapidly excreted while the β-isomer, which is a central nervous system depressant, accounts for most of HCH residues in the organism. There are few reports of mild acute intoxication following occupational exposure to HCH or lindane: it is characterised by headache, tremors, vomiting, prostration, and convulsion (Smith 1991); studies on workers exposed to lindane for many years did not demonstrate any long-term adverse effect. Abnormal EEG patterns have been reported in workers exposed to HCH (Müller *et al.* 1981) but not to lindane (Baumann *et al.* 1981).

Chlorophenoxy compounds

2,4-Dichlorophenoxyacetic acid (2,4-D), 2,4,5-trichlorophenoxyacetic acid (2,4,5-T), 4-chloro-2-methylphenoxyacetic acid (MCPA) and 2-(2,4,5-trichlorophenoxy)propionic acid (silvex) are the most used chlorophenoxy herbicides (see Figure 13.1). The use of 2,4,5-T is restricted in several countries because commercial formulations might contain 2,3,7,8-tetrachlorodi-benzo-p-dioxin (TCDD) as an impurity. However, TCDD contamination has greatly diminished over the years due to improved industrial technology (IARC 1986). These herbicides act against growth hormones in plants, but there is no known hormonal effect in mammals. Both 2,4-D and 2,4,5-T have moderate oral toxicity. Signs of toxicity include nausea and vomiting, myalgia, muscular hypertonia, headache, and changes in the electrocardiogram (WHO 1984a). Chloracne has also been observed in workers exposed to 2,4,5-T, but this effect was thought to be due to 2,3,7,8-TCDD (see Table 13.5). Peripheral neuropathy in workers exposed to 2,3-D or 2,3,5-T has also been reported with slowed nerve conduction velocity in sural and median nerves (Murphy 1987; WHO 1984a).

Case–control studies have shown an association between cancer and occupational exposure to chlorophenoxy herbicides. However, elevated odd ratios for cancer at some sites were not consistent in independent studies as reviewed by the IARC (IARC 1986). Subsequently, more studies have been published showing variations in relative risk estimates for soft-tissue sarcomas, non-Hodgkin's lymphomas and Hodgkin's disease (Pearce and Reif 1990). Recently, in a mortality study among farmers, a significant dose–response relationship was found between risk of non-Hodgkin's lymphoma and the use of some herbicides (Wigle et al. 1990). The percentage of non-Hodgkin's lymphoma attributable to herbicide exposure was estimated to be 22% in this study. A cancer mortality study on an international cohort of workers exposed to chlorophenoxy herbicides showed a significant excess of mortality for soft-tissue sarcomas, but not for non-Hodgkin's lymphomas occurring in the period 10–19 years after the beginning of exposure (Saracci et al. 1991). A more recent review of the available studies also concluded that data were not sufficient to support a causal relationship between exposure to 2,4-D and non-Hodgkin's lymphomas (US EPA 1994).

Phenoxy acids can be absorbed through the skin and are mainly excreted unchanged in the urine (Lavy and Mattice 1989). Concentration of 2,4-D or 2,4,5-T in urine after occupational exposure ranged from 1–32 mg/l in the absence of signs or symptoms of poisoning (IARC 1986). Urinary excretion of 2,4-D is higher at the end of the work-week due to its long kinetics (Knopp 1996).

Bipyridyl compounds: paraquat and diquat

Paraquat (1,1'-dimethyl-4,4'-bipyridylium salts) and diquat (6,7-dihydrodipyrido-1,1'-ethylene-2,2'-bipyridylium salts) are nonselective contact herbicides (see Figure 13.1). In cells, they undergo a single electron addiction to form a free radical that reacts with molecular oxygen to reform the parent compound and concomitantly produce a superoxide anion. This oxygen radical may then be responsible for cell death.

Table 13.5 Pesticides causing orthoergic dermatitis (OD) or allergic contact dermatitis (ACD)

Type of pesticide	Type of dermatitis	Comment
Insecticides		
Organophosphorus compounds		
Dichlorvos	ACD	
Naled	OD/ACD	
Thiometon	ACD	
Organochlorine compounds		
Aldrin	OD	
DDT (1,1,1-trichloro-2,	OD	Due to solvents, mainly xylenes
2-bis(4-chlorophenyl ethane)	ACD	
Dicofol		
Lindane and BHC (benzene		
hexachloride)	OD/ACD	Probably due to contaminants
Pyrethrum	ACD	
Herbicides		
2,4-Dichlorophenoxyacetic acid	OD/ACD	Probably due to contaminants
2,4,5-Trichlorophenoxyacetic acid	OD/ACD	Chloracne was also described but it was likely caused by the contaminant TCDD (2,3,7,8-tetrachlorodibenzo-*p*-dioxin)
Alachlor	ACD	From Won *et al.* 1993
Allidochlor	ACD	
Atrazine	ACD	
Cynazine	ACD	
Dichlobenyl	OD	
Diquat	OD	Nail lesions also described
MCPA (4-chloro-2-methylphenoxyacetic acid)	OD	
Metholachlor	ACD	
Nitrofen	OD	
Paraquat	OD	Nail lesions also described
Phenmedipham	ACD	
Propanyl	OD	Also chloracne caused by impurities
Propazine	OD	Probably caused by intermediates in the synthesis
Simazine	OD	

continued

Table 13.5 continued		
Type of pesticide	Type of dermatitis	Comment
Fungicides		
1-Chlorodinitrobenzene	ACD	
Benefin	ACD	From van Ginkel and Sabapthy 1995
Benomyl	ACD	
Captan	ACD	
Captafol	OC/ACD	
Chlorthalonil	ACD	
Dinocap	ACD	
Fluazinam	ACD	From Pentel *et al.* 1994
Imazalil	ACD	
Mancozeb	ACD	
Maneb	OD/ACD	
Pentachlorophenol	OD	Chloracne was also described but it was likely caused by the contaminant TCDD (2,3,7,8-tetrachlorodibenzo-*p*-dioxin)
Thiophanate-methyl	OD/ACD	
Thiram	ACD	
Trifluralin	ACD	From van Ginkel and Sabapthy 1995
Zineb	OD/ACD	
Ziram	OD	
Rodenticides		
ANTU (α-Naphthylthiourea)	ACD	
Solvents and fumigants		
Kerosene	OD	
Tetralin	OD	
Xylene	OD	
Inorganic and organometallic pesticides		
Arsenic and its compounds	OD	
Chromium (sodium dichromate)	OD/ACD	
Copper (cupric sulfate)	OD	
Zinc (zinc chloride)	OD	
Miscellaneous pesticides		
Chlorfenson	ACD	
Propargite	ACD	

Source: summarised from Hayes and Laws (1991) except where otherwise indicated.

Several cases of suicidal or accidental acute paraquat poisoning have been reported. The clinical picture is characterised by kidney and liver necrosis that is followed within a few weeks by pulmonary fibrosis. Lung toxicity is related to selective paraquat accumulation in the lung by an energy-dependent process (Smith *et al.* 1979). Fatal poisoning was also associated with ingestion of the liquid concentrate that caused ulceration of the upper digestive tract. A small number of occupational poisonings have been described and were due to skin absorption (Hart 1987; Smith 1988; WHO 1984b). Biological monitoring of exposure may be carried out by measuring urinary paraquat excretion. Paraquat levels up to 0.76 mg/l have been found in exposed workers (Chester and Woollen 1981; Swan 1969). These values are more than one order of magnitude lower than those found in cases of acute poisoning (Hart 1987). Pulmonary function in those workers was not affected.

Diquat poisoning is much less common than paraquat poisoning and causes kidney and liver necrosis, diarrhoea also being a prominent feature (WHO 1984b). Lung fibrosis was never reported. Urinary diquat lower than 0.047 mg/l was found in workers with a dermal exposure up to 1.82 mg/h (Wojeck *et al.* 1983).

Triazines

Figure 13.1 shows the chemical structure of the substituted *S*-triazine. The most used triazines are atrazine, cynazine, propazine and simazine. All compounds have low acute toxicity and no case of systemic poisoning has ever been described. In exposed workers, atrazine is rapidly excreted both unmodified and as dealkylated metabolites, the latter in much greater amounts (Catenacci *et al.* 1993).

Amides

Alachlor and metholachlor (see Figure 13.1) are the most used herbicides. They have a very low acute toxicity (rat oral LD_{50} values are >1800 mg/kg) and cutaneous absorption is negligible (Stevens and Sumner 1991). Cases of accidental acute, nonlethal metholachlor intoxication have been described; symptoms were nausea, vomiting or diarrhoea, and headache (Stevens and Sumner 1991).

Arsenical herbicides

Dimethylarsenic acid and methylarsenic acid are the only two arsenical compounds still in use as pesticides (Tomlin 1994). Their toxicity is low and neither acute nor long-term effects have been described in workers exposed to these compounds (WHO 1981). These two arsenic derivatives are not carcinogens. However, they may contain inorganic arsenic (a recognised human carcinogen) as an impurity.

Dicarboximides

Captan, captafol and folpet belong to this group of fungicides. The use of captafol has been restricted or banned in many countries. Because of some structural similarities with thalidomide, these compounds were suspected teratogens but animal studies were negative (Edwards *et al.* 1991). Dermatitis was reported as a consequence of single exposures (Table 13.5). Although epidemiological studies are not available, captafol is classified in the group 2A by IARC (IARC 1991) because of its carcinogenic effects in mice and rats. Captan and folpet cause duodenal tumours in rodents probably through a nongenotoxic mechanism (proliferative stimulation) (WHO 1996b).

Hexachlorobenzene

Hexachlorobenzene (HCB) (see Figure 13.1) is still used in some countries as a fungicide for seed treatment. Cases of poisoning following occupational exposure have not been reported. The epidemic of HCB poisoning (porphyria turcica) in Turkey during the 1950s was due to ingestion of wheat seedlings treated with HCB (Morris and Cabral 1985).

Anticoagulant rodenticides

Anticoagulants used as rodenticides are antimetabolites of vitamin K and inhibit the synthesis of prothrombin (Ecobichon 1996). Warfarin was the first compound to be introduced; other anticoagulants include coumafuryl, diphacinone and chlophacinone. The risk for workers is associated with accidental ingestion of massive doses.

Fluoroacetic acid and its derivatives

The most common compounds are (see Figure 13.1) sodium fluoroacetate (R=ONa), 2-fluoroacetamide (R=NH$_2$) and MNFA (R=N-methyl-N-1-naphthyl). These are among the most potent rodenticides (rat oral LD$_{50}$ 0.20 mg/kg for fluoroacetate and 4–15 mg/kg for fluoroacetamide) which are also highly toxic to man. The active compound is fluoroacetate that inhibits the citric acid cycle thereby lowering energy production. The heart and central nervous system are the target organs and symptoms of severe poisoning include cyanosis, convulsions and alteration of heart rate. The cause of death is usually ventricular fibrillation or respiratory failure. There are no reports of occupational poisoning, but nonspecific symptoms such as nausea, weakness, fatigue and minor electrocardiogram (ECG) alterations (bradycardia, prolonged PQ segment, U waves) in the absence of evident overexposure to MNFA have been reported (Pelfrene 1991).

ANTU (α-naphtylthiourea)

ANTU (see Figure 13.1) is used as a rodenticide. Poisoning by ANTU is characterised by massive pulmonary oedema and pleural effusion. There is great variability in species sensitivity to ANTU and humans appear to be relatively resistant. Neither death nor permanent sequelae have been described after accidental or intentional human poisoning with ANTU (IARC 1983; Pelfrene 1991). In some countries it has been withdrawn from use because of the presence of carcinogenic impurities such as β-naphthylamines (Tomlin 1994).

DBCP (1,2-Dibromo-3-chloropropane)

DBCP (CH_2Br-$CHBr$-CH_2Cl) is a soil fumigant and nematocide. DBCP is readily absorbed through the skin. Male rats repeatedly exposed for 7 hours to 5 ppm of DBCP developed degeneration of seminiferous tubules, increased number of Sertoli cells, reduced sperm counts and abnormal sperm morphology (Torkelson et al. 1961). A group of workers involved in DBCP production became aware that only a few of them had fathered children: 13% were found to be azoospermic, 17% oligospermic and 16% had low normal counts (Whorton et al. 1979). Studies performed in another factory (Egnatz et al. 1980) and in agricultural workers (Takahashi et al. 1981) confirmed the effect of DBCP exposure on testicular function. The levels of DBCP in air of the two factories were reported to be 1 ppm or less. A 7-year follow-up showed that no major improvements in testicular function of the azoospermic/oligospermic workers took place (Eaton et al. 1986).

1,2-Dibromoethane

1,2-Dibromoethane (CH_2Br-CH_2Br), also known as ethylene dibromide (EDB), is used as a crop and soil fumigant. EDB is readily absorbed from the skin, lung and gastrointestinal tract. EDB is metabolised either to bromoacetaldehyde, or to (2S)-bromoethylglutathione that is responsible for hepatotoxicity (White et al. 1983).

Workers of the papaya fumigation industry exposed to EDB had statistically significant decreases in sperm count and motility, and increased morphological sperm abnormalities (Ratcliff et al. 1987). This is consistent with experimental data (Short et al. 1979). Workers had an average duration of exposure of 5 years and a geometric mean breathing zone exposure to EDB of 88 p.p.b. (8 hours time-weighted average) with peaks up to 262 p.p.b.. In five other epidemiological studies no significant reproductive effect of exposure to EDB was demonstrated, but the potency of these studies was questioned (Dobbins 1987).

Hydrogen cyanide

Hydrogen cyanide (HCN) is an insecticidal fumigant. Toxicity is due to the block of cellular respiration (Smith 1996). Occupational exposure occurs through inhalation, while accidental and intentional poisoning usually involves ingestion of a cyanide salt. Lower doses cause at first stimulation of respiration and heart rate; venous blood remains oxygenated and the patient is not cyanotic. Then respiration becomes slow and gasping until it stops. In the past, some applicators died after careless handling of cyanide, but when used by trained applicators the safety record is good. Biomonitoring of cyanide exposure can be carried out by measuring urinary excretion of thiocyanates (Lauwerys 1983). However, thiocyanates are present in variable amounts in nonexposed subjects, and higher levels are found in smokers (Bertelli et al. 1984). The suggested limit for postshift urinary thiocyanates is 2.5 mg/g of creatinine for nonsmokers (Lauwerys 1983).

Dinitrophenols

Derivatives of 2,4-dinitrophenol ($R_1=R_2=H$ in Figure 13.1) are used as herbicides, fungicides, or insecticides. Although all share the same mechanism of toxicity in mammals, they may differ in their ability to be absorbed through the skin (Gasiewicz 1991). These compounds exert their toxic effect by uncoupling oxidative phosphorylation. The most used compounds are 2,4-dinitrophenol (wood preservative and woodworm insecticide), DNOC ($R_1=H$, $R_2=CH_3$, insecticide and fungicide; no longer used in many western countries), dinocap ($R_1=H$, $R_2=1$-methyl-n-heptyl, acaricide and fungicide), and dinoseb ($R_1=COCH=CHCH_3$, $R_2=sec$-butyl, herbicide). The best known of these compounds is DNOC which is easily absorbed by the skin: signs and symptoms of acute DNOC and 2,4-dinitrophenol intoxication include fatigue, restlessness and excessive sweating. Fatal hyperthermia may occur. Biological monitoring of workers exposed to DNOC involves the measurement of the parent compound in the blood. A 10 ppm threshold in blood has been suggested (Lauwerys 1983). Symptoms appear at or above 30 ppm, while 50 ppm is considered dangerous (Gasiewicz 1991). Dinocap and dinoseb have a lower acute toxicity.

Pentachlorophenol and related compounds

The structural formula of pentachlorophenol (PCP) is shown in Figure 13.1. The technical product is a mixture of PCP, tetrachlorophenol and traces of chlorinated dibenzo-p-dioxins, polychlorinated dibenzofurans and polychlorobenzenes (WHO 1987b). In recent years, improved technologies led to a decreased concentration of impurities (Gasiewicz 1991; IARC 1986). PCP and its salts (mainly sodium-PCP) have a variety of applications in industry and in agriculture as algaecides, bactericides, fungicides, herbicides, insecticides, and molluscicides but their main use is as wood preservatives. Recently, strict limits on PCP uses other than wood preservative have been established in many countries because of concern for the effects both to humans and to the environment (especially on aquatic organisms) (WHO 1987b). Several cases of human poisoning have been described, mainly after skin exposure. In workers exposed to a mixture of tetrachlorophenol (20%) and PCP (3%), it was calculated that 30–100% of the dose deposited on the skin was absorbed (Fenske et al. 1987). Acute signs of intoxication resemble those elicited by nitrophenolic compounds, being characterised by a marked increase in metabolic rate due to the uncoupling of oxidative phosphorylation. Fatal cases are always associated with a marked rise in temperature and terminal spasm (WHO 1987b; Wood et al. 1983). Reversible and nonspecific liver and kidney functional changes have been described after exposure to PCP (WHO 1987b). Monitoring of exposure can be performed by measuring urinary excretion of PCP; postshift levels of 1 mg/g of creatinine (Lauwerys 1983) or 2 mg/g of creatinine prior to the last shift of the workweek (ACGIH 1996) have been proposed as permissible values. The ACGIH (1996) also proposed an end of shift-free PCP plasma concentration of 5 mg/l.

Triphenyltin compounds

Triphenyltin (fentin) acetate and triphenyltin hydroxide (see Figure 13.1) are used as fungicides and molluscicides. Few cases of human poisoning have been described. Symptoms were dizziness, nausea, skin irritation and liver damage. A case of acute dermal poisoning by fentin acetate has also been described (Colosio *et al.* 1991). The prominent features were urticaria on trunk and arms and genital oedema. The patient suffered from periodic urticaria during the following 5 months. Patch tests with the commercial formulation and the pure component of the formulation were negative. During the acute phase EEG alterations were also observed. Urinary excretion and blood levels of tin showed biphasic kinetics suggesting the possibility of accumulation.

Sources for further information

Books

The most comprehensive book on pesticide toxicology is *The Handbook of Pesticide Toxicology* (Hayes, W.J. Jr, and Laws, E.R. Jr (eds), Academic Press, San Diego 1991). The edition published in 1991 is the updated combination of two previously published books, namely *Toxicology of Pesticides* (Hayes, W.J. Jr (ed.), Williams and Wilkins, Baltimore 1975) and *Pesticides Studied In Man* (Hayes, W.J. Jr (ed.), Williams and Wilkins, Baltimore 1982). The book reports most of the animal and human toxicological data available in the open literature.

Other reference books include:

- *The Pesticide Manual* (Tomlin, C. (ed.) British Crop Protection Council, Surrey, UK, 1997, 11th edn) which lists all commercially available compounds and for each of them essential information on nomenclature, development, physicochemical properties, uses, toxicology, formulation and method of analysis (with references) is given. It also includes biological agents. A new edition is published at intervals of 3–4 years.
- *Farm Chemicals Handbook* (Poplyk, J. (ed.) Meister Publishing Company, Willoughby, OH) contains a pesticide dictionary where compounds are listed with their common and commercial names, and essential information on chemistry, uses, toxicology. It also contains an extensive list of company addresses. It is published yearly.
- *Pesticides. Synonyms and Chemical Names* (published by the Commonwealth Department of Health, Australia). It lists alphabetically all common names (including rejected and discontinued ones) and trade names, and reports their chemical name.
- *Casarett and Doull's Toxicology. The Basic Science of Poisons* (Klaassen, C.D. (ed.) McGraw-Hill, Health Professions Division, New York, 1996, 5th edn) in which a whole chapter by D. Ecobichon is devoted to the toxicology of pesticides.

Series

The United Nations Environment Programme, the International Labour Organization, and the WHO sponsor, through the International Programme on Chemical Safety, the publication of reports on chemicals, group of chemicals or general toxicological problems in a series called *Environmental Health Criteria*. Some of the volumes deal with a single or a group of pesticides and review all available data on properties, analytical methods, environmental distribution, effect of residues, kinetics and metabolism, effects on animals and on man. The books may be obtained from WHO in Geneva, Switzerland.

The same organisation also publishes a series called *Health and Safety Guide* where essential information (20–30 pages) on single chemical substances (including many pesticides) is given. The booklets may be obtained from WHO in Geneva, Switzerland.

Each year a Joint Meeting of the FAO Panel of Experts on Pesticides Residues in Food and the Environment and the WHO Expert Group on Pesticide Residues is held and monographs on residue aspects and on toxicology are published afterwards. The toxicological monographs summarise for each pesticide considered the safety data (both published in the open literature and confidentially provided by the manufacturer) on which a decision regarding the acceptable daily intake is made. The monographs are published by FAO (Rome, Italy) until 1998, and subsequently by WHO in Geneva, Switzerland.

The International Agency for Research on Cancer publishes the *IARC Monographs on the Evaluation of Carcinogenic Risks to Humans*. The results of the evaluation of pesticides have been published in volumes 4, 5, 12, 15, 30, 53 and Supplement 7. The monographs may be obtained from IARC, Lyon, France.

References

ACGIH (1996) *1996–1997 Threshold Limit Values for Chemical Substances and Physical Agents and Biological Exposure Indices*. Cincinnati, OH: The American Conference of Governmental Industrial Hygienists.

Aldridge, W.N. (1990) An assessment of the toxicological properties of pyrethroids and their neurotoxicity. *Crit. Rev. Toxicol.* 21: 89–104.

Baker, E.L. Jr, Zack, M., Miles, J.W. *et al.* (1978) Epidemic malathion poisoning in Pakistan malaria workers. *Lancet* i: 31–34.

Baumann, K., Behling, K., Brassow, H.L. and Stapel, K. (1981) Occupational exposure to hexachlorocyclohexane. III. Neurophysiological findings and neuromuscular function in chronically exposed workers. *Int. Arch. Occup. Environ. Health* 48: 165–172.

Bertelli, G., Berlin, A., Roi, R. and Alessio, L. (1984) Acrylonitrile. In: Alessio, L., Berlin, A., Boni, M. and Roi, R. (eds) *Biological Indicators for the Assessment of Human Exposure to Industrial Chemicals*. Brussels/Luxembourg: Commission of the European Communities, pp. 1–16.

Catenacci, G., Barbieri, F., Bersani, M., Ferioli, A. and Cottica, D. (1993) Biological monitoring of human exposure to atrazine. *Toxicol. Lett.* 69: 217–222.

Chen, S., Zhang, Z., He, F. *et al.* (1991) An epidemiological study on occupational acute pyrethroid poisoning in cotton farmers. *Br. J. Ind. Med.* 48: 77–81.

Chester, G. and Woollen, B.H. (1981) Studies of the occupational exposure of Malaysian plantation workers to paraquat. *Br. J. Ind. Med.* 38: 23–33.

Coats, J.R. (1982) Structure–activity relationships in DDT analogs. In: Coats, J.R. (ed.) *Insecticide Mode of Action*. New York: Academic Press, pp. 29–43.

Colosio, C., Tomasini, M., Cairoli, S. *et al*. (1991) Occupational triphenyltin acetate poisoning: a case report. *Br. J. Ind. Med*. 48: 136–139.

Currie, K.L., McDonald, E.C., Chung, L.T.K. and Higgs, A.R. (1990) Concentrations of diazinon, chlorpyrifos, and bendiocarb after application in offices. *Am. Ind. Hyg. Assoc. J*. 51: 23–27.

Davis, J.E., Staiff, D.C., Butier, L.C. and Stevens, E.R. (1981) Potential exposure to dislodgable residues after application of two formulations of methyl parathion to apple trees. *Bull. Environ. Contam. Toxicol*. 27: 95–100.

De Bleeker, J., van den Neucker, K. and Colardyn, F. (1993) Intermediate syndrome in organophosphorus poisoning: a prospective study. *Crit. Care Med*. 21: 1706–1711.

Dobbins, J.G. (1987) Regulation and the use of 'negative' results from human reproductive studies: the case of ethylene dibromide. *Am. J. Ind. Med*. 12: 33–45.

Eaton, M., Schenker, M., Whorton, M.D., Samuels, S., Perkins, C. and Overstreet, J. (1986) Seven-year follow-up of workers exposed to 1,2-dibromo3-chloropropane. *J. Occup. Med*. 28: 1145–1150.

Ecobichon, D.J. (1996) Toxic effects of pesticides. In: Klaassen, C.D. (ed.) *Casarett and Doull's Toxicology. The Basic Science of Poison*, 6th edn. New York: McGraw-Hill, pp. 643–689.

Edwards, R.I., Ferry, D.H. and Temple, W.A. (1991) Fungicides and related compounds. In: Hayes, W.J. Jr and Laws, E.R. Jr (eds) *Handbook of Pesticide Toxicology*. San Diego: Academic Press, pp. 1409–1470.

Egnatz, D.G., Ott, M.G., Townsend, J.C., Olson, R.D. and Johns, D.B. (1980) DBCP and testicular effects in chemical workers: an epidemiological survey in Midland, Michigan. *J. Occup. Med*. 22: 727–732.

Elliot, M. (1976) Properties and applications of pyrethroids. *Environ. Health Perspect*. 14: 3–13.

Fenske, R.A., Horstman, S.W. and Bentley, R.K. (1987) Assessment of dermal exposure to chlorophenols in timber mills. *Appl. Ind. Hyg*. 2: 143–147.

Gallo, M.A. and Lawryk, N.J. (1991) Organic phosphorus compounds. In: Hayes, W.J. Jr and Laws, E.R. Jr (eds) *Handbook of Pesticide Toxicology*. San Diego: Academic Press, pp. 917–1123.

Gasiewicz, T.A. (1991) Nitro compounds and related phenolic pesticides. In: Hayes, W.J. Jr and Laws, E.R. Jr (eds) *Handbook of Pesticide Toxicology*. San Diego: Academic Press, pp. 1191–1269.

Hart, T.B. (1987) Paraquat: a review of safety in agricultural and horticultural use. *Human Toxicol*. 6: 13–18.

Hayes, W.J. Jr and Laws, E.R. Jr (eds) (1991) *Handbook of Pesticide Toxicology*. San Diego: Academic Press.

He, F., Sun, J., Han, K. *et al*. (1988) Effects of pyrethroid insecticides on subjects engaged in packaging pyrethroids. *Br. J. Ind. Med*. 45: 548–551.

He, F., Wang, S., Liu, L., Chen, S., Zhang, Z. and Sun, J. (1989) Clinical manifestations and diagnosis of acute pyrethroid poisoning. *Arch. Toxicol*. 63: 54–58.

Hodgson, M.J., Biock, G.D. and Parkinson, D.K. (1986) Organophosphate poisoning in office workers. *J. Occup. Med*. 28: 434–437.

Hueto, P., Bindoula, G. and Hoffman, J.R. (1995) Ethylenebisdithiocarbamates and ethylenethiourea: possible human health hazards. *Environ. Health Perspect*. 103: 568–573.

IARC (1983) *Monographs on the Evaluation of Carcinogenic Risks to Humans. Volume 30, Miscellaneous Pesticides*. Lyon: International Agency for Research on Cancer.

IARC (1986) *Monographs on the Evaluation of Carcinogenic Risks to Humans. Volume 41*,

Some Halogenated Hydrocarbons and Pesticide Exposures. Lyon: International Agency for Research on Cancer.

IARC (1987) *Monographs on the Evaluation of Carcinogenic Risks to Humans. Overall Evaluations of Carcinogenicity: an Updating of IARC Monographs Volumes I to 42.* Lyon: International Agency for Research on Cancer.

IARC (1991) *Monographs on the Evaluation of Carcinogenic Risks to Humans. Volume 53, Some Pesticides and Occupational Exposures in Insecticide Application.* Lyon: International Agency for Research on Cancer.

Jager, K.W., Roberts, D.V. and Wilson, A. (1970) Neuromuscular function in pesticide workers. *Brit. J. Ind. Med.* 27: 273–278.

Jeyaratnam, J. (1990) Acute pesticide poisoning: a major health problem. *World Health Stat. Q.* 43: 139–144.

Jusic, A., Jurenic, D. and Milic, S. (1980) Electromyographical neuromuscular synapse testing and neurological findings in workers exposed to organophosphorus pesticides. *Arch. Environ. Health* 35: 168–175.

Keplinger, M.L. and Deichmann, W.B. (1967) Acute toxicity of combinations of pesticides. *Toxicol. Appl. Pharmacol.* 10: 586–595.

Knopp, D. (1996) Assessment of exposure to 2,4-dichlorophenoxy acetic acid in the chemical industry: results of a five-year biological monitoring. *Occup. Environ. Med.* 51: 152–159.

Kurttio, P. and Savolainen, K. (1990) Ethylenethiourea in air and in urine as an indicator of exposure to ethylenebisdithiocarbamate fungicides. *Scand. J. Work Environ. Health* 16: 203–207.

Kurttio, P., Vartiainen, T. and Savolainen, K. (1990) Environmental and biological monitoring of exposure to ethylenebisdithiocarbamate fungicides and ethylenethiourea. *Br. J. Ind. Med.* 47: 203–206.

Lauwerys, R. (1983) *Industrial Chemical Exposure: Guidelines for Biological Monitoring.* Davis: Biomedical Publications.

Lavy, T.L. and Mattice, J.D. (1989) Biological monitoring techniques for humans exposed to pesticides. Use, development, and recommended refinements. In: Wang, R.G.M., Franklin, C.A., Honeycutt, R.C. and Reinert, J.C. (eds) *Biological Monitoring for Pesticide Exposure. Measurement, Estimation, and Risk Reduction (ACS Symposium Series 382).* Washington, DC: American Chemical Society, pp. 192–205.

Le Quesne, P.M., Maxwell, I.C. and Butterworth, S.T.G. (1980) Transient facial sensory symptoms following exposure to synthetic pyrethroids: a clinical and electrophysiological assessment. *Neurotoxicology* 2: 1–11.

Le Quesne, P.M. and Maxwell, I.C. (1981) Effect of metrifonate on neuromuscular transmission. *Acta Pharmacol. Toxicol. Scand.* 49 (Suppl. 5): 99–104.

Lotti, M. (1989) Neuropathy target esterase in blood lymphocytes. Monitoring the interaction of organophosphates with a primary target. In: Wang, R.G.M., Franklin, C.A., Honeycutt, R.C. and Reinert, J.C. (eds) *Biological Monitoring for Pesticide Exposure: Measurement, Estimation and Risk Reduction (ACS Symposium Series No. 382).* Washington, DC: American Chemical Society, pp. 117–123.

Lotti, M. (1991a) Treatment of acute organophosphate poisoning. *Med. J. Aust.* 154: 51–55.

Lotti, M. (1991b) Central neurotoxicity and neurobehavioural toxicity of anticholinesterases. In: Ballantyne, B. and Marrs, T.C. (eds) *Basic and Clinical Toxicology of Organophosphates and Carbamates.* London: Butterworths, pp. 75–83.

Lotti, M. (1992) The pathogenesis of organophosphate-induced delayed polyneuropathy. *Crit. Rev. Toxicol.* 21: 465–487.

Lotti, M., Becker, C.E. and Aminoff, M.J. (1984) Organophosphate polyneuropathy: pathogenesis and prevention. *Neurology* 34: 658–662.

Lotti, M., Becker, C.E., Aminoff, M.J. et al. (1983) Occupational exposure to the cotton defoliants DEF and Merphos. A rational approach to monitoring organophosphorus-induced delayed neurotoxicity. J. Occup. Med. 25: 517–522.

Maranelli, G., Perbellini, L. et al. (1991) Solfuro di carbonio ematico nel monitoraggio biologico dell'esposizione a ditiocarbammati. In: Giuliano, G. and Paoletti, A. (eds) Proceedings of the LIV National Meeting of the Società Italiana di Medicina del Lavoro e Igiene Industriale. Bologna: Monduzzi Editore, pp. 1251–1256.

Maroni, M. (1988) Carbamate pesticides. In: Alessio, L., Berlin, A., Boni, M. and Roi, R. (eds) Biological Indicators for the Assessment of Human Exposure to Industrial Chemicals. Luxembourg: Commission of the European Communities, pp. 27–54.

Mehler, L., Edmiston, S., Richmond, D., O'Malley, M. and Krieger, R.I. (1990) Summary of Illnesses and Injuries Reported by California Physicians as Potentially Related to Pesticides – 1988, HS-1541. Sacramento: California Department of Food and Agriculture.

Miyamoto, J. (1976) Degradation, metabolism, and toxicity of synthetic pyrethroids. Environ. Health Perspect. 14: 15–28.

Moretto, A. (1991) Indoor spraying with the pyrethroid insecticide λ-cyhalothrin: effects on spraymen and inhabitants of sprayed houses. Bull. World Health Org. 69: 59–94.

Moretto, A. and Lotti, M., (1998) Poisoning by organophosphorus insecticides and sensory neuropathy. J. Neurol. Neurosurg. Psych. 64: 463–468.

Morgan, D.P., Hetzler, H.L., Siach, E.F. and Lin, L.I. (1977) Urinary excretion of paranitrophenol and alkyl phosphates following ingestion of methyl or ethyl parathion by human subjects. Arch. Environ. Contam. Toxicol. 6: 159–173.

Morris, C.R. and Cabral, J.R.P. (1985) Hexachlorobenzene: Proceedings of an International Symposium. Lyon: IARC.

Müller, D., Klepel, H., Macholz, R. and Knoll, R. (1981) Electroneurographic and electroencephalographic finding in patients exposed to hexachlorocyclohexane. Psychiatr. Neurol. Med. Psychol. 33: 468–472.

Murphy, S.D. (1987) Toxicology of chlorophenoxy herbicides and their contaminants. In: Costa, L.G., Galli, C.L. and Murphy, S.D. (eds) Toxicology of Pesticides: Experimental, Clinical and Regulatory Perspectives. NATO ASI Series. Berlin: Springer-Verlag, pp. 49–61.

Nigg, H.N., Stamper, J.H. and Queen, R.M. (1984) The development and use of a universal model to predict tree crop harvester pesticide exposure. Am. Ind. Hygiene Assoc. J. 45: 182–186.

Pearce, N. and Reif, J.S. (1990) Epidemiologic studies of cancer in agricultural workers. Am. J. Ind. Med. 18: 133–148.

Pelfrene, A.F. (1991) Synthetic organic rodenticides. In: Hayes, W.J. Jr and Laws, E.R. Jr (eds) Handbook of Pesticide Toxicology. San Diego: Academic Press, pp. 1271–1316.

Pentel, M.T., Andreozzi, R.J. and Marks, J.G. Jr (1994) Allergic contact dermatitis from the herbicides trifluralin and benefin. J. Am. Acad. Dermatol. 31: 1057–1058.

Plestina, R. (1984) Prevention, Diagnosis and Treatment of Insecticide Poisoning. WHO/VBC/84.889. Geneva: WHO.

Popendorf, W.J. and Leffingwell, J.T. (1982) Regulating OP pesticide residues for farmworker protection. Res. Rev. 82: 125–201.

Ratcliff, J.M., Schrader, S.M., Steenland, K., Clapp, D.E., Turner, T. and Hornung, R.W. (1987) Semen quality in papaya workers with long-term exposure to ethylene dibromide. Br. J. Ind. Med. 44: 317–326.

Roan, C., Morgan, D. and Paschal, E.H. (1971) Urinary excretion of DDA following ingestion of DDT and DDT metabolites in man. Arch. Environ. Health 22: 309–315.

Roberts, D.V. (1976) EMG voltage and motor nerve conduction velocity in organophosphorus pesticide factory workers. Int. Arch. Occup. Environ. Health 36: 267–274.

Rosenstock, L., Keifer, M., Daniell, W.E., McConnel, R., Claypoole, K. and The Pesticide

Health Effects Safety Group (1991) Chronic central nervous system effects of acute organophosphate pesticide intoxication. *Lancet* 338: 223–226.

Rowley, D.L., Rab, M.A., Hardjotanojo, W. *et al.* (1987) Convulsions caused by endrin poisoning in Pakistan. *Pediatrics* 79: 928–934.

Salem, H. and Olajos, E.J. (1988) Review of pesticides: chemistry, uses and toxicology. *Toxicol. Ind. Health* 4: 291–321.

Saracci, R., Kogevinas, M., Bertazzi, P.A. *et al.* (1991) Cancer mortality in workers exposed to chlophenoxy herbicides and chlorophenols. *Lancet* 338: 1027–1032.

Senanayake, N. and Karalleide, L. (1987) Neurotoxic effects of organophosphorus insecticides. An intermediate syndrome. *N. Engl. J. Med.* 316: 761–763.

Short, R.D., Winston, M., Hong, C.B., Minor, J.L., Lee, C.C. and Seifter, J. (1979) Effects of ethylene dibromide on reproduction in male and female rats. *Toxicol. Appl. Pharmacol.* 49: 97–105.

Smith, A.G. (1991) Chlorinated hydrocarbon insecticides. In: Hayes, W.J. Jr and Laws, E.R. Jr (eds) *Handbook of Pesticide Toxicology*. San Diego: Academic Press, pp. 731–795.

Smith, J.G. (1988) Paraquat poisoning by skin absorption: a review. *Human Toxicol.* 7: 15–19.

Smith, L.L., Rose, M.S. and Wyatt, I. (1979) The pathology and biochemistry of paraquat. In: *Oxygen Free Radicals and Tissue Damage. Ciba Foundation Series* 65 (New Series). Amsterdam: Excerpta Medica, pp. 321–347.

Smith, R.P. (1996) Toxic responses of the blood. In: Klaassen, C.D. (ed.) *Casarett and Doull's Toxicology. The Basic Science of Poison*. New York: McGraw-Hill, pp. 335–354.

Spear, R.C., Jenkins, D. and Milby, T.H. (1975) Pesticide residues and field workers. *Environ. Sci. Technol.* 9: 308–313.

Spear, R.C., Popendorf, W.J., Leffingwell, J.T., Milby, T.H., Davies, J.E. and Spencer, W.F. (1977) Fieldworker's response to weathered residues of parathion. *J. Occup. Med.* 19: 406–410.

Stålberg, E., Hilton-Brown, P., Kolmodin-Hedman, B., Holmstedt, B. and Augustinsson, K.B. (1978) Effect of occupational exposure to organophosphorus insecticides on neuromuscular function. *Scand. J. Work Environ. Health* 4: 255–261.

Steenland, K., Jenkins, B., Ames, R.G., O'Malley, M., Chrislip, D. and Russo, J. (1994) Chronic neurological sequelae to organophosphate pesticide poisoning. *Am. J. Publ. Health* 84: 731–736.

Stevens, H.T. and Sumner, D.D. (1991) Herbicides. In: Hayes, W.J. Jr and Laws, E.R. Jr (eds) *Handbook of Pesticide Toxicology*. San Diego: Academic Press, pp. 1317–1468.

Stokes, L., Stark, A., Marshall, E. and Narang, A. (1995) Neurotoxicity among pesticide applicators exposed to organophosphates. *Occup. Environ. Med.* 52: 648–653.

Swan, A.A.B. (1969) Exposure of spray operators to paraquat. *Br. J. Ind. Med.* 26: 322–329.

Takahashi, W., Wong, L., Rogers, B.J. and Hale, R.W. (1981) Depression of sperm counts among agricultural workers exposed to dibromochloropropane and ethylene dibromide. *Bull. Environ. Contam. Toxicol.* 27: 551–558.

Talcott, R.E., Mallipudi, N.M., Umetsu, N. and Fukuto, T.R. (1979) Inactivation of esterases by impurities isolated from technical malathion. *Toxicol. Appl. Pharmacol.* 49: 107–112.

Tomlin, C. (ed.) (1997) *The Pesticide Manual. A World Compendium*, 11th edn. Surrey, UK: The British Crop Protection Council.

Torkelson, T.R., Sadek, S.E., Rowe, V.K. *et al.* (1961) Toxicologic investigations of 1,2-dibromo-3-chloropropane. *Toxicol. Appl. Pharmacol.* 3: 545–559.

Tracey, J.A. and Gallagher, H. (1990) Use of glycopyrolate and atropine in acute organophosphorus poisoning. *Hum. Exp. Toxicol.* 9: 99–100.

US EPA (1994) *An SAB report. Assessment of potential 2,4-D carcinogenicity. Review of the* epidemiological and other data on potential carcinogenicity of 2,4-D by the SAB/SAP Joint Committee (EPA-SAB-EHE-94-005). Washington DC: US Environmental Protection Agency.

van Ginkel, C.J.W. and Sabapathy, N.N. (1995) Allergic contact dermatitis from the newly introduced fungicide fluazinam. *Contact. Derm.* 32: 160–162.

Verberk, M.M. and Sallé, H.J.A. (1977) Effects on nervous function in volunteers ingesting mevinphos for one month. *Toxicol. Appl. Pharmacol.* 42: 351–358.

Wadia, R.S., Chitra, S., Amin, R.B., Kiwalkar, R.S. and Sardesai, H.V. (1987) Electrophysiological studies in acute organophosphate poisoning. *J. Neurol. Neurosurg. Psych.* 50: 1442–1448.

White, R.D., Gandolfi, A.J., Bowden, G.T. and Sipes, I.G. (1983) Deuterium isotope effect on the metabolism and toxicity of 1,2-dibromoethane. *Toxicol. Appl. Pharmacol.* 69: 170–178.

WHO (1979) *Environmental Health Criteria 9. DDT and its Derivatives.* Geneva: WHO.

WHO (1981) *Environmental Health Criteria 18. Arsenic.* Geneva: WHO.

WHO (1984a) *Environmental Health Criteria 29. 2,4-Dichlorophenoxyacetic Acid (2,4-D).* Geneva: WHO.

WHO (1984b) *Environmental Health Criteria 39. Paraquat and Diquat.* Geneva: WHO.

WHO (1986a) *Environmental Health Criteria 63. Organophosphorus Insecticides: A General Introduction.* Geneva: WHO.

WHO (1986b) *Environmental Health Criteria 64. Carbamate Pesticides: A General Introduction.* Geneva: WHO.

WHO Regional Office for Europe (1987a) *Environmental Health 27, Drinking-water Quality: Guidelines for Selected Herbicides.* Copenhagen: WHO.

WHO (1987b) *Environmental Health Criteria 71. Pentachlorophenol.* Geneva: WHO.

WHO (1988) *Environmental Health Criteria 78. Dithiocarbamate Pesticides, Ethylenethiourea, and Propylenethiourea: A General Introduction.* Geneva: WHO.

WHO (1990) *Public Health Impact of Pesticides used in Agriculture.* Geneva: WHO.

WHO (1996a) *The WHO Recommended Classification of Pesticides by Hazard and Guidelines to the Classification 1996–1997, WHO/PCS/96.3.* WHO: Geneva.

WHO (1996b) *Pesticide Residues in Food. 1995. Evaluations. Part II. Toxicological and Environmental.* Geneva: WHO.

Whorton, D., Milby, T.H., Krauss, R.M. and Stubbs, H.A. (1979) Testicular function in DBCP exposed pesticide workers. *J. Occup. Med.* 21: 161–166.

Wigle, D.T., Semenciw, R.M., Wilkins, K., Riedel, D., Ritter, L., Morrison, H.I. and Mao, Y. (1990) Mortality study of Canadian male farm operators: non-Hodgkin's lymphoma mortality and agricultural practices in Saskatchewan. *J. Natl Cancer Inst.* 82: 575–582.

Wojeck, G.A., Price, J.F., Nigg, H.N. and Stamper, J.H. (1983) Worker exposure to paraquat and diquat. *Arch. Environ. Contam. Toxicol.* 12: 65–70.

Won, J.H., Ahn, S.K. and Kim, S.C. (1993) Allergic contact dermatitis from the herbicide Alachlor. *Contact Dermatitis* 28: 38–39.

Wood, S., Rom, W.M., White, G.L. and Logan, D.C. (1983) Pentachlorophenol poisoning. *J. Occup. Med.* 25: 527–530.

Zhang, Z., Sun, J., Chen, S., Wu, Y. and He, F. (1991) Levels of exposure and biological monitoring of pyrethroid in spraymen. *Br. J. Ind. Med.* 48: 82–86.

Chapter 14

Toxicity of organic solvents

N.H. Stacey and C. Winder

Introduction

A **solution** is a homogeneous mixture of substances. Components present in minor proportions are called **solutes**; the major component is called the **solvent**. As it can contain a number of chemicals in solution, water is a solvent.

A solvent can be defined as a substance that has the power of dissolving, and the solute is the substance that has been dissolved in the solvent. Most often when we refer to a solution we are considering a liquid dissolved in a liquid, a solid dissolved in a liquid or a gas dissolved in a liquid.

The extent to which substances can form solutions depends upon the kind and strength of the attractive forces between the molecules involved. These forces are exerted by species of the following types:

- nonpolar molecules
- polar molecules (which contain permanent dipoles)
- ions
- metallic atoms.

The function of a solvent is to **disperse** solids, liquids or gases into solution and to **separate** the solute molecules. In an **ideal solution**, the different molecules have the same attractive fields and mix without change in volume or heat content. **Real solutions** are discussed in terms of their departures (sometimes considerable) from the idealised model.

In occupational toxicology discussion of solvents almost invariably relates to organic rather than aqueous solvents. Accordingly, only organic solvents will be further considered in this chapter.

The chemistry of organic compounds

Organic compounds like solvents are generally structures with varying numbers of hydrogen and carbon atoms (hydrocarbons). Other atoms (notably oxygen, nitrogen, sulfur and the halogens) may also be present.

The properties of organic solvents depend on four main factors:

- the number of carbon atoms present
- the presence of only single bonds (saturated molecules) or double or triple bonds (unsaturated molecules) between carbon atoms
- the configuration of the solvent molecule, such as straight chain (aliphatic), branched chain or ring (aromatic) structures
- the presence of functional groups.

Turpentine and alcohol were some of the earliest solvents used until the late 19th century. Today most widely used solvents are hydrocarbon solvents and oxygenated solvents distilled from crude oil or coal tar. Other solvents have more specialised applications (Wypych 2000).

Hydrocarbon solvents contain only carbon and hydrogen and there are four types of hydrocarbon in commercially available solvents:

- linear paraffins, characterised by straight chain, saturated alkanes
- isoparaffins, characterised by branched, saturated alkanes
- naphthenes or cycloparaffins (saturated cycloalkanes)
- aromatics (unsaturated arenes).

Oxygenated solvents contain oxygen in addition to carbon and hydrogen and there is a wide range being used in industry. The main types and their typical examples include:

- ketones: methyl ethyl ketone
- esters: n-butyl acetate
- ethers: diethyl ether
- alcohols: isopropanol
- glycol ethers: ethylene glycol monobutyl ether
- glycol ether acetates: propylene glycol monomethyl ether acetate.

Oxygenated solvents are largely manufactured from chemical precursors. In general, these solvents are more powerful solvents than hydrocarbon solvents, in that they dissolve a wider range of materials. Hence, they are most used in many industrial applications.

Still other groups of solvents contain carbon, hydrogen, possibly oxygen and other elements, such as nitrogen, sulfur and the halogens. Examples include:

- amines: dimethyl amine
- nitrate solvents: nitromethane
- halogenated solvents: trichloroethylene.

Many of these solvents have specialised uses. For example, the halogenated solvents were in widespread use until recently, as they are not as flammable as other solvent groups. Some (such as 1,1,1-trichloroethane) are now being phased from use as many halogenated solvents appear to add to global greenhouse warming and are involved in thinning of the ozone layer.

It is also possible for a solvent to be a combination of two or more of these chemical types, such as ethanolamine or chlorobenzene.

Physical properties of solvents

The physical properties of organic solvents have a great bearing on the degree of fire and explosion hazard. Some properties of hydrocarbon solvents that are determinants of hazard are:

- boiling point (or range)
- vapour pressure
- flash point
- fire point
- flammable (or explosive) limits
- autoignition temperature
- vapour density
- in hydrocarbon mixtures, the content of particularly hazardous components such as benzene and n-hexane (likely to be present in substantial amounts only if the boiling points of the pure compounds fall within the boiling range of the mixture).

Examples of common organic solvents

There are over 100 different organic solvent compounds in common use in industry and commerce. Examples are shown in Table 14.1.

Organic solvents have a wide range of uses which are summarised in the Table 14.2, which also shows the proportion of solvent use in the various major use categories.

As can be partly predicted from the above pattern of solvent use some of the main workplaces/work activities using solvents include automotive with spray painting, woodworking/furnishing, metal trades, printing, plastics/petrochemical, dry cleaning, laboratories, building and agriculture.

A wide range of solvents is used in the surface-coating industry because of the many different types and requirements of products. Toluene and xylene figure prominently in spray painting in the automotive industry for example. Adhesives also call on the varying properties of the different solvents for various product applications. Metal cleaning is a little more specific in its requirements with chlorinated solvents such as trichloroethylene and 1,1,1-trichloroethane finding considerable application in vapour degreasing. In dry cleaning tetrachloroethylene (perchloroethylene) has been widely used as a solvent.

It should be noted that for many of the applications to which solvents are put it is their abilities to vaporise readily that make them desirable. This very property

Table 14.1 Common solvents

Class of solvent	Example	Molecular formula
Acetamide solvents	N,N-Dimethylacetamide	
Alcohol solvents	Amyl alcohol n-Butanol iso-Butanol Cyclohexanol Diacetone alcohol Ethanol Glycerol Methanol 4-Methyl-2-pentanol Propanol	
Aldehydes	Formaldehyde Acetaldehyde	
Aliphatic solvents	n-Heptane n-Hexane Octane	
Amide solvents	Hexamethylphosphoramide	
Amine solvents	Diethylenetriamine Dimethylamine Ethanolamine Ethylenediamine Triethyamine	
Aromatic solvents	Benzene 4-tert-Butyltoluene Chlorobenzene Cumene 1,2-Dichlorobenzene Ethylbenzene Mesitylene Pseudocumene Styrene Toluene Xylene	

continued

Table 14.1 continued

Class of solvent	Example	Molecular formula
Cyclic aliphatic solvents	Cyclohexane Methylcyclohexane	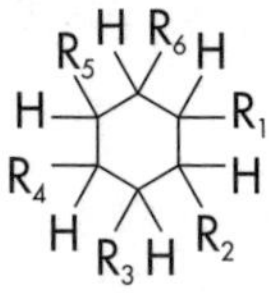
Ester solvents	Amyl acetate n-Butyl acetate iso-Butyl acetate Ethyl acetate Ethyl formate Methyl acetate Methyl formate iso-Propyl acetate n-Propyl acetate	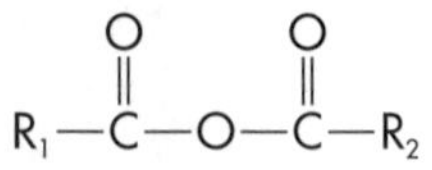
Ethers	Dichloroethyl ether Diethyl ether Diisopropyl ether Methylal	R_1—O—R_2
Formamides	N,N-Dimethylformamide	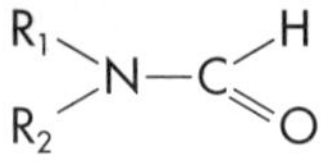
Furan solvents	Furfural Furfural alcohol Tetrahydrofuran	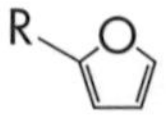
Glycol ethers	Diethylene glycol Ethylene glycol Ethylene glycol monobutyl ether Ethylene glycol monoethyl ether Ethylene glycol monomethyl ether Hexylene glycol Propylene glycol	
Glycol ether acetates	Ethylene glycol monoethyl ether acetate Ethylene glycol monomethyl ether acetate	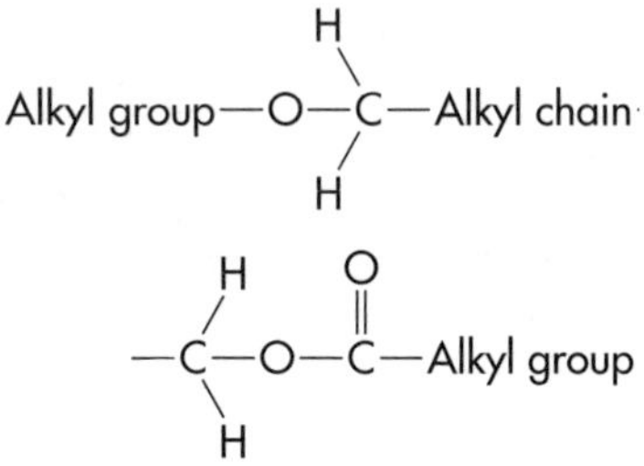

continued

Table 14.1 continued

Class of solvent	Example	Molecular formula
Halogenated solvents	Bromoform Carbon tetrachloride Chloroform 1,1-Dichloroethane 1,2-Dichloroethane 1,2-Dichloroethylene Dichloromethane 1,2-Dichloropropane Tetrachloroethane Tetrachloroethylene 1,1,1-Tricloroethane 1,1,2-Tricloroethane Trichloroethylene Trichlorofluoromethane 1,1,2-Trichloro-trifluoroethane	R—Cl or F or Br or I
Ketone solvents	Acetone Cyclohexanone Diisobutyl ketone Mesityl oxide Methyl *n*-amyl ketone Methyl butyl ketone Methylcyclohexanone Methy lethyl ketone Methyl *iso*-butyl ketone	$R_1-\overset{\overset{\displaystyle O}{\|}}{C}-R_2$
Nitrate solvents	Nitrobenzene Nitroethane Nitromethane Nitropropane	R—NO$_2$
Solvent mixtures	Ligroin Mineral turpentine Naphtha Petroleum spirit Stoddard solvent Varnish maker's and painter's naphtha White spirits Wood turpentine	

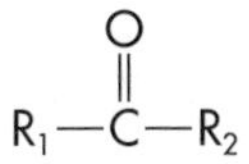

continued

Table 14.1 continued

Class of solvent	Example	Molecular formula
Miscellaneous solvents	Acetonitrile	
	Caprolactam	
	Carbon disulfide	
	Dimethyl sulfoxide	
	1,4-Dioxane	
	Dioxolane	
	Epichlorohydrin	
	Isophorone	
	Limonene	
	N-Methylpyrrolidone	
	Morpholine	
	Sulpholane	
	Tetralin	

Note: In molecular formulae, the letter R denotes any organic group.

and use means that they readily get into the air and unless contained or removed by ventilation then they will be found in the workplace air. This is probably why they form a group of chemicals that are of considerable and continuing concern in the workplace. That is, there is substantial scope for the worker to be exposed which is a prerequisite to the toxicity of a chemical having the opportunity to manifest its actions.

To provide an indication of the extent of solvent production figures are provided in Table 14.3 for solvent consumption.

Hazards of solvents

Organic solvents are hazardous. These hazards can be divided into danger hazards and toxic hazards.

Table 14.2 Uses of solvents

Application	Percent use
Surface coatings	43.3
Metal cleaning (degreasing)	10
Household products	8.1
Adhesives	6.7
Pharmaceuticals	6.1
Dry cleaning	3.9
Other	20

Source: adapted from Collings and Luxon 1982.

Table 14.3 Consumption of solvents

Solvent class	Tonnes (thousands)
Aliphatic hydrocarbons	1200
Aromatic hydrocarbons	900
Chlorinated hydrocarbons	800
Alcohols	600
Ketones	450
Esters	300
Glycol ethers	150

Source: adapted from Collings and Luxon 1982.

Danger hazards

Apart from the ability of a solvent to vaporise, which is related to its vapour pressure, there are other physical properties that need to be considered when using them in the workplace. These relate to safety aspects particularly and include, for example, boiling range, flash point, fire point and flammable range.

The most important hazard of organic solvents relate to flammability. Many organic solvents have low flash points, and will burn if ignited. One reason that the chlorinated solvents became so widespread was that they have quite high flash points and are not usually flammable under conditions of normal use. The dangerous goods classification criterion for flammability is defined as a liquid with a flash point of 61°C or below.

Toxic hazards

Most solvents are volatile. The property of quick evaporation makes solvents available for exposure and absorption into the body. In general, the toxicity of solvents is related to their ability to dissolve fatty materials (the property of lipophilicity).

Before the toxicity can be discussed, it is necessary to understand biotransformation and disposition processes, as outlined in Chapter 2.

The biotransformation of some solvents has been well studied. These include ethanol, benzene, n-hexane and chlorinated solvents.

Biotransformation of ethanol

The toxic effects of alcohol on the liver are directly related to its metabolism. Initially, ethanol is broken down to acetaldehyde by the enzyme alcohol dehydrogenase. The acetaldehyde is then broken down to acetate by the enzyme acetaldehyde dehydrogenase (see Figure 14.1).

Biotransformation of benzene

The toxic effects of benzene are caused by biotransformation to benzene oxide (an epoxide), and then hydroxylated derivatives such as phenol, hydroquinone and

$$CH_3CH_2OH \quad + \quad NAD \quad \rightarrow \quad CH_3CHO \quad + \quad NADH$$

Ethanol   coenzyme   Acetaldehyde   reduced coenzyme

Alcohol dehyrogenase

$$CH_3CHO \quad + \quad NAD \quad \rightarrow \quad CH_3COO^- \quad + \quad NADH$$

Acetaldehyde   coenzyme   Acetate   reduced coenzyme

Acetaldehyde dehyrogenase

Figure 14.1 The biotransformation of ethanol.

catechol. These metabolites are available for conjugation reactions. However, it is epoxide formation that is the probable intermediate involved in carcinogenicity (see Figure 14.2).

Metabolism of benzene proceeds through the mixed function oxidase system to form benzene epoxide, as an initial product. Conversion of the epoxide to 1,2-dihydrodiol by the enzyme epoxide hydrase occurs, leading to the formation of catechol. Benzene epoxide also undergoes nonenzymatic rearrangement to phenol, which is further biotransformed to hydroquinone. Metabolites such as quinones or semiquinones derived from catechol or hydroquinone are possible genotoxicants. As rearrangement of benzene oxide to phenol proceeds at a rate that is slow enough for reaction with DNA to be feasible, benzene oxide has also been

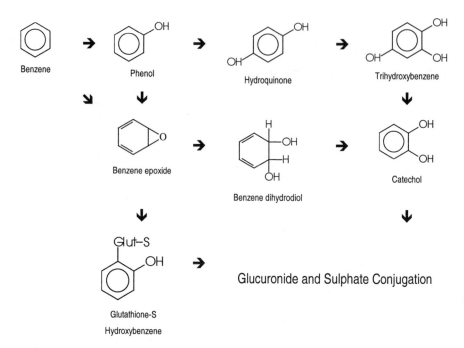

Figure 14.2 The biotransformation of benzene.

considered as a possible candidate as a carcinogen, although its reactivity *in vivo* has not been established.

The generalisation that strong electrophilic reactivity is a basic requirement for the initiation of carcinogenesis and detailed knowledge of the enzymatic biotransformations involved in metabolic activation and deactivation of many chemicals are providing important tools for the recognition of carcinogens.

Biotransformation of *n*-hexane

n-Hexane is biotransformed to 2-hexanol, further to 2,5-hexanediol, and then to the major metabolite 2,5-hexadione. These three substances are solvents in their own right. All produce similar polyneuropathies on exposure, although 2,5-hexadione is many times more potent than *n*-hexane or 2-hexanone in producing neuropathy (see Figure 14.3).

The neurotoxicity of 2,5-hexadione is fairly specific to its γ-diketone structure (that is, two ketone groups on an aliphatic chain separated by two carbon atoms), as 2,3-, 2,4-hexadione and 2,6-heptanedione are not neurotoxic, while 2,5-heptanedione, 3,6-octadione and other γ-diketones are neurotoxic (see Table 14.4).

One further important interaction is that of *n*-hexane and methyl ethyl ketone (2-butanone). This interaction is synergistic in nature, with exposed animals showing earlier onset and greater severity of neurotoxic signs, and produced greater concentrations of 2,5-hexadione than animals exposed to 2-hexanone alone.

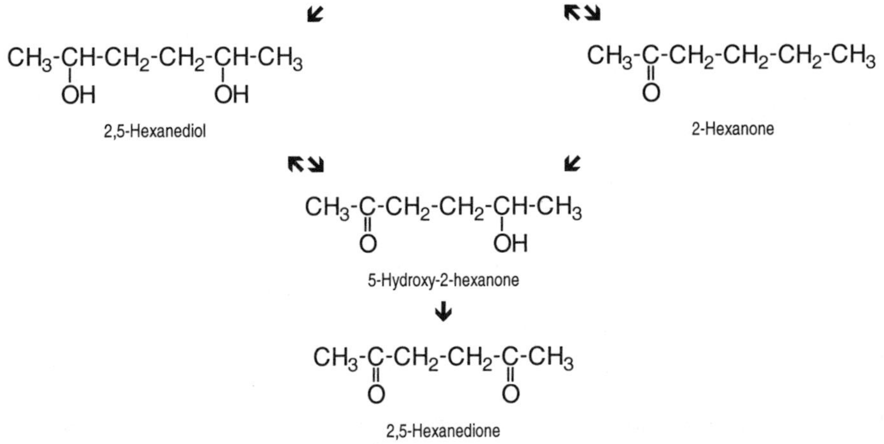

Figure 14.3 The biotransformation of *n*-hexane.

Table 14.4 Structure of neurotoxic di-ketones

Chemical	Chemical formula	γ-Diketone	Neurotoxic?
2,3-hexadione	$CH_3-\underset{\underset{O}{\|\|}}{C}-\underset{\underset{O}{\|\|}}{C}-CH_2-CH_2-CH_3$	✗	✗
2,4-hexadione	$CH_3-\underset{\underset{O}{\|\|}}{C}-CH_2-\underset{\underset{O}{\|\|}}{C}-CH_2-CH_3$	✗	✗
2,5-hexadione	$CH_3-\underset{\underset{O}{\|\|}}{C}-CH_2-CH_2-\underset{\underset{O}{\|\|}}{C}-CH_3$	✓	✓
2,5-heptadione	$CH_3-\underset{\underset{O}{\|\|}}{C}-CH_2-CH_2-\underset{\underset{O}{\|\|}}{C}-CH_2-CH_3$	✓	✓
2,6-heptadione	$CH_3-\underset{\underset{O}{\|\|}}{C}-CH_2-CH_2-CH_2-\underset{\underset{O}{\|\|}}{C}-CH_3$	✗	✗
2,5-octadione	$CH_3-\underset{\underset{O}{\|\|}}{C}-CH_2-CH_2-\underset{\underset{O}{\|\|}}{C}-CH_2-CH_2-CH_3$	✓	✓
2,6-octadione	$CH_3-\underset{\underset{O}{\|\|}}{C}-CH_2-CH_2-CH_2-\underset{\underset{O}{\|\|}}{C}-CH_2-CH_3$	✗	✗
3,6-octadione	$CH_3-CH_2-\underset{\underset{O}{\|\|}}{C}-CH_2-CH_2-\underset{\underset{O}{\|\|}}{C}-CH_2-CH_3$	✓	✓

Excretion

The main routes of excretion of absorbed solvents are in the faeces, in the urine, or in exhaled air. Again, the route of excretion depends on individual chemical characteristics, and also the amount of biotransformation that has occurred in the body.

In faeces

Though generally not a major route of excretion for solvents, some may be directly excreted from the body into the gastrointestinal tract. One possible further route of excretion is through bile – when there is considerable transformation activity in the liver, biliary excretion can be quite considerable. Bile is directly passed into the faeces in the intestine (where the possibility exists that some of this excretion may be reabsorbed further down the intestinal tract). However, this can be a significant route of excretion for biotransformed chemicals.

In urine

Urine is the main route of excretion for many absorbed chemicals. The ability of a solvent to be excreted in urine depends on its water solubility properties, and the water solubility properties of its biotransformed breakdown products. The organ which carries out urinary excretion is the kidney, which filters the blood and removes unwanted materials from the body. The product of these processes is urine, which collects in the bladder and is eventually excreted. Again, the kidney may be damaged if sufficient concentrations of chemicals or breakdown products are reached. An example of a kidney toxicant is chloroform, although this solvent is not in widespread use nowadays.

In exhaled air

Solvents are some of the more volatile chemicals and there may be considerable excretion of nonbiotransformed chemicals through exhaled air. Also, biotransformation of solvents may produce metabolic products such as carbon dioxide, which will also be exhaled.

Toxic effects of solvents

The effects of solvents on health are both general, in that they are common to all substances in this class, or specific to individual compounds. Observed effects will depend on a number of factors, including:

- individual solvent properties
- exposure level
- frequency of exposure
- exposure to other materials
- subject sensitivity.

Toxic effects of various solvent classes are listed in Tables 14.5–14.7. However, some organic solvents have an individual toxicity that cannot be predicted from trends in other similar structures.

General effects of solvents

The health effects of the solvents will be described below by virtue of the body systems they affect. The main body systems affected are the lungs, the skin, the central nervous system, the liver, and the kidneys. There is also some concern regarding the potential of the chemicals in these products to cause birth defects and cancer. These are outlined below.

The lungs

The primary function of the lung is to provide a means for getting gases into and out of the body. To do this, the lung has evolved into an organ with a structure

Table 14.5 Toxicity of the aliphatic solvents

Ethene	$CH_2{=}CH_2$	Depression of CNS
Ethyne	$CH{\equiv}CH$	Depression of CNS, metabolic effects
Propane	$CH_3{-}CH_2{-}CH_3$	Depression of CNS, cardiac effects
Propene	$CH_3{-}CH_2{=}CH_2$	Depression of CNS
Butane	$CH_3{-}CH_2{-}CH_2{-}CH_3$	Depression of CNS, cardiac effects
1,3 Butadiene	$CH_2{=}CH\text{-}CH{=}CH_2$	Carcinogen
Hexane	$CH_3{-}CH_2{-}CH_2{-}CH_2{-}CH_2{-}CH_3$	Neuropathy

Table 14.6 Toxicity of the aromatic solvents

Benzene	(benzene ring)	Depression of CNS, haematological effects, leukaemia
Toluene	(benzene ring)$-CH_3$	Depression of CNS (more so than benzene), dermatitis
Xylene	(benzene ring with $-CH_3$ and H_3C)	Depression of CNS, dermatitis
Styrene	(benzene ring)$-CH_2{=}CH_2$	Depression of CNS, mucous membrane effects, dermatitis
Phenol	(benzene ring)$-OH$	Depression of CNS, respiratory effects

Table 14.7 Toxicity of the halogenated solvents

Halogenated solvent	Formula	Effects
Carbon tetrachloride	CCl_4	Hepatoxicant
Chloroform	$CHCl_3$	Hepatoxicant, narcosis
Chloroprene	$CH{=}CHCl{=}CH_2$	Carcinogen, irritant
Ethylene dibromide	CH_2BrCH_2Br	Carcinogen
Ethylene dichloride	CH_2ClCH_2Cl	Carcinogen
Methyl bromide	CH_3Br	Skin, respiratory toxicity
Tetrachloroethene	$CHCl_2{=}CHCl_2$	CNS depressant
Trichloroethane	CCl_3CH_3	CNS depressant
Trichloethylene	$CCl_2{=}CHCl$	CNS depressant
Vinyl chloride	$CH_2{=}CCl_2$	Carcinogen

which is ideal for the uptake of oxygen (from the air), as well as other inhaled materials. The lung also excretes waste gases (mainly carbon dioxide), although in some cases it can excrete absorbed chemicals, either as unchanged or biotransformed products. The key part of the respiratory system is the thin membrane between the air and the blood, where minute blood vessels called capillaries come into close proximity with air where gas exchange (oxygen in, carbon dioxide out) occurs. The lungs also have a respiratory defence system, which is directed at maintaining its gas exchange function.

For many chemicals in the occupational environment (including solvents) the lungs are the main route of exposure. Once inhaled, solvents may also directly affect the lung and its defence systems (Schenker and Jacobs 1996).

Solvents and the respiratory system

There are a number of respiratory effects encountered in solvent exposure.

Asphyxiation

This can be either due to dilution of oxygen concentration in air by high concentrations of solvent vapours to asphyxiant levels (below 10–12% oxygen) such as from exposure to the smaller alkanes, or chemical asphyxiation, for example carbon monoxide production from biotransformation of methylene dichloride (dichloromethane) vapours.

Irritation

This can occur at a number of levels in the entire respiratory tract, from the nasal passages and upper airways, to deep in the gas exchange areas (pulmonary component). In the nasal passages and upper airways, the main response is hypersecretion of mucous and obstruction. The effects of irritants in the lower airways are similar, with hypersecretion with bronchitis (cough and sputum) and some dyspnoea (shortness of breath) caused by constriction of the airways. Many solvents can act as irritants, such as xylene, some alcohols and alkanes. Irritants may also trigger attacks of asthma. In the gas exchange areas of the pulmonary tissues, the responses to irritants are different. Pulmonary oedema involves a breakdown of the blood–air barrier with an increase in fluid over the gas exchange tissues. Pneumonitis is a condition where the gas exchange tissues become filled with inflammatory cells. Both these conditions will reduce the ability of the lungs to absorb or release gases. In general, oedema or pneumonitis is caused by massive short-term exposures. However, chemical pneumonitis can be produced following the vomiting of swallowed materials containing petroleum solvents and through this, their accidental entry into the lung (aspiration).

Sensitisation and allergy

The main allergic response of the lung is a reflex bronchoconstriction known as asthma. The mechanism of effect is wholly or partly through the body's immune system, where the body makes antibodies to the chemical that then react to subsequent exposures. This means that once formed, very small amounts of exposure may precipitate quite debilitating asthma. While exposure is obviously an important determinant in sensitisation, other factors, such as predisposition, the status of the immune system and exposures to other sensitisers, will also have a bearing. It is important to realise that the body's ability to become sensitised is not based on orthodox dose–response relationships, and therefore very careful consideration must be given to the choice of control measures to be employed. Most solvents are not well-known inducers of asthma, though organic chemicals such as phthalic anhydrides and other cyclic ethers have hypersensitivity effects.

Cancer

The production of cancer is elaborated on below, though some specific points should be made about cancer in the lung. Three occupations where solvent exposure is significant also report higher incidences of lung cancer. These are: (i) oil and petrol refining; (ii) painting as an occupation; and (iii) spraying insecticides. However, with the probable exception of benzene exposures in oil refining, the likely cancer-causing agents are as yet identified, and evidence of this nature merely highlights the problems of interpretation of scientific and medical evidence, and its application to the workplace.

The skin

The skin is the body's largest organ. It comes into contact with many chemical substances, some of which will be hazardous. Fortunately, the skin is not highly permeable, and is therefore a relatively good barrier separating the body from the environment. However, some chemicals can be absorbed by the skin in sufficient quantities to cause systemic effects. Skin disease itself is one of the most common occupational health problems.

Standards of personal cleanliness and the condition of the skin are predisposing factors which may have an effect on the ability of chemicals to pass through the skin. Continually wet or oily (due to exposure to oils) or cracked and dry skins may be more permeable to chemicals than intact, healthy skin.

The commonest occupational skin disorders are contact dermatitis, which may be of irritant or allergic types. Other occupational skin disorders of relatively low occurrence are outlined in the chapter on skin disorders. Irritant contact dermatitis is the most common and significant condition from exposure to solvents, followed by allergic contact dermatitis, contact urticaria and scleroderma.

Irritant contact dermatitis

This is the most common form of skin damage, caused by direct skin contact. About 90% of all cases of occupational skin damage are of this type. The response of the body is redness and inflammation of the skin, which progresses to itching, swelling, cracking and possibly blisters. The skin can become thick, rough and cracked on prolonged and repeated exposure.

Allergic contact dermatitis

As with the induction of asthma in the lung (see above), some solvents can pene-trate the skin and provoke a sensitisation reaction, with the formation of specific antibodies to the chemical. On subsequent exposure, often to very small amounts, a contact dermatitis reaction can occur. The production of allergic contact der-matitis is variable; sometimes with months or years of exposure occurring before a sensitisation response is provoked. However, once sensitisation occurs it is often permanent. As with asthma, induction of allergic reactions in the skin is not subject to orthodox dose–response relationships, and selection of appropriate control measures is very important.

Importantly, irritant and allergic contact dermatitis can occur together.

Solvents and the skin

While solvents represent a fairly diverse group of chemical substances, their effects on skin are similar enough to allow them to be grouped together. Many solvents have been identified as dermal irritants, the most potent irritants from the solvents in thinners, paints, adhesives, surfactants and detergents. Of all the skin irritants, solvents are the commonest. The reason for this is that the skin contains quite high levels of fats, and solvents, on contact with the skin, will remove the fat (this is called defatting). Defatting renders the skin dry, scaly and eventually cracked. Some solvents, such as white spirit (also called Stoddard solvent), acetone, trichloroethylene, defat the skin but do not penetrate it to any great extent. Other solvents, such as toluene, xylene, butanol and styrene (used in structural poly-esters) may cause skin irritation and irritant contact dermatitis. Some workers have also reported dermatitis to the materials (notably rubber) used in protective gloves and shoes.

Solvents and organic compounds that cause allergic contact dermatitis include wood turpentine, water-based paint preservatives (such as formaldehyde), and epoxy resins (particularly the low molecular weight epoxys). However, this con-dition is much less common than irritant contact dermatitis.

Solvents are also able to penetrate some of the materials used to make protec-tive equipment, so extra caution should be taken in the choice of gloves, aprons, and so on.

Solvents may also increase the toxicity of other materials by virtue of their ability to increase absorption. An example of this is that some resins are only mildly damaging to skin in their pure state (where they are absorbed into skin relatively slowly), whereas once dissolved in a solvent, their effects are increased because greater quantities will be absorbed across the skin.

The nervous system

The nervous system is comprised of the central nervous system (CNS) which is normally considered to be the brain and spinal cord, and the peripheral nervous system (PNS) which is considered to be those neuronal structures outside of the CNS (nerves, ganglia and support cells).

Further subdivisions of the brain include the cerebral cortex (which regulates many higher brain functions, including body movement and long-term memory), the hippocampus and limbic systems (which control 'strong emotions' and short-term memory) and the cerebellum (which coordinates movement). The main product of the central nervous system is integration of body functions and behaviour. These have been shown to be sensitive to the effects of a number of chemicals.

Solvents and the CNS

Of the volatile solvents, most have been shown to cause effects on behaviour and nervous system function. Most volatile solvent vapours typically produce signs of central nervous system disturbance.

Short-term effects

The short-term effects of solvents on the nervous system are tiredness, disorientation, a sense of intoxication, drowsiness, euphoria, dizziness, confusion and eventually, unconsciousness. Generally, these symptoms follow moderate to high short-term exposure, and will wear off following cessation of exposure. Extremely high levels of exposure can produce paralysis, convulsions, coma and eventually death.

A dramatic example of the acute effects of solvents on the central nervous system can be appreciated from the following situation. Two workers entered a tank to paint it. Being a confined space meant that there was a concentration of solvent from the paint in the air as there was poor ventilation. One of the workers climbed out because he felt dizzy. He then noticed that his workmate had collapsed on the floor of the tank. During a subsequent rescue operation 17 more people were overcome by the toluene vapours. On more than one occasion there were three men lying unconscious or semiconscious on the floor of the tank. Fortunately in this case rescue was eventually effected by drilling holes in the tank and inserting air hoses to disperse the vapours and no fatalities were recorded. This is a common scenario when the combination of solvent vapours and confined spaces occurs and has been the cause of many deaths. As can be appreciated this is another example of the mix of toxicity and exposure being required before a toxic response can be realised.

Long-term effects

The neuropsychological effects of long-term low-level exposure to solvents has been an area of attention, contention and concern, since this phenomenon was first reported in the early 1980s. There is now an appreciable literature on this topic. Often, the first symptoms are nonspecific or subjective, themselves difficult to critically evaluate. Effects including memory impairment, coordination impairment, deterioration of personality and depression have been indicated as important features of this condition. The study of the long-term effects of solvents on nerve function uses neurological, neurobehavioural or psychological evaluation. These specialised tests measure components of behaviour such as language comprehension, logical and spatial thinking, power of observation, coordination and memory. Results of these tests indicate that some functions are more affected by solvents than others.

In the last few years, the term 'solvent neurotoxic syndrome' has been coined to describe the major symptoms of nervous system effects caused by exposure to solvents. These symptoms include difficulties in concentration, forgetfulness, headaches, irritability, insensitivity, personality disorders, mental disabilities and even suicidal tendencies. Importantly, exposures in many reports suggest that exposures associated with this condition are apparently below recommended exposure standards.

While workplace exposure to solvents remains an issue, programs to reduce exposure as a result of research in the 1970s and 1980s has resulted in a reduction of solvent-related encephalopathy and other health effects (Baker 1994). More recent reports suggest the possibility of more than additive neurotoxic effects of solvents, especially in relation to potentiation and synergistic effects by exposure to ketones.

Solvents and the peripheral nervous system

Many of the neuronal processes of the CNS affected by solvents show similar effects in the PNS. However, one structure, the axon, is particularly at risk. The axon is a long protoplasmic process that extends from the neuronal cell body, and enervates various structures remote from the cell body. Many axons are short, but some are quite long. Nutrient supply down axons is by diffusion, but this is inadequate for long axonal processes. Therefore nutrient supply through the axon is also through specific neurotubule active-transfer mechanisms. If these processes are damaged, nutrient transfer is impaired, and the structures at the end of the axon can become nutrient deficient, and in extreme cases, can undergo impairment of function and neuropathological change, leading to death of the axon. Many toxic agents can affect these processes and cause axonopathy, including some solvents. The archetypal peripheral neurotoxicant is n-hexane.

Example: the toxicity of n-hexane

This aliphatic solvent will also be used to illustrate the importance of understanding various aspects of the toxicity of chemicals so that appropriate decisions can be made about their use and the use of those chemicals that are structurally related to them.

The uses of n-hexane have been varied; it is found in the printing industry, in vegetable oil extraction, in glues, paints, varnishes, plastics and rubber. Large numbers of workers have been exposed in the course of their work and the reports of human peripheral neuropathy are summarised in Table 14.8.

There is good correlation between the clinical observations and the observed experimental pathology. The disease process is well defined with a gradual onset and initial findings in the lower extremities. There is a swelling of the axons, demyelination and degeneration of the nerve fibres of the peripheral nervous

Year	Work activity
1964	Laminating process
1969	Glue in shoemaking
1971	Cabinet making
1980	Footwear manufacture
1999	Automotive technicians

Table 14.8 Reports of *n*-hexane peripheral neuropathy in workers

system. Once removed from exposure to *n*-hexane there is generally a recovery but it may take months or years. The biotransformation of *n*-hexane is an essential component in the eventual expression of toxicity. This was outlined in Figure 14.2.

The key features for the toxicity of this solvent are found in the γ-diketone structure of the metabolite 2,5-hexanedione. Other experimental studies have shown that it is this structural characteristic that is associated with this toxic effect. In Table 14.4 it can be seen that it is only when animals are treated with chemicals bearing this structure that the peripheral neuropathy is observed. Thus the 5- and 7-chain aliphatic hydrocarbons, pentane and heptane, do not result in peripheral neuropathy because they do not form γ-diketone metabolites. Therefore, a deeper appreciation of the structure involved in this toxic response is required in order that the associated issues are properly understood. Consideration of the biotransformation pathway also indicates that methyl *n*-butyl ketone, another chemical with considerable occupational use, would be expected to cause peripheral neuropathy, which is indeed the case.

From this example of *n*-hexane it can be seen that knowledge of the biotransformation of the chemical is essential to be able to understand what is responsible for the peripheral neuropathy and to be able to adequately predict which structurally related chemicals would or would not be expected to cause similar effects.

The liver

The liver is the largest internal organ in the body. As noted in the earlier section on biotransformation, the liver has a number of important functions, including metabolism of absorbed nutrients by the digestive system and biotransformation of exogenous chemicals. The most important reaction here is to convert what are generally fat-soluble substances into more water-soluble substances. By doing this, absorbed chemicals can be more readily excreted in urine. It is this last point that is critical for an understanding of the interaction of liver function with solvent exposure, as it is vulnerable to damage.

While a number of solvents have been shown to induce liver damage (hepatotoxicity) such chemicals are now all but withdrawn from use (for example, carbon tetrachloride, chloroform). The liver contains many biotransformation pathways, and absorbed chemicals may be changed to less toxic or more toxic products. Sufficient exposure to toxic chemicals and their biotransformation products can produce effects in the liver. This damage may impair the capacity of the liver to do its job properly (Dossing and Skinhoj 1985).

Many chemicals cause liver damage in a manner consistent with the dose–response relationship (that is, the higher the dose or exposure, the greater the effect). However, other chemicals do not (Zimmerman 1982). This is referred to as an idiosyncratic response, and may be due to immunological or metabolic abnormalities.

Liver damage can be detected with a range of procedures, from liver function tests, to external palpation to enlargement and tenderness, through to needle biopsy.

Elevated levels in one or more of these enzymes may indicate liver damage. Other enzymes and enzyme reaction products (such as bile acids and bilirubin) may also be used, although this is a specialised research area. Other measures of liver function include levels of uric acid, cholesterol or triglycerides, although these more likely indicate metabolic dysfunction, of which liver dysfunction is only one type.

Occupational exposure to solvents has been associated with significant liver toxicity. Also, the interaction of solvents with each other, from interoccupational exposure or nonoccupational exposure to solvents (such as alcohol consumption), may cause liver damage in solvent-exposed workers.

The kidneys

The kidneys form part of the urinary system. They have a number of functions concerning filtration of blood and excretory fluid (urine) production; they keep the body's fluids and electrolytes in balance; and they help to keep the acidity of the blood at a steady level.

For solvent exposure, the first point is important. Exposure to some chemicals will cause damage to the kidney (nephrotoxicity), which may in turn jeopardise excretory efficiency. This may be seen as a poorer ability of the body to excrete toxicants and their breakdown products or the inability of the body to retain chemicals that it needs (such as glucose, albumin and amino acids).

Solvents and the kidney

The most well-established solvent risk factors for kidney damage are the glycol ethers and some chlorinated solvents.

Reproductive effects

Two or three of every hundred babies born have some form of congenital anomaly at birth. While known environmental and occupational factors account for about 6%, the cause of about 70% of these congenital anomalies is unknown. The birth of an infant represents a significant point of the reproductive process, but a number of other factors (loss of libido, infertility, miscarriage, and so on) indicate the multifactoral and complex nature of the developmental and reproductive processes.

Solvents and reproduction

Of the chemicals that have been shown to cause reproductive effects, those commonly encountered as solvents and the effects they cause are shown in Table 14.9.

Of these, benzene is now still encountered as a minor contaminant in some thinners; formaldehyde is a minor component of some water-based paints and is therefore rarely encountered; and the propellant use of vinyl chloride in spray packs was discontinued in the early 1970s.

Evidence is also available from animal studies to indicate that:

- 2-ethoxyethanol, monomeric methacrylates and methyl ethyl ketone cause teratogenic effects
- dichloromethane, styrene, 1,1,1-trichloroethane, trichloroethylene, tetrachloroethylene and xylene cause foetotoxicity.

Cancer

The production of cancer by chemicals is a very special aspect of a toxicological reaction. While some cancer-causing chemicals (carcinogens) show a dose–response relationship, it is generally accepted that the uncertainties of the evidence and the seriousness of the outcome indicate that it is not possible to establish acceptable exposures for agents that cause cancer.

Also, while animal evidence is not necessarily indicative of an effect in humans, in the absence of data on humans, it is biologically plausible and prudent to regard agents for which there is sufficient evidence of carcinogenicity in experimental animals as if they represented a carcinogenic risk to humans.

Solvents and cancer

Organic solvents

The key body for the evaluation of carcinogenic risk to humans is the International Agency for Research into Cancer (IARC). Since 1971, IARC have published a substantial number of monographs detailing the published evidence for cancer-causing factors. IARC classify carcinogenic risk (see Chapter 11) as follows:

Table 14.9 Reproductive effects of solvents

Reproductive effects	Solvent(s)
Menstrual disorders	Toluene, styrene, benzene
Abortion or infertility	Formaldehyde, benzene
Testicular atrophy	2-Ethoxyethanol
Decreased libido and impotence	Vinyl chloride
Decreased foetal growth, low birth weight	Toluene, formaldehyde, vinyl chloride

- Group 1   agents which are carcinogenic to humans
- Group 2A   agents which are probably carcinogenic to humans
- Group 2B   agents which are possibly carcinogenic to humans
- Group 3   agents which are not classifiable
- Group 4   agents which are probably not carcinogenic to humans.

Solvents that have been subject to an IARC evaluation are shown in Table 14.10.

Of these, the most important is benzene. The use of this solvent is now virtually discontinued, though benzene may appear as an impurity in small concentrations in some thinners, and petrol can contain up to 5% benzene.

Table 14.10 IARC classifications for some solvents

Solvent	IARC group
Acrylonitrile	Group 2B
Benzene	Group 1
Bromoform	Group 3
Carbon tetrachloride	Group 2B
Chloroform	Group 2B
bis-Chloromethyl ether	Group 1
Cyclohexanone	Group 3
1,2-Dichloroethane	Group 2B
Dichloromethane	Group 2B
1,2-Dichloropropane	Group 2B
Dimethylformamide	Group 3
Dimethyl sulfate	Group 3
1,4-Dioxane	Group 2B
Ethanol (alcohol drinking)	Group 1
Ethyl acrylate	Group 2B
Ethylene	Group 3
Formaldehyde	Group 2A
Hydrogen peroxide	Group 3
Methyl acrylate	Group 3
Methyl chloride	Group 3
Methyl methacrylate monomer	Group 3
Mineral oils (untreated)	Group 1
Styrene	Group 2B
Tetrachloroethylene	Group 2A
Toluene	Group 3
1,1,1-Trichloroethane	Group 3
1,1,2-Trichloroethane	Group 3
Trichloroethylene	Group 2A
Vinyl chloride monomer	Group 1
Xylene	Group 3

The categorisation of ethanol as a Class 1 carcinogen is based principally on the production of cancer (mainly in the liver) in chronic alcohol abusers. This is a noteworthy example, as it raises issues of nonthreshold carcinogens, acceptable exposures and mechanisms of carcinogenesis (see Chapter 11).

Occupations involving solvent exposure

The IARC also evaluated the evidence regarding the association of some occupations with cancer, and noted that occupational exposure as in the petroleum refining industry, occupation as a painter, and occupation as a pesticide applicator are associated with human cancer (IARC 1989a, 1989b, 1991).

Interactions

Evidence is also available to suggest that exposures to mixtures of solvents have additive (for example, toluene and xylene) or synergistic (such as trichloroethylene and styrene) interactions. While many effects from exposure to solvents are additive, the effects from such exposures can cause considerably more problems than can be predicted from a combination of individual effects. Further, this enhancement of each other's effects may occur in ways that are not yet fully understood.

Specific toxic effects of solvents

As well as the general toxic effects of solvents there are also several important specific effects of particular solvents. In the past, solvents have been shown to cause a number of significant health effects, including:

- central nervous system damage (carbon disulfide)
- permanent peripheral nerve damage (n-hexane)
- liver and kidney damage (carbon tetrachloride)
- birth defects (cellosolve acetate)
- ocular effects (methanol)
- cancer in animals (chlorinated solvents, dioxane) or humans (alcohol)
- aplastic anaemia and leukaemia (benzene).

Apart from the general and specific toxicities listed above there are other toxic effects of particular groups of solvents that need to be considered. These are listed in Table 14.11 for easy reference. It should be noted that while the effects are generally related to the group some of them will be found with only some of the solvents within that group. More details can be found in the reference texts provided at the end of this chapter.

While the categories of solvent covered in this chapter outline the toxicity of the majority of solvents found in workplaces there are some others that bear mention as they have considerable toxicity. Amongst these are carbon disulfide and dioxane. Carbon disulfide has been clearly associated with neurotoxicity in humans producing classic symptoms including irritability, violence, sexual dif-

Table 14.11 Toxic effects of groups of solvents

Solvent group	Effects		
	Examples	Acute	Chronic
Aliphatic hydrocarbons	Petrol, kerosene, diesel, *n*-hexane	Nausea, pulmonary irritation, ventricular arrhythmia	Weight loss, anaemia, proteinuria, haematuria, bone-marrow hypoplasia
Aromatic hydrocarbons	Toluene, xylene, benzene	Nausea, ventricular arrhythmia, respiratory	Headache, anorexia, lassitude
Halogenated hydrocarbons	Carbon tetrachloride, dichloromethane, trichloroethane, trichloroethylene, tetrachloroethylene	Irritant, liver, kidney, heart	Fatigue, anorexia, liver, kidney, cancer*
Ketones	Acetone, methyl ethyl ketone, methyl *n*-butyl ketone	Irritant, respiratory depression	
Alcohols	Methanol, ethanol, isopropanol	Irritant, gastrointestinal	Liver, immune function
Esters	Methyl formate, methyl acetate, amyl acetate	Irritant, liver, palpitations	
Glycols	Ethylene glycol, diethylene glycol, propylene glycol	Kidney	Kidney
Ethers	Diethyl ether, isopropyl ether	Irritant, nausea	
Glycol ethers	Ethylene glycol monomethyl ether, ethylene glycol, monoethyl ether, propylene glycol monomethyl ether	Irritant, nausea, anaemia, liver, kidney, reproductive system	

*In experimental animals, nongenotoxic.

ficulties, insomnia and bad dreams. Dioxane is a cyclic ether which produces damage to various organs at high levels of exposure as well as some of the more common symptoms including irritation, headache and nausea. It has been found to be carcinogenic in studies with experimental animals but the applicability of these findings to humans remains contentious.

Summary

The solvents are a large group of commercially important chemicals. Many of the properties that make solvents useful products assist in making them harmful. The property of high evaporation rates causes high concentrations to be found in industrial environments; the solubility of solvent vapours in biological tissues causes a general toxicity which is generic to most solvents. Further, some individual solvents have toxic effects that cannot be predicted from structurally similar compounds.

There are a number of problems with solvent exposure:

- They are a heterogeneous group of chemicals.
- They have a widespread use.
- Because they have different applications and the method of use is basically through dispersion of vapours in workplace atmospheres they are not generally amenable to control measures (there are many different products, many solvents are highly volatile, there is a high potential for inhalational exposure, and there is a high potential for skin contact).
- Generally, there is poor availability of information on hazard (poor label information, disregard of label information, nonavailability of material safety data sheets or other information, variability of training within and across industry).
- There is also under-reporting of solvent-related disease (there are difficulties in diagnosis, there is a lack of recognition by health personnel, a variable onset of disease, and lack of reporting).

Lastly, there is the traditional problem of worker overfamiliarity with hazards and the perception that if no effects occurred acutely, then control measures are considered unnecessary.

References and further reading

Albers, J.W., Wald, J.J., Garabrant, D.H., Trask, C.L. and Berent, S. (2000) Neurologic evaluation of workers previously diagnosed with solvent-induced toxic encephalopathy. *J. Occup. Environ. Med.* 42: 410–423.

Baker, E.L. (1994) A review of recent research on health effects of human occupational exposure to organic solvents. *J. Occup. Med.* 36: 1079–1092.

Baker, E.L. (1988) Organic solvent neurotoxicity. *Ann. Rev. Pub. Health* 9: 223–232.

Bernstein, D.I. (1997) Allergic reactions to workplace allergens. *J. Am. Med. Ass.* 278: 1907–1913.

Boman, A. and Maibach, H.I. (2000) Percutaneous absorption of organic solvents. *Int. J. Occup. Environ. Health* 6: 93–95.

CEC (1986) *Organochlorine Solvents: Health Risks to Workers.* Commission of the European Communities/Royal Society of Chemistry, Brussels.

Chen, J.D., Wang, J.D., Jang, J.P. and Chen, Y.Y. (1984) Exposure to mixtures of solvents among paint workers and biochemical alterations of liver function. *Biochim. Biophys. Acta* 777: 75–83.

Collings, A.J. and Luxon, S.G. (eds) (1982) *Safe Use of Solvents.* New York: Academic Press.

Cone, L.E. (1986) Health hazards of solvents. *State Art Rev. Occup. Med.* 2: 297–330.

Dossing, M. and Skinhoj, P. (1985) Occupational liver injury. Present state of knowledge and future perspective. *Int. Arch. Occup. Environ. Health* 56: 1–21.

Grasso, P. (1988) Neurotoxic and neurobehavioural effects of organic solvents on the nervous system. *State Art Rev. Occup. Med.* 3: 525–539.

Harris, R.L. (ed.) (2000) *Patty's Industrial Hygiene*, 5th edn (4-volume set). New York: John Wiley and Sons.

IARC (1991) *IARC Monographs on the Evaluation of the Carcinogenic Risk of Chemicals to Man 53. Occupational Exposures in Insecticide Application and Some Pesticides.* International Agency for Research on Cancer, Lyons.

IARC (1989a) *IARC Monographs on the Evaluation of the Carcinogenic Risk of Chemicals to Man 45. Occupational Exposures in Petroleum Refining; Crude Oil and Major Petroleum Fuels.* International Agency for Research on Cancer, Lyons.

IARC (1989b) *IARC Monographs on the Evaluation of the Carcinogenic Risk of Chemicals to Man 47. Some Organic Solvents, Resin Monomers and Related Compounds, Pigments and Occupational Exposures in Paint Manufacture and Painting.* International Agency for Research on Cancer, Lyons.

IPCS (1992) *1,1-Trichloroethane. Environmental Health Criteria* 136: International Program on Chemical Safety, WHO, Geneva.

IPCS (1993) *Benzene. Environmental Health Criteria* 150: International Program on Chemical Safety, WHO, Geneva.

IPCS (1987) *Butanols, Four Isomers. Environmental Health Criteria* 65: International Program on Chemical Safety, WHO, Geneva.

IPCS (1994) *Chloroform. Environmental Health Criteria* 163: International Program on Chemical Safety, WHO, Geneva.

IPCS (1985) *Dimethyl Sulphate. Environmental Health Criteria* 48: International Program on Chemical Safety, WHO, Geneva.

IPCS (1989) *Formaldehyde. Environmental Health Criteria* 89: International Program on Chemical Safety, WHO, Geneva.

IPCS (1997) *Methanol. Environmental Health Criteria* 196: International Program on Chemical Safety, WHO, Geneva.

IPCS (1992) *Methyl Ethyl Ketone. Environmental Health Criteria* 143: International Program on Chemical Safety, WHO, Geneva.

IPCS (1990) *Methyl iso-Butyl Ketone. Environmental Health Criteria* 117: International Program on Chemical Safety, WHO, Geneva.

IPCS (1984) *Methylene Chloride. Environmental Health Criteria* 32: International Program on Chemical Safety, WHO, Geneva.

IPCS (1991) *n-Hexane. Environmental Health Criteria* 122: International Program on Chemical Safety, WHO, Geneva.

IPCS (1983) *Styrene. Environmental Health Criteria* 26: International Program on Chemical Safety, WHO, Geneva.

IPCS (1984) *Tetrachloroethylene. Environmental Health Criteria* 31: International Program on Chemical Safety, WHO, Geneva.

IPCS (1986) *Toluene. Environmental Health Criteria* 52: International Program on Chemical Safety, WHO, Geneva.

IPCS (1985) *Trichloroethylene. Environmental Health Criteria* 50: International Program on Chemical Safety, WHO, Geneva.

IPCS (1997) *Xylenes. Environmental Health Criteria* 190: International Program on Chemical Safety, WHO, Geneva.

Iregren, A. (1996) Behavioral methods and organic solvents: Questions and consequences. *Environ. Health Persp.* 104 (Suppl. 2): 363–366.

Juntunen, J. (1993) Neurologic solvent poisoning in industry: A review. *Environ. Res.* 60: 98–111.

Klaassen, C.D. (ed.) (2001) *Casarett and Doull's Toxicology. The Basic Science of Poisons*, 6th edn. New York: McGraw-Hill.

MacFarland, H.N. (1986) Toxicology of solvents. *Am. Ind. Hygiene Ass. J.* 47: 704–707.

Meeks, R.G., Harrison, S.D. and Bull, R.J. (1991) *Hepatotoxicity*. London: CRC Press.

Mikkelsen, S. (1997) Epidemiological update on solvent neurotoxicity. *Environ. Res.* 73: 101–112.

Morrow, L.A., Ryan, C.M., Hodgeson, M.J. and Robin, N. (1990) Alterations in cognitive and psychological functioning after organic solvent exposure. *J. Occup. Med.* 32: 444–450.

Noraberg, J. and Arlien-Soborg, P. (2000) Neurotoxic interactions of industrially used ketones. *Neurotoxicology* 21: 409–418.

Riihimaki, V. and Ulfvarson, U. (eds) (1986) *Safety and Health Aspects of Organic Solvents. Progress in Clinical and Biological Research, Volume 220.* New York: Alan R. Liss.

Rinsky, R.A. (1987) Benzene and leukemia: An epidemiologic risk assessment. *N. Engl. J. Med.* 316: 1044.

Salvaggio, J.E., Butcher, B.T. and O'Neil, C.E. (1986) Occupational asthma due to chemical agents. *J. All. Clin. Immunol.* 78: 1053–1058.

Schenker, M.S. and Jacobs, J.A. (1996) Respiratory effects of organic solvent exposure. *Tuberculos. Lung Dis.* 77: 4–18.

Wedeen, R.P. (1997) Occupational and environmental renal disease. *Sem. Nephrol.* 17: 46–53.

White, R.F. and Proctor, S.P. (1997) Solvents and neurotoxicity. *Lancet* 349: 1239–1243.

Williamson, A.M. and Winder, C. (1993) A prospective cohort study of the chronic effects of solvent exposure. *Environ. Res.* 62: 256–271.

Wypych, G. (ed.) (2000) *Handbook of Solvents*. Toronto: ChemTech Publishing.

Zimmerman, H.J. (1982) Chemical hepatic injury and its detection. In: Plaa, G.A. (ed.) *Toxicology of the Liver*. New York: Raven Press.

Chapter 15

Toxicity of gases, vapours and particulates

C. Winder

Introduction

There is a huge range of airborne contaminants in workplace air – vapours, fumes, mists, dusts and fibres. In common usage, these terms require definition:

- **gas** is a term usually applied to a substance that is in the gaseous state at room temperature and pressure
- **vapour** is applied to the gaseous phase of a material that is ordinarily a solid or liquid at room temperature and pressure
- **aerosol** is applied for a relatively stable suspension of solid particles in air, liquid droplets in air or solid particles dissolved or suspended in liquid droplets in air
- **mists** and **fogs** are aerosols of liquid droplets formed by condensation of liquid droplets on particulate nuclei in the air
- **fumes** are solid particles formed by combustion, sublimation or condensation of vaporised material (conventionally, the size of fume particles are considered to be less that 0.1 μm, though the size can increase by aggregation or flocculation as the fume ages)
- **dusts** are solid particles in air formed by grinding, milling or blasting
- **fibres** are solid particles with an increased **aspect ratio** (the ratio of length to width); they have special properties because of their ability to be suspended in air for longer periods than dusts and other aerosols.

These can be divided into two main types of contaminant:

- those that are **dissolved** in air, such as gases and vapours
- those that are **suspended** in air, such as fumes, dusts, mists, aerosols, fibres (also called particulates).

Dissolved contaminants will reach, and can have effects in, all parts of the respiratory system.

Suspended particles may be inspired into the respiratory system, although the depth that they can reach is a function of their size, measured as the mean aerodynamic diameter of the particle.

The toxicity of atmospheric contaminants

There are a number of factors that can effect the severity of inhaled gases and vapours, fumes, mists and particles:

- size of the inhaled contaminant (for particulates)
- solubility (ability to dissolve in tissue fluids)
- reactivity (ability to react with tissue components); included in reactivity is the general property of oxidative burden, caused by the generation of free radicals
- conditions of exposure, such as concentration and duration of exposure
- lung defences, genetic or acquired
- exercise/rest state
- immunological status
- tissue water content.

Of these, size and solubility are the most important intrinsic properties, and conditions of exposure relate to risk.

Figure 4.10 (in Chapter 4) shows the size of various atmospheric contaminants. Only those below a size of 5 μm are considered respirable, that is they can reach into the deep portions of the respiratory system, where the alveoli are located.

Gases and vapours (dissolved airborne contaminants)

Gases are one of the three fundamental states of matter, with fundamental properties of no fixed form or volume. They share this property with vapours (the gaseous phase of a liquid). Some vapours have significant potential to cause problems in workplaces, because they can reach high concentrations. It is therefore appropriate to give such vapours the same consideration as gases.

Gases have the property of being extremely easy to handle. They can be enclosed, compressed, liquefied, containerised, piped, and so forth, without too much technological difficulty. While contained, they present little risk to users or other potentially exposed individuals.

However, in that there is no physical barrier between the air and the various structures that make up the human respiratory system, gases can enter the body very easily and quickly. So risks can arise once gases are released and available for exposure (whether occupational, domestic or environmental).

Toxicity of gases and vapours

A significant number of gases that can be encountered in the workplace environment, including asphyxiants, irritants, sensitisers and toxic gases. The toxic effects of gases and vapours are manifold.

Asphyxiation

This is the ability of a gas or vapour to displace oxygen from air by dilution (simple asphyxiation) or by interfering with the ability of the body to transport oxygen (toxic asphyxiants).

Irritation

This is the ability of a gas or vapour to cause local symptoms of irritation upon exposure or contact with the tissues of the respiratory system.

Sensitisation

The ability of a gas or vapour to cause an immune response upon exposure or contact with the tissues of the respiratory system leading to respiratory sensitisation and asthma.

Toxicity

The ability of a gas or a vapour to produce a toxic effect either at the site of contact (that is, the respiratory system), or following absorption and distribution to other tissues or organ systems.

While the target organ/system is the respiratory system, these gases can cause a range of health effects in a number of body systems. These effects include effects in the nervous system, blood-forming tissues, and the liver (see Table 15.1).

Asphyxiants

Simple asphyxiants

- **Carbon dioxide (CO_2)** is produced by living biological materials as part of normal respiration. It is also formed from the complete combustion of carbon-containing materials:

$$C + O_2 \Rightarrow CO_2$$

Carbon dioxide is a common cause of asphyxiation, although other toxic effects are observed at lower concentrations. The most notable of these is respiratory acidosis, where the carbon dioxide dissolves in biological fluids to form a weak solution of carbonic acid (H_2CO_3):

$$CO_2 + H_2O \Rightarrow H_2CO_3$$

Table 15.1 Gases and vapours, and the body systems they affect

Organ systems affected	Examples
Respiratory system	
Irritants	Chlorine (Cl_2)
	Ammonia (NH_3)
	Oxides of nitrogen (NO_x)
	Sulfur dioxide (SO_2)
	Sulfur trioxide (SO_3)
	Fluorine (F_2)
	Phosphine (PH_3)
	Phosgene ($COCl_2$)
	Ozone (O_3)
	Formaldehyde (CH_2O)
	Acrolein
Corrosives	Acid mists
	Caustic mists
Asphyxiants	Nitrogen (N_2)
■ Simple	Hydrogen (H_2)
	Methane (CH_4)
	Helium (He_2)
	Argon (Ar_2)
	Neon (Ne_2)
	Ethylene (C_2H_4)
	Ethane (C_2H_6)
■ Toxic	Carbon monoxide (CO)
	Hydrogen cyanide (HCN)
	Hydrogen sulfide (H_2S)
Sensitisers	Isocyanates ($-N{=}C{=}O$)
	Amines ($-C-NH_2$)
Central nervous system	Carbon disulfide (CS_2)
	Aliphatic hydrocarbons
	Solvent vapours
Blood-forming system	Arsine (AsH_3)
Carcinogens	Vinyl chloride (C_2H_3Cl)
	Nickel carbonyl (NiCO)
	Formaldehyde (CH_2O)
	Sulfuric acid mists (H_2SO_4)

- **Nitrogen (N_2)**
- **Inert gases** such as helium, argon and neon.
- **Low molecular weight, straight chain hydrocarbons** (C_1 to C_4) including aliphatic compounds, such as methane CH_4, ethane CH_3CH_3, ethene CH_2CH_2, propane $CH_3CH_2CH_3$, or butane $CH_3CH_2CH_2CH_3$.

Toxic asphyxiants

Carbon monoxide

Carbon monoxide (CO) is produced from the partial combustion of carbon-containing materials, for example, fuel, coal, coke, solvents:

$$2C + O_2 \Rightarrow 2CO$$

The gas has a strong affinity for the oxygen-containing protein haemoglobin, and will inhibit the normal oxygen-carrying capacity of the blood. It is a chemical asphyxiant. It is the single largest cause of poisoning deaths in the USA, accounting for up to 4000 fatalities annually. Low levels may affect the cardiovascular system, and CNS effects can occur owing to cerebral hypoxia.

Cyanides (typically HCN)

Hydrogen cyanide (HCN) is a gas associated with the smell of bitter almonds. HCN and salts of cyanide (CN^-) are encountered in electroplating, ore extraction, photography, and the chemical industry. Cyanide is another toxic asphyxiant. The mechanism of action is different to that of carbon monoxide in that it binds to the cytochrome enzyme chain, inhibiting the process of cellular metabolism. It is one of the most rapidly acting lethal poisons.

Hydrogen sulfide

Hydrogen sulfide (H_2S) is a gas commonly associated with the smell of rotten eggs. Though it has excellent odour-warning characteristics, these habituate very quickly and a slowly rising concentration may hardly be noticed. The gas is encountered in poorly ventilated spaces such as underground sewers or liquid manure tanks. Hydrogen sulfide is another chemical asphyxiant, disrupting energy metabolism similar to (but more potent than) hydrogen cyanide. Encephalopathy from hypoxia is possible.

Irritants

Gases

Ammonia

Ammonia (NH_3) (CAS number 7664-41-7) is a colourless gas (boiling point −34°C) with a characteristic pungent smell. Salts of ammonia have been used as 'smelling salts'.

Ammonia is produced using a variety of methods, including condensation of nitrogen compounds extracted from fossil fuels and decomposing ammonia compounds. However, the Haber process is the one most often used industrially.

In this process, nitrogen (N_2) and hydrogen (H_2) gases are mixed under high temperature (500°C) and pressure (250 atmospheres):

$$N_2 + 3H_2 \Rightarrow 2NH_3$$

Ammonia is used in fertiliser production, nitric acid manufacture, refrigeration, explosives manufacture, plastics production and oil refining.

The hazards of ammonia are related to its extreme solubility in water. Ammonia and water combine to make the alkaline solution ammonium hydroxide (NH_4OH):

$$NH_3 + H_2O \Rightarrow NH_4OH$$

In biological systems, ammonia will dissolve in the first water it encounters. In people, this will be tears in the eye, and the mucous lining in the upper respiratory system. As an alkaline material is being formed, the basic response is of extreme irritation.

Therefore ammonia is essentially an upper airway irritant. Effects from single intense exposures include intense pain in the eyes, mouth and throat, loss of voice and development of stridor (high-pitched noise in breathing). Longer single exposures will produce ulceration of the mouth and nose, cyanosis and pulmonary oedema. Death usually results from asphyxiation. No long-term effects have been reported.

Ammonia has an odour threshold of about 5 ppm, and an exposure standard of 2.5 ppm.

Chlorine

Chlorine (Cl_2) (CAS number 77282-50-5) is a heavy greenish-yellow gas or boiling yellow liquid with a characteristic pungent irritating odour.

It is a highly reactive gas with many uses. Large-volume industrial uses include chemical manufacturing, water sterilisation, paper manufacturing and as a disinfectant and bleaching agent. Chlorine is now made almost entirely by the electrolysis of alkali chlorides (for example, in brine) using diaphragm or mercury cells. The reaction also yields sodium hydroxide (NaOH) and hydrogen gas:

$$2NaCl + H_2O \Rightarrow 2NaOH + H_2 \nearrow + Cl_2 \nearrow$$

Chlorine is only moderately soluble in water, producing a weak solution of hypochlorous acid (HOCl) and hydrochloric acid (HCl):

$$Cl_2 + H_2O \Rightarrow HOCl + HCl$$

The hypochlorous acid dissociates into hypochlorite ions ($-OCl^-$) which readily penetrate cell walls. Once in the cell, these ions decompose to hydrochloric acid derivatives that gradually destroy the cell.

Chlorine is a severe respiratory and skin irritant. Exposure–response relationships exist:

- About 5 ppm   Eye irritation
- Above 15 ppm   Throat irritation
- 15–30 ppm   Cough, choking, burning
- Above 50 ppm   Pneumonitis
- 430 ppm   Death from 30 minutes' exposure
- Above 1000 ppm   Death within minutes.

Other effects from short-term exposure include chest pain, dyspnoea (irregular breathing), production of white or pink sputum, sore and reddened conjunctiva, coarse wheezes and crackles when breathing. Pathological changes include bronchospasm, swelling and ulceration of mucosa with desquamation and pulmonary oedema (either immediate or delayed several hours). No long-term effects have been reported, even from severe overexposure.

Chlorine has an odour threshold of about 0.5–1 ppm, and an exposure standard of 1 ppm

Other halogen gases

Fluorine

Fluorine (F_2) (CAS number 7782-41-4) is a very soluble, heavy greenish-yellow gas with a low boiling point of $-188°C$. It is produced by the electrolysis of a solution of hydrogen fluoride (HF – hydrofluoric acid) and potassium fluoride (KF).

Fluorine is a highly reactive material, which is used as an oxidising agent and in uranium extraction. The gas has no major uses, and has been replaced by hydrofluoric acid wherever possible.

The major hazard is that the gas is highly corrosive and reactive. The gas is extremely soluble, and dissolves in water to form hydrofluoric acid (HF):

$$2F_2 + 2H_2O \Rightarrow 4HF + O_2$$

The dose–response relationship for fluorine is:

- 0.5 ppm   Eye/throat irritation
- 1.5–3 ppm   Choking
- 2–5 ppm   Chest pain
- 4–8 ppm   Severe upper respiratory tract irritation
- 30 ppm   Death.

Other effects include skin burns and ulceration. Long-term effects have been reported, and include nervous system and thyroid disorders. Fluorosis has also been suggested.

Fluorine has an odour threshold of 0.14 ppm, and an exposure standard of 1 ppm.

Bromine

Bromine (Br_2) (CAS number 7726-95-6) is a slightly reddish brown fuming liquid with a melting point of $-7°C$. It is used in the production of petrol antiknock enhancers (ethylene dibromide), in chemical manufacture, in dyes, photography and gold extraction.

Vapours of bromine are corrosive and severely irritating. Bromine dissolves in water to form bromide ions (Br^-), which then displace physiological chloride ions (Cl^-). The exposure–response relationship looks like:

- 5–20 ppm   Eye/throat irritation
- 25 ppm   Chest pain
- 100 ppm   Bronchospasm
- 100 ppm   Delayed onset oedema.

Long-term effects include liver and kidney damage.

Bromine has an odour threshold of 0.05 ppm, and an exposure standard of 0.1 ppm.

Nitrogen oxides

The oxides of nitrogen (NO_x) are usually variable in chemical composition, but usually comprise mixtures of NO, NO_2 and N_2O_4. The generic term NO_x is often used to describe such mixtures. The oxides of nitrogen are listed in Table 15.2. Oxides of nitrogen are produced from a number of processes, including:

- reactions involving nitric acid or nitrates
- explosives manufacture
- welding
- from diesel-powered machinery, especially in confined spaces (for example, in mines)
- agricultural silos (as a side reaction in anaerobic environments)
- from decomposing organic matter.

The major occupational hazard relates to nitrogen dioxide (NO_2), which is a heavy red–brown gas with a pungent odour. This dissolves in water to produce a solution of nitric (HNO_3) and nitrous (HNO_2) acids:

$$N_2O_4 \Leftrightarrow 2NO_2 + H_2O \Rightarrow HNO_3 + HNO_2$$

The nitrous acid subsequently decomposes to more nitric acid.

Nitrogen dioxide is a respiratory irritant. Commonly reported symptoms include mild upper respiratory tract symptoms at low concentrations, progressing to intense coughing and choking at moderate concentrations (sufficient to force the worker to leave).

Clinically, people receiving short-term intense exposures report mucoid or frothy sputum, increasing irregular breathing, reduction in lung function, wheezes

Table 15.2 The oxides of nitrogen

Name	Formula	Properties	Reactivity
Nitric oxide CAS No. 10112-43-9	NO	Colourless gas MP $-164°C$ BP $-152°C$	Produced in many reactions involving the reduction of nitrogen-containing materials; moderately reactive
Nitrogen dioxide CAS No. 10102-44-0	NO_2	Brown gas MP $-11°C$	Exists in a temperature-dependent equilibrium with N_2O_4; reactive
Nitrous oxide CAS No. 10024-97-2	N_2O	Colourless gas MP $-91°C$ BP $-89°C$	Unreactive; nonirritating; used as an anaesthetic
Dinitrogen trioxide	N_2O_3	Dark blue gas FP $-101°C$	Extensively dissociated as gas, dissociates to NO and NO_2
Dinitrogen tetroxide CAS No. 10544-72-6	N_2O_4	Colourless gas BP $-21°C$	Dissociates to NO_2 as gas and partly as liquid
Dinitrogen pentoxide	N_2O_5	Colourless solid MP $30°C$ BP $-152°C$	Unstable as vapour; ionic solid
Nitrogen trioxide $\Leftrightarrow$ dinitrogen pentoxide	$NO_3 \Leftrightarrow N_2O_6$	MP $-164°C$ BP $-152°C$	Unstable, and not well characterised

and crackles in breathing, frank pulmonary oedema within 24 hours (with a nodular component on X-rays, cyanosis, tracheitis). Pathologically, there is extensive damage to the mucosal cells, inflammatory cell exudation, dilated alveolar capillaries, and alveoli full of fluid and blood cells. Deaths usually occur from pulmonary failure. Sometimes slight cases that have recovered relapse after 2–3 weeks to severe cases, including fever and chills.

There is some suggestion of emphysema to prolonged low-level exposure.

The odour threshold for NO_2 is 0.4 ppm Exposure standards exist for NO (25 ppm) and NO_2 (3 ppm).

Oxides of sulfur

There are two – sulfur dioxide (SO_2; CAS number 7446-09-5) and sulfur trioxide (SO_3). Mixtures of the two are called SO_x.

Sulfur dioxide is a heavy colourless gas with a suffocating odour, a melting point of $-76°C$ and a boiling point of $-10°C$. The gas is produced from many processes, either natural or synthetic, when sulfur-containing materials are burned. Industrial emissions arise from a number of principal sources:

- combustion of fossil fuels, such as coal and oil for power generation and other purposes
- smelting or ores containing sulfur (particularly those containing copper, lead, mercury, silver, cadmium and zinc)
- oil refinery operations (mainly ore roasting)
- coke production
- sulfuric acid manufacture.

Sulfur trioxide is obtained by the slow reaction of sulfur dioxide and oxygen:

$$2SO_2 + O_2 \Rightarrow 2SO_3$$

Small quantities of SO_3 accompany SO_2 emissions (up to 4–5% of the sulfur burnt). Sulfur trioxide will then combine with water to produce sulfuric acid (H_2SO_4):

$$SO_2 + H_2O \Rightarrow H_2SO_4$$

This can occur in industrial situations, where the acid mist can corrode ducting, baghouses and other equipment, although the main concern of this reaction is in the environment, where the production of acid rain in areas down wind from emission sources can cause significant damage.

The oxides of sulfur are fairly readily soluble in water, and therefore are primarily upper airway irritants. A small exposure–response relationship has been noted:

- 1–5 ppm   Mild bronchoconstriction
- 5 ppm   Alteration in lung function
- 5–10 ppm   Some cases of bronchospasm.

Other short-term effects include rapid and extreme irritation of the eyes, nose, mouth and upper airways (sufficient to force the worker to leave), cough, and marked vasoconstriction. Inhaled SO_2 is removed slowly from the respiratory tract (takes more than a week).

There is evidence that long-term low-level exposure increases airways resistance, and exacerbates chronic obstructive pulmonary disease at prolonged exposure at low levels. (It has been estimated that low-level exposure to three irritants, NO_x, SO_x and possibly ozone, is a significant cause of chronic bronchial disease in the general public.)

Halogen acid gases

HCl, HF, HBr are covered in acid mists (see below).

Ozone

Ozone (O_3) (CAS number 7782-44-7) is a colourless gas with a characteristic smell. It is produced from heavy electrical equipment such as photocopiers and X-ray machines, and in arc welding. It is also found at high altitudes.

The gas is only slightly soluble. The breakdown of ozone to oxygen produces short-lived, but highly reactive free radicals. These free radicals contribute to per-oxidation reactions, which contribute to the senescence process. Short-term effects include cough, nose and eye irritation, chest tightness, headache. Severe cases have not been reported, and while pulmonary oedema has been observed from long-term exposure in animal experiments, similar effects have not been reported in humans.

Ozone has an odour threshold of 0.05 ppm and an exposure standard of 0.1 ppm.

Phosgene

$$O=C=C \begin{matrix} \diagup Cl \\ \diagdown Cl \end{matrix}$$

First used in trench warfare in World War I, phosgene ($COCl_2$) (CAS number 75-44-5) is a heavy colourless gas (boiling point 8°C) with a faint odour of newly mown hay. The gas is produced from the reaction of carbon monoxide and chlorine and, for all its hazards, is still used as a reaction intermediate in the chemical industry to make dyes, isocyanates and pesticides. The gas may also be produced in the combustion of chlorine-containing organic material (for example, welding containers that contain chlorinated solvents, or smoking in an atmosphere of vapours of chlorinated hydrocarbons).

Phosgene has a low solubility in water, and is broken down to form hydrochloric acid and carbon dioxide. The main hazard of phosgene is that it is an irritant.

Phosgene has an odour threshold of 1 ppm, and an exposure standard of 0.1 ppm.

Vapours

Aldehydes

The principal chemical property of the aldehydes is the presence of at least one aldehyde (—CHO) group.

Aldehydes normally have good warning properties, are lacrimators (cause the eyes to water), are directly irritating on contact, are generally lipid soluble, and cause toxicity by altering cell membrane lipid function or by denaturing membrane proteins.

As a general group, aldehydes cause:

- contact dermatitis (either of the allergic or irritant types)
- bronchospasm in sensitive individuals
- strong conjunctival, corneal and respiratory tract irritation
- tracheobronchitis and pulmonary oedema in severe inhalational exposures.

Formaldehyde

$$O=C\overset{\displaystyle H}{\underset{\displaystyle H}{}}$$

Formaldehyde (CHO) (CAS number 50-00-0) is a colourless gas with an unpleasant odour. It is used for a variety of purposes, including embalming, as a biocide, and in chemical and plastics manufacture. Formaldehyde is very soluble in water.

Formaldehyde is an irritant and sensitiser, and information on carcinogenicity continues to increase. Its main short-term effects include tears, nasal irritation, cracking of nasal tissue, sneezing, sore throat, chest constriction and headache. Long-term effects include attenuation of short-term symptoms due to decreased sensitivity. Formaldehyde also produces asthma and dermatitis.

The odour threshold for formaldehyde is 0.8 ppm, the exposure standard is 1 ppm

Other aldehydes include materials such as acetaldehyde, glutaraldehyde and acrolein.

Acetaldehyde

$$O=C-C-H$$ (with H substituents)

Glutaraldehyde

$$O=C-C-C-C-C=O$$ (with H substituents)

Acreolin

$$O=C-C=C$$ (with H substituents)

Acid mists

While strictly speaking, acid mists are not dissolved in air, their effects are of the local irritation type. Basically, they exert their effect by contact of an acid with tissue, and their effects can be predicted from the pH of the acid and conditions of exposure. The main direct effect of contact with acids is caused by the high levels of hydrogen ions present. At high concentrations the effects are related to corrosion of tissue. This will abate as the level of acidity falls to tissue pH levels. At lower exposures there are a range of effects, including membrane breakdown, protein denaturing, cell death and necrosis. Anions of some acids may also have a toxicity of their own (for example, fluorosis from contact with dilute solutions of hydroflouric acid).

Long-term exposure will lead to tissue damage, such as desquamation, loss of mucosal layer, perforation of the nasal septum and tissue corrosion.

Exposure to sulfuric acid mists was categorised as a human carcinogen by the International Agency for Research on Cancer (IARC) in 1992.

Caustic mists

As with acid mists, irritancy and corrosion effects can be predicted from the pH of the alkali and conditions of exposure. Some alkalis, such as caustic soda (NaOH) and caustic potash (KOH) have the ability to diffuse through tissue and cause deep wounds.

Irritant and toxic fumes

- metal fumes that produce metal fever
- polymer fumes that produce polymer fume fever
- metal fumes producing bronchitis and pneumonitis
- cadmium lung.

Sensitisers

Isocyanates

The isocyanates, of which toluene di-isocyanate (TDI) is the most common, are widely used in the production of polyurethane. They contain the reactive $-N{=}C{=}O$ group which combines with other organic structures. A common reaction is the reaction of an isocyanate with the hydroxyl group of an alcohol to form a urethane:

$$R-N + C + O + R'OH \Rightarrow R-NHCOOR'$$

For example, in the production of a polyurethane resin, the reaction is generally between a diisocyanate and polymeric hydroxyl compounds such as polyether or polyester resins:

$$R-(N{=}C{=}O)_2 + 2\,R'-(OH)_2 \Rightarrow R-(NHCOOR'-OH)_2$$

Examples of isocyanates include:

- Hexamethylene diisocyanate (HDI) (asthma)
- Isophorone diisocyanate

- Methylene bisphenyl isocyanate (MDI) (asthma)
- Methyl isocyanate (Bhopal disaster)
- Methyl isothiocyanate (CNS seizures, coma, death)
- 1,5-Naphthalene diisocyanate (NDI)
- Toluene diisocyanate (TDI) (irritant, sensitiser, asthma)

The asthma induced by isocyanates generally resolve following removal from exposure, though it can be persistent in some sensitised workers.

Aldehydes

Aldehydes (see above) may also have sensitisation properties.

Toxic gases

Carbon disulfide

Carbon disulfide (CS_2) is a clear colourless liquid with a characteristic aromatic odour. It has been in use for over 100 years, and the incidence of chronic carbon disulfide toxicity is now rare. However, neuropsychiatric disorders, peripheral neuropathies, and so on, may occur following long-term low-level exposure.

Oxygen

One nonoccupational (and remarkable) toxicity is that of high levels of oxygen (O_2) as part of oxygen therapy in infants. This causes diffuse pulmonary damage characterised by damage to the alveoli. In animals exposed to 95–100% O_2 the damage is sufficient to cause death within 3–4 days.

Summary: gases and vapours

Gases and vapours have both local effects in the respiratory tract and systemic effects after inhalation. Therefore, the respiratory tract is both a point of entry and a target organ of toxicity. Local adverse effects relate mainly to irritation, although increased susceptibility to allergy, infection, structural disease and cancer also occur in the respiratory tract in response to gas/vapour/fume exposure. There are a number of systemic effects, although the main effect is asphyxiation.

At the level of exposure to gases, the main controls are by somehow placing a barrier between the gas and the respiratory system. As with the control of all hazards, the hierarchy of controls should be applied, and elimination, substitution and isolation are preferable over measures that dilute gas concentrations (such as ventilation). The final control that should be considered is the use of respiratory protection.

Particulates (suspended airborne contaminants)

In this chapter, particulate is a term used to describe all materials (whether liquid or solid) suspended in air.

Classification of particulates

There are many different types of particulates found in workplaces, and they can be broadly classified as being either organic or inorganic. Organic dusts originate from plant and animal materials, but also from synthetic material of an organic nature (for example, chemical intermediates or plastics). Inorganic particulates can be further subdivided as metallic or nonmetallic, and they can be divided by whether they have silica present.

A more fuller classification is shown in Figure 15.1, with a list of common particulates in Table 15.3.

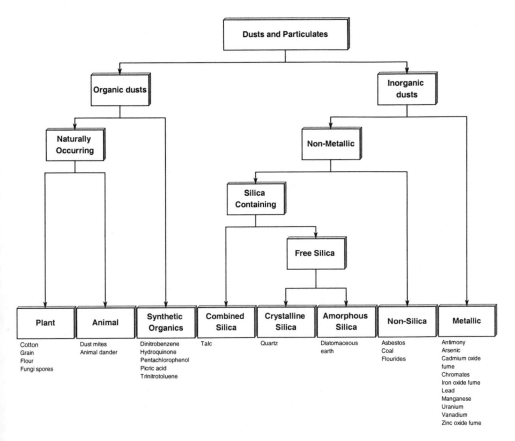

Figure 15.1 Classification of particulates.

Table 15.3 Common particulate toxicants

Particulate	Sources	Toxic effects/diseases
Asbestos	Mining, manufacture of asbestos products, construction, ship building	Asbestosis, pleural plaques, lung cancer, mesothelioma
Aluminium dust and abrasives	Manufacture of aluminium products, ceramics, paints, electrical goods, fireworks, abrasives	Aluminosis, alveolar oedema, interstitial fibrosis
Beryllium	Mining and extraction, alloy manufacture, ceramics	Berylliosis, pulmonary oedema, pneumonia, granulomatosis, lung cancer, cor pulmonale
Cadmium (oxide)	Welding, manufacture of electrical goods, pigments	Pneumonia, emphysema, cor pulmonale
Chromium[VI]	Manufacture of chromium compounds, pigment manufacture, tanneries	Bronchitis, fibrosis, lung cancer
Coal dust	Coal mining	Fibrosis, coal miner's pneumoconiosis
Cotton dust	Textile manufacture	Byssinosis
Iron oxides	Hematite mining, iron and steel production, welding, foundry work	Siderosis, diffuse fibrosis-like pneumoconiosis
Kaolin	Pottery manufacture	Kaolinosis, fibrosis
Manganese	Alloy production, chemical industry	Manganism, manganese pneumonia
Nickel	Mining, production, electroplating	Pulmonary oedema, lung cancer, nasal cavity cancer
Silica	Mining and quarrying, stone cutting, construction, sand blasting	Silicosis, fibrosis, silicotuberlosis
Talc	Rubber industry, cosmetics	Talcosis, fibrosis
Tantallum carbide	Manufacture and sharpening of cutting tools	Hard metal disease, hyperplasia of bronchial epithelium, fibrosis
Tin	Mining, tin production	Stanosis
Tungsten carbide	Manufacture and sharpening of cutting tools	Hard metal disease, hyperplasia of bronchial epithelium, fibrosis
Vanadium	Steel manufacture	Irritation, bronchitis

The toxicity of particulates

While the main risk of exposure to particulates is to the respiratory system, some particulates (for example, nickel) can also cause skin problems on contact. Still other dusts can have soluble components that can cause systemic toxicity (such as kidney damage from cadmium or nervous system damage from lead).

The size of an airborne contaminant is important in inhalational toxicology. Airborne particles can have a range of masses, densities, shapes and sizes, and these will bear on the amount of time the particle will be airborne. From the time it is generated, the particle will be subject to forces of:

- intrinsic kinetic energy from any forces that arose during the process of particle generation (for example, particles from a grinding wheel)
- air movement at the source of particle generation (such as air flow from ventilation systems)
- diffusion
- air resistance
- gravity.

Dimensional considerations in particulate size characteristics

All dimensions increase (radius and diameter, surface area, volume) as particulate size increases. However, they do not all increase at the same rate. Area is the square of the radius of a spherical particulate ($4\pi r^2$), and volume is its cube ($4/3\pi r^3$). These changes in dimensions are shown in Figure 15.2.

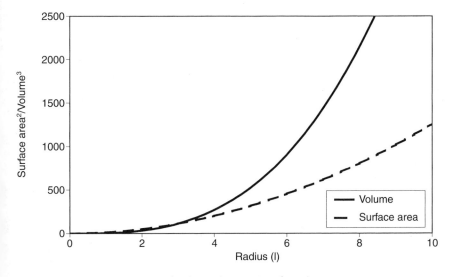

Figure 15.2 Linear, area and volume dimensions of a sphere.

These dimensions are for a sphere, and they change for different shaped particles. However, a sphere is the most ideal solid structure: all other shapes will have higher surface areas and volumes than a sphere for an equivalent mass.

In estimating the toxicity of a particulate, volume (mass) and surface area are more important than any linear dimension. While the volume increases at a higher rate than surface area, this occurs for single spheres.

Most particulate concentrations are expressed in terms of mass/volume (for example, mg/m³). There is no measure of the number of particulates in such gravimetric values. If the gravimetric concentration and the density of a particulate are kept constant, and all the particulates in any sample are the same size, then as the radius decreases, the number of particles increases, as a function of the cube of the increase in the radius (see Figure 15.3).

Under such circumstances the total surface area actually increases, because there are more particles at any particle size. This is shown in Figure 15.4, which shows the surface area for a single particle (the dashed line), and for all the particles (solid line).

Mathematically, this is seen as:

$$\text{Mass} = \frac{4}{3}\pi r^3 D N \quad \text{and} \quad \text{Area} = 4\pi r^2 N$$

where r is the length of the radius, D is the density and N is the number of particles. Therefore, for a given mass, the total surface area increases linearly as r decreases:

$$\frac{\text{Total area}}{\text{Total mass}} = \frac{4\pi r^2 N}{\frac{4}{3}\pi r^3 D N} = \frac{3}{rD}$$

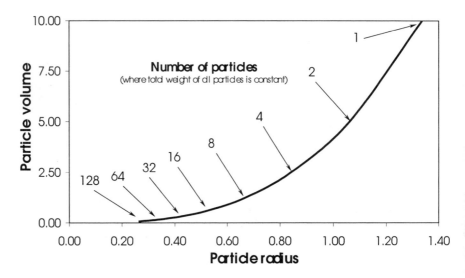

Figure 15.3 Increasing number of particulates for decreasing radius size.

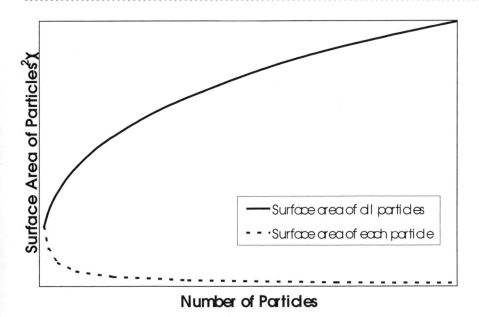

Figure 15.4 Surface area/volume relationship for single and total particulate.

Therefore, where a particulate expresses its toxicity through factors related to its surface, the surface area becomes a critical factor when the number of particles is large, but their size is small. For example, in recent years, a growing literature has suggested that the 'superfines' component of particulate matter makes a significant contribution to the toxicity of the particulate.

For example, particulate matter below $10\,\mu m$ (PM_{10}) is now known to be a cause of significant respiratory problems, and perhaps particulate matter below $2.5\,\mu m$ ($PM_{2.5}$) may be even more significant.

Terminal velocity

The critical factor for determining the time a particle will be airborne is the terminal velocity. Large particles will fall out of the air quite quickly, becoming dusts. Small particles will fall under the force of gravity, but will be subject to air resistance. The smaller the particle, the greater the effect of air resistance, and the longer a particle will stay in air. Eventually, the forces of gravity and air resistance will reach equilibrium, and terminal velocity will be reached.

Experiments with 'standard spherical' particles show that terminal velocity is fairly linear, until the particle diameter falls below 20–$30\,\mu m$, when air resistance has a greater effect (see Figure 15.5).

An important concept arising out of this figure is that the terminal velocity of particles becomes very low with very small diameters.

Put into perspective, a particle with a diameter of $300\,\mu m$ will reach terminal velocity (at $1\,m/s$) very quickly, and fall one metre in one second. Such particles do not represent a risk to exposed individuals as they are not generally available to be inhaled. Because of the experimental basis of this data (using spherical objects

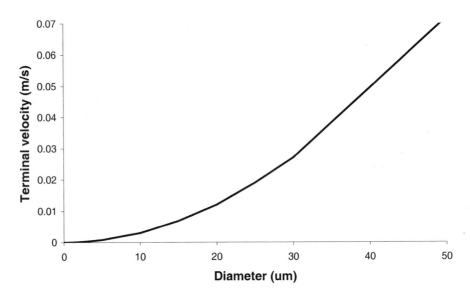

Figure 15.5 Particle size and terminal velocity.

unlikely to be encountered in the occupational environment), it is likely that terminal velocities would be lower for workplace dusts, fumes and other particles. Indeed, only particles with a diameter below about 50–100 μm are relevant in inhalational toxicology.

Further, it is generally accepted that particles with an aerodynamic diameter below 50 μm, with a terminal velocity of below 7 cm/s (or 0.07 m/s), do not remain airborne for too long. This should be contrasted with a dust particle with a diameter of 1 μm, which has a terminal velocity of 0.03 mm/s (or 0.00003 m/s), where settling due to gravity is negligible for practical purposes, and movement in air is more important than settling through it.

Inhaled aerosols are rarely of constant dimensions – most contain particles of varying sizes and shapes. The size distribution of many aerosols encountered in workplaces approximates to a log normal distribution, which can be statistically characterised by the median or geometric mean and the geometric standard deviation. The relative mass of small particles to large particles depends on the cube of the radius of the particle, and a toxic effect caused by inhalation of a particulate may be due to a very small proportion of the total particles present in the aerosol.

For this reason, dusts and particulates are divided into size-specific fractions, based on their ability to enter and penetrate into the respiratory system. The main fractions of a total aerosol are shown in Figure 15.6.

Total dust is divided into 'noninspirable' and 'inspirable' (or inhalable) fractions. For a particulate to have a toxic effect in the respiratory system, it must be inhaled.

The largest inhaled particles, with aerodynamic diameters greater than 30 μm, are deposited in the upper airways of the respiratory system, that is the air passages from the point of entry at the nostrils or lips to the larynx (the *nasopharyngeal* fraction).

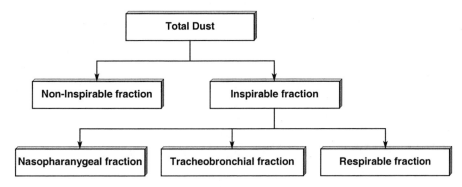

Figure 15.6 Size-selective components of particulates.

During nasal breathing, particles are captured by nasal hairs and by impaction onto the mucosal lining of the turbinates and upper airways. Of the particles that fail to deposit in the nasopharyngeal fraction, the larger particles will deposit in the tracheobronchial airway region and can either be absorbed (if soluble or capable of dissolution) or be cleared by mucociliary clearance. This is the *tracheobronchial* fraction. The remaining particles not deposited in the nasopharyngeal and tracheobronchial regions can penetrate to the alveolar region, the region of gas exchange with the body. This is the 'respirable' fraction.

A great deal of work has gone into establishing the limits of regional and total deposition of particulates in the respiratory system based on size. It is generally considered that the nasopharyngeal region collects particles of above about 7 (to 30) μm, that particulates of 2–30 μm are captured in the tracheobronchial region, and particles of less than 7 μm can penetrate to the alveolar regions. However, these values are approximate, and there is a discernible range of particle sizes that fall into nasopharyngeal, tracheobronchial and respiratory fractions (see Figure 15.7).

Following on from this work is the effort given to developing occupational hygiene monitoring methods that allow estimation of total, inspirable and respirable fractions on workplace aerosols.

An additional toxicity factor: adsorption of gases and vapours onto particulate surfaces

As noted above, the toxicity of some particulates is based on surface area factors, as opposed to factors related to particulate volume. One additional feature with this phenomenon is that irritant or toxic gases and vapours may become adsorbed onto the surface of some particulates. The process of adsorption may increase the concentration of the gases or vapours in relative terms, so that if they are desorbed from the surface of the particulate, for example in the respiratory tract, they may attain higher concentrations in critical tissues that may not be reached through inhalation of an atmosphere of a lower contaminant concentration. Examples here include irritant gases adsorbed to soot particles in combustion processes and toxic fluorides adsorbed to inorganic particulate in aluminium smelting.

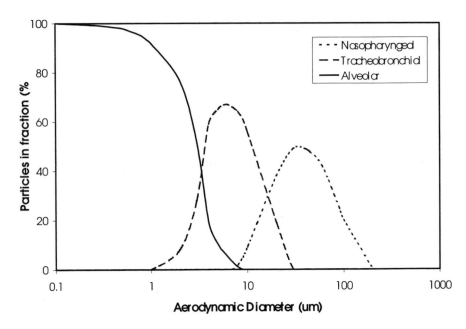

Figure 15.7 Particulate deposition in the lungs by size and region.

Pathological responses to inhaled materials

In the respiratory system

A full range of pathological responses in the respiratory system to inhaled materials (whether gases/vapours or particulates) is summarised in Table 15.4. The main lung diseases from occupational exposure to particulates are shown in the chapter on respiratory toxicology.

Respiratory effects from systemic toxicity

The respiratory system is also a target organ for toxicants from other sources of exposure. For example, ingestion and absorption of the bipyridylium herbicide paraquat produces extensive lung injury, characterised by diffuse interstitial and intra-alveolar fibrosis. The mechanism of toxicity is cellular uptake through the polyamine system and intracellular lipid peroxidation. This produces cellular necrosis and fibroblast proliferation, leading to progressive lung lesions that are invariably fatal. These pathological changes are seen especially in the presence of oxygen, therefore the lungs are a target organ.

Other toxicants may also target the respiratory system as an organ of systemic toxicity, for example:

■   the chemotherapeutic agents bleiomycin and 1,3-*bis*(2-chloroethyl)-1-nitro-surea

Table 15.4 Pathological responses of the respiratory system to inhaled materials

Response	Mechanism
Absorption	Absorption of agents can have toxic effects, either in the lung, or elsewhere in the body.
Asphyxiation	Either: by a reduction in the concentration of oxygen in inspired air by physical displacement (simple asphyxiation); or by a reduction in oxygen transport in the body by chemical reaction (chemical asphyxiation).
Local irritation	Related to the solubility of the substance onto moist surfaces and mucous membranes of nose, eyes, mouth and upper respiratory tract.
Irritation of airways/ bronchoconstriction	Irritation of the airways leads to bronchoconstriction; more extensive and smaller airway constriction occurs with exercise or exertion than at rest.
Increase in the secretion of mucous	Increase in secretion of mucous will slow down ciliary movement, and may block smaller airways.
Cell damage/oedema	Damage to cellular components of airways and alveoli results in increased permeability, loss of compliance, necrosis and intraluminal (within the airways) rather than interstitial (within cells of the airways) oedema; pulmonary oedema may, in turn, be compounded by secondary infection.
Macrophage cytotoxicity	Alteration in function or destruction of alveolar macrophages will alter clearance processes, which can lead to collection of respired particles in a given area.
Sensitisation and allergy	Dependent on immunological status and disposition to asthma.
Lung overload by particles	When lung burdens of particulates are sufficient to exceed physiological clearance mechanisms such as macrophage phagocytosis, lung burdens of such particulates will persist, and completely nonphysiological mechanisms of disease pathogenesis may occur.
Emphysema	Abnormal presence of air: in the lungs, emphysema is an overdistension of the alveoli, and in parts, destruction of their walls, giving rise to the formation of large sacs from the rupture and running together of a number of contiguous air vesicles; another form, acute interstitial emphysema, is the infiltration of air beneath the pleura and between the pulmonary air cells.

continued

Table 15.4 continued	
Response	Mechanism
Granulomatous reactions	Granuloma is a new growth made up of granular cells, caused by chronic inflammation.
Fibrogenesis	The growth of fibrous tissues, comprising fibres of collagen and elastin. Between these cells lie star-shaped cells or fibroblasts, from which collagen or elastin is formed. Elastin has elastic properties, and is used in the walls of arteries, and so forth. Normally, collagen is grouped into bundles which are held together by other fibres, used to make ligaments, tendons and sinews. It is also the substance laid down in the repair of wounds, or as a result of inflammation, and forms scar tissue. Fibrogenesis is the growth of collagen fibres in response to cellular inflammation and damage.
Cancer	Oncogenesis leading to primary lung tumours.

- the anticancer and immunosuppressive agent cyclophosphamide
- a toxin of mouldy potatoes – 4-imopemeanol
- the pyrrolizidine alkaloid monocrotaline
- cationic amphophilic drugs such as amiodarone and chlorphentermine.

Summary

Airborne contaminants represent a major category of safety, health and environmental risks. The respiratory system is a major source of exposure of contaminants into the body, by virtue of the fact that there is no barrier between the external environment and the air sacs involved in the entry of gases into the body.

Atmospheric contaminants will affect the respiratory system on contact, causing a range of effects related to the structure or function of the various elements from which it is composed. Contaminants will affect airways, air sacs, and defence mechanisms.

As well as a system in the front line of exposure and effect, the respiratory system is a major route of entry of contaminants into the body.

Much of the study of airborne contaminants has concentrated on occupationally exposed individuals. However, public health studies have investigated respiratory health problems from environmental contamination, such as the 'pea soup' fogs in Europe in the 1950s and smog in the USA in the 1970s and 1980s.

There are many different types of contaminants – however, they should be considered as being of two types – gases and vapours that are *dissolved* in air, or various types of particulates that are *suspended* in air. These properties also indicate the ways in which contaminants can be collected – gases and vapours need to be absorbed onto adsorbents and particulates need to be captured onto filter media, and controlled.

References and further reading

Amoore, J.E. and Hautala, E. (1983) Odor as an aid to chemical safety: odor thresholds compared with threshold limit values and volatilities for 214 industrial chemicals in air and water dilution. *J. Appl. Toxicol.* 3: 273.

Annotation (1981) Health hazards in formaldehyde. *Lancet* i: 926.

Barrow, C.S. (1986) *Toxicology of the Nasal Passages.* Washington: Hemisphere Publishing.

Beach, F.X.M., Jones, E.S., Scarrow, G.D. *et al.* (1969) Respiratory effects of chlorine gas. *J. Occup. Med.* 4: 152.

Beckett, W.S. and Bascom, R. (eds) (1992) Occupational lung disease. *Occup. Med. State Art Rev.* 7(2): 189–370.

Costa, D.L. (2001) Air pollution. In: Amdur, M.O., Doull, J. and Klaassen, C.D. (eds) *Cassarett and Doull's Toxicology: the Basic Science of Poisons*, 5th edn. New York: Macmillan, pp. 854–871.

Dawson, S.V. and Schenker, M.B. (1979) Health effects of ambient concentrations of nitrogen dioxide. *Am. Rev. Resp. Dis.* 120: 281.

Demeter, S.L. and Cordasco, E.M. (1994) Occupational asthma. In: Zenz, C., Dickerson, O.B. and Horvath, E.P. (eds) *Occupational Medicine*, 3rd edn. St Louis: Mosby, pp. 213–228.

Ellenhorn, M.J. and Barceloux, D.G. (1988) Respiratory tract irritants. In: *Medical Toxicology: Diagnosis and Treatment of Human Poisoning.* New York: Elsevier, pp. 871–894.

Ellenhorn, M.J. and Barceloux, D.G. (1988) Odor threshold values and 1985–1986 threshold limit values. In: *Medical Toxicology: Diagnosis and Treatment of Human Poisoning.* New York: Elsevier, pp. 1412–1414.

Everett, E.D. and Overholt, E.L. (1968) Phosgene poisoning. *J. Am. Med. Ass.* 205: 103.

Gardner, D.E, Crapo, J.D. and Massaro, E.J. (eds) (1988) *Toxicology of the Lung.* New York: Raven Press.

Haldane, J.S. and Priestley, J.G. (1935) *Respiration.* Oxford: Clarendon Press.

Higgins, I.T.T. (1971) Effects of sulphur oxides and particulates on health. *Arch. Environ. Health* 22: 584.

IARC (1992) *Occupational Exposures to Mists and Vapours from Strong Inorganic Acids; and Other Industrial Chemicals.* International Agency for Research in Cancer Monographs on the Evaluation of Carcinogenic Risks to Humans, Lyon, 54: 41–119.

Jeffery, P.K. and Reid, L.M. (1977) The respiratory mucous membrane. In: Brain, J.D., Proctor, D.F. and Reid, L.M. (eds) *Respiratory Defence Mechanisms*, Part I. New York: Marcel Dekker.

Jones, R.N. *et al.* (1981) Longitudinal changes in pulmonary function following single exposure to chlorine gas. *Am. Rev. Resp. Dis.* 123: 125.

Kennedy, G.L. and Trochimowicz, H.J. (1982) Inhalation toxicology. In: Hayes, A.W. (ed.) *Principles and Methods of Toxicology.* New York: Raven Press.

Mauderly, J.L. (1989) Effect of inhaled toxicants on pulmonary function. In: McClellan, R.O. and Henderson, R. (eds). *Concepts in Inhalation Toxicology.* New York: Hemisphere, pp. 347–401.

Morgan, W.K.C. and Seaton, A. (1984) *Occupational Lung Diseases*, 2nd edn. Philadelphia: Saunders.

Ng, T.P and Chan, S.L. (1991) Factors associated with massive fibrosis in silicosis. *Thorax* 46: 292–332.

NICNAS (1994) *Glutaraldehyde: Full Public Review.* National Industrial Chemicals Notification and Assessment Scheme/AGPS, Sydney.

Paivon, J.C., Brochard, P., Jaurand, M.C. and Bignon, J. (1991) Silica and lung cancer, a controversial issue. *Eur. Resp. J.* 4: 730–744.

Quakenboss, J.J. *et al.* (1986) Personal exposure to nitrogen dioxide: Relationship to indoor/outdoor air quality and activity patterns. *Environ. Sci. Technol.* 20: 775–783.

Salvaggio, J.E., Butcher, B.T. and O'Neil, C.E. (1986) Occupational asthma due to chemical agents. *J. Allergy Clin. Immunol.* 78: 1053–1058.

Schlesinger, R.B. (1990) The interaction of inhaled toxicants with respiratory clearance mechanisms. *Crit. Rev. Toxicol.* 20: 257–286.

Schlueter, D.P. (1994) Silicosis and coal worker's pneumoconiosis. In: Zenz, C., Dickerson, O.B. and Horvath, E.P. (eds) *Occupational Medicine*, 3rd edn. St Louis: Mosby, pp. 171–178.

Stokinger, H.E. (1965) Ozone toxicity: a review of research and industrial experience 1954–1964. *Arch. Environ. Health* 10: 719.

Terenski, M.F. and Cheremisinoff, P.N. (1973) *Industrial Respiratory Protection*. Ann Arbor: Ann Arbor Science.

Walton, M. (1973) Industrial ammonia gassing. *Br. J. Ind. Med.* 30: 78.

West, J.B. (1985) *Respiratory Physiology – The Essentials*. Baltimore: Williams and Wilkins.

WHO PACE (1999) *Hazard Prevention and Control in the Work Environment: Airborne Dust*. World Health Organisation Prevention and Control Exchange, Geneva.

Witschi, H.R. and Brain, J.D. (1985) *Toxicology of Inhaled Particles*. Berlin: Springer-Verlag.

Witschi, H.R. and Last, J.A. (2001) Toxic responses of the respiratory system. In: Klaassen, C., Doull, J. and Amdur, M.O. (eds) *Cassarett and Doull's Toxicology: The Basic Science of Poisons*, 4th edn. New York: Pergamon, pp. 383–406.

Fields interfacing with toxicology

Chapter 16

Occupational hygiene – interface with toxicology

C. Gray

The scope of occupational hygiene

Assessment and control of toxic substances in the workplace is one of the main roles of the occupational hygiene profession. Although this is by no means the exclusive responsibility of occupational hygienists and is shared with other professionals as well as with workers and management, occupational hygienists have a special combination of skills and knowledge, gained by formal training and/or extensive experience, which equips them for this role. This requires some knowledge of toxicology, occupational diseases, epidemiology, standard setting, occupational health law and management, as well as a thorough understanding of methods of exposure assessment and control. Occupational hygienists are also concerned with physical hazards such as noise and vibration, nonionising radiation, heat stress and to some extent with biohazards and ergonomics.

They often work in a team along with occupational physicians, occupational health nurses, safety officers, toxicologists and others. Although there is some overlap between the skills of these various professionals their roles are usually quite distinct. The greatest overlap might be found between the occupational hygienist and safety officer. Ideally, an all-round safety officer would be able to carry out at least some of the functions of an occupational hygienist. The distinction is usually to be found in the scientific background of the occupational hygienist which enables him/her to understand more deeply the physical, chemical and biological principles that underlie factors which are a threat to health and wellbeing.

Traditionally, occupational safety was concerned primarily with reducing the risk of traumatic injury due to machine accidents, falls, fire, explosion, etc. Occupational hygiene grew out of the recognition of the more insidious effects of repeated exposure to chemicals or other hazards in the workplace. The demarcation which has grown up between these two professions is to some extent artificial, but there is a general and growing recognition of the need for specialists who have sound scientific backgrounds and the appropriate training and experience to equip them to understand and control the increasingly complex hazards in our technically advanced workplaces.

In assessing hazards in the workplace, occupational hygienists are able to draw upon a large store of knowledge built up over many years by the efforts of occupational physicians, toxicologists, epidemiologists and other occupational hygienists. By reference to this knowledge they are able to recognise hazards that may need control or further evaluation.

Measurements can be made of worker exposure to these hazards, for example the airborne concentration of a toxic gas in a worker's breathing zone can be measured and the results can be compared with criteria which allow the risk of disease or injury to be assessed. Appropriate control measures can then be instituted to reduce the risk to the exposed workers. Such controls might involve the substitution of a toxic substance used in a process, or the provision of engineering controls such as ventilation.

This sequence of *recognition, evaluation,* and *control* represents the traditional formula for occupational hygiene practice. A fourth component in the sequence is now recognised and that is *anticipation.* This can involve deliberate consideration of new technology, processes, equipment and materials before they are introduced into the workplace. It can also involve taking account of the consequences of unplanned process changes; the failure of control equipment, the inappropriate use of process chemicals for cleaning purposes and the shortcuts in work practice which result in unnecessary exposures.

Occupational hygienists concerned with the assessment and control of chemicals in the workplace would hardly be able to operate without access to toxicological information. They would not know which substances required controlling, nor to what extent. However, this flow of information is not just one way. Occupational hygiene data are an essential part of the toxicological appraisal of industrial substances and are utilised both in epidemiology and in setting of exposure standards. Occupational epidemiology involves examining the relationship between health and workplace exposures and occupational hygienists, or others with relevant experience, are involved in providing assessments of current or historical exposures, depending upon the design of the epidemiological study.

Setting of exposure standards, such as the American Threshold Limit Values (TLVs) and the Australian NOHSC National Exposure Standards, usually draws heavily upon workplace exposure data gathered by occupational hygienists. In fact, the TLVs were developed by technical committees of the American Conference of Governmental Industrial Hygienists (ACGIH). Similarly, various biological limits have been developed based upon biological monitoring techniques such as the measurement of toxic substances or their metabolites in urine. These seek to impose limits on uptake of chemicals by workers rather than on concentrations of substances in the workplace air to which workers are exposed. The ACGIH Biological Exposure Indices (BEIs) are examples of such biological limits. The BEIs are not derived directly from toxicological data as might be expected, but are based on the TLVs and are derived by comparison of occupational exposures and biological monitoring results taking into account the relevant biotransformation and elimination kinetics.

It should be apparent from the foregoing that occupational health involves a complex interplay between toxicology, epidemiology, occupational hygiene and occupational medicine. This is illustrated in a simplified way in Figure 16.1.

Exposure to chemicals in the workplace

The term 'exposure' implies contact with a substance which can lead to uptake and when we measure exposure in some way we are attempting to measure this potential for uptake.

Airborne gases and vapours and inhalable dusts can enter via the respiratory tract and this is often the most important route of uptake in the occupational setting. Some toxic substances can also be absorbed via the skin and most via the gastrointestinal tract. Lipid-soluble substances including organic solvents and some pesticides readily pass through the intact skin and for some dusty materials hand to mouth contact can result in significant uptake via the gastrointestinal tract. This is thought to be of some importance in the lead industry for example.

Although it is acknowledged that gastrointestinal and skin uptake can be important, exposure limits such as the Australian exposure standards refer only to airborne concentrations of substances which result in uptake via the respiratory tract. The skin notation which is applied to some substances, indicates that there is a potential for systemic toxicity as a result of skin uptake, but there is no exposure limit for this route of entry. Although we have no way of directly quantifying personal exposure by the skin or gastrointestinal routes, biological monitoring can sometimes be used to measure the uptake by all routes including these.

Toxic substances may be present in the air in several physical forms as shown in Table 16.1, the major divisions being: (a) dusts and other particulates; and

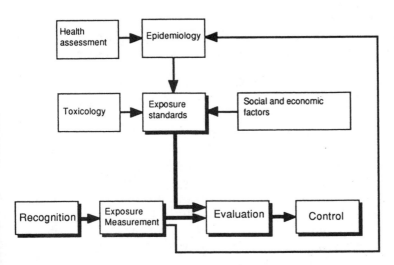

Figure 16.1 Simplified diagram of the relationship between occupational hygiene and occupational epidemiology and toxicology from the occupational hygiene perspective. It should also be borne in mind that clinical or toxicological investigations may be instigated as a result of occupational hygiene or epidemiological observations.

Table 16.1 Types of air contaminant	
Gases	Individual molecules of substances which cannot be liquefied at normal temperatures, e.g. carbon dioxide
Vapours	Individual molecules of substances which are liquid or solid at normal temperatures, e.g. benzene
Total dust	All particles in the air
Inhalable dust	Can be drawn into upper airways
Respirable dust	Can penetrate to alveoli of lungs
Fibres	Elongated particulates (length:breadth >3:1)
Mist	Airborne liquid droplets
Fume	Solid particles formed by evaporation–condensation processes (e.g. welding fume) or by gas phase reactions (e.g. ammonium chloride)
Smoke	Solid particles or liquid (usually tar-like) droplets formed by combustion mechanisms leading to evaporation–condensation, or gas phase reactions leading to condensation

(b) gases and vapours. Dusts, fumes, fibres, mists and smokes are referred to as 'particulate matter' or 'particulates', whilst a more or less stable suspension of particles in air is referred to as an 'aerosol'.

Preliminary assessment

Assessment involves both recognition and evaluation of workplace hazards. Recognition might seem a simple task, after all we could easily observe the use of solvent-based adhesive for example. However, workplaces are usually complex and contain many potential hazards and the problem is to identify all the important ones and to prioritise them for immediate corrective action, further evaluation or for attention at some later date.

In the recognition stage, an initial 'walk-through' or preliminary survey is often carried out to assess the potential hazards associated with the workplace. Generally the walk-through survey is carried out without the use of measuring instruments and involves careful observation of processes, raw materials, intermediates, products, by-products, wastes and significant contaminants. By linking a substance inventory with toxicity data and information on the processes and work procedures, hazards can be systematically identified and prioritised. The preliminary survey can also help to identify physical hazards such as noise, heat stress and poor illumination, and adverse ergonomic factors.

The use and adequacy of control measures such as ventilation, personal protective equipment or special work practices can also be noted. As a result of the preliminary surveys, many hazards which may require a full occupational hygiene evaluation may be identified.

Exposure monitoring

The second phase of the assessment procedure is evaluation and commences with monitoring of personal exposures or workplace air concentrations of chemicals identified in the preliminary assessment phase. There are two main reasons for monitoring chemicals in the workplace:

- to estimate personal exposure to hazardous substances
- to assist in the design or selection of controls to reduce exposure.

The first of these two reasons for monitoring is usually related to assessing compliance with exposure standards. The second is to provide information for control purposes that will help reduce the exposures in some way. This might involve more detailed investigations of exposure patterns in processes in order to identify which process steps give rise to the greatest exposures so that controls can be designed which are most effective.

For example, personal monitoring might have shown that average exposures to solvent vapour in a manufacturing operation are above the exposure standard. Further investigations might reveal that only one or two process steps, such as loading a solvent tank or equipment cleaning, make a major contribution to the overall exposure. Controls (elimination, substitution, ventilation or personal protection) can then be focused on those process steps rather than on the entire manufacturing operation.

Monitoring might also be required for other reasons including:

- Checking confined spaces prior to entry as they might contain toxic, inflammable or explosive atmospheres or be deficient in oxygen.
- To ensure that processes involving highly dangerous substances are under continuous control; that leaks have not developed and that control equipment is working. This usually requires the use of direct-reading instruments.

Exposure data might also be used for the purposes of epidemiology, to investigate a possible association between medical outcomes and exposures.

There are many methods of monitoring chemicals in air and these fall into two major divisions:

- Air sampling techniques in which a sample of the contaminant is collected for subsequent analysis in the laboratory.
- Direct-reading methods in which an analytical instrument is taken into the workplace to obtain 'real-time' measurements.

Most frequently, air samples are taken to determine the exposure levels of a worker or group of workers. This is most effectively achieved by measuring concentrations of substances in the worker's 'breathing zone' where the air is representative of that which the individual worker is breathing; sometimes this is defined as within 300 mm of the nose or mouth. Often it is satisfactory to collect an air sample at the location of the lapel or breast pocket, but in some situations it is necessary to sample as close to the nose as possible. This is the case where a worker wears a visor (e.g. a welder's helmet) but is also necessary when a worker is very close to the source of the emission and the concentration gradients are very steep. To evaluate engineering controls or determine sources of contamination, samples are normally collected in the vicinity of the operation itself.

The duration of sampling is determined principally by the objective of monitoring and must be selected to take account of the requirements of the analytical method. A full-shift personal sample is preferred when results are to be compared to 8-hour time-weighted average exposure standards. Where a substance has a short-term limit it is more difficult to demonstrate compliance. In principle, serial 15-minute measurements are necessary, although it is possible to employ less frequent sampling with appropriate statistical analysis. It can be even more difficult to satisfactorily demonstrate compliance with a ceiling limit. This requires continuous monitoring using a direct-reading instrument, but sometimes short-term sampling of up to 15 minutes' duration has to be used instead.

It is also important to realise that exposures can be highly variable from worker to worker and from day to day, and it can be misleading to put great reliance in a single, or just a few, exposure measurements. Monitoring should follow a carefully considered strategy aimed at answering clearly stated questions about exposure in a given group of workers and the statistical validity of the results should be assessed.

Dust monitoring

Personal sampling most commonly makes use of small portable battery-operated pumps which can be worn by the worker and are used to draw air through a sampling device which will collect the contaminant. Dust can be collected in several ways, the most common of which is to draw a measured volume of air through a suitable filter.

Airborne dusts can have a wide range of particle size and very coarse dust which cannot be inhaled is of little toxicological significance. What is important is the fraction of the dust that can be drawn into the airways. For systematically poisonous substances any dust which enters the nose is important; this is the 'inhalable' fraction. If the lung is the target organ, as in the case of fibrogenic dusts such as quartz, then only that fraction of the dust which can penetrate to the deep regions of the lung is important; this is the 'respirable' fraction.

When assessing the inhalability of dusts, the important property of the dust particles is their 'aerodynamic diameter'. This is the diameter measured by the rate of fall of the particles in air rather than their geometric size or shape.

The inhalable dust fraction is defined by the ACGIH as corresponding to the curve shown in Figure 16.2 and can be sampled using the Institute of Occupational Medicine, Edinburgh (IOM) inhalable dust monitor shown in Figure 16.3 or, more approximately, using the modified UKAEA (seven-hole) sampling head.

The respirable dust fraction is defined by the International Standards Organisation (ISO) as corresponding with the curve shown in Figure 16.4 and is sampled using a suitable cyclone (Figure 16.5) which rejects the nonrespirable dust fraction. A different sampling head is used for asbestos monitoring, consisting of a downward facing open-face filter holder with a cowl on the inlet to prevent very large particles being collected.

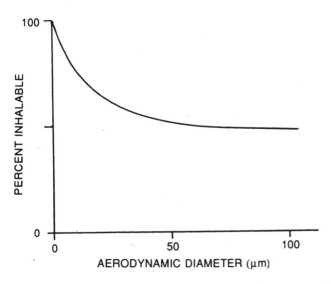

Figure 16.2 The inhalable fraction of airborne dust (ACGIH definition).

Figure 16.3 The IOM inhalable dust monitor shown assembled (left) and dismantled (right).

Once a dust sample has been collected it is then measured or analysed. For example, this might involve accurate weighing, metal analysis by atomic absorption spectrometry (AAS), or asbestos fibre counting by phase contrast optical microscopy (PCOM).

In order to monitor a worker's exposure to dust, a filter is mounted in the sampling head which is then connected to a sampling pump by a length of plastic

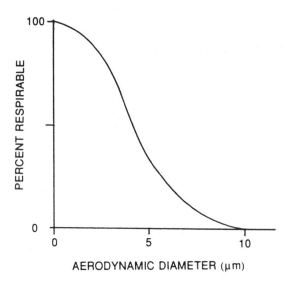

Figure 16.4 The respirable fraction of airborne dust (ISO definition).

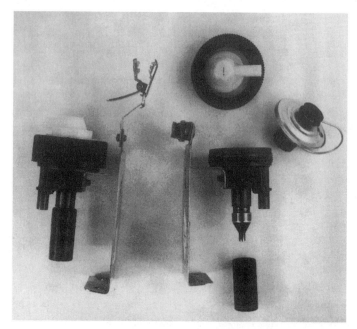

Figure 16.5 A cyclone respirable dust monitor shown (left) and dismantled (right). The filter is mounted in the cassette shown held closed by a spring clip.

tubing and the pump is set to the required flow rate (2 l/min in the case of an inhalable dust monitor or 1.9 l/min for a cyclone). The sampling head is clipped in the breathing zone of the worker, for example near the lapel, and the pump is hung from a belt or waist band (Figure 16.6). Sampling is then carried out for an accurately timed period which may be a full working day or a shorter interval.

Figure 16.6 A worker wearing a cyclone respirable dust monitor and personal sampling pump.

Monitoring gases and vapours

Gases and vapours may be sampled by a variety of different techniques, depending on their physical and chemical properties. It is possible to sample any gas or vapour into a container such as a plastic bag, and special sampling bags made of Teflon (polytetrafluoroethylene) or Tedlar (polyvinylfluoride) are available commercially for this purpose. These have low permeabilities and adsorptive properties and allow many vapours to be stored for a day or so with minimal loss. Samples collected in this way are bulky and contain only small amounts of contaminant for chemical analysis, and for these reasons bag sampling is not often used by occupational hygienists. Other techniques of sampling gases and vapours remove the contaminant from the air by trapping it in a sampling device and thereby allow a larger amount to be collected for analysis. Probably the two most important such sampling techniques are absorption in a liquid or adsorption onto a porous solid such as charcoal.

For many gases and some vapours it is necessary to collect the analyte using liquid absorption. This commonly involves bubbling an air stream through a solution that will dissolve the contaminant or trap it by chemical reaction.

For example, alkaline gases such as amines can be trapped in acid solutions and formaldehyde can be trapped as an addition product in aqueous sodium bisulfite solution. The liquid is placed in a glass wash bottle or impinger (Figure 16.7) which can be mounted in the breathing zone of the worker while the pump is attached to the belt or waist band. Contaminants which have been collected in liquids may be analysed by a variety of methods depending on the nature of the contaminant and the adsorbing solution. Commonly used techniques are ultra-violet/visible (UV/VIS) spectrometry and liquid chromatography.

For organic vapours, adsorbent sampling onto porous solids such as charcoal is the standard method. A measured volume of air is drawn through a sampling tube made of glass or metal packed with a solid adsorbent and the contaminant vapours are adsorbed from the air stream onto the surface of the adsorbent (Figure 16.8). A typical solid adsorbent tube for organic vapours contains 100 mg of activated char-coal and 50 mg as a back-up section (Figure 16.9). Other solid adsorbent materials include silica gel and a wide range of commercially available porous polymers.

Some examples of solid absorbents are:

- Activated charcoal: this is by far the most commonly used adsorbent. It has a very large surface area/weight ratio and a high adsorptive capacity. It is non-polar and therefore preferentially absorbs organic vapours rather than polar molecules such as water.
- Silica gel: this is a highly polar adsorbent and efficient collector of polar vapours such as alcohols and amines. It is also hygroscopic and this can cause problems in humid conditions.

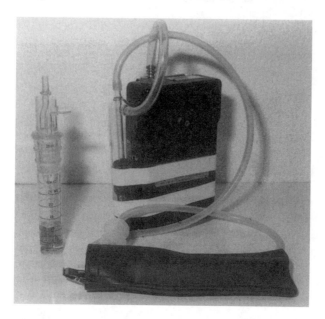

Figure 16.7 Sampling pump and impingers. An impinger is shown strapped to the pump with adhesive tape. The pouch at the forefront is designed to hold and protect an impinger.

Figure 16.8 A low flow sampling pump and charcoal tubes.

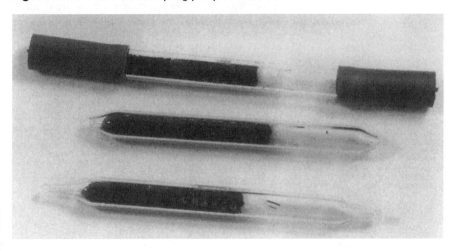

Figure 16.9 Charcoal tubes. Bottom: unused. Middle: tips broken off. Top: capped after use.

■   Porous polymers: these, like charcoal are generally nonpolar but have lower adsorptive capacity. They are suitable for thermal desorption.

■   Molecular sieves: these include zeolites and carbon molecular sieves which retain adsorbed species according to their molecular size.

The adsorbent sampling tube is usually mounted in the wearer's breathing zone, connected to a pump worn on a belt.

After sampling, the tubes are sealed with plastic caps before the tube is sent for analysis. Many variations on geometry and packing are possible. Larger amounts of adsorbent enable sampling to be conducted over longer time periods. Polar adsorbents, such as silica gel, can be used in special applications for polar gases or vapours such as methanol.

A potential pitfall of adsorbent sampling is the phenomenon of 'breakthrough' which occurs after a certain volume of contaminated air has been drawn through the sampling tube. When breakthrough occurs the sample is gradually swept out of the tube with the air stream and is lost. Breakthrough can be avoided if the volume of air that can be sampled before breakthrough occurs (the 'breakthrough volume') is known in advance. Breakthrough volumes for many compounds are published by manufacturers of adsorbent tubes and the back-up section in charcoal tubes is to help indicate when breakthrough has occurred.

Passive sampling

Since the early 1980s, a number of 'passive' sampling devices for gases and vapours have been introduced which do not make use of a pump to collect the sample. These are devices in which the rate at which the gas is absorbed or adsorbed is governed by its rate of diffusion across a well-defined diffusion path. Two general types of passive device are available commercially which are known as 'badges' and 'tubes', respectively. Passive samplers for organic vapours contain an adsorbent material such as charcoal cloth or a porous polymer. For gases a chemically reactive sampling material is used. The adsorbent is located in a small container which is open at one end to allow contaminant molecules in the air to diffuse into the sampler and become trapped on the adsorbent. Tube-type passive samplers such as those marketed by Perkin Elmer have a small frontal area and a relatively small sampling rate. Badge-type samplers have a much larger frontal area and a sampling rate that is considerably higher than for tubes. The effective sampling rate depends on the molecular weight of the contaminant that is being sampled.

Passive samplers are usually calibrated by the manufacturer against standard atmospheres in the laboratory, and users should satisfy themselves that the calibration is suitable for the concentration range and sampling conditions that they are dealing with. Sometimes this requires extensive validation work in the workplace.

Preparation for analysis

Where a vapour has been sampled on to a solid adsorbent, it must be removed from the adsorbent before analysis can take place. The simplest way of removing the vapour is to wash it off with a suitable solvent.

Carbon disulfide is widely used to remove organic vapours from charcoal. The charcoal tube is cut open and the charcoal granules are transferred to a small bottle. A small amount of carbon disulfide is added and the bottle, once capped, is tumbled or otherwise agitated for a short period. A measured amount of the carbon disulfide, now containing the contaminant from the charcoal, is then removed using a microsyringe and injected directly into the analytical instrument (usually a gas chromatograph).

A convenient alternative technique for desorption is to heat the adsorbent tube so that the adsorbed vapour is desorbed into a gas stream and passes directly into the analytical instrument. Alternatively the thermally desorbed vapour can be collected in a chamber or by freezing and analysed later. A fully automated thermal desorber is manufactured by Perkin Elmer for desorbing their own design of adsorber tubes.

Some occupational hygienists carry out their own chemical analysis, filter weighing and occasionally asbestos counting, but it is increasingly the case that occupational hygiene samples are analysed by analysts or technicians. When samples must be sent to external laboratories for analysis, the hygienist must be satisfied that suitable analytical methods are used and that adequate quality control is exercised.

One safeguard that is available for some analyses is to ensure that the laboratory is appropriately registered. The National Association of Testing Authorities (NATA) is the accrediting body for Australian laboratories, and in 1987 set up a technical committee to develop quality standards specifically for laboratories carrying out occupational hygiene measurements.

Direct-reading instruments

Direct-reading instruments provide an immediate indication (within a fraction of a second up to a few minutes) of the concentration of air contaminants and are available for airborne dusts including fibres, and for many gases and vapours. Direct-reading instruments should not be seen simply as an alternative to air sampling techniques. They can provide details of the variation of air concentrations over time and space which can help with the efficient selection of controls and are particularly useful where measurements are required quickly; some can also be used to operate alarms. They also have a number of limitations which can yield misleading results unless these are recognised and taken into account.

The simplest form of direct-reading 'instrument' is perhaps the indicator tube (or detector tube). Detector tubes consist of a glass tube initially sealed at both ends and filled with porous granules of an inert material impregnated with a chemical reagent which changes colour in the presence of the contaminant. The actual concentration of the gas is determined either by the length of the stain or by comparing the intensity of the colour with standards. Air is pumped through the tube by means of a small hand-operated pump which may be of a bellows or piston design (Figure 16.10).

Although detector tubes are easy to use with little training, interpretation of the results may require considerable training and experience. Detector tubes may be designed to detect one contaminant or a group of chemically related compounds, but either way the tube may also respond to other contaminants which may be present in the air. This is known as 'cross-sensitivity' and can cause misleading results. For example, benzene detector tubes equally respond to other aromatic hydrocarbons such as toluene and xylene. Before detector tubes are used it is important that the manufacturer's instructions and warnings on cross-sensitivity are studied and account is taken of possible interfering contaminants that may be present.

Another potential problem with most detector tubes is that they provide a measurement at only one time and place while air concentrations to which workers are exposed may fluctuate widely. It is not possible to have great confidence that a short-term measurement is representative of average conditions or that even a

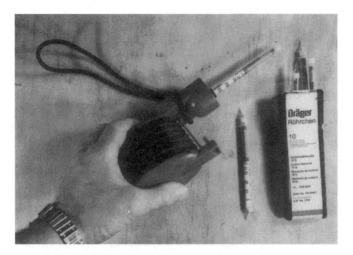

Figure 16.10 Detector tube and bellows-type pump.

short operation has been adequately assessed. Nevertheless, short-term detector tubes have an important part to play in the assessment of exposures and process emissions. A number of long-term detector tubes are also available which can be operated over an extended period with a small battery-operated pump. These provide a time-weighted average measurement of contaminant concentration and can be inspected periodically to obtain an updated result. Additionally, a number of passive long-term detector tubes are available which operate without a pump and provide a convenient means of personal monitoring for some gases.

Direct-reading dust monitors

A variety of direct-reading instruments for dusts and other particulates is available. These generally measure the airborne concentration of the particulates but not their composition. They utilise some physical property of airborne particles such as light scattering, although some measure the mass of dust collected on an oscillating quartz crystal.

The main detection principles used in commercially available direct-reading dust monitors are:

- optical: light scattering or obscuration
- electrical: e.g. ion interception
- piezoelectric: oscillating quartz crystal
- β attenuation: attenuation of β radiation by deposited dust.

Examples of direct-reading dust monitors commonly used in Australia are the miniature real-time aerosol monitor (MINIRAM) and the fibrous aerosol monitor (FAM).

The MINIRAM is a small forward light-scattering dust monitor. Dust-laden air diffuses into the sensing zone through which a light beam passes. Dust particles in this zone scatter light which is measured by a photocell and the dust concentration is shown instantaneously on a liquid crystal digital display. Alternatively, the average concentration over a period can be displayed and the data collected during a whole week (40 hours) can be stored in internal memory. The MINIRAM is calibrated by the manufacturer against an atmosphere of standard Arizona road dust and this calibration often does not hold for other particulates. In such cases the user must calibrate the instrument for the particulate to be monitored by comparison with a pump and filter technique.

The FAM is also a light-scattering dust meter but is designed to detect and count fibrous particles. The air sample is drawn to a sensing zone through which a laser beam passes. A rotating high-voltage electrical field causes any fibrous particles present to rotate and they are detected by the varying pattern of scattered light that results. The FAM can be used both to count and measure the size of airborne fibres and is usually calibrated against amphibole asbestos for which it is most reliable.

Direct-reading gas and vapour monitors

A large number of types of direct-reading instruments for gases and vapours are available. Some of these are highly specific and measure only one contaminant, whilst some are nonspecific and will detect any of a wide range of contaminants.

The main principles of detection employed in reading gas and vapour monitors include:

- colorimetric: detector tubes and paper tape monitors
- electrical/electrochemical: conductivity, reactions at an electrode
- combustion: oxidation of organic compounds on a catalytic surface
- flame ionisation: measurement of the electrical conductivity of a flame in which the air sample is burned
- photoionisation: measurement of the conductivity of air due to ions produced from the contaminant by ultraviolet light
- infrared: absorption of infrared radiation by molecules of the contaminant.

In addition, portable versions of standard laboratory analytical instruments such as gas chromatographs are available.

Two common examples of direct-reading gas/vapour monitors are the mobile infrared analyser (MIRAN) and the total ionisables present (microTIP) photoionisation meter.

The MIRAN is a single-beam gas-phase infrared spectrometer with a large sample cell containing a pair of concave mirrors which, by multiple reflections, provide a long optical path length and therefore high sensitivity. The instrument can be set to different wavelengths characteristic of different types of contaminant. This gives some selectivity; for example, ketones could be detected in the presence of hydrocarbons without much cross-sensitivity, however, different ketones and other carbonyl compounds would interfere. The MIRAN can be calibrated by the user for a wide range of contaminants.

The microTIP is a small photoionisation monitor employing an ultraviolet light source to produce ionisation in the contaminant which is then measured electrically. Nearly all organic compounds and many inorganic compounds are detected but with different sensitivities and the instrument must be calibrated for each separately. Methane is not detected and can actually suppress the response due to other compounds present.

Caution must be employed in the use of direct-reading instruments and in the interpretation of their results. Many of the instruments are nonspecific and the occupational hygienist may often have to verify findings by supplementary sampling and laboratory analyses. Nevertheless, direct-reading instruments can provide valuable information supplementary to that which can be obtained by air sampling.

Biological monitoring

Analysis of tissue or body fluids is employed to measure toxic chemicals in the body or to assess biochemical changes which might provide the early signs of toxic damage. These two different objectives may be termed 'biological exposure monitoring' and 'biological effect monitoring' (Table 16.2).

Biological exposure monitoring assesses the uptake of chemicals by measuring the chemical or its metabolites in the body or excreted fluids. These may include blood, fat, bone or organs, urine, faeces, breath, sweat, nails, hair, sputum and breast milk. As discussed earlier, the American Conference of Governmental Industrial Hygienists (ACGIH) has set a number of biological exposure indices (BEIs) which are concentrations of substances or principle metabolites in specified body fluids which would indicate exposures around the corresponding threshold limit value (TLV) concentration.

Biological effect monitoring must be clearly distinguished from biological moni-

Table 16.2 Examples of biological monitoring techniques for exposure and effects of various substances

Substance	Biological exposure	Biological effect
Lead	Lead in blood Lead in urine	Zinc protoporphyrin in red blood cells
Cadmium	Cadmium in urine	Low molecular weight proteins in urine
Carbon monoxide	Carbon monoxide in exhaled breath	Carboxyhaemaglobin in blood
Nitrobenzene	p-Nitrophenol in urine	Methaemaglobin in blood
Toluene	Toluene in blood or exhaled breath Hippuric acid in urine	
Organophosphate insecticides		Cholinesterase in red cells or plasma

toring of exposure. Whereas the latter technique attempts to quantify the body burden of a substance, biological effect monitoring aims to identify individuals with early signs of adverse health effects, e.g. increased levels of cytoplasmic enzymes in the case of exposure to hepatotoxic chemicals, or increased levels of protein in the urine in cases of exposure to nephrotoxic chemicals. However, not all biological effects can necessarily be described as adverse.

In principle, biological monitoring provides a more accurate assessment of health risk than measurement of air concentrations in the workplace. A biological parameter reflecting the internal dose is necessarily more closely related to systemic adverse effects than is any environmental measurement. Unfortunately, biological monitoring can present practical difficulties and the results are sometimes difficult to interpret. Nevertheless, it has a number of major advantages over exposure monitoring.

In particular, it can provide a measure of uptake by all routes including the skin and gastrointestinal tract, whereas exposure monitoring can assess only uptake via the respiratory tract. Additionally where workers are using personal protective equipment such as respirators, and monitoring cannot provide a true measure of respiratory exposure, biological monitoring may still give a useful indication of uptake.

When a biological monitoring method is based on the determination of the chemical or its metabolite in biological media, it is necessary to know how the substance is distributed to the different compartments of the body, biotransformed and finally eliminated if we are to correctly interpret the results. For most substances the toxicokinetics are best described by a compartment model consisting of various tissue types with different capacities, perfusion rates and partition coefficients.

Following exposure, the chemical is eliminated from the various compartments with different half-lives. For example, for volatile organic compounds elimination from blood and alveolar air is generally rapid, whereas that from adipose tissue is relatively slow. Consequently, elimination curves from body fluids are composites of several curves with increasing half-life, reflecting the disappearance from the main body compartments (e.g. vascular, extravascular, and fat in the case of a lipid-soluble solvent). When a toxicant is substantially metabolised before elimination the toxicokinetics are also dependent upon the rate of biotransformation. If the toxicokinetics are known, then it is relatively simple to select an appropriate time after exposure when the biological sample should be taken such as end of shift, end of the working week or 16 hours' postexposure.

The majority of methods for biological monitoring of exposure rely on the determination of the chemical or its metabolite in blood, urine, or alveolar air.

Blood

The amount of chemical absorbed can sometimes be assessed by measuring its concentration in blood. This technique is suitable for a variety of metals and many organic compounds. In the case of organic compounds, it may be appropriate to measure the unchanged chemical in the blood or a metabolite.

The concentration of a solvent in blood frequently reflects either the most recent exposure, if blood is collected during exposure, or the integrated exposure during the preceding day, if blood collection is performed after the end of exposure (i.e. before the next shift). However, the concentration in blood of some cumulative organic chemicals, such as polychlorinated biphenyls or hexachlorobenzene, mainly reflects stored body burden; the blood level of these chemicals being related to the duration of exposure or to the concentration in the storage compartment (e.g. adipose tissue).

Blood samples may be collected by syringe or a skin prick. Normally venous blood is collected with an addition of an anticoagulating agent and analysed whole or after separating the cells from the plasma. Whole blood may be analysed if the toxic substances are to be found in solution in the plasma. Cell fractions may be analysed if the toxicant is protein bound and is found predominantly in the red blood cells (e.g. lead). If the cellular portion is required then the blood is centrifuged so that the plasma and cells are separated.

Substances commonly assessed in blood samples include cadmium, lead, mercury, styrene, and toluene.

Urine

The most commonly used test in occupational health to assess the internal dose of a chemical is the measurement of a metabolite or the chemical itself in urine. For many organic chemicals, the urinary excretion of the metabolites increases progressively during the workshift to reach a maximum at, or sometimes after, the end of the exposure period; it then declines in an exponential fashion. Depending on the exposure conditions, the biological half-life of the metabolite, and the sensitivity of the analytical method, the metabolite concentration can be determined in the urine collected during the shift, at the end of the shift, or before the next shift.

Urine analysis presents two advantages. First, it is noninvasive and not unduly inconvenient for the worker to provide a sample. Second, when a metabolite is monitored, its level in urine is much less influenced by peak exposures than the concentration of the unchanged chemical in blood or alveolar air. Urine analysis is suitable for many metals and pesticides as well as a range of other substances.

Because urine volume is subject to considerable variation concentrations of excreted substances may be subject to random errors. Two techniques are commonly used to correct for such variations in urine volume.

- Specific gravity correction: the specific gravity of the urine is measured and the concentration of the chemical/metabolite is adjusted to a standard specific gravity of 1.02.
- Creatinine correction: the creatinine concentration in urine is measured as well as the chemical/metabolite and the results are expressed as grams of substance per gram of creatinine or moles of substance per mole of creatinine.

Chemicals commonly assessed in urine include cadmium, chromium, fluorides, lead, MEK, phenol and trichloroacetic acid for trichloroethylene exposure.

Exhaled breath

For volatile compounds, such as organic solvents, the concentration in alveolar air can be used to estimate current or recent exposure. Several factors influence the concentration of a solvent in alveolar air, the most important being the blood–air partition coefficient. The less blood-soluble the solvent is, the higher will be its concentration in alveolar air compared with that in the blood. Exhaled breath analysis is an attractive method as it is easy to perform and is non-invasive.

Exhaled breath analysis can be performed using direct-reading instruments. However, it is more common to collect a breath sample in an inert plastic bag and analyse this later. A one-way exhalation valve is often used to collect the sample and a charcoal filter on the inlet prevents contamination. Air contaminants may be transferred onto charcoal tubes for storage.

Chemicals assessed in exhaled breath include benzene, carbon monoxide, n-hexane, methylchloroform, styrene and toluene.

Biological monitoring results can often be interpreted on an individual basis and used to identify overexposed workers. In the lead industry high individual blood lead values indicate persons who should be removed from the source of exposure until their blood lead concentration returns to a low value.

It must be kept in mind that in biological monitoring, personal human data are obtained and the ethical aspects must receive attention. In particular, the monitoring procedure itself must be without health risk, sufficient information must be given to the subjects before and after monitoring, and individual results must remain confidential.

Exposure assessment for epidemiology

In epidemiological investigations actual exposure measurements made by occupational hygienists for individuals in the cohort are rarely available. Historical exposure measurements are usually difficult to reconstruct due to insufficient data, incomplete work histories, and changes in manufacturing processes, control technology, and industrial hygiene measurement techniques. Instead, estimates of exposure have to be made in a variety of ways which have considerably less precision than exposure measurements.

If no quantitative exposure data are available, epidemiological investigations are still possible by using exposure surrogates such as:

- ever/never employed in the industry or in a specific job category
- length of service in the industry or in a specific job category
- years since first exposure.

These measures are not quantitative and usually do not permit exposure–response analysis.

Some occupational exposure data may be available for the study population, although this is rarely comprehensive. In other circumstances exposures must be estimated in some way.

Historical exposure data

Occupational hygiene exposure data may be available for groups of workers, linked for example to job categories. As an approximation an individual's exposure might be assumed to be equal to the average for his job group. There are several potential sources of error in this approach, including:

- Intentional bias: occupational hygienists usually investigate situations they believe give high exposures in order to direct control measures efficiently. Historical exposure data might, therefore, not be representative of average exposures.
- Changes in technology: exposure data obtained prior to or following changes to processes or control equipment might be invalid in the study period.
- Personal protection: measurements made where the workforce uses respiratory protection are not measures of exposure but only of environmental concentration.
- Other routes of uptake: occupational hygiene measurements take no account of skin or gastrointestinal uptake.

These last two problems might be overcome if biological monitoring data are available, but this is not very common.

Exposure estimation

Where no reliable measured exposure data are available for the study population, exposure intensities are often estimated on the basis of job descriptions or simply job title. Such exposure estimates can be obtained in many ways, including:

- interviews with subjects with the disease being studied, their relatives, and/or their employers
- measurements of current exposure (suitable where process and control technology and work practice have not changed greatly)
- simulation (repeating historical work practice so that exposure measurements can be performed)
- exposure panels (teams of experienced hygienists and physicians familiar with the industries being studied) to estimate exposure intensities for industries or job categories
- job exposure matrices (a table listing typical exposures against job titles).

Exposures estimated in these ways are usually very imprecise and can often only be used for ranking (e.g. low, medium, high).

Control

It must be remembered at all times that the primary responsibility of the occupational hygienist is to do what is necessary and reasonable to control hazardous exposures, not just to assess them. Control might be achieved in many different ways, ideally at the source of the emission by eliminating or substituting processes or materials, or by process modification. A second option is to try to contain or capture the emission after it has been produced. This might be achieved by ventilation of some sort, or by segregation of the process from the workforce. Generally the least satisfactory option is to try to protect the workers by the use of protective clothing and respirators after their working environment has become contaminated.

All possible routes of exposure (respiratory, skin and gastrointestinal tract) must be identified if adequate control is to be achieved. For example, during the application of some pesticides the major uptake is via skin contact, and efforts to control the airborne substance will not be rewarded by greatly reduced overall exposures. In such cases the control strategy might involve reduced handling, avoidance of spills, suitable clean-up procedures, waste disposal arrangements, protective clothing, washing facilities, provision of work clothes and changing rooms, and a range of operational rules covering food, drink and tobacco use at work.

Several hierarchical schemes are possible for the control of substances in the workplace, of which the following is an example. The relative order of importance of the various control options can be different in different processes.

- Control at source:
  (a) eliminate or substitute the process or the hazardous substances
  (b) modify the process to reduce the emissions.
- Transmission control:
  (a) segregation of the process from the workers
  (b) good housekeeping to prevent re-release from, or contact with, surface contamination
  (c) local exhaust ventilation
  (d) dilution ventilation.
- Personal protection:
  (a) personal hygiene
  (b) respirators
  (c) gloves
  (d) impervious clothing, etc.

Summary

This chapter provides an overview of the contribution of occupational hygiene to industrial toxicology in terms of recognition, evaluation and control, and in epidemiology and standard setting. Readers who desire more detailed information on any of the topics are referred to the many excellent texts that are available, some of which are listed in the Bibliography to this chapter, and to the various short training and postgraduate courses which are available.

Bibliography

General occupational hygiene

Occupational Health and Hygiene Guidebook for the WHSO (1992) Brisbane: David Grantham.
Plog, B.A. (ed.) (1988) *Fundamentals of Industrial Hygiene*, 3rd edn. Chicago, USA: National Safety Council.

Monitoring

Australian Standards AS 2986–1987 Workplace Atmospheres – Organic Vapours – Sampling by Solid Adsorption Techniques. Canberra.
Gill, F.S. and Ashton, I. (1992) *Monitoring for Health Hazards at Work.* Oxford: Royal Society for the Prevention of Accidents/Blackwell Scientific.
Gray, C.N. and Thomson, J.M. (1984) Passive and active atmospheric monitoring. In: Harrington, J.M. (ed.). *Recent Advances in Occupational Health*, Volume 2. Edinburgh: Churchill Livingstone.
Hering, S.V. (ed.) (1979) *Air Sampling Instruments*, 7th edn. ACGIH. (For evaluation of atmospheric contaminants.) Cincinnati.
Ness (1991) *Air Monitoring for Toxic Exposures.* New York: Nostrand.

Quality control in analysis

NIOSH (1973) *The Industrial Environment, its Evaluation and Control*, Chapter 22. Washington, DC: US Government Printing Office.
NIOSH (1984) Introduction, Section C. In: *Manual of Analytical Methods*, 3rd edn. Cincinnati, OH: NIOSH.
Whitehead, T.P. (1977) *Quality Control in Clinical Chemistry.* New York: Wiley.

Biological monitoring

Baselt, R.C. (1980) *Biological Monitoring Methods for Industrial Chemicals.* Davis: Biomedical Publications.
Documentation of the threshold limit values and biological exposure indices (1991) In: *American Conference of Government Industrial Hygienists*, 6th edn. American Conference of Governmental Industrial Hygienists, Cincinnati.
Hinch, A.L. (1974) *Biological Monitoring for Industrial Chemical Exposure Control.* Boca Raton, FL: CRC Press.
Ho, M.H. and Dillan, H.K. (1987) *Biological Monitoring of Exposure to Chemicals – Organic Chemicals.* New York: Wiley.

Exposure assessment for epidemiology

ACGIH (1991) *Exposure Assessment for Epidemiology and Hazard Control.* Michigan: Lewis.
Gerin, M., Siemiatycki, J., Kemper, H. and Begin, D. (1985) Obtaining occupational exposure histories in epidemiological studies. *J. Occup. Med.* 27: 420–426.
Kauppinen, T. and Partinen, T. (1988) Use of plant- and period-specific job-exposure matrices in studies on occupational cancer. *Scand. J. Work Environ. Health* 14: 161–167.
Rosenstock, L., Logfero, J., Heyer, N.J. and Carter, N. (1984) Development and validation of a self-administered occupational health history questionnaire. *J. Occup. Med.* 26: 50–54.

Chapter 17

Occupational medicine – interface with toxicology

W.-O. Phoon

Background

Although Bernardino Ramazzini (considered the father of occupational health) lived and died a few centuries ago, the subject with which his name is associated is still relatively new. It is rapidly evolving and has developed more subdisciplinary areas than perhaps most other health disciplines.

Most of the original pioneers of occupational health, like Ramazzini, were physicians; for example, Paracelsus, Percival Potts, Charles Thackrah, Alice Hamilton, and Thomas Morrison Legge. Ramazzini himself was a professor of internal medicine at different times in the universities of Padua and Modena in Italy. Cases of poisoning from mercury, lead and other substances and other occupational diseases led him to go out from his consulting room in the hospital to the workplaces from which his patients came. This led to the pioneering studies on occupational diseases encapsulated in his famous textbook on the subject.

The early years of the development of occupational health focused on the diagnosis and treatment of overt occupational diseases, including those from chemicals and physical agents. This stage of development was not very different from similar periods in the development of internal medicine. A rich repository of new knowledge about the symptoms and signs of occupational diseases then rapidly accumulated. The 19th and 20th centuries and the early years of the present century could be termed the 'classical age of clinical occupational medicine' in the sense that the symptom complexes of conditions such as lead poisoning, mercury poisoning and silicosis all became fully documented.

The input of other sciences, such as nursing, psychology, physiology, biochemistry, pharmacology, toxicology, chemistry, physics and engineering greatly enhanced the development of occupational health as a scientific discipline. Epidemiology and statistics gave added strength to the systematic detection and quantification of occupational conditions. More recently, the appearance of the newer (and largely applied) professions, such as physiotherapy, vocational therapy and ergonomics, have added to the repertoire of skills needed to control and prevent occupational hazards and to rehabilitate victims from them.

Nowadays, occupational health practitioners come from a wide and expanding spectrum of different basic disciplines. Each category of such practitioners has both a unique part to play, drawing upon his or her special skills, as well as a collective part to play as a member of a multi- or interdisciplinary team in the prevention, management or rehabilitation of occupational illnesses (Phoon 1992a).

Definition of occupational medicine

As with most other subjects, there are many definitions of occupational medicine. A simple definition is that occupational medicine encompasses the role played by medical practitioners in occupational health. As such, occupational medicine focuses on the prevention, diagnosis and management of occupational injuries and diseases, including health surveillance.

Training for occupational medicine

Occupational medicine should be regarded as an extension of the work of a basic practitioner. The usual period of training of a doctor at medical school is 6 years or longer. After that there is normally a compulsory period of supervised work experience ('housemanship' or 'internship') of at least 1 year. Most occupational physicians undergo a few more years of supervised training work in hospital, community or general practice before embarking upon formal training in occupational health. Professional training for the latter subject is generally relatively brief. In many universities, doctors and others can obtain their Master of Occupational Health and Safety degree or equivalent qualification after only approximately 10 months of full-time work, followed by the part-time preparation of a treatise. Treatise preparation may last 1–2 years or more. In general, however, occupational physicians all over the world require additional (usually 2) years of supervised training in practical situations to qualify for registration as a specialist occupational physician, and only after an 'exit' examination. Even then, the duration of formal training in occupational health (as distinct from experience) is appreciably shorter than the duration of formal training for medicine, as a whole. This is little different from training in other medical specialisations, for example surgery or psychiatry.

Training to be a medical doctor is fairly homogeneous throughout the world, though the standards may vary. Occupational health, however, as a subject in the curriculum for medical students, varies immensely. Medical schools often provide as few as 10 hours of training throughout the entire course.

The postgraduate training for medical practitioners in occupational health is far more diverse. In certain countries (e.g. the UK) occupational medicine is generally regarded as a 'clinical discipline'. Consequently there is strong emphasis on clinical teaching. In some other countries there is next to no clinical teaching, and occupational health in such locations, including the role of medical practitioners, is probably best described as a subsegment of public health (Phoon 1992b).

Occupational medicine as applied to toxicology

The medical doctor needs to have at least a reasonable level of competence in general medicine to diagnose and manage toxicological problems. Skill in taking a good medical history, both general and occupational, is essential. Sometimes an exhaustive history is taken, but at all times proper judgement is necessary to sort out 'the wheat from the chaff'. As has been said – with some truth – the most important part of a stethoscope is between the earpieces; good history taking is an art requiring a high level of deductive reasoning sharpened by knowledge and experience.

Taking a history, however competently, is insufficient in itself. The doctor needs to be able to perform a clinical examination to confirm or refute the provisional diagnosis in his or her mind after the taking of the history. This is especially important when it is not easy to find organ system medical specialists (such as dermatologists, liver specialists or respiratory physicians) who are particularly knowledgeable in occupational health. This situation is often aggravated by over-specialisation (especially in developed countries).

Often toxicological problems may fall between different areas of internal medicine because such problems may affect several organ systems in the same case; 'general' internal medicine has lost prestige and become rather unfashionable in many developed countries. The danger of missing the 'woods' because of overemphasis on individual 'trees' is not very conducive to the detection and management of toxicological problems.

Intoxications or diseases from chemical substances, which an occupational physician may encounter, fall into the following main categories:

- exposure to a single chemical
- exposure to multiple chemicals
- acute or short-term exposure
- subacute or long-term (chronic) exposure; for some effects (e.g. malignant mesothelioma from asbestos) the latency period (between exposure and the development of symptoms and signs) could last as long as 35–40 years.

Occupational physicians should be aware that many factors could predispose to ill effects from chemicals, including the following:

- Allergy to chemicals; in many cases, this allergy is increased with repeated exposures (e.g. latex or nickel allergy).
- Hereditary or constitutional deficiencies; for example, glucose-6-phosphate deficiency, a hereditary enzyme defect, predisposes red blood cells to haemolysis, and may therefore aggravate the damage from chemical haemolysins (e.g. arsine).
- Pre-existing organ-system damage; for example liver damage from viral hepatitis or ethyl alcohol increases the ill effect from other hepatotoxicants, such as some of the chlorinated solvents.

The role of the physician for workers exposed to chemicals involves medical examination and diagnosis and management of chemical intoxication.

Medical examinations include:

- Pre-employment or pre-exposure examinations which are invaluable in making sure that susceptible workers are not exposed to chemicals (e.g. pregnant women to lead), or to establish pre-exposure 'baselines' (e.g. cholinesterase blood levels for seasonal sprayers of pesticides on crops).
- Periodic examinations of exposed workers: in most countries of the world, workers exposed to chemicals (or other hazards) are required by law to undergo such periodic examinations (e.g. those exposed to solvents or heavy metals). Biological monitoring is often required. This usually consists of measuring the amount of a potentially toxic chemical or its metabolite or metabolites in blood, urine or expired air in the exposed workers.

'Health surveillance' is the term usually applied to these two procedures of physical examinations and biological monitoring in such exposed workers. The findings from the two procedures enable the physician to determine the degree of chemical exposure in each worker and the risk to health.

Diagnosis and medical management of cases of chemical intoxication: Through having a sound knowledge of chemical risks and by taking a medical history, physical examination, and ordering appropriate laboratory tests, the physician can usually make a diagnosis of chemical intoxication. In other cases, tests for chemical concentrations in the work environment or special investigations (e.g. neurobehavioural, nerve conduction, pulmonary function tests or radiological examinations) may be required before a definitive diagnosis is reached.

Medical management of cases of occupational poisoning is described briefly towards the end of this chapter.

To illustrate the principles in the diagnosis and management of toxicological problems in occupational situations, two case studies are discussed here. They relate to inorganic lead and organic solvents, respectively.

Case study 1: inorganic lead

Lead is, of course, a metal well known to humanity for a very long time. Even in modern times inorganic lead is still fairly extensively used, for the manufacture of accumulator storage batteries, as alloys (usually mixed with tin), and so on. Organic lead is used, though now to a diminishing extent, as an antiknock agent in petrol fuel.

The diagnosis of inorganic lead poisoning is based on the parameters described below.

History of exposure

Whenever possible, a history of exposure to significant amounts of inorganic lead should be established. It is pointless to talk merely of 'lead exposure', because in almost any environment, especially in urban areas, the exposure to small amounts of the metal is inevitable.

For practical purposes, not only the amount of exposure but also the duration of exposure should be significant in occupational circumstances. It is very rare in those circumstances for a worker to suffer from lead poisoning after an exposure of less than several weeks. A duration of at least several months is more usual.

Possibility of absorption

A worker can be working in an atmosphere heavily laden with lead but may not absorb any because he or she is using completely effective personal protective equipment throughout exposure. However, the use of personal protective equipment is not comfortable, fraught with possibilities of human error (for example, a respirator not fitting the facial contours well because of wearing a beard), the filter may not be changed regularly, or the person may simply not wear the device except very intermittently. Therefore, personal protective equipment must not be relied on as the main means of preventing lead poisoning. The use of other methods is more dependable, for example elimination or exhaust ventilation to control the concentration in the work environment.

High exposure occupations (e.g. lead smelting (primary and secondary), burning off of lead paints, the assembly of lead accumulator batteries, and lead disking) are usually attended by some degree of lead absorption, even where good standards of occupational hygiene apply. Other exposure situations (e.g. mixing of lead in paints and lead soldering) are associated with lower but not absent risks of lead absorption.

A consistent or suggestive history

The 'classical' syndrome of lead intoxication, with symptoms of weight loss, anorexia, epileptiform fits, constipation, abdominal colic, wrist drop, wasting, anaemia and Barton's line on the gums, is nowadays extremely rare; much more common would be a history of general ill health with rather ill-defined symptoms, although the more severe and dramatic symptoms and signs mentioned above can still be seen in some less developed countries.

Of particular significance is a history of usually several months after taking employment in a job situation with apparently appreciable exposure to the metal. Poisoning can occur as a result of changes in either the work environment or work conditions. For example, a person might absorb more lead than usual when the exhaust ventilation system is not working properly; or a worker might work overtime so the exposure proportionately increases with the same ambient concentration of the substance.

Deterioration in health can often be elicited by asking a worker whether he is as well as he was, say, 6 to 12 months ago, or what his weight was then compared with now. Vague symptoms such as lassitude, inability to sleep or disturbed sleep, a metallic taste in the mouth, undue tiredness and muscle aches could provide useful clues.

Obviously, the physician or nurse taking the history has to be careful not to ask too many leading questions! It is advisable to let the patient describe his or her medical complaints first without any or very little prompting and then follow up with questions that address specific points if necessary.

Physical examination

Signs such as mild anaemia, slight degree of muscle wasting, and evidence of peripheral neuropathy may be present. Depending on the symptoms, the exclusion of other disease not related to lead exposure should also be borne in mind by the examiner.

Laboratory investigations

A wide spectrum of laboratory investigations can be performed in cases of suspected lead poisoning. They usually fall into the following categories.

- Concentrations of lead in body fluids (e.g. lead in blood or urine).
- Depression of concentrations of enzymes in the blood relating to haem synthesis (e.g. γ-aminolaevulinic acid dehydratase (ALAD)).
- Excess of excretion in the urine of substances normally used for haem synthesis in the urine because of interference with haem synthesis (e.g. excess of aminolaevulinic acid or zinc protoporphyrin in the urine).
- Effects of lead on body systems (e.g. anaemia and reduced packed cell volume due to interference with haem synthesis, reduction in nerve velocity, etc.).

In children there is sometimes the presence of a 'lead line' in the shafts of long bones near the epiphyses, but this is not seen in adults. What laboratory investigations are made depends on what the symptoms and signs indicate. The laboratory investigation generally agreed to be the most useful, and indeed the most used (subject to quality control), is the lead concentration in blood (PbB).

The use of the laboratory investigations described in the first three of those listed is commonly referred to as 'biological monitoring', which can be defined as a regular measuring activity in which selected, validated indicators of the uptake of toxic substances are determined in order to prevent health impairment. The regular use of biological monitoring in combination with medical examination, as well as investigative methods for the detection of early effects of toxic substances upon body systems (such as the last investigation mentioned above) is usually collectively referred to as 'health surveillance' (of workers who are exposed to such substances).

Environmental monitoring

For single cases seen at the usual clinic, it is not always practicable to inspect the workplace or to measure the lead concentration there. For epidemiological studies or for establishing surveillance for groups of workers, however, measurements of lead levels in the work environment are essential. There is a 'lag period' in the correlation of environmental lead to blood lead levels, as the latter reflects absorption 1–2 weeks previously rather than at the present time.

Case study 2: organic solvents

Most of the substances used as solvents are organic chemicals employed in a large variety of purposes such as degreasing, painting, thinning, extraction and cleaning. As is well known, solvents can cause a wide spectrum of health effects.

- Solvents are one of the leading causes of occupational dermatitis. The majority of organic solvents are depressants of the central nervous system.
- Acute exposure may cause narcosis, excitation, headache, dizziness, nausea and confusion. Severe exposure may cause convulsions, coma and failure of the vital brain centres, sometimes with fatal results. Chronic exposure may lead to undue fatigue, headache, disturbed sleep, intellectual impairment, poor memory, impaired vision, reduced manual dexterity and motor weakness. Certain solvents, such as carbon disulfide, methyl bromide, triorthocresyl phosphate, n-hexane and methyl-n-butyl ketone, may cause peripheral neuropathy.
- Effects on the respiratory system are mainly irritation to mucous membranes of the throat and respiratory passages. Several solvents, such as carbon disulfide, halogenated carbon compounds and benzene can cause effects on the cardiovascular system. The risk of angina pectoris increases with carbon disulfide exposure. Benzene, toluene and halogenated carbon compounds can cause cardiac arrhythmias, especially ventricular fibrillation.
- Hepatotoxicity is a possibility with sufficient exposure to any organic solvent. Carbon tetrachloride and tetrachlorethane are the most toxic solvents in this regard. Other hepatotoxic solvents include chloroform, tetrachlorethylene, trichlorethylene, diethylene dioxide and diethylene glycol.
- Aromatic hydrocarbons, halogenated carbon substances, and glycols can cause damage to the kidneys, in the form of acute renal failure. Aromatic hydrocarbons can cause glomerulonephritis. Acute renal failure due to toluene has been described in glue sniffers.
- Bone-marrow depression and leukaemia have been associated with benzene.
- Carbon disulfide and carbon tetrachloride have been implicated in the reduction of fertility in both sexes in humans. Chromosomal aberrations have been detected following exposure to benzene.

The diagnosis of solvent poisoning can be extremely complex, especially since in many occupational situations several solvents can be used together. It is not always possible to identify what solvents were actually used by patients who are seen long after exposure to solvents has ceased.

The diagnosis of organic solvent poisoning is based on the parameters described below.

History of exposure

The 'index of suspicion' of the medical examiner should be raised if the patient comes from one of the many trades or handles one of the many processes known to be associated with solvent exposure.

Possibility of exposure

Organic solvents are used for many purposes in modern life. Workers can be exposed while using these chemicals as cleaners, degreasers and thinners, in industries related to engineering, printing, painting, construction and repairs. Solvents are also used extensively in the manufacture of different chemicals. They also occur in a wide range of domestic chemicals. Often a distinction must be made as to whether solvent exposure has occurred from occupational or domestic circumstances.

A consistent or suggestive history

Unfortunately, there are no really pathonogmonic symptoms for solvent intoxication. Patients presenting with mild psychotoxic symptoms referable to the nervous system are often misdiagnosed as cases of psychoneurosis or even malingering. In acute cases, there may be a fairly well-defined correlation between work exposure and the onset and duration of the symptoms. The patient may be quite well when he or she arrives at the workplace, but may begin to feel more tired and have headache as the working day proceeds. When the worker returns home at the end of the day, he or she may feel so tired that they do little else except eat the end-of-shift meal and go to bed early. If the exposure is intermittent, for example, the worker may not do spray painting every day, he or she may be able to volunteer the information that only on spray-painting days do these symptoms arise. In instances where different substances are used for different forms of spray painting, the patient may sometimes be able to identify those substances that are more prone to cause symptoms. Another useful clue is whether or not the patient had similar symptoms before starting the present job. However, enquiry then has to be made about whether any previous jobs carried similar risks of solvent exposure.

Tiredness, or rather undue tiredness, can be very subjective. Two individuals can show very different degrees of tiredness after the same amount of mental or physical exertion. In such circumstances it is useful to ask the patient whether the volume of work has changed recently. If not, does he or she feel the same extent of tiredness as, say, 6 to 12 months ago? Does the tiredness persist on weekends or long vacations? It must be remembered, in this connection, that workers sometimes engage in more frenetic activities during their own time than at work. Moreover, the use of organic solvents in domestic or recreational activities must also be considered, for example, DIY activities or hobbies involving the use of glues, thinners, and so on.

Sometimes, solvent intoxication may be precipitated by an increased volume of work, the introduction of a new substance containing another solvent, or the breakdown or cessation of usage of a normally effective exhaust ventilation system or other preventive measures like a respirator. Therefore questions directed at eliciting these kinds of information are often necessary.

The chronic effects of solvents on the neurological system do not necessarily have time relationships to exposure and its cessation. Effects such as memory loss and intellectual deficit can persist long after cessation of exposure, and if exposure has been sufficiently intense or for long periods, may even be permanent. If the presentation relates to the liver, a provisional diagnosis of solvent poisoning is generally made on the basis of a history consistent with exposure to a hepatotoxic solvent and the exclusion of other known toxic substances affecting the liver. It must be borne in mind, however, that other agents injurious to the liver, for example, alcohol and hepatitis virus, could operate synergistically either to provoke or aggravate the liver disease.

The process of excluding other causes of reported symptoms applies to the diagnosis of cases presenting with skin, respiratory, renal or cardiovascular complaints.

Physical examination

A comprehensive and thorough examination must be done, since solvent exposure can affect so many organ systems and assume so many forms. In general, special concentration should be on the neurological system and liver, although other systems may be affected at the same time. Like most other cases of occupational dermatitis, solvent dermatitis is a form of contact dermatitis. Typically it is on the exposed parts of the body, for example, hands and forearms. Most cases of solvent dermatitis are due to irritation. A minority of cases are allergic in nature, and in those patch testing can be positive.

Laboratory investigations

There are very few specific tests. The usual laboratory investigations fall into the following categories:

- tests for increased levels of solvents (e.g. in expired air or for slowly metabolised solvents, in the urine)
- tests for increased levels of metabolites excreted (e.g. in the urine, phenol levels after benzene exposure, trichloroacetic acid and similar compounds after trichlorethylene exposure, hippuric acid after toluene exposure, and methyl-hippuric acid after xylene exposure)
- tests for functions of target organs: central nervous system (electroencephalography, psychomotor tests of memory and reaction times, etc.); peripheral nervous system (nerve conduction velocity, electromyography, and use of visual and somatosensory evoked potential).

Environmental monitoring

This could be useful but is not always practicable. Moreover, results only refer to the present situation and not necessarily to the past when the concentrations of the solvent (or solvents) could have been quite different.

Medical management of cases of occupational poisoning

The management of occupational poisoning varies not only with the cause of the poisoning but also with other factors such as:

- severity of the health effects
- whether the health effects are acute or chronic in nature
- the availability of specialist advice, such as poisons advisory and treatment centres
- the level and nature of nearby health services
- the skill and expertise of the occupational health personnel concerned.

In some cases of occupational poisoning there are fairly specific remedies such as antidotes or chelating agents. In others, general measures such as supporting treatment, exchange transfusion, dialysis, and so on, may be used. Management should include proper attention to electrolyte balance, nutrition and oxygenation. Particularly for severe cases, careful watch should be kept for complications such as chest infections and pulmonary oedema, and liver and kidney failure, and the appropriate treatment promptly instituted.

In the present context, it is not appropriate to discuss in detail the medical management of cases of occupational poisoning. In many countries of the world, poisons advisory centres exist, which can often provide invaluable information concerning antidotes to specific poisons. Just as important perhaps as specific antidotes is supportive treatment, such as the maintenance of vital functions in the victim by resuscitation and assisted respiration, proper nutrition, electrolyte balance, and prevention of secondary infection. Further, advice on contraindicated treatments and interventions may be crucial (such as oxygen treatment for paraquat poisoning).

For all serious cases, it is best to admit the patient to hospital for observation or treatment. Pulmonary oedema from chemicals may not manifest itself until several hours after exposure. Therefore it may be wise to err on the side of caution and admit all cases of acute exposure to respiratory poisons for at least 24 hours for observation and whatever treatment may be necessary (Phoon 1992c). In cases of chronic poisoning (e.g. lead poisoning) it is prudent, moreover, to treat patients with chelating agents as inpatients. Chelating agents can give rise to severe electrolyte imbalance and may precipitate attacks of acute poisoning from mobilisation of metal stores from the bones or other organs or tissues into the blood stream.

Rehabilitation of cases of occupational poisoning should be planned and implemented as soon as possible and not just before the patient is ready for discharge from the outpatient department or ward. The duty of the occupational physician includes continual assessment of either temporary or permanent disability of the affected worker, and advice to the workplace or family doctor on what duties that worker could undertake. Often modified duties or shorter hours may be advised, at least initially after a long period of absence from work. Such rehabilitation programmes may require the expertise and cooperation of specialists in rehabilitation medicine. Of course, on return of the rehabilitated worker to the workplace, it is absolutely vital to ensure that the exposures which precipitated the health effects have been eliminated or controlled, so that the same problem does not recur.

In cases of permanent disability, the occupational physician often has the duty of assessing the extent of that disability, although he or she often shares that responsibility with specialists in the particular organ system involved (for example, thoracic physicians).

Conclusions

The role of the physician in occupational health (occupational medicine) is of vital importance to the prevention, diagnosis, management and rehabilitation of workers exposed to toxic chemicals or suffering ill effects from them. To perform that role, the physician has to have a good grasp of knowledge about chemicals and to be at the same time competent in clinical medicine. The physician also has to:

- work closely with his or her colleagues in the occupational health team, such as nurses, hygienists and toxicologists;
- be able to relate well to workers, supervisors and managers in the workplace;
- cooperate with other health personnel, such as family doctors, organ-system medical specialists, and members of the rehabilitation team.

Finally, the physician must keep abreast of the plethora of new knowledge about chemicals and their possible effects upon the human body.

Bibliography

He, F.S. (1987) Occupational neurotoxicology: current problems and trends. *Keynote Addresses of XXII International Congress on Occupational Health*. Sydney: XXII ICOH Organising Committee, pp. 48–77.

Lauwerys, R. (1989) *Detection of the Nephrotoxic Effects of Industrial Chemicals: Assessment of the Current Screening Tests (Lucas Lecture)*. London: Faculty of Occupational Medicine, Royal College of Physicians.

Phoon, W.-O. (1992a) Occupational medicine: principles and practice (special article). *CME Rev.* 2: 16–18.

Phoon, W.-O. (1992b) Education and training in occupational health in developing countries. In: Jeyaratnam, J. (ed.) *Occupational Health in Developing Countries*. Oxford: Oxford University Press, pp. 413–431.

Phoon, W.-O. (1992c) Incidence, presentation and therapeutic attitudes to anticholinesterase poisoning in Asia. In: Ballantyne, B. and Marrs, T.C. (eds) *Clinical and Experimental Toxicology of Organophosphates and Carbamates*. Oxford: Butterworth-Heinemann, pp. 482–489.

Phoon, W.-O. (1996) Cardiovascular diseases. In: Jeyaratnam, J. and Koh, D. (eds) *Textbook of Occupational Medicine Practice*. Singapore: World Scientific, pp. 60–73.

Phoon, W.-O. (1998) Training of health and safety professions. In: Stellman, J. (ed.) *Encyclopaedia of Occupational Health and Safety*. Geneva: International Labour Office, pp. 18.22–18.26.

Phoon, W.-O. and Ong, C.N. (1982) Lead exposure patterns and parameters for monitoring lead absorption among workers in Singapore. *Ann. Acad. Med. Singapore* 1: 593–600.

Ramazzini, B. (1983) *De Morbis Artificum Diatriba*. Typis Antonii Capponi, Mutinae, 1700. Translated by Wright, W.C. and Ramazzini, B. *Diseases of Workers*. New York: Classics of Medicine Library, Gryphon Publications.

WHO (1986) *Environmental Health Criteria, 63. Organophosphorus Insecticides. Report of a WHO Task Group on Organophosphorus Insecticides*. Geneva: World Health Organization.

Chapter 18

Epidemiology – interface with toxicology

T. Driscoll and C. Winder

Introduction

Epidemiology underpins a great deal of the research carried out in occupational toxicology. However, while epidemiological data is commonly consulted in occupational toxicology, the process of collecting data in such a way so as to be able to make clear conclusions is invariably more problematic. The process of designing, conducting, analysing and interpreting the results of such studies is part of epidemiology.

It is not possible to interpret the results of toxicological studies in workers unless an appreciation of the underlying principles of study design is known. This chapter covers a short introduction to occupational toxicological epidemiology, and is an overview of the topic, covering some basic principles, terms and measures.

Definition of epidemiology

The traditional meaning of epidemiology has denoted the study of human illness (Greek *epidemos* means 'upon the population') (Hernberg 1992).

However, several definitions of epidemiology have been suggested. The simplest is that epidemiology is the study of the occurrence of disease or other health-related characteristics in human populations (Olsen *et al.* 1991). It has also been classically defined as the study of the distribution and determinants of health-related events in human populations and the application of this study to the control of health problems (Karvonen *et al.* 1986; WHO 1989). Hence, in epidemiological research, the primary units of concern are groups of persons, rather than separate individuals (Friedman 2003).

For that matter, epidemiology usually requires the comparison of the occurrence of diseases in two different groups of people. The number of new diseased individuals in a particular group of people depends upon the number of people in the group and the time period over which new cases are recorded. Consequently, the size of the group and the duration of the study is important (Olsen *et al.* 1991).

A substantial proportion of epidemiological studies attempt to establish a relationship between the occurrence of a health problem and some preceding exposure. Part of this demonstration is to estimate the magnitude of the relationship and to assess the possible existence of a causal association between exposure and health outcome (Christie 1988).

Epidemiology: assessing risks to health

Early epidemiology

As a concept, epidemiology dates back to Hippocrates, in that his writing describes the occurrence of diseases in various populations. Epidemiology as a discipline has no precise beginnings, although through the ages, physicians concerned with outbreaks of cholera, typhoid, bubonic plague, influenza and other infectious diseases were probably the first epidemiologists. Epidemiology grew out of medicine, but more importantly, it grew out of a need to understand what caused deaths in a population. Initially, epidemiology looked at the occurrence of infectious diseases.

To collect information about deaths in London from plague and so forth, a weekly tally (Bills of Mortality) was initiated around 1532 to collect information on causes of death. In 1662, a London haberdasher (John Graunt) analysed records from over 100 years of these Bills of Mortality, and observed that deaths in populations exhibited a numerical regularity from year to year, and laid the foundation for some critical epidemiological concepts, including the development of the life table.

Sir Percival Pott reported the first occupational cancer in the 1780s when he noted the increase of scrotal cancer in chimney sweeps. It was not until the introduction of gas chromatography in the 1930s that the polyaromatic hydrocarbon association with this condition was identified, suggesting cause and effect.

William Farr, a physician working in London as the Superintendent of the Statistical Department of the General Register Office in the mid-1800s, developed recording systems for data on death and began looking for patterns and associations. The basic approaches developed in these systems are still used today (although pen and paper approaches are a thing of the past). Farr was also one of the first investigators to analyse data on the occupational causes of mortality.

The most well-known person regarded as the founder of epidemiology is John Snow, a physician practising in Victorian London. In 1855, a cholera epidemic occurred in London. Using some of William Farr's data on cholera cases, and doing some investigation of his own (including mapping 500 cases) he identified that nearly all the deaths had taken place within a short distance of the water pump at Broad Street. He made representations to the Board of Guardians of St James parish, and the pump was disabled by removing the handle. The incidence of cholera diminished.

While this would have been enough for most physicians, Snow went further. He collected data on the water supply to various districts around London, and compared them to numbers of cases of cholera. By developing a measure that related the number of deaths to the number of persons in the population, he was able to show that a high incidence of cholera was observed in certain areas. He was able to match these areas to particular suppliers of water. The Lambeth Company supplied water from the north of London, where it was free of the sewage of London; the Southwark and Vauxhall water supply was not.

Again, this would have been enough for most investigators, but not for John Snow. It was possible to suggest that the differences in cholera incidence between various areas could be something other than water supply (for example, social class). Snow identified that one area was supplied by both sources, and went from house to house where cholera deaths had occurred, asking for the name of the water supplier. Analysis of the data showed that houses supplied with Lambeth Company water had a death rate of 37 per 10,000 houses; while the corresponding rate for houses supplied from Southwark and Vauxhall was 315 per 10,000 houses (with a background London rate of 59 per 10,000 houses).

Snow's work developed a model of epidemiology still in use today:

- recognition of an association between exposure and effect (water supply and death from cholera)
- formation of a research question (sewage in water causes cholera)
- collection of information to substantiate (or repudiate) the research question (deaths from cholera were higher in districts supplied by one particular company)
- recognition of alternative explanation (social class was associated with the different areas as well)
- development of a method to minimise effects of alternative explanation (looking at cholera death rates within, not between, areas)
- minimisation of collection of false information (identifying which water company supplied water to which house).

Of course, Snow was trying to prevent cholera. This he did. But perhaps unseen in this story, is that by taking such detailed steps to prove beyond doubt that association between water supply and death from cholera, Snow's work developed epidemiology into purposeful research.

Epidemiology continued to develop slowly through the 19th and early part of the 20th century, with attention on the theoretical study of how epidemics spread.

One major impetus to epidemiology occurred after the Second World War when a significant increase in lung cancer in men was reported (before 1940 lung cancer was virtually a medical curiosity). Perhaps not surprisingly, cigarette smoking was suspected.

Studies investigating this problem attempted a range of approaches:

- studying people with lung cancer and collecting information on past habits (these became known as case studies)
- investigating whether people with lung cancer differed from people without lung cancer and matching for various factors (the case–control study)
- smokers and nonsmokers were followed and the rates of a variety of health conditions, including lung cancer, were measured (the cohort study).

These formed the basis of a more intensive phase of epidemiological development. Other types of study evolved to deal with the methodological or interpretive problems arising from these studies. Also, study of other exposures or other effects began to spread.

For example, in 1960 a paper was published in the *British Medical Journal* reporting cases of an apparently rare cancer of the lining of the lung (the pleura) in South African asbestos miners (Wagner *et al.* 1960). Asbestos exposure was associated with all but one of 33 cases of pleural mesothelioma. This descriptive study of a new occupational cancer did little in the way of 'proving' that asbestos caused cancer; however, only the foolhardy would ignore its findings.

One relatively recent development that has assisted in the development of extraordinarily complex associations has been the development of the computer, and refinement of biostatistical techniques that allow systematic analysis of data.

Epidemiology: some important concepts

Epidemiology is the study of health-related phenomena in groups of people. It attempts to identify the distribution and determinants of health-related phenomena (see Figure 18.1).

The results of epidemiological analyses are used to better identify, control or prevent health-related phenomena.

Epidemiology often involves observation without intervention, although intervention followed by observation is also commonly used in public health and observation followed by intervention is common in occupational health and safety.

Unlike scientists involved in laboratory investigations, epidemiologists are rarely able to exert significant control over the parameters of their studies. Instead,

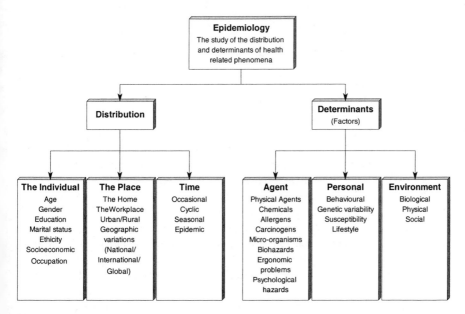

Figure 18.1 Factors in the study of epidemiology.

they usually attempt to control the study through careful selection of study subjects, rigorous methods of data collection, and appropriate techniques of data analysis.

Why bother with epidemiology? Although a number of conditions (or 'outcomes') and exposures have an obvious cause–effect relationship (for example, acute lung damage from inhalation of chlorine gas), the relationship between exposure and outcome may not be obvious for a number of reasons. There may be a long delay between exposure and development of the outcome (for example, 30 or 40 years for asbestos and mesothelioma). The outcome may be rare (for example, liver angiosarcoma from vinyl chloride monomer exposure) or the exposure has not been seen before (for example, HIV). There may be many causes of a particular outcome, thus obscuring the role of individual exposures (for example, smoking obscuring the role of chromium in causing lung cancer). To overcome these problems, and discover the true relationships between various exposures and outcomes, it is necessary to study groups of people rather than individuals. Therefore, a competence in epidemiological concepts is required.

The underlying principle of epidemiology is that certain outcomes (for example, liver disease, noise injury or lung cancer) are caused by certain exposures (such as carbon tetrachloride, noise or nickel carbonyl), and that the relationship between exposure and outcome follows some sort of logical and predictable pattern. Note that epidemiology does not address the problem of idiosyncratic reactions, except on a group basis.

The basic practical principles of epidemiology are to:

- identify causes
- identify sources
- identify exposures or means of transmission
- identify mechanisms of effect
- identify trends
- evaluate the effectiveness of interventions, controls and treatments.

To understand epidemiological studies there is a need to be able to understand key epidemiological concepts. Some of these are outlined below.

Epidemiological terms

- **Study question:** The problem that the study has been designed to answer (e.g. does workplace exposure to dust/chemicals cause lung disease?).
- **Study base:** The population (or population-time) who are eligible for selection for the study and about whom inferences are to be made. A properly defined study base must be defined in terms of both time and place, and the study base is fundamental to the success or failure of an epidemiological study (e.g. all workers in a factory).
- **Study sample:** The fraction of the study base being studied (e.g. all production workers).
- **Study factor:** The exposure or factor being studied (e.g. workplace exposure to dust/chemicals).

- **Outcome factor:** The outcome (disease/condition) believed to be associated with the study factor (e.g. all lung diseases excluding asthma).
- **Selection:** How subjects are selected for inclusion or exclusion in the study (e.g. all workers with over 5 years' exposure).
- **Measurement:** The way relevant parameters are determined in the study. Method(s) used to measure or assess the study factor, the outcome factor, and any confounders or bias (e.g. interview questionnaire for employment and lifestyle factors, lung function tests, workplace assessment).
- **Cases:** Subjects that have the outcome factor (e.g. workers with bronchitis, emphysema or lung cancer).
- **Controls:** Subjects that do not have the outcome factor (e.g. workers without any lung disease).
- **Rate:** The number of affected people, usually expressed in terms of the population in the study (e.g. 17 cases of back pain in a company with 380 employees: rate $= 17/380 \times 100 = 4.47\%$).
- **Confounders:** Factors that distort (falsify) the relationship between study factor and the outcome factor (e.g. smoking, age, gender).
- **Bias:** Distortion of the true relationship between the study factor and the outcome factor due to some systematic error. Bias may result from problems with subject selection or with measurement (e.g. prompting in questions, recall bias by affected workers).

Example

Consider a study of workers at a chemical manufacturing plant designed to determine the influence of exposure to one or more chemicals in causing liver disease:

- the **study question** could be 'Does exposure to one or more chemicals at this plant cause liver disease?'
- the **study base** is all employees at the plant
- whilst the **study sample** would be the workers who agreed to participate in the study
- with the **study factor** being exposure to the chemical(s)
- the **outcome factor** being the presence of liver dysfunction
- subjects might be **selected** randomly (for example, every fifth worker), or by asking everyone and including the volunteers in the study
- the study factor might be **measured** by taking samples of the air where the subject works
- the outcome factor might be **measured** using standard tests of liver function
- possible **confounders** include alcohol use and current or previous liver infection
- potential confounders might be **measured** using a questionnaire
- the study might be **biased** because all of the exposed workers had the blood for their liver function tests collected in one type of tube, whereas all of the unexposed workers had their blood collected in a different type of tube that affected the liver function test results.

Measures used in epidemiology

Incidence and prevalence

Incidence and prevalence are two terms commonly used in epidemiology. Many definitions are available, but they are all basically variations on a theme. Strictly speaking, the incidence and prevalence refer to the number of persons with a particular attribute (usually called 'cases'), but more commonly they are used to refer to the rate of cases.

When expressed in terms of rate, incidence and prevalence have a numerator (number of cases) and a denominator (number of persons in the population), and some indication of the time over which the measurement applies. The choice of an appropriate denominator depends on the question of interest. Usually the population at risk is used. This is the group of persons who are susceptible to developing the outcome.

Incidence is the number of new cases occurring in a certain population over a given period of time. As a rate, this is calculated using the formula:

$$\text{Incidence (I)} = \frac{\text{Number of new cases in a given time period}}{\text{Population at risk during the given time period}}$$

If a population is continually changing at the time when cases are determined (through developing the outcome, being cured, leaving the population, entering the population or dying) the 'person-year at risk' can be used. This is calculated by adding together the length of time during the measuring period at which each individual member of the population is at risk.

$$\text{Incidence rate (IR)} = \frac{\text{Number of new cases in a given time period}}{\text{Person-years at risk during the given time period}}$$

Prevalence is the number of cases existing in a particular population. Usually point prevalence is used, which is the prevalence at a particular point in time. As a rate, this is calculated using the formula:

$$\text{Prevalence (P)} = \frac{\text{Number of cases at a given point of time}}{\text{Population at the given point of time}}$$

An approximation to the point prevalence is given by

$$\text{Prevalence (P)} = \text{Incidence density} \times \text{Duration of course of disease}$$

Therefore, prevalence is the product of incidence multiplied by duration. This relationship is most apparent in a group with a stable, chronic disease. When this occurs, the incidence may be derived, providing that the prevalence and duration are known.

The relationship between incidence and prevalence is shown in Figure 18.2. Factors affecting the observed prevalence rate are shown in Table 18.1.

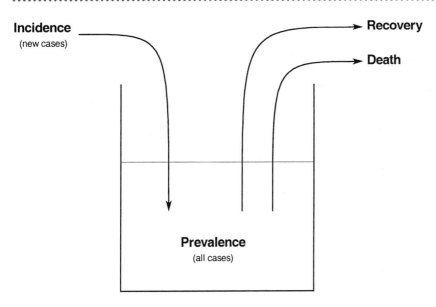

Figure 18.2 The relationship between incidence and prevalence.

Table 18.1 Factors affecting the observed prevalence rate

Increased by ↑	↓ Decreased by
Longer duration of the disease	Shorter duration of the disease
Longer lifespan of cases without cure	High case fatality rate from disease
Increase in number of new cases	Decrease in new cases
Removal of healthy people from the study population	Removal of cases from the study population
Addition of susceptible people	Removal of susceptible individuals (producing a survivor population)
Improved diagnostic facilities (leading to better reporting)	Improved cure rate of cases

Example

Consider a study of hepatitis B in 1100 hospital workers over 5 years. If 100 workers are immune due to previous infection, active disease lasts for 3 months, and the number developing the disease is 2 in each of the 5 years, we can make the following statements.

The **incidence** is 10 cases in 1000 workers per 5 years, which is 2 cases per 1000 workers per year (2/1000/y) or 200 cases per 100,000 per year (200/100,000/year). (Note that the whole population is not used, because not all workers were at risk of getting the disease.) This is called the **cumulative incidence**. We can become even more refined by including in the denominator only those subjects who are at risk, and only for the period of time that they are at risk. Each person would therefore contribute an amount of time between 0 and 5 years, with the sum of these contributions forming the denominator for the incidence. This method, which calculates the **incidence density**, is the most accurate way of determining incidence.

The **point prevalence** would depend on when individuals developed the disease and how long it lasted, but if 3 individuals had hepatitis B half way through the second year, then the prevalence at that time would be 3/1000, or 3000/100,000, or 0.3%.

Measures of risk

Relative risk (RR) is the ratio of the incidence in an exposed group to the incidence in an unexposed group. It is used in random controlled trials and cohort studies. It is measured using the formula:

$$RR = \frac{IR_{exposed}}{IR_{nonexposed}}$$

The relative risk is calculated by establishing four populations of individuals (for example, workers):

- exposed individuals who have the outcome
- exposed individuals who do not have the outcome (normally obtained by subtracting the exposed individuals with the outcome from the total number in the group)
- nonexposed individuals who have the outcome
- nonexposed individuals who do not have the outcome (normally obtained by subtracting the nonexposed individuals who do not have the outcome from the total number in the group).

This can be expressed in a table such as that shown in Table 18.2.

The rate (risk) of the outcome among the exposed ($IR_{exposed}$) is given by:

$$\frac{A}{A + B}$$

Table 18.2 Calculation of relative risk

	Outcome present	Outcome absent	
Exposed	A	B	A + B
Nonexposed	C	D	C + D
	A + C	B + D	A + B + C + D

The rate (risk) of the outcome among the nonexposed ($IR_{nonexposed}$) is given by:

$$\frac{C}{C+D}$$

The relative risk (RR) is given by:

$$\frac{IR_{exposed}}{IR_{nonexposed}} \text{ or } \frac{\dfrac{A}{A+B}}{\dfrac{C}{C+D}}$$

Example

Ninety out of a group of 493 workers in the processing area of a cement factory report skin problems compared with 70 out of 1201 workers in a saw mill. This is represented in Table 18.3.

The relative risk (RR) is therefore:

$$\frac{\dfrac{90}{493}}{\dfrac{70}{1271}} = 3.31$$

There are problems applying the relative risk estimation with knowing what the size of the groups are and what confidence can be given to the data.

If the outcome is not common, then A will be a lot less than B, and C will be a lot less than D, and A/B/C/D is an approximation of the relative risk.

Attributable risk is the rate of the health condition in the exposed group minus that rate in the nonexposed group:

$$AR = IR_{exposed} - IR_{nonexposed}$$

This can be re-expressed as

$$\frac{A}{A+B} - \frac{C}{C+D} = \frac{90}{493} - \frac{70}{1604} = 0.13$$

Table 18.3 Calculation of relative risk from outcomes

	Outcome present	Outcome absent	
Exposed	90	403	493
Nonexposed	70	1201	1271
	160	1604	1764

The **odds ratio** is used to approximate the relative risk in studies where incidence cannot be easily calculated (case-referent and cross-sectional studies). In many case–control studies, data on disease rates are difficult to measure (for various reasons), and the rate ratio must be approximated. The odds ratio is used here. This measures the odds of disease in the exposed group divided by the odds of disease in the nonexposed group.

$$\text{The odds of outcome among the exposed} = \frac{A}{B}$$

$$\text{The odds of outcome among the nonexposed} = \frac{C}{D}$$

$$\text{The outcome odds ratio (OR}_O) = \frac{\dfrac{A}{B}}{\dfrac{C}{D}} = \frac{AD}{BC}$$

$$\text{The odds of exposure among those with the outcome} = \frac{A}{C}$$

$$\text{The odds of exposure among those without the outcome} = \frac{B}{D}$$

$$\text{The exposure odds ratio (OR}_E) = \frac{\dfrac{A}{C}}{\dfrac{B}{D}} = \frac{AD}{BC}$$

$$\text{Therefore, OR}_E = \frac{AD}{BC} = \frac{90 \times 1201}{403 \times 70} = 3.83$$

It should be noted the formula for the OR_O and the OR_E are the same.

While the odds ratio provides a measure of risk, it can be seen that the relative risk is a more accurate calculation.

Precision is the degree of agreement between repeated measures of the same phenomena.

Accuracy is the degree of agreement between repeated measures of the same phenomena and the truth.

Note that a measurement can be very precise (that is repeatable) but inaccurate. However, a measure that repeatably and correctly determines the truth is both accurate and precise.

Standardised mortality ratio (SMR) is the ratio of the number of deaths observed to the number of deaths expected, based on the experience of a comparison population.

For example, the observed deaths in a worker group compared to expected deaths in the group based on the death rate in the whole population.

Many studies use death from one or more causes as the outcome of interest, and compare the rates of outcome between two or more groups. This comparison can be in the form of a relative risk, but can also be expressed in other ways. A number of factors can affect whether or not someone develops a particular condition and, if they do, whether they die as a result. As already mentioned, these factors are called potential confounding factors and must be taken into account in any analysis. There are a number of ways to do this, but a commonly used method in occupational studies is to calculate a standardised rate.

This standardisation is usually made on the basis of age and/or gender. Basically, a calculation for the standardised mortality ratio (SMR) is performed as follows:

- A rate of death is calculated for each level of the variable on which the standardisation is performed. For example, separate rates for males and females, or separate rates for each 5 year age group.
- The rate from one group is then applied to the number of people in the comparison group to calculate an expected number of deaths. In effect we are asking: if the rate of death from one group was applied to the population in the other group, how many deaths would we expect to occur?
- The observed number of deaths (the number that actually occurred) is then compared to the expected number of deaths (the number calculated on the basis of the rate from the other group) by dividing the observed number by the expected number. Usually this fraction is multiplied by 100 for ease of expression. If the observed deaths are the same as the expected number, the value will be 100. If the observed deaths are greater than the expected (that is, the population has a higher risk of death than the comparison population) then the value will be greater than 100. If the observed deaths are less than the expected deaths, then the value will be less than 100.

Usually the rate from the comparison population is applied to the number of people in the study population. This is because the study population is usually small and so the rates calculated in each strata can vary a lot just due to chance. The standardised ratio that is calculated in this manner is called the standardised mortality ratio or SMR. That is:

$$SMR = (Observed\ deaths/Expected\ deaths) \times 100$$

SMRs provide a convenient way to compare populations with a single measure. However, they can be criticised on the basis that a single measure can obscure differences in risk in different subgroups. For example, if the risk of developing a disease as a result of exposure to something was four times for men but only two times for women, the calculated SMR would depend on how many men and women were in the population. If there were equal numbers, the SMR would be 300. If the population was nearly all male the SMR would be near 400. If nearly all female, the SMR would be near 200. So SMRs can be misleading.

The healthy worker effect

This is a commonly observed phenomenon by which lower death rates (or injury or disease rates) are observed in workers relative to the general population (McMichael 1976). From a general perspective, populations are made up of two groups, the healthy and the sick. Those that tend to be healthy tend to be employed, those who are sick tend not to work. So from first principles, workers have a tendency to be more healthy than the population from which they arise. People of good health are selected for employment, and as good health is associated with the outcome (for example, a lower risk of injury or disease) and the exposure (by virtue of employment) it meets the definition of confounding.

Sample sizes and power

This is a difficult area to cover in detail without some statistical background. However, it is possible to consider the principles in a fairly general manner.

Power is the ability of a study to detect an association (real effect). For example at least x cases are needed before a positive association in a group of y workers can be established – fewer cases may suggest a false negative.

Most studies use only a sample of the population that the results are supposed to represent. The underlying population has a precise value for any particular parameter (for example, mean weight). The study attempts to determine that parameter, but can only make an estimate of it because it only contains a sample of the underlying population. This estimate is unlikely to be exactly the same as the true value, because random selection factors would probably prevent the sample exactly representing the underlying population.

Having calculated the estimate from our study, and presuming the study has been conducted properly, we can calculate (using accepted statistical assumptions and formulae) a range in which the true result probably lies. This range is usually called the confidence interval, and its width depends upon two factors, the desired degree of certainty and the number of persons in the study.

If we want to be 60% sure that we know what range the true result lies in, then the range will be a certain width. If we want to be more certain, then the range must become wider. The generally accepted (but arbitrary) degree of uncertainty is 95%, which in effect means that if it is accepted that the true result lies within a certain range, there is less than a 5% chance (or 1 in 20) that the true result lies outside this range (you will usually see this written as probability (p) below 0.05). It is apparent that this level of probability is quite generous to the non-true result. It means that making the association can be quite arduous, but that once it is established, it has a reasonable likelihood of being true. It should be noted that scientific certainty is different from other forms of certainty, such as may be required in criminal law (beyond a reasonable doubt) or civil cases (on the balance of probabilities, where p is less than 0.5, not less than 0.05).

The more subjects in the study sample (that is, the bigger the sample size) the more likely it is that the estimate made from those subjects is close to the true value for the population. Therefore, the range in which we can be 95% sure that the true result lies becomes narrower, until when the whole of the population is included in the study, our estimate is the same as the true result and the confidence interval becomes 0. So the more subjects in the study, the more likely the study is to find the true results, and so the more powerful the study is.

In most epidemiological studies, most effort is spent in finding comparisons between groups rather than in measurement of a particular factor in only one group. The above principles can be applied to determining whether any differences have occurred due to random factors (or chance), or whether they are due to real differences in the underlying groups.

For example, consider an investigation of FEV_1 (forced expiratory volume in 1 second, a measure of lung function) in a group of miners, with a group of clerks from the mine's office as the comparison group. We find that the mean FEV_1 in miners is 3.2 litres and the mean FEV_1 in clerks is 3.6 litres. Did this difference arise through chance or is it present because the two groups really do have different lung function? In practice, a statistical test of comparison such as a t test or ANOVA would be used to answer this question, but the basis of this test is the comparison of the ranges (or confidence intervals) surrounding the estimates.

Confidence interval is the range in which there is some confidence that the true result lies. For example relative risk $= 1.87 \ (0.56 - 5.02)$.

Consider the two possible outcomes: if the confidence intervals do not cross one another, then the results are considered statistically significantly different and the populations are presumed to truly have different lung function.

However, this does not mean that the populations truly do not have the same lung function. If the results for the two groups are statistically different from each other this could be due to three factors. Perhaps there truly is a difference between the lung function in the two groups. Alternatively, there may truly be no difference, but the study may have a flawed design, in which case the conclusions may be incorrect and a new study may be required.

There may truly be no difference, and the results may have been due to a chance occurrence. Remember, the basis of the statistical tests is probability, and if there is less than a 5% probability that the results occurred due to chance, we are still accepting that nearly five times in a hundred we will conclude that there is a real difference when in fact there is not. We can minimise the chance of this happening by saying we want to be more sure of the true results before we will accept any differences that are detected (for example, by decreasing the p value to 0.01). The smaller the p value, the more certain we are that, if we detect a difference, the difference truly exists.

However, we achieve this increased certainty at the cost of an increased likelihood of concluding that there is not a real difference when in reality there is. Alternatively, we can include more subjects in the study, thereby narrowing our confidence intervals and making it easier to detect any true differences between the groups.

If the confidence intervals do cross one another, then the estimates would not be considered statistically significantly different and the populations can not be presumed to have different lung function.

However, this does not mean that the populations truly have the same lung function. If the results for the two groups are not statistically different from each other this could be due to three factors. Perhaps there truly is no difference between the lung function in the two groups. Alternatively, there may truly be a difference but the study may have a flawed design, in which case you need a new study! Finally, there may truly be a difference, but the number of subjects may have been insufficient to give the study enough statistical power to detect the difference (because the confidence intervals were too wide). Thus the power of a study relates to its ability to detect a difference between results if such a difference truly exists.

As noted above, the confidence intervals can be narrowed, and the power improved, by including more subjects in the study. Table 18.4 shows an example in Swedish fishermen.

The incidence of deaths from drowning in fishermen compared with those of a regional population was 13 observed as against 1.1 expected. This gave an SMR of 11.82, with 95% confidence intervals (95% CI) of 6.29 to 20.2. While drowning is an obvious risk with working on water, what are the possible problems of interpreting cancer of the breast in this study, which reports an SMR of 4.17? By looking at the 95% CI (which ranges from 0.1 to 23.2), the increased SMR can be put into context – the wide range in these CI suggests extraordinary variability. Further, it is only when the data on observed/expected cases is assessed (1 observed, 0.2 expected) that the results can be considered appropriately. In this case, the SMR and CI values are based on too few data, and the reported SMR can be considered anomalous.

Study types

A number of different study designs or study types are used in epidemiology. Each has its own advantages and disadvantages. The particular design used at any particular time is influenced by factors such as the study question, the study base that is available, the expected relationship between the study and outcome factors, financial and personnel resources, and how quickly a result is needed.

In choosing a study design, the main question to decide is 'Why is the study being conducted?' Do you want to attempt to determine if the presence of one factor is likely to be associated with or cause the presence of another factor (a question of causality), or merely to determine what proportion of a population has a given factor (a question of magnitude)?

Table 18.4 A study of risks to Swedish fishermen

Cause of death	Observed	Experimental	Standardised mortality ratio	Confidence level
All causes	419	484.5	0.86	0.78–0.95
All malignant neoplasms	90	107.4	0.84	0.68–1.03
Alimentary organs	31	40.9	0.76	0.52–1.09
Lung larnyx	16	20.6	0.78	0.46–1.28
Breast	1	0.2	4.17	0.10–23.2
Leukaemia	5	3.6	1.38	0.45–3.22
Multiple myeloma	7	2.3	3.08	1.24–6.35
Cardiovascular diseases	224	255.7	0.88	0.77–0.99
Accidents/poisoning/violence	58	38.5	1.50	1.15–1.96
Drowning	13	1.1	11.82	6.29–20.2
Alcohol	8	11.0	0.73	0.31–1.43

Source: incidence data from Svensson *et al.* 1994, reproduced with permission.

Study types generally fall into two categories, descriptive or analytic. Descriptive studies are only useful for questions of magnitude. Analytic studies can usually answer questions of both magnitude and causality.

Descriptive studies

A descriptive study involves surveying a population regarding the presence or absence of certain factors of interest. There is no attempt to relate the presence of one factor with the presence (or absence) of another. Descriptive studies therefore are useful for giving an overall picture of a population, but not for investigating a causal relationship between two factors.

Analytic studies

A number of study types can be used to answer questions of causality, with some study types being better, or more reliable, than others. In (approximate) descending order of reliability they are randomised control trials, cohort studies, case–control studies, cross-sectional studies, before–after studies and ecological studies.

Randomised control trial (RCT)

This study type involves selecting the study sample and then randomly allocating subjects to exposure groups. The groups are then followed over time to determine whether different exposure groups have different proportions of subjects developing the outcome, or subjects who develop different levels of the outcome. The key advantage of the randomised control trial (RCT) is that subjects become members of an exposure group purely on a random basis. Therefore, individual attributes of subjects that might influence the outcome (potential confounders) should be equally spread between the exposure groups, thereby reducing the likelihood of confounding occurring. This randomisation should control for both known and unknown potential confounders, an attribute which no other study type has. The RCT is not often used in the occupational setting because of the practical or ethical difficulties in randomly allocating persons to particular work exposures. However, in the clinical setting it is of great value in assessing the efficacy of various, hopefully beneficial, interventions (for example, drugs). Use of RCTs in assessing the worth of new interventions is also desirable and often possible, but is not as commonly used. The design of a RCT is shown in Figure 18.3.

Cohort study

Cohort studies are similar to RCTs in that groups of subjects with different exposures are followed over time and then their outcome determined. However, unlike in RCTs, the allocation of subjects to exposure groups is not random. Instead, it is usually determined by unknown factors. For example, a study of the relationship between smoking and lung cancer will probably have smoking and nonsmoking groups. Subjects will be allocated to the different exposure groups because they

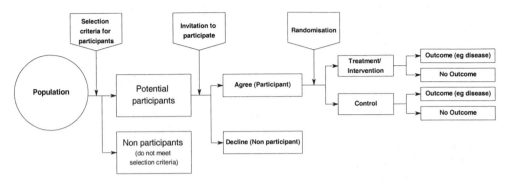

Figure 18.3 Design of a randomised cohort study.

have chosen to smoke or chosen not to smoke, rather than because the study designer has told them what they should do. Potential confounding factors (for example, age; income; level of health) might be associated with the decision of an individual to smoke or not, and so may not be equally spread between exposure groups. If these factors influence whether the subject develops the outcome of interest, or is identified as having the outcome of interest, then this could influence the final results unless the effect of the potential confounders is controlled when the study is analysed. However, unknown confounders cannot be controlled because by definition the study designer is unaware of them (see Figure 18.4). This is the main reason why RCTs are preferable to cohort studies.

Advantages of cohort studies are:

- They can provide a picture of the baseline situation.
- Sample selection by exposure is essential for rare exposures.
- They can minimise losses to follow-up and assess their effect.
- If concurrent, they can protect against detection biases.

Disadvantages of cohort studies are:

- They are poorly suited to the study of rare outcomes.
- Prolonged follow-up is required for long latency conditions.
- Careful follow-up is labour intensive and costly.
- Subject attrition rates require large initial sample sizes.

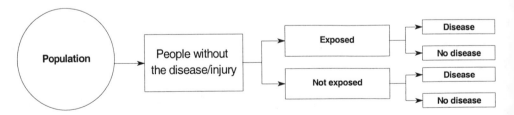

Figure 18.4 Design of a cohort study.

Case–referent studies

The case–referent study type (more commonly known as a case–control study) has a very different design to RCTs and cohort studies. It involves identifying a study base and identifying all the members of the study base who have the outcome (see Figure 18.5). These persons are called cases and usually all cases are included in the study if possible. Next, a sample of all members of the study base who do not have the outcome are chosen. These persons are known as the referents or controls. Ideally all the referents would be included in the study, but this is usually impractical (because the study would become too large) and unnecessary (because the validity and statistical efficiency of the study are not improved to any marked extent). The previous exposure characteristics of the cases and the referents are then compared.

If the study is well designed, the referents theoretically should have the same characteristics as the cases, including the exposure characteristics, unless the exposure truly is associated with the outcome.

The key to a successful case–referent study is the study base. If it is well defined the study should be valid, presuming the other aspects of the study are conducted properly. If it is not well defined, then the cases and the referents may not be comparable and the study may therefore be invalid.

It can be seen that case–referent studies and cohort studies approach a problem from opposite directions. In the first, subjects are selected on the basis of the presence or absence of the outcome, with previous exposure being determined. In the second, subjects are selected on the basis of exposure, with the subsequent development of the outcome determined. A good case–referent study is likely to be as valid as most good cohort studies, but the case–referent study is more likely to run into selection and measurement problems than is a cohort study.

A typical simple analysis of a case–referent study is shown in Table 18.5,

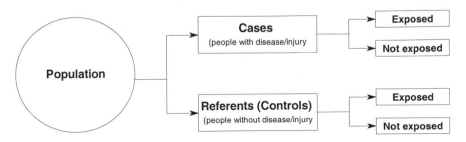

Figure 18.5 Design of a case–referent study.

Table 18.5 Two-by-two table for a case–control study

		Worked in factory		
		Yes	No	Total
Wheezing	Yes	52	9	61
	No	12	57	69
	Total	64	66	130

examining work in a particular factory where a machine generating fugitive emissions operates 7 hours a day, and where cases of wheezing are occurring. The data is entered onto a two-by-two table. The odds ratio can be calculated from this table.

Advantages of case–referent studies are:

- Generally there is a short study period.
- They can be used to study rare health-related phenomena.
- They are relatively inexpensive to carry out.
- They can be used to study several risk factors.
- They are useful for the study of new health-related phenomena.

Disadvantages of case–referent studies include:

- Sometimes it is difficult to find appropriate controls.
- Sometimes it is difficult to determine if exposure preceded the disease.
- They are prone to bias in subject selection.
- Usually it is not possible to calculate incidence rates.

Cross-sectional studies

Unlike the previous study types, subjects in cross-sectional studies are not selected on the basis of either exposure or outcome. Instead, they are included in the study purely because they are present in the study base at the time of the study. Exposure and outcome are determined at the same time, and the association between the two determined. This is a convenient (and usually inexpensive) way to conduct a study, but has two major shortcomings. Firstly, subjects who have developed the outcome, and therefore are of major interest in the study, may have left the population prior to the study being conducted (for example, a study of lung disease in miners, in which miners with severe lung disease may be missed because they were unable to continue working). Secondly, because exposure and outcome are being determined at the same time, it is not always clear which came first (for example, in a study of exercise and obesity, did the lack of exercise cause obesity, or did the obesity make exercise difficult?).

Advantages of cross-sectional studies include:

- They are useful for descriptive studies, such as describing the clinical spectrum of disease, or prevalence surveys.
- They are fairly quick and inexpensive to carry out, can identify early clues for later, more detailed study.
- They are less prone than case control studies to recall error and bias.

Disadvantages of cross-sectional studies include:

- They may be unable to sort out temporal sequence of exposure and outcome.
- There is a greater potential than cohort studies for sample distortion bias.
- Careful follow-up is labour intensive and costly.
- Subject attrition rates can require large initial sample sizes.

Ecological studies

Ecological studies involve determining the proportion of various populations with certain factors, and attempt to relate one factor to another at a population level (for example, comparison of heart disease rates in countries with different average intakes of salt, fats or vegetables). There is no information on factors at an individual level, and thus no way of properly relating exposure to outcome, nor to take account of potential confounders (except at a population level).

Before–after studies

Before–after studies are like repeat descriptive studies, where some aspect of the outcome of interest is determined before and after a particular exposure or intervention. Again, there is usually no information on individual exposure, outcome or potential confounders. This study type is usually used in studies of large population interventions (for example, safe sex campaigns).

Causation

The final part of carrying out of epidemiological study is its conclusions. Because of the nature of epidemiology, the attribution of proof is extremely difficult. At best, most epidemiological studies will report an association between the study factor and the outcome factor, with some level of probability and confidence limits.

Association

The study and outcome factors are associated if one is more (or less) common in the presence of the other

As a concept, establishment of an association will need to consider (among other things) the impact that errors may have on the findings.

Type I error: A false positive. This is where an association is found, but is not real. Occasionally, such an association may arise as an anomaly in the analysis. Other causes include systematic errors or inappropriate attribution of factors.

Type II error: A false negative. That is, the study does not have the power to detect a positive association. False negative conclusions can also be reached when:

- the measures are crude; the morbidity measures are irrelevant or wrong
- there are random errors
- nondifferential classification occurs
- the study design is too small
- the study design is inefficient
- incorrect exposure categories are used
- exposures are too low and/or too short
- follow-up is too short or incomplete
- allowance is not made for a latency period.

In thinking about associations or causation, other factors should also be considered. The **guidelines for causation** are:

- Temporal relation: does the cause precede the effect? (essential)
- Plausibility: is the association consistent with other knowledge? (mechanism of action, laboratory evidence)
- Consistency: have similar results been shown in other studies?
- Strength: what is the strength of the association between cause and effect? (relative risk)
- Dose–response: is increased exposure to the possible cause associated with increased effect?
- Reversibility: does the risk decrease with removal of a possible cause?
- Study design: is the evidence based on a strong study design?
- Judging the evidence: how many lines of evidence lead to the conclusion?

Occupational epidemiology

The connection between occupation and a variety of diseases dates back to the 5th century BC in Hippocrates' message to his contemporary physicians to explore patients' environmental, lifestyle, and vocational backgrounds as determinants of aetiology and treatment. In this undertaking, epidemiological surveys directed towards general and high-risk populations are needed to generate sufficient numbers of affected and nonaffected persons to sort out the role of confounding variables, demonstrate the strength of associations between occupational exposures and diseases, and propose causal inferences for further study (White 1985).

In promoting epidemiological studies of occupationally defined populations, Gaffey stated: 'If one is concerned with exposure to a man-made substance, it is a truism that there exists, somewhere, a group of persons who manufactured that substance, and whose exposure in most cases was dramatically greater than that of the general population' (Gaffey 1988).

Indeed, epidemiology is essential to occupational health and safety because it helps to answer whether there is a health-related problem among workers and if so, what is causing it, whether it could be some exposure in the work environment, and how big the risk of this exposure is (Christie 1988). Emphasising the importance of epidemiological research in the occupational setting, Christie (1988) stated that: 'Epidemiology, far from being esoteric research carried out in ivory towers, is concerned with seeking solutions to real time problems in the field of occupational health and safety.'

Thus, when applied to occupational health, epidemiology has the dual task of describing the distribution of deaths, incidents, illnesses, and their precursors in the workforce and of searching for the determinants of health, injury, and disease in the occupational environment, as conceptualised in Figure 18.6 (Karvonen and Mikheev 1986).

Other disciplines in occupational health, such as occupational hygiene, toxicology, ergonomics, safety engineering, occupational rehabilitation and so on, also contribute to this process.

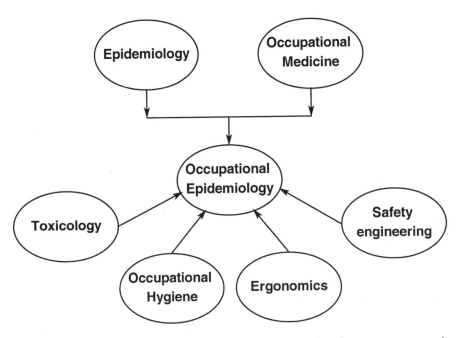

Figure 18.6 Relationship of epidemiology and occupational medicine to occupational epidemiology.

Epidemiology in the toxicological setting

Occupational toxicological epidemiology is a term used to describe the application of epidemiological principles to chemical exposures in the workplace. The range of factors contributing to the causation of occupational diseases and injuries relates to factors in the worker (health status, genetic factors, personal and lifestyle factors, as well as previous exposure) and factors relating to the workplace (exposure to risks and risk control measures). These then interact to produce occupational disease or injury, which are themselves affected by injury/disease initiation and progression and repair mechanisms (see Figure 18.7). As well as being critical in occupational health and safety activities (for example, workplace environmental monitoring, health surveillance and treatment of affected workers), these factors need to be explored in occupational epidemiological studies.

Methods of occupational epidemiology

The methods used in occupational epidemiology differ only in emphasis from those used in acute or chronic disease and injury epidemiology. Consequently, most of this chapter has looked at the methods of epidemiology. Obviously, community trial-type studies do not often form part of occupational epidemiology where chemical exposures are involved. The most common studies of working populations are shown in Table 18.6.

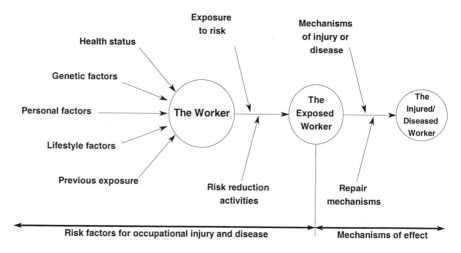

Factors in the Development of Occupational Disease and Injury

Figure 18.7 Factors in the development of occupational disease and injury.

Table 18.6 Epidemiological studies used to study occupational populations

Type of study	Also known as	Unit of study
Observational studies		
Descriptive studies	Case studies	Affected individuals
Analytical studies		
▪ Correlational	Ecological	Populations
▪ Cross-sectional	Prevalence	Individuals
▪ Case–control	Case–referent	Individuals
▪ Cohort	Follow-up	Individuals
Interventional studies		
Randomised control trials		Individuals

Most epidemiological studies are of one or other of the studies listed in this table, although many studies combine aspects of several study types.

Occupational medicine has developed over the years, along with other forms of medicine. Occupational physicians and other occupational health professionals began looking at the injuries and other health problems caused by short-term exposure to physical and chemical agents. Their emphasis was on treatment of the affected individual, but ultimately the focus was on prevention. No elaborate scientific methodology had to be developed to decide whether exposure to dust produced acute bronchitis – efforts were directed to treating the disease and reducing the exposure.

However, working populations in particular occupations tend to be small, workers can change the type of work they do in a lifetime, and some risk factors in working populations cause conditions at low frequencies. A practising physician may not notice, in a career spanning 50 years, for example, that kidney disease is common in plumbers. The 1974 report by Creech and Johnson that exposure to vinyl chloride was associated with development of angiosarcoma of the liver was made from just four cases. This was an anomaly due to the low incidence of a rare cancer. In most cases it is difficult to see the wood for the trees.

In fact, the number of associations between workplace and disease is not small (Alderson 1983; Checkoway *et al.* 1989; Hernberg 1992; Marsh 1994; Monson 1990; WHO 1989). Table 18.7 outlines just some of these associations.

The list is representative only. A full list of such associations would be much longer, if not virtually endless. Many of these associations would have been revealed through the observations of toxicologists, occupational health and safety professionals and the techniques of occupational epidemiology.

Occupational epidemiology is therefore an important discipline in the occupational health and safety field.

Summary

In summary, epidemiology can be loosely described as the study of epidemics. This requires:

- careful selection of sample and control populations
- use of suitable methodology that allows for stringent data collection, identification and control of confounding and avoidance of the introduction of bias
- careful selection of study design
- choice of appropriate statistical analysis.

With regard to the epidemiology of chemical exposures in working populations, toxicology, occupational medicine and epidemiology have a contribution to make to the detection, monitoring and prevention of chemically related occupational diseases and injury. However, there are many pitfalls associated with their application. The use of particular methods of investigation such as epidemiological surveys and medical screening should be carefully scrutinised for their applicability to a given situation, so that maximum use can be made of these tools.

The scope of epidemiological research is continually expanding. Initially concerned with the study of occupationally related diseases in workers exposed to specific high-risk agents such as dye workers (Case *et al.* 1954), asbestos miners (Wagner *et al.* 1960), or of rubber workers (McMichael *et al.* 1974), the field of occupational epidemiology has grown. This has been associated with advances in the methodology of epidemiology and biostatistics, and in advances in the related occupational health and safety disciplines.

Further, while the information that occupational epidemiology provides is useful when derived from individual workplaces, exposed populations or risk factors, it is in the collection and analysis of such information into competent and comprehensive reviews that really useful information arises.

Occupational epidemiology has one other function. The potential to produce effects in workers from exposure to occupational risks is not limited to working

Table 18.7 Associations of workplace and disease

Risk factor	Health problem
Abrasive materials	Dermatitis
Acids and alkalis	Chemical burns
Aluminium smelting	Occupational asthma
4-Aminobiphenyl	Bladder cancer
Anaesthetic gases	Miscarriage
Animals	Zoonoses
Arsenic	Skin cancer
Asbestos	Asbestosis
Asbestos	Lung cancer
Asbestos	Mesothelioma
Benzene	Leukaemia
β-Naphthylamine	Bladder cancer
Cadmium	Kidney problems
Carbon disulfide	Psychosis
Carbon tetrachloride	Liver disease
Cement	Dermatitis
Chloromethyl ethers	Cancer
Coal dust	Pneumoconiosis
Coal gasification	Cancer
Colophony	Occupational asthma
Confined spaces	Asphyxiation
Dibromochloropropane	Male sterility
Electricity	Electrical shock
Epoxy resins	Dermatitis
Ethylene oxide	Miscarriage
Formaldehyde	Dermatitis
Formaldehyde	Nasal cancer
Glycol ethers (low MW)	Reproductive problems
Grain dust	Respiratory problems
Hexavalent chromium	Lung cancer
Hexavalent chromium	Skin problems
Infectious agents	Infection
Ionising radiation	Genetic defects
Ionising radiation	Cataracts
Ionising radiation	Cancer
Iron/steel production	Cancer
Isocyanates	Occupational asthma
Isopropyl alcohol	Cancer
Lead (inorganic)	Reproductive problems
Leatherwork	Cancer
Manual handling	Musculoskeletal problems
n-Hexane	Peripheral neuropathy

continued

Table 18.7 continued

Risk factor	Health problem
Nickel	Lung cancer
Noise	Deafness
Oils	Dermatitis
Overuse	Musculoskeletal problems
Painting (occupation)	Cancer
PCBs	Reproductive problems
Pesticide application	Cancer
Pitches and tars	Skin cancer
Platinum salts	Occupational asthma
Polyaromatic hydrocarbons	Cancer
Radon	Lung cancer
Rubber manufacture	Cancer
Rubella	Birth defects
Shampoos	Dermatitis
Shiftwork	Psychological problems
Silica	Silicosis
Solvents	Dermatitis
Solvents	Impaired CNS function
Soots	Skin cancer
Sunlight	Melanoma
Sunlight	Dermatitis
Tobacco smoke, passive	Lung cancer
Toxic chemicals	Poisoning
Tricresyl phosphate	Peripheral neuropathy
Vibration	White finger disease
Vinyl chloride	Liver cancer
Violence at work	Trauma
Western red cedar	Occupational asthma

populations. Where exposure to such risks can occur outside the working environment, for example to discharges or pollutants, the possibility of adverse health problems to nonoccupational populations can arise.

The methods of occupational epidemiology can be adapted (perhaps not without some difficulty) for the study of associations between environment and health. This allows identification of a fuller spectrum of effects, and provides better protection for the health of workers, their families, neighbours and communities.

References and further reading

Alderson, M. (1983) *An Introduction to Epidemiology*, 2nd edn. London: Macmillan Press.
Beaglehole, R., Bonita, R. and Kjellstrom, T. (1994) *Basic Epidemiology*. Geneva: World Health Organisation.
Case, R.A.M., Hosker, M.E., McDonald, D.B. and Pearson, J.T. (1954) Tumours of the

urinary bladder in workmen engaged in the manufacture of certain dyestuff intermediates in the British chemical industry: Part I. *Br. J. Ind. Med.* 11: 75–104.

Chan-Yeung, M., McMurren, T., Cationo-Begley, F. and Lam, S. (1993) Occupational asthma in a technologist exposed to glutaraldehyde. *J. Allergy Clin. Immunol.* 91: 974–978.

Checkoway, H., Pearce, N.E. and Crawford-Brown, D.J. (1989) *Research Methods in Occupational Epidemiology.* Oxford: Oxford University Press.

Christie, D. (1988) *A Guide to Occupational Epidemiology.* Australia: CCH International.

Coggon, D., Rose, G. and Barker, D.J.P. (1997) *Epidemiology for the Uninitiated,* 4th edn. Plymouth: BMJ Publishing Group.

Creech, J.L. and Johnson, M.N. (1974) Angiosarcoma of the liver in the manufacture of polyvinyl chloride. *J. Occup. Med.* 16: 150–151.

Driscoll, T.R. (1993) Are work-related injuries more prevalent than disease in the workplace? *Occup. Med.* 43: 164–166.

Fletcher, R.H., Fletcher, S.W. and Wagner, E.H. (1988) *Clinical Epidemiology – the Essentials,* 2nd edn. Baltimore: Williams and Wilkins.

Friedman, G.D. (2003) *Primer of Epidemiology,* 4th edn. New York: McGraw-Hill.

Gaffey, W.R. (1988) Quantification of risk in defined populations. In: Gordis, L. (ed.) *Epidemiology and Health Risk Assessment,* Oxford: Oxford University Press, pp. 189–193.

Garabrant, D.H., Held, J., Langholz, B., Peters, J.M. and Mack, T.M. (1992) DDT and related compounds and risk of pancreatic cancer. *J. Nat. Cancer Inst.* 84: 764–771.

Hernberg, S. (1987) 'Negative' results in cohort studies – how to recognise fallacies. *Scand. J. Work Environ. Health* 7 (Suppl. 4): 121–126.

Hernberg, S. (1994) Epidemiology in occupational health. In: Zenz, C. (ed.) *Developments in Occupational Medicine.* Chicago: Year Book Medical Publishers.

Hernberg, S. (1992) *Introduction to Occupational Epidemiology.* New York: Lewis Publishers.

Jeyaratnam, J., Lun, K.C. and Phoon, W.-O. (1987) Survey of acute pesticide poisoning among agricultural workers in four Asian countries. *Bull. World Health Org.* 65: 512–257.

Karvonen, M. (1986) Epidemiology in the context of occupational health. In: Karvonen, M. and Mikheev, M.I. (eds) *Epidemiology of Occupational Health.* Copenhagen: World Health Organisation Regional Office in Europe, European Series No. 20.

Karvonen, M. and Mikheev, M.I. (eds) (1986) *Epidemiology of Occupational Health.* Copenhagen: World Health Organisation Regional Office for Europe, European Series No. 20.

Kodama, A.M. and Dollar, A.M. (1983) Indoor air fires as a cause of cancers of the upper respiratory and digestive systems in certain underdeveloped countries. *J. Environ. Health* 46: 88–90.

Marsh, G.M. (1994) Epidemiology of occupational diseases. In: Rom, W.N. (ed.) *Environmental and Occupational Medicine,* 2nd edn. Boston: Little, Brown and Co, pp. 725–731.

McMichael, A.J. (1976) Standardised mortality ratios and the 'healthy worker effect': Scratching beneath the surface. *J. Occup. Med.* 18: 165.

McMichael, A.J., Spirtas, R. and Kupper, L.L. (1974) An epidemiologic study of mortality within a cohort of rubber workers, 1964–72. *J. Occup. Med.* 16: 458–464.

Monson, R.R. (1990) *Occupational Epidemiology,* 2nd edn. Boca Raton: CRC Press.

Morrell, S., Kerr, C., Driscoll, T., Taylor, R., Salkeld, G. and Corbett, S. (1998) Best estimate of the magnitude of mortality due to occupational exposure to hazardous substances. *Occup. Environ. Med.* 55: 634–641.

Olsen, J., Merletti, F., Snashall, D. and Vuylsteek, K. (1991) *Searching for Causes of Work-Related Diseases: An Introduction to Epidemiology at the Work Site.* Oxford: Oxford University Press.

Riegelman, R.K. (1981) *Studying a Study and Testing a Test. How to Read the Medical Literature.* Boston: Little, Brown and Co.

Rothman, K.J. and Greenland, S.G. (1998) *Modern Epidemiology*, 2nd edn. Philadelphia: Lippincott-Raven.

Svensson, B.G., Mikoczy, Z., Stromberg, U. and Hagmar, L. (1995) Mortality and cancer incidence among Swedish fishermen with high dietary intake of persistent organochlorine compounds. *Scand. J. Work Environ. Health* 21: 106–115.

Wagner, J.C., Sleggs, C.A. and Marchand, P. (1960) Diffuse pleural mesothelioma and asbestos exposure in the North Western Cape Province. *Br. J. Ind. Med.* 17: 260–271.

White, K.L. (1985) Epidemiology comes of age. *Med. J. Aus.* 142: 84–86.

WHO (1989) *Epidemiology of Work-Related Diseases and Accidents. Tenth Report of the Joint ILO/WHO Committee on Occupational Health. Technical Report Series 777.* Geneva: World Health Organization.

Winder, C. (1991) Trends in occupational cancer in Australia 1979–1989. *J. Occup. Health Safety – Aust. N.Z.* 7: 355–363.

Wong, O., Morgan, R.W., Bailey, W.J., Swencicki, R.E., Claxton, K. and Kjeifets, L. (1986) An epidemiological study of petroleum refinery employees. *Br. J. Ind. Med.* 43: 6–17.

Zhong, L., Goldberg, M., Parent, M.-E. and Hanley, J.A. (1999) Risk of developing lung cancer in relation to fumes from Chinese style cooking. *Scand. J. Work Environ. Health* 25: 309–316.

Uses of
toxicological data

Chapter 19

Chemicals, workplaces and law

C. Winder and J. Barter

Introduction

Over 100,000 chemicals and polymers are in commercial use worldwide (NTP 1984). Approximately half are thought to be harmful to health (ILO 1990). In Australia, at least 20,000 toxic and hazardous chemicals are imported each year and a further 700 are manufactured domestically. Compilation of the Australian Inventory of Chemical Substances (AICS) lists approximately 40,000 industrial chemical entities in Australia (CEPA 1992b), to which should be added those agricultural and veterinary chemicals, therapeutic substances, cosmetics and food additives not presently listed on AICS. The numbers of chemical products that these raw materials are formulated into is not known, but is likely to be considerably over 100,000.

While the benefits of chemicals have become important in many economic activities, increased evidence suggests that some chemicals can contribute to health and environmental problems. It is now recognised that chemicals need to be managed properly in order to achieve a sustainable level of industrial and agricultural development with high levels of human health and environmental protection (Nichols and Crawford 1983).

In 1992, the United Nations Conference on Environment and Development (the Rio Earth Conference) gave rise to the *Agenda 21 Report* (UNCED 1992). This report outlined the responsibilities of states towards the achievement of sustainable development, and was adopted by heads of government in over 150 countries.

Chapter 19 of *Agenda 21* addresses the environmentally sound management of toxic chemicals, including basic programs for:

- adequate legislation
- information gathering and dissemination
- capacity for risk assessment and interpretation

- establishment of risk management policy
- capacity for implementation and enforcement
- capacity for rehabilitation of contaminated sites and poisoned persons
- effective education programs
- capacity to respond to emergencies.

Chapter 19 of *Agenda 21* recommends that national programs for the environmentally sound management of chemicals should be in place in all countries by the year 2000. It also called for the formation of an intergovernmental forum to improve coordination and management of chemicals, and the International Conference on Chemical Safety duly met in Stockholm in 1994. This conference considered mechanisms for the development and implementation of recommendations of Chapter 19. The Stockholm Conference established the Intergovernmental Forum on Chemical Safety (IFCS) as a means for discussing and exchanging information.

The IFCS recognised the establishing of the nature and complexity of chemicals management infrastructure in different countries, of different levels of development, and with different national and cultural aspirations. The IFCS sought information on the chemical management infrastructure of each nation:

- to provide a baseline of information on chemicals management infrastructure in legal, institutional, administrative and technical areas
- to identify infrastructure-related strengths, weaknesses and gaps
- to identify priorities for national action.

Most developed countries and many other nations prepared national profiles of chemicals management infrastructure following publication of *Agenda 21*.

Political and geographic structures

There is a diverse range of political structures throughout nations and states across the world. In many political units with democratic structures, there is a parliament elected by the people. Each political unit also usually has its own independent judiciary.

As well as national government, other tiers of government can exist at the state, province, regional or local government level. These government bodies vary considerably in size, population, income, range and complexity of functions.

The term 'law' generally falls into two broad categories:

- legislation in the form of statutes passed by parliament (statute law)
- rules derived from decisions of the courts (common law) which include decisions of courts and judicial interpretation of statute law.

Societal strategies for the management of chemicals: evolution of laws and voluntary initiatives

Evolution of chemical management laws

In many countries, legislative and administrative measures have been introduced to deal with chemical hazards. Whilst the origin of such measures can be traced back to the development by the courts of common law principles such as the law of nuisance, and to certain ancient statutes, the subject is essentially of recent origin. This, combined with the development of legislation in response to local as well as international developments (for example, thalidomide, asbestos, persistent bioaccumulative toxic chemicals) has meant that the legislative control of chemicals has developed of its own accord. As a result, it is a highly complex area.

Most nations take their obligations in controlling chemicals very seriously. There has been intense activity in many nations over the past 30–40 years to identify and deal with problems arising out of the use of chemicals. In turn, this has produced a greater emphasis on regulatory control.

Chemicals control regulation is a highly complex area, in which scientific and legal issues are brought together. This is further complicated by historical precedents or jurisdictional subtleties.

Chemical control legislation operates at different levels, with different jurisdictional demarcations and different administrative arrangements. This complexity of legislation is not limited to one state or one nation. For example, in the USA, four federal agencies are primarily responsible for regulating exposures to chemicals. These agencies administer over two dozen statutes which have been enacted over time, and all have protection of health as their main goal. Table 19.1 summarises the stepwise development of the major chemical control laws at the federal level in the USA.

Evolution of voluntary initiatives

In response to the increasing awareness over time of the need to manage hazards associated with the use of chemicals, voluntary initiatives were developed to deal with various aspects of the manufacture, transportation, use, and disposal of chemicals. The most prominent example of such voluntary efforts is the Responsible Care® initiative developed and implemented by the chemical manufacturers in North America and subsequently implemented by manufacturers, distributors, and users in various countries globally. This initiative incorporates industry best practices into a codified set of management practices which are utilised across the spectrum of activities associated with use of chemicals. In some countries, implementation of the Responsible Care® codes has provided the impetus for development of regulatory legislation incorporating these concepts into law.

Legal concepts

The theoretical or philosophical foundations of law are often obscured by the detail of legislation. However, there are a number of important legal concepts

Table 19.1 Major US chemical control laws

Act or law	Responsible Federal Agency
Food and Drugs Act 1906	FDA
Food, Drug and Cosmetic Act 1938	FDA
Food Additives Amendment 1958	FDA
Pesticide Residue Amendment 1954	EPA
Medical Devices Amendment 1976	FDA
Federal Hazardous Substances Act 1960	CPSC
Labelling of Hazardous Materials Amendment 1988	CPSC
Occupational Health and Safety Act 1970	OSHA
Consumer Products Safety Act 1972	CPSC
Federal Water Pollution Control Act 1972	
Amendments 1977, 1987	EPA
Safe Drinking Water Act 1974	
Reauthorised for 5 years 1986	EPA
Clean Air Act 1970	
Amended 1977, 1990	EPA
Toxic Substances Control Act 1976	
Amended 1981, 1984, 1986	EPA
Resource Conservation and Recovery Act 1976	
Amended 1980, 1984	EPA
Federal Insecticide, Fungicide and Rodenticide Act 1947	
Amended 1972, 1988	EPA
Comprehensive Environmental Response, Compensation and Liability Act 1980	EPA
Hazard Communication Standard 1983	
Amended 1988	OSHA
Superfund Amendments and Reauthorisation Act 1986	EPA

CPSC, Consumer Product Safety Commission; EPA, Environmental Protection Agency; FDA, Food and Drug Administration; OSHA, Occupational Safety and Health Administration.

which through natural justice or equity, find their ways into the basic operating principles of much law, including that for the management or control of chemicals. Some of these concepts are outlined below.

Duty of care

The principles of duty of care are basically similar for neighbours, employers, suppliers and polluters, and is based on the 1932 judgement of the UK House of Lords, which established the general principle of liability for negligent conduct. For example, the employer duty of care to workers and to others is explicitly outlined in Australian State occupational health and safety legislation (Wyatt and Oxenburg 1996).

Right to know

The principles of right to know include worker, community and consumer right to know (Fuller 1995). Although legislation dealing with all of these stakeholders is well developed in North America and the United Kingdom (Strange 1990), only worker right to know exists in legislation in Australia (Gunningham and Cornwall 1994), where all State occupational health and safety legislation contains provisions for the employer to provide such information, instruction, training and supervision to employees as may be necessary to ensure the health and safety at work of employees (Winder 1995). In North America, the US Federal Hazard Communication Standard and the Canadian Workplace Hazardous Materials Information System contain similar provisions for the identification and notification of hazards in order to protect workers.

Risk minimisation

The concept of reducing risks to health is a general duty in various legislation, which often also includes the concept of **as low as reasonably practicable** (Winder 1993). The concept of risk minimisation needs to take into account:

- the severity of the hazard or risk
- the state of knowledge regarding: the hazard or risk; the ways of removing or mitigating that hazard or risk; the availability and suitability of ways of removing the hazard or risk
- the cost of removing the hazard or risk.

Important in the risk minimisation concept is the idea of 'reasonable practicability'.

No risk

The concept of no or zero risk assumes that no use of a particular chemical is likely to produce benefits sufficient to outweigh the potential risk. This approach is epitomised by the Delaney Clause, enacted as part of the 1958 Food Additives

Amendment to the US Food, Drug and Cosmetic Act. Whereas the amendment requires that a food additive must be found to be safe before FDA can approve its use in food, the Delaney Clause stipulates that this finding cannot be made for a food additive that has been shown to induce cancer in humans or in experimental animals. Over the years, the US FDA has developed administrative rationales for dealing with this issue as a *de minimis* or negligible risk.

De minimis

Simply put, *de minimis* means that the law does not bother with inconsequential matters. This is a concept that allows, for example, the public to be exposed to a carcinogen found in food as a residue, because the residue is at such a low level that it is considered to be inconsequential. In application of this principle, the benefits of significant regulatory impact are weighed against the likely impact.

The Precautionary Principle

This concept articulates an approach to chemicals risk management in circumstances of scientific uncertainty, reflecting the need to take prudent action in the face of potentially serious risks without having to await the completion of further scientific research. The most broadly accepted definition of the **Precautionary Principle** is derived from the *Declaration of the Rio Conference on Environment and Development* (June 1992): 'in order to protect the environment, the precautionary approach shall be widely applied by States according to their capabilities. Where there are threats of serious or irreversible damage, lack of full scientific certainty shall not be used as a reason for postponing cost-effective measures to prevent environmental degradation.'

There is considerable debate about circumstances when it is appropriate to apply the Precautionary Principle rather than the accepted and widely utilised Risk Assessment/Risk Management paradigm. The key issue is the extent of scientific uncertainty in establishing a credible risk assessment that demonstrates reasonable evidence of 'threats of serious or irreversible damage to health or the environment'.

The Precautionary Principle is incorporated into some chemicals management legislation, such as the NSW Environmental Protection Agency Act.

Intergenerational equity

A relatively new principle, which asserts that the present generation should ensure that the health diversity and productivity of the environment is maintained or enhanced for the benefit of future generations.

Types of law

Traditionally, all governments have bodies of law applying within areas of their own jurisdiction. In many countries with legal systems based on the British model, this body of law is made up of:

- common law (derived from operation of the courts)
- statutory (or government-derived legislation) law, comprised of primary legislation (Acts) and regulations
- nonstatutory sources of law.

Common law

Basic duties and rights of suppliers, employers and workers have developed over time from case decisions in the courts. This is called common law, and at its broadest terms it places a responsibility on each person of legal age to act with care to each other. The operation of this responsibility is called 'duty of care'.

Manufacturers, importers and suppliers

Under common law, suppliers of chemical substances owe their customers a duty of care relating to product safety and liability. This points to the supplier being the initial source of information on hazard and safe handling. The general duty of care of any supplier certainly includes notification to the purchaser that further information is available, and that accurate information is supplied when required. Examples of regulations based on this duty are the US Hazard Communication Standard promulgated by OSHA and various right to know laws developed in many political jurisdictions globally.

Employers

At the workplace, employers owe their employees a duty of care, and matters related to this duty of care are tested in the courts. Historically, under duty of care, an employer has an obligation to provide:

- a safe place of work
- safe materials, processes and tools to perform the work
- knowledge of work hazards, including those that are not immediately apparent but that may be encountered during performance of work
- competent supervisors and fellow employees
- rules and procedures by which all workers could perform safely, and means for their enforcement.

Polluter pays

The polluter pays principle relates to the concept that activities leading to wealth generation which can cause harm or damage should be appropriately rectified by the wealth generator. For example, if the process of extraction of minerals produces environmental damage in the form of mining wastes or environmental contamination, the generator of such damage should be responsible for fixing the problems they cause. It is recognised that if the polluter pays principle is not applied, then environmental problems persist or governments (and ultimately the community) pay for pollution, either at the cost of health, environment or resources.

Statute law

As well as common law, there is statute law, which are acts of parliament, regulations and orders, which are laid down by governments specifying certain responsibilities and rights.

Statute law for the control of chemicals falls into four broad categories:

- legislation outlining general obligations
- regulations which outline more specific duties
- legislation aimed at specific chemicals or groups of chemicals
- legislation for the notification and assessment of chemicals.

Legislation outlining general obligations

Historically, restrictive-type legislation has been used to regulate chemicals. However, modern legislation has been moving towards legislation that involves the introduction of enabling legislation outlining general rather than prescriptive obligations.

For example, all jurisdictions in Australia have now enacted performance-based occupational health and safety legislation which enshrine in legislation many of the common law precedents established over many years. While the general duties are the same, the scope of the legislation and the precise wording of each state act varies. These general duties are similar from state to state, and include:

- the duty to ensure health, safety and welfare at work of employees
- the provision and maintenance of safe plant and systems of work without risks to health
- the provision of information and training
- the provision and maintenance of a working environment that is safe and without risks to human health
- adequate facilitates for employee welfare at work.

Regulations which outline more specific duties

Because general duties legislation is sometimes too imprecise for accurate interpretation, ancillary legislation, in the form of regulations to the main Act, is required. For example, the Australian Hazardous Substances Regulation is subordinate legislation to the Occupational Health and Safety Act which outlines in more precise (but still performance-based) detail the control of chemicals at work.

Legislation aimed at specific chemicals

Acts of Parliament and regulations have been passed which deal with:

- Specific chemicals (such as regulations for asbestos, lead or liquid petroleum gas (LPG)). These tend to address the specific major hazards of these materials and how they should be contained or controlled.

- Groups of chemicals (such as legislation for poisons, dangerous goods, radio-active materials, pesticides). These tend to address specific matters such as labelling, transportation, storage and packaging, though they may cover other issues, such as point of sale considerations, security, and so on.
- The location of chemicals in use or disposal (such as legislation for environmental protection, waste disposal, factories and shops, occupational health and safety, construction and safety, clean air or clean waters). These tend to address more general issues, relating to areas of scope or jurisdictional demarcation.

Legislation of this type tends to be developed in an *ad hoc* fashion by various government departments with responsibilities for various matters, including public health, occupational health and safety, environmental protection, pollution control, transport, and so on.

Legislation for the notification and assessment of chemicals

In the past, notification and assessment of certain categories of chemicals has been introduced in response to specific public health or environmental concerns. For example, the Therapeutic Goods Act 1966 was introduced in response to the thalidomide tragedy, and much pesticides legislation was introduced in the 1970s in response to increasing concern about pesticides in the environment. Similarly, regulations have established inventories of approved or notified chemicals in several countries, for example:

- United States Toxic Substance Control Act (TSCA)
- European Union (EU) Inventory of Existing Commercial Chemical Substances (EINECS)/European Union List of Notified Chemical Substances (ELINCS)
- Australian Inventory of Chemical Substances (AICS)
- Japanese Existing and New Chemical Substances (ENCS)
- Canadian Domestic Substances List/Canadian Non-Domestic Substances List (DSL/NDSL)
- Korean Existing Chemicals List (ECL)
- Switzerland Giftliste 1 (List of Toxic Substances 1)/Inventory (Newly Notified Substances).

It can be argued that chemicals should be assessed for their potential to damage health or the environment prior to their introduction, as a number of well-publicised examples (thalidomide, asbestos, DDT, PCBs, and so forth) indicate the folly of introducing new chemicals without a suitable assessment of risk. Accordingly, the US TSCA legislation includes a requirement for a Pre-manufacturing Notification (PMN) to the Environmental Protection Agency (EPA) prior to introduction of a new chemical into commerce. In the European Union, a Screening Information Data Set (SIDS) must be developed and reviewed prior to commercialisation.

Obtaining information and conducting toxicological assessments relating to risks to health and safety and/or the environment for all categories of chemicals is important because:

- therapeutic substances are deliberately ingested
- cosmetics are applied to the skin, mucous membranes or hair
- food additives are present in food, sometimes in large quantities
- agricultural chemicals often find their way into food chains either as contaminants or by design
- industrial chemicals can affect the health of workers and in some cases, be significant environmental pollutants.

Therefore, the potential for exposure varies significantly, and in some cases, may be substantial.

Nonstatutory sources of law

The process of developing and adopting legally enforceable standards for particular materials is usually cumbersome, requires a long time, often requires detailed consultation, and if enacted, is slow to respond to new information. A good case in point is the generic carcinogen policy proposed by the US OSHA in 1977, finalised in 1980 and never implemented.

In occupational settings, employers frequently control workplace practices and exposure levels on the basis of voluntary nonenforceable guidance recommendations. Such guidance may be developed by independent groups of health professionals, for example, the Australian National Health and Medical Research Council, the Deutsche Forschungsgemeinschaft (maximum concentration value in the workplace; MAK values), the American Conference of Government Industrial Hygienists (threshold unit values; TLVs), or the American Industrial Hygiene Association (WHEELs). For materials that have not been considered by such expert groups, health professionals from industry groups or individual companies frequently develop such guidance. Occasionally, sources of detailed technical requirements on material relevant to a law can be incorporated into legislation.

Other examples of nonstatutory sources include:

- Standards for permissible exposure levels to contaminants, such as: acceptable levels of contaminants in soil of contaminated sites (ANZECC/NHMRC 1992); the NHMRC recommendations for lead levels in children (NHMRC 1993); occupational exposure standards (NOHSC Australia 1995); water quality guidelines (ANZECC/ARMCANZ 2002).
- Performance standards, such as the National Environmental Health Strategy (Woodruff and Guest 1997) or the NOHSC Australia National Strategy for the Management of Chemicals at Work (NOHSC Australia 1989).
- Codes of practice and guidance (such as AS 2865 Safe Working in a Confined Space or the NOHSC Australia Code of Practice Vinyl Chloride). These outline general provisions that will assist in reaching a desirable standard.
- Materials, equipment and process specifications (for example, AS 1692 Tanks

for Flammable and Combustible Liquids or AS 2161 Industrial Safety Gloves and Mittens), which are more specific recommendations which tightly define how to reach an appropriate level of compliance.

All these documents can be considered under the general descriptions of 'standards' (see Figure 19.1).

Life-cycle concepts

Chemicals management encompasses a range of inter-related activities (see Figure 19.2).

Not all of these steps are amenable to legislation, but many have legislative controls. This makes the regulatory control of chemicals quite complex.

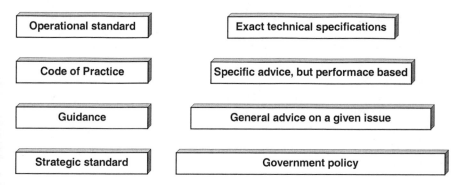

Figure 19.1 Standards and codes of practice.

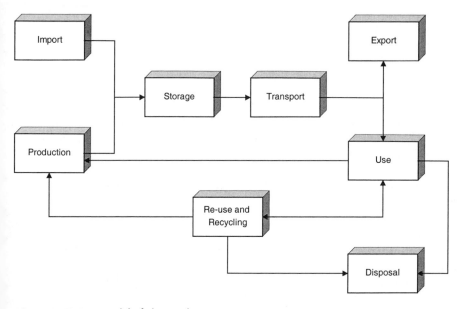

Figure 19.2 A model of chemicals activities.

Summary description of key legal instruments relating to chemicals

The life-cycle approach to chemicals management allows the identification of relevant components of chemicals management and use. As shown in Figure 19.2, the key life-cycle components identified for chemicals management are:

- import
- manufacture
- storage
- transport
- use
- disposal
- export.

These life-cycle components must then be contrasted against the jurisdictional variability that exists in governments. For example, there are a large number of distinct groupings, including:

- environment
- planning
- waste management
- health
- occupational health and safety
- consumer protection
- transport
- agriculture
- trade and industry
- legal
- waterways/marine environments
- local government.

The identification of these groupings varies from government to government, so there is likely to be some variability in interpretation of operational legislation.

These different life-cycle or jurisdictional components can then be used to identify who regulates, or is interested in, the various aspects of chemicals management. These are identified with a tick in Figure 19.3. What then emerges is a highly complex mosaic of jurisdictional activities and overlaps.

Legislation for notification, assessment and registration of chemicals

It is self-evident that regulation of chemicals at the individual nation or political unit level is inefficient, and that centralised or coordinated processes would assist in removing administrative complexity. Nowhere is this more apparent than in notification, assessment and registration of chemicals. Over the past two decades, there has been significant regulatory effort to harmonise procedures for assessing chemicals.

Life cycle component	Import	Manufacture	Storage	Transport	Use	Disposal	Export
Portfolio Environment	✔	✔	✔	✔	✔	✔	
Planning		✔	✔	✔	✔		
Waste Management			✔	✔	✔	✔	
Health		✔		✔	✔	✔	
OHS		✔	✔	✔	✔	✔	
Consumer protection		✔			✔		
Transport			✔	✔			
Agriculture (Pesticides)		✔	✔		✔	✔	
Trade and Industry	✔	✔			✔		✔
Legal		✔			✔	✔	
Waterways/Marine				✔		✔	
Local Government		✔		✔	✔	✔	

Figure 19.3 The regulatory mosaic for chemicals control legislation.

Notification, assessment and registration procedures for chemicals are usually treated under a number of categories, such as industrial chemicals; agricultural/veterinary products; pharmaceutical substances; cosmetics; and food additives. This division has arisen largely out of actions relating to microeconomic reform, in particular to avoid duplication, to decrease costs through centralised assessment and/or registration programs, and to reduce the burden on industry.

Notification and assessment legislation has evolved reflecting these divisions. For example, there are four main pieces of federal legislation which relate to the notification and assessment of chemicals in Australia. These are shown in Table 19.2.

Cosmetics and consumer products fall by default into the legislation for industrial chemicals in Australia.

Table 19.2 Notification and assessment legislation in Australia

Chemical category	Legislation	Coordinating agency
Industrial chemicals	Industrial Chemicals (Notification and Assessment) Act 1989	National Occupational Health and Safety Commission (NOHSC) Commonwealth Department of Health and Ageing (from 2002)
Agricultural and veterinary chemicals (Code) Act 1994	Agricultural and Veterinary Chemicals Veterinary Chemicals	National Registration Authority for Agricultural and
Pharmaceutical chemicals	Therapeutic Goods Act 1989	Therapeutic Goods Administration
Food additives and contaminants	National Food Authority Act 1991	Australia New Zealand Food Authority

Notification and assessment of pharmaceuticals

While systems for the notification and assessment of therapeutic substances really were in place from the 1950s, the thalidomide tragedy brought intense focus to the control of drugs. National systems are now in place regulating the safety, efficacy and quality of therapeutic goods (these include devices as well as medicines).

Notification and assessment of agricultural and veterinary chemicals

Legislation for agricultural and veterinary chemicals began getting quite detailed throughout the 1970s and onwards. One reason for this was Rachel Carson's book *Silent Spring*, which described the impact of extensive pesticide use on fauna in the USA in the late 1960s. National legislation can be quite varied across various nations, but nowadays there are quite stringent requirements in place. There are two objects for such legislation:

- to make provision for the evaluation, approval, and control of the supply of active constituents for proposed or existing agricultural and veterinary chemical products
- to make provision for the evaluation, registration and control of the manufacture and supply of agricultural and veterinary chemical products.

Notification and assessment of industrial chemicals

Industrial chemicals legislation, such as the US Toxic Substances Control Act 1976, the EC Sixth Amendment or the Australian Industrial Chemicals (Notification and Assessment) Act 1989 provides for national schemes for the notification and assessment of industrial chemicals. This legislation operates for the purpose of:

- aiding in the protection of the workers, the public and the environment by finding out the risks to occupational health and safety, to public health and to the environment that could be associated with the importation, manufacture or use of the chemicals
- providing information, and making recommendations, to regulatory agencies with responsibilities for the regulation of industrial chemicals
- giving effect to national obligations under international agreements relating to the regulation of industrial chemicals
- collecting statistics in relation to the chemicals.

Notification and assessment of cosmetics

The notification and assessment of cosmetics may be incorporated into consumer product or industrial chemicals legislation, depending on the country of jurisdiction. A major feature of systems for the control of cosmetics is the 'generally regarded as safe' (GRAS) list, which lists products that have been in use for many years, that are assumed to hold no health problems.

Notification and assessment of food additives and contaminants

Pure foods legislation is also longstanding legislation, in some cases dating back centuries (for example, the Roman edict to prohibit the use of lead nitrate as a sweetener in wine). Legislation for the notification and assessment of food additives, and legislation for the control of residues in foods are in place in many countries.

The listing of permissible additives, preservatives and colours and the setting of tolerances or residue limits are important regulatory functions. Many countries also have GRAS lists for established food additives.

Notification and assessment of consumer products

Consumer safety legislation is often quite patchy on the notification and assessment of the ingredients of consumer products, and in any event, is usually policed reactively.

Legislation by use category

Notwithstanding this complexity and uncertainty, there are numerous pieces of legislation which have a significant impact on chemicals.

The breadth and scope of the legislation in each of these regulatory mosaics is immense, and it would be difficult to provide a complete analysis. However, there is some consistency in many nations with legislation for the control of chemicals, and common themes are present in the 'big picture' legislation.

Customs acts

Customs acts regulate the export and import of chemicals and hazardous waste to ensure that:

- imported chemicals are (or have been) properly assessed under notification and assessment legislation (see above)
- advice is given to other countries receiving exported chemicals with known toxicity
- exported or imported hazardous waste is disposed of safely, so that human beings and the environment, both within and outside nations, are protected from the harmful effects of waste.

Clean air acts

This legislation is concerned with preventing and minimising air pollution from premises, mobile equipment and motor vehicles. It usually provides for licensing of premises scheduled under the act that will be allowed to discharge pollution to air (which actually makes such legislation a 'dirty air act'). A basic premise of such acts is the prescription of emission standards and control equipment. Where no standards exist, the 'best practicable means' approach is usually adopted.

Clean waters acts

These acts usually complement clean air acts, and provide for the prevention, control or mitigation of water pollution. The definition of 'water pollution' is crucial in this legislation, as it may be quite broad, for example the introduction of any matter into fresh or marine water to cause a change in its physical, chemical or biological condition. This legislation also usually allows for the licensing of premises that discharge to sewers, drains, watercourse or groundwater.

Pesticides acts

Of all the legislation to control chemicals, pesticides acts tend to be the most variable in the state jurisdictions, depending on jurisdictional needs and political influences. These acts are aimed at agricultural chemicals and (sometimes) veterinary drugs (these may have their own controls, such as Stock Medicine or Animal Husbandry Acts) and control a number of pesticide activities including possibly:

- the assessment of pesticides for human or animal health and for environmental protection
- determination of allowable levels of pesticide residues in foods
- the registration of pesticides
- approval of labels
- approval of containers
- issue of permits for the sale of unregistered pesticides
- control of the supply, sale, date marking and possession of pesticides
- the use of pesticides, including storage, personal controls during use, recommended application rates, disposal of used products or containers
- re-entry periods to areas where pesticides have been applied
- development of analytical methods to assure that food items do not contain higher that allowable levels of pesticide residues
- the prohibition or prevention of certain food items containing prohibited residues from being available for consumption.

Aerial spraying control acts

These control the licensing of pilots and aircraft for pesticide spraying. The legislation normally covers orders for the concentration of chemical applied, the type of aircraft and application equipment used, the land where the chemical is to be applied, and the climatic conditions during spraying. Some, but not all, statutes also contain provisions to cover hazard communication to third parties.

Dangerous goods acts

These are acts for the transport of hazardous chemicals, which have a high degree of conformity worldwide. The reason for this is that they are, in the main, based on the recommendations of the United Nations Committee of Experts on the Transport of Dangerous Goods covering the classification of hazardous materials, standards for labelling and packaging, minimal information and documentation requirements, and so on. These include a numbering system for identification of specific chemicals or groups of chemicals with the same hazards (the UN Number), a classification system according to the predominant hazard of the material (the Dangerous Goods Class) and a system for recognising the degree of danger or risk (the Packaging Group). Most dangerous goods transport systems use the UN system, or derivatives of it (the UN recommendations are incorporated into regulations and standards of a number of international and national organisations, including air transport, maritime transport and road transport agencies). National legislation follows accordingly.

Poisons acts

This legislation usually provides for the classification (scheduling), labelling, packaging and sale of poisons, including:

- natural poisons
- therapeutic poisons (prescription drugs, drugs of dependence and drugs of abuse)
- domestic, agricultural and industrial poisons.

Regulations under this sort of legislation may control for certain groups of poisons, such as carcinogens or teratogens.

Therapeutic goods and/or cosmetics acts

Aimed at particular sectors of the chemical industry, rather than a chemical or group of chemicals, this legislation aims to regulate the manufacture, distribution and advertising of therapeutic goods or devices and/or cosmetics. The legislation also usually imposes quality and quantity standards on these products.

Factories, shops and/or industries acts

These tend to be older, broad-based acts, covering a range of industrial activities. In the area of chemicals control, these acts often have a number of specific regulations, including those for the control of:

- lead
- asbestos
- abrasives
- welding
- electroplating
- spray painting
- synthetic mineral fibres
- pest controllers
- foundries.

Occupational health and safety acts

Occupational health and safety legislation, if introduced after 1973, tends to follow a model of enabling statute, which puts general obligations, duties and rights on a legal basis. This is a fundamentally different approach to the older style 'prescriptive' legislation, such as the Factories and Shops Acts (see above), which only sought to regulate specified industries and processes. Occupational health and safety legislation requires employers to ensure the health, safety and welfare of their employees and other persons at work. This duty of care sometimes (but not always) extends to manufacturers and suppliers of materials and production equipment to workplaces.

There are also usually a number of chemically related regulations or nonstatutory codes of practice to such acts, similar to those found under the Factories and Shops Acts (see above). However, one regulation that has a major impact on the control of chemicals in the workplace is the Hazardous Substances regulation, which explicates employer obligations to workers.

Hazardous substances legislation

The Hazardous Substances Regulation to Occupational Health and Safety Legislation is based on the risk management approach of identify, assess, control (for example, NSW Work Cover 2001). The regulation outlines a range of responsibilities, including:

- obligations of manufacturers and importers (suppliers), employers and employees
- requirements for record keeping
- criteria for determining a hazardous substance
- requirements for hazard communication on chemical hazards
- procedures for workplace assessment
- requirements for exposure control, including permissible exposure standards
- consideration of the need for workplace monitoring
- consideration of the need for health surveillance
- requirements for education and training
- systems for the emergency services.

This regulation represents an attempt to manage chemicals in the workplace, from purchase to disposal.

Planning legislation

Many developments are now known to pose substantial risks. Planning legislation aims to ensure that, in relation to such developments, that matters affecting the environment significantly are fully examined and taken into consideration.

Consumer protection/trade practices acts

These provide protection for consumers of, among other things, chemicals. Under such acts, products can only be supplied which comply with approved safety standards. These standards include contents, packaging, warnings, and disclosure of product information. It is likely that chemicals falling into the jurisdiction of one or more pieces of legislation (cosmetics, drugs, domestic pesticides, foods) will also be covered by legislation of this type.

Radiation control acts

This legislation is concerned with the regulation of radioactive materials and radiation apparatus, including:

- licensing of premises (such as nuclear reactors and facilities, but also possibly X-ray installations)
- the transport, storage, use and safe disposal of any radioactive waste products from manufacture, production, treatment and storage of radioactive substances.

Major hazards acts

These require the control of major hazards (rather than the siting or location of major hazards, which tend to be controlled under planning legislation).

Waste minimisation acts

This legislation covers chemicals or chemical wastes considered to be environmentally hazardous, and may be subject to control by licensing or prohibition. It also sometimes provides for the restoration of premises contaminated by chemicals or chemical wastes.

Ozone protection legislation

Ozone protection legislation institutes a system of controls on the manufacture, import and export of substances that deplete stratospheric ozone.

Sea dumping legislation

Sea dumping legislation provides for the protection of the environment regulating the dumping into the sea, and incineration at sea, of wastes and other matter.

Chemical weapons legislation

Chemical weapons legislation should be consistent with the Chemical Weapons Convention (CWC). The CWC is an international treaty that bans the development, production or use of chemical weapons, and requires the destruction of existing weapons. This treaty requires the identification of facilities that produce or use chemicals that have been identified as useful in the production of chemical weapons. The treaty requires the reporting of quantities of listed chemicals imported, used, produced or exported and also contains provisions for inspection of such facilities by international teams.

Banned or severely restricted chemicals legislation

The WHO produces a list of banned or severely restricted chemicals based on regulatory activities in a diverse number of countries. This may be used to further the control of listed chemicals under relevant legislation. In recent years, considerable international activity has focused on efforts to develop appropriate regulation to control the release of persistent bioaccumulative toxic (PBT) chemicals.

Ministerial, coordinating and consultative mechanisms

Chemicals management infrastructure is not carried out without significant consultation of relevant issues and possible impacts.

There are a range of coordinating and consultative mechanisms which assist in the process, from Ministerial Councils to public fora. Examples include:

- councils of ministers
- peak councils of government and nongovernment individuals
- policy committees of senior government officials
- coordinating committees of government officials of different agencies
- consultative committees of government and/or nongovernment members on specific jurisdictional matters, such as the pesticides or pollutants
- technical committees of government and/or nongovernment members on specific jurisdictional matters
- public consultative committee.

In all, the possible mechanisms for input into many of the processes of chemicals management is quite varied.

Nonregulatory mechanisms for managing chemicals

Nonregulatory mechanisms for managing industrial chemicals are focused on education and information dissemination and research. In these areas a diverse group of stakeholders are actively involved (for example, government agencies, chemical industry, other employers, unions, environmental groups, community groups, and the public are active). Education and information dissemination ranges from conducting community awareness campaigns to integrating information on chemical safety into formal courses, such as tertiary courses.

In the area of research, a range of bodies fund or encourage research into the effects of chemicals on humans and the environment, and into safe handling and use systems.

The best example of a voluntary, nonregulatory chemicals management program is the Responsible Care® program, an initiative of the chemical industry. This initiative was developed and implemented by the chemical manufacturers in North America and has subsequently been adopted by chemical manufacturers, distributors and users in a number of international settings. Responsible Care® incorporates industry best practices into a codified set of management practices that cover the entire life-cycle of chemicals. These practices cover the range of activities associated with the use of chemicals:

- community awareness and community response
- process safety
- employee health and safety
- distribution
- waste and release reduction
- waste management
- product stewardship.

In some countries, introduction of Responsible Care® has provided the impetus for the development of laws and regulations to cover certain aspects of chemical management. In other countries, it has become nearly as important as regulations in influencing chemical management.

Discussion

With regard to chemicals and the law, it can be seen that there are four different approaches to the application of the law in the control of chemicals:

1 Primary legislation or subordinate regulations

Recourse to legislation is the primary means of controlling chemicals in many nations. This can take place at three separate levels.

Political unit level

Where governments (state or province) can enact specific legislation as part of their jurisdictional responsibilities, such as Poisons or Dangerous Goods Acts.

National level

In countries which do not have a federal or provincial system of government (such as the UK or New Zealand) this is a fairly straightforward process, subject to the parliamentary systems in place in those countries. In countries that have a less centralised structure, such as the USA, Canada, Australia and Germany, systems for enacting legislation are dependent on constitutional demarcations and administrative arrangements between national and local levels. Nationally imposed law can be picked up or modified by other jurisdictions within the country, depending on priorities and political imperatives. In cases where the legislative function is a federal responsibility, national governments can (and do) introduce legislation to control chemicals. For example, in Australia, the federal government has the responsibility for chemicals assessment, and legislation such as the Notification and Assessment (Industrial Chemicals) Act 1989 or the Agricultural and Veterinary Chemicals Code Act 1994 is in place to deal with such functions. However, where responsibilities are devolved to other jurisdictions, they can produce a chronic problem of nonuniformity of legislation within the different states.

International level

Here, national governments can pick up internationally derived recommendations. For example, many nations are signatory to a significant number of conventions for the control of chemicals, such as the Basel Convention for the Transboundary Movement of Hazardous Wastes, or the Vienna Convention for the Protection of the Ozone Layer. However, in other cases, ratification of international conventions is only possible if complementary legislation is available in all jurisdictions.

2 Legislative review

From time to time it becomes necessary to re-examine the operation of the law, and for this, the processes of legislative review are required. There are a number of review mechanisms available.

Judicial review

This form of review has not been not used for matters of review of chemicals management infrastructure.

Parliamentary review

This has been used a number of times, such as the House of Representatives Standing Committee on Environment and Conservation Inquiry into Hazardous Chemicals (SCEC 1982), the Senate Select Committee on Agricultural and Veterinary Chemicals in Australia (1990) or the New South Wales Parliament Inquiry into Domestic Lead Pollution (1994).

Review by government agency

Such as the Industry Commission report into Occupational Health and Safety (IC 1995), or the NSW Chemicals Inquiry (WorkCover 1990).

Review by independent body

Such as the review of pesticide residues in agricultural products conducted by the Australian Science and Technology Council (ASTC 1989).

The impact of such reviews depends on their terms of reference, their scope and the ability or will of parliaments or government departments to adopt and implement recommendations stemming from such reviews. Normally, governments act with commendable speed to adopt recommendations that stem from high-level reviews. However, this is not always the case.

3 Consensus approaches

Consensus approaches are worthy of consideration, in which development of traditional restrictive legislation-based controls are moving away to a less prescriptive approach, incorporating multipartite development of standards and preventive strategies with more emphasis on consultation. These may or may not be incorporated into statute.

4 Recourse to common law

Recourse by common law is made through the courts. Common law can be a powerful means of establishing precedents, by which the operation of existing legislation can be examined and modified. Over the past decade, courts in Australia have made decisions on a number of cases relating to the use of chemicals. For example:

- imprisonment of a company director for knowingly discharging hazardous wastes to a wetlands;
- recognition of exposure to tobacco smoke as a workplace contaminant for which employers should control;
- establishment of the boundaries of due diligence for environmental management or workplace safety;
- the copyright status of workplace material safety data sheets;
- establishment of multiple chemical sensitivity as a compensatable condition.

However, common law actions have a number of problems. Firstly, they require a plaintiff to mount a common law action, a process that is time-consuming, frustrating and expensive. Secondly, they rely on a legal profession who are knowledgeable about the law, but largely ignorant about chemistry, toxicology, public health and environmental protection. There is often a problem of interpretation, in that the elusiveness of scientific certainty contrasts poorly by the yes/no, black/white demands of the law. Thirdly, judicial decisions are subject to the whims and vagaries of judges or juries. Lastly, common law judgements can be so complex that they may be misinterpreted.

In the USA there have been recent attempts to develop procedures concerning the admissibility of technical evidence in lawsuits, in order to prevent use of 'junk science' in the courtroom. As a result of a lawsuit (*Daubert vs. Merrill Dow*), guidelines on qualifications of technical experts and the scientific acceptability of technical information to be used as evidence were developed for use in federal courts. This guidance has been used in several additional cases and seems to be a growing practice.

Phases of regulation

Taking a broad view of developments in OECD countries, Nichols and Crawford (1983) outlined various phases of regulatory approach:

- the minimalist period of the pre-1960s
- the fragmented approach of the late-1960s and early 1970s, in response to potential or actual damage to health or the environment. Governments reacted to known hazards in recognised situations, with the emphasis – in the main – on reactive, prescriptive and corrective methods (these tended to fit into existing institutional arrangements)
- the sectoral approach of the 1970s and 1980s of various agencies and government departments charged with the responsibility of a specific area, such as environment, workplace, public health or transport.

Development of chemicals management infrastructure worldwide have mirrored these phases.

All these can impact on the law. One further point should be introduced. As noted above, chemicals control legislation has been introduced in discrete 'waves' of the minimalist, fragmented, and sectoral approaches. This leads to the question: what is likely to be the next wave?

In many respects, criticism of the current regulatory system (PIAC 1992) is based on a misunderstanding of ministerial and legislative responsibilities and interjurisdictional demarcations. For chemicals, it makes little sense if a chemical exposure causes problems in one (or more than one) of the current jurisdictional demarcations (environment, public health, occupational health and safety, transport, and so on). Perhaps what is needed next for consistent regulation of chemicals is a transectoral 'whole of government' approach, where chemicals are regulated because they are chemicals, not because they are found in the environment or workplaces or anywhere else.

Chemicals assessment and control legislative systems in many countries are comprehensive, although some chemicals may be covered by more than one piece of legislation (for example, they may be used for both industrial and agricultural purposes). However, coordination systems are in place to avoid duplication of assessment work and conflicting application of controls. A number of labelling systems are in place and Australia actively participates in international work on harmonisation of classification and labelling systems.

Enforcement programs are conducted by the range of regulatory authorities with responsibility for chemicals legislation. Programs are adapted as required by authorities to address the objectives of the legislation. The effectiveness of enforcement programs of such legislation is sorely needed.

Summary

In government activities, the management of chemicals requires activities in the areas of:

Assessing the risks from chemicals:
- premarket assessment (for new chemicals)
- existing chemicals assessment (for existing chemicals considered to be a risk)
- land use planning guidance for new facilities where chemicals are to be used
- land use planning guidance for existing facilities where the risks from the use of chemicals is changing (and the risk is increasing)
- identification and assessment of hazardous wastes.

Controlling sources of chemicals:
- import and export
- mining or other forms of extraction
- chemicals manufacturing.

Controlling 'downstream' use of chemicals:
- industry
- agriculture
- pharmaceutical and cosmetics
- food
- consumer products.

Controlling impacts of chemicals:
- public health
- occupational health and safety
- environment.

Controlling various activities:
- import/export
- storage
- transport and distribution
- emergency response management.

Controlling the disposal of chemicals, containers and their wastes:
- to air
- to land
- to water
- to the marine environment.

Many of these activities, such as mining, are only indirectly associated with chemicals, so their legislative controls are not comprehensive. Further, in countries with discrete political units such as state or provinces, the various levels of government make allocation of responsibilities complex.

However, there are a number of common themes that are apparent in this legislation, as shown in Table 19.3.

Table 19.3 Themes in chemicals management legislation

Activity		Type of legislation	Examples
Identification	Individual chemicals	Notification and assessment legislation	Toxic Substances Control Act or sixth amendment
	Places where chemicals are used	Planning legislation	City or state planning acts
Infrastructure	Government agencies	Legislation which establishes regulatory bodies	Australian National Environment Protection Council (NEPC) legislation
	Courts	Legislation which establishes legal systems and courts	Environment court legislation
Assessment of impacts	Sectoral responsibilities	Legislation for environment, health, workplace and so forth	Environment protection legislation Public health acts Occupational health and safety acts
	Specific activity responsibilities	Legislation for transport, agriculture, wastes and so forth	Hazardous wastes legislation
Control of impacts	Sectoral responsibilities	Legislation for environment, health, workplace and so forth	Poisons acts Hazardous substances legislation
	Specific activity responsibilities	Legislation for transport, agriculture, wastes and so forth	Pesticides legislation Dangerous goods legislation

Controls for assessment

The assessment of chemicals for human health or environmental effects is a national function. This optimises resources by removing the possibility of duplication of assessments and diminishes the likelihood of variable risk assessments by other agencies.

Obviously, the data and information requirements of, and assessment mechanisms for, each piece of legislation vary. Further, each assessment agency can introduce technical and consultative mechanisms by which the assessment process proceeds and is made accountable. These can range from ministerial councils, who set policy and make decisions, to technical working groups where various matters are considered. In some cases, consultation with industry, community or union groups is part of the assessment process.

Controls of chemical sources

Chemicals can be sourced by import, extraction and manufacturing.

Import

Legislation for the control of import of materials exists in the customs legislation. However, the materials covered under the legislation as illegal imports or contraband tend to be illegal for other than chemical safety reasons, such as drugs of abuse or addiction or animal products. However, customs agents are reluctant to impound unassessed chemicals, which tend to be in large quantities (compared with contraband) as they may:

- not necessarily know the hazards or risks
- not necessarily not know the value
- have to adopt a liability for which they lack resources or capability.

Therefore, common with other legislation (such as the Dangerous Goods Act), chemicals imported into countries are generally allowed one trip from the port of entry to a depot where any noncompliance with legislation can be dealt with.

Mining

Extraction of chemicals by mining or other forms of extraction (such as petro-chemicals from oil wells) is covered under mining legislation. Mining activities are covered under mining or resources legislation and may be general to all mines, or may cover specific types of mines such as coal mines.

Chemicals manufacturing

Chemicals manufacturing is considered an industry like many others. Many countries have not considered this industry to be different enough from other industries in the manufacturing sector to warrant its own legislation, as has occurred for other sectors, such as mining or construction. Therefore chemicals manufacturing is controlled under general legislation for such things as environment, occupational health and safety, waste disposal and so forth. Other legislation, such as poisons acts and dangerous goods legislation also has a major impact in this industry – indeed more so than any other. However, again this legislation is not specific for the chemical industry sector.

Planning legislation

There is no specific legislation for the identification and assessment of sites where chemicals are planned to be located. Such administrative arrangements are incorporated in the general requirements of planning legislation. Planning legislation covers a number of issues, such as land use, land zoning and there are significant differences in scope and application of the various pieces of planning acts across nations.

However, overlap is such that it is now not possible to develop a new facility, or significantly expand an existing facility, in those jurisdictions where planning legislation is in place where chemicals are to be manufactured or used, without having to comply with planning requirements.

Further, major projects in many jurisdictions are required to undergo environmental impact assessment procedures, whereby they can be scrutinised even further.

Infrastructure legislation

While much legislation deals with chemicals and their effects, or deals indirectly with the consequences of activities involving chemicals, other legislation has a direct effect on how a country manages chemicals by virtue of the regulatory bodies and the administrative infrastructure it establishes, such as government departments, statutory bodies, courts and so forth. No analysis of chemicals management infrastructure should ignore such structures.

Chemicals handling and use

It is in this area that the important legislation for controlling chemicals in many countries is found. These include important legislation such as:

- environmental and pollution control legislation, such as clean waters and clean air acts
- poisons legalisation
- dangerous goods legislation
- occupational health and safety legislation, including regulations for hazardous substances
- occupational health and safety legislation for day to day activities involving chemicals
- environmental legislation where such activities may impact on the environment
- public health legislation where such activities may affect the health of the public
- waste minimisation or disposal legislation if chemicals must be disposed.

Also, for some groups of chemicals (such as pesticides or radioactive substances) or industry sectors (such as agriculture, pest treatment or asbestos removal) there may be specific controls in other legislation which incorporate responsibilities in all these areas (as well as in other areas). In such cases, the potential for comprehensive chemicals management can exist.

References and bibliography

AAVCC (1989) *Requirements for Clearance of Agricultural and Veterinary Chemical Products.* Australian Agriculture and Veterinary Chemical Council/AGPS, Canberra.

ACIC (1987) *Emergency Response Manual.* Australian Chemical Industrial Council, Melbourne.

ACIC (1993) *Responsible Care: Code of Practice – Community Right to Know*. Australian Chemical Industrial Council, Melbourne.

ACIC (1991) *Responsible Care: Code of Practice – Emergency Response and Community Awareness (ERCA)*. Australian Chemical Industrial Council, Melbourne.

ACIC (1993) *Responsible Care: Code of Practice – Manufacturing*. Australian Chemical Industrial Council, Melbourne.

ACIC (1993) *Responsible Care: Code of Practice – Research and Development*. Australian Chemical Industrial Council, Melbourne.

ACIC (1992) *Responsible Care: Code of Practice – Warehousing and Storage*. Australian Chemical Industrial Council, Melbourne.

ACIC (1992) *Responsible Care: Code of Practice – Waste Management*. Australian Chemical Industrial Council, Melbourne.

ACIC (1992) *Waste Management Code of Practice and Implementation Guide*. Australian Chemical Industrial Council, Melbourne.

AEC (1986) *National Guidelines for the Management of Hazardous Waste*. Australian Environment Council/AGPS, Canberra, November 1986.

ANZECC (1985) *Air Emission Inventories for Australian Capital Cities*. Australian and New Zealand Environment and Conservation Council, Canberra.

ANZECC (1994) *Guide to Environmental Legislation in Australia and New Zealand*, 4th edn. In: Fabricus, C. P. (ed.) *ANZECC Report No. 2*. Australia and New Zealand Environment and Conservation Council/AGPS, Canberra.

ANZECC (1994) *National Guidelines for the Management of Wastes: National Manifest System and Classification System*. Australia and New Zealand Environment and Conservation Council/AGPS, Canberra.

ANZECC (1992) *National Strategy for Ecologically Sustainable Development*. Australia and New Zealand Environment and Conservation Council/AGPS, Canberra.

ANZECC/ARMCANZ (2002) *Australian and New Zealand Water Quality Guidelines*. Australia and New Zealand Environment and Conservation Council/Agriculture and Resource Management Council of Australia and New Zealand, Canberra.

ANZECC/NHMRC (1992) *Australian and New Zealand Guidelines for the Assessment and Management of Contaminated Sites*. Australia and New Zealand Environment and Conservation Council/National Health and Medical Research Council/AGPS, Canberra.

ASTC (1989) *Health, Politics, Trade: Controlling Chemical Residues in Agricultural Products*. Australian Science and Technology Council/AGPS, Canberra.

Brickman, R., Jasanoff, S. and Ilgen, T. (1985) *Controlling Chemicals: The Politics of Regulation in Europe and the United States*. Ithaca: Cornell University Press.

Brooks, A. (1994) *Guidebook on Australian Occupational Health and Safety Law*, 4th edn. Sydney: CCH Australia.

BRRU (1986) *Social Regulation: Issues Concerning Industrial Chemicals. Information Paper No. 3*. Business Regulation Review Unit/AGPS, Canberra.

Carson, R. (1962) *Silent Spring*. New York: Houghton Mifflin.

CEC (1987) *Legislation on Dangerous Substances: Classification and Labelling in the European Communities*. Commission of the European Communities, Luxembourg.

CEPA (1992a) *Australian Inventory of Chemical Substances*. Commonwealth Environment Protection Agency/AGPS, Canberra.

CEPA (1992b) *National Pollutant Inventory: Public Discussion Paper*. Commonwealth Environment Protection Agency/AGPS, Canberra.

CEPA (1992c) *National Waste Minimisation and Recycling Strategy*. Commonwealth Environment Protection Agency/AGPS, Canberra.

Crawford, B. (1994) *Australian and New South Wales Regulation and Control of Hazardous Materials*, 3rd edn. Sydney: Alken Press.

Eisner, H. (1988) Safety: the perils of self regulation. *Safety Aust.* 14.

FORS (1997) *Australian Dangerous Goods Code*, 6th edn. Canberra: Federal Office of Road Safety/Australian Government Printing Service.

Fuller, J. (1995) *Chemicals, Communication and the Community: Recognising a Right to Know in the Chemical World.* Public Interest Advocacy Centre, Sydney.

Gun, R. (1995) The NOHSC Model Regulations for chemical safety: How much benefit? *J. Occup. Health Safety Aust. N.Z.* 10(6): 523–528.

Gunningham, N. and Cornwall, A. (1994) *Toxics and the Community: Legislating the Right to Know.* Australian Centre for Environmental Law/Australian National University, Canberra.

HSE (1988) *Control of Substances Hazardous to Health Regulation.* UK Health and Safety Executive/Her Majesty's Stationary Office, London.

IARC (1966) *Overall Evaluations of Carcinogenicity to Humans* (to May 1996). International Agency for Research on Cancer, Lyon. At: http://www.iarc.fr/crthall.htm.

IC (1995) *Work, Health and Safety: Inquiry into Occupational Health and Safety*, Volumes 1 and 2. Industry Commission, Melbourne.

ILO (1990) *Safety in the Use of Chemicals at Work. Convention No. 170.* International Labour Organisation, Geneva.

IOMC (1996) *Strengthening National Capabilities for the Sound Management of Chemicals.* Inter-Organization Programme for the Sound Management of Chemicals, Geneva.

Lewis, S.M., Moore, H., Winder, C. and Mooney, C. (1993) *A Survey of Industrial Product Use in the Rockdale Area.* NOHSC Australia/NSW WorkCover Authority, Sydney.

Moore, S.J. and Thompson, G. (1997) Implementing the Basel Convention on Transfrontier Movement of Hazardous Wastes, Australian experience. *Poll. Control* 97.

Moore, S.J. and Worrall, M.J. (1991) Waste management plans for major industries. In: *Waste Technology and Management.* DITAC/AGPS, Canberra.

NHMRC (1985) *Code of Practice for the Disposal of Radioactive Wastes by the User.* Canberra: National Health and Medical Research Council, Radiation Health Committee Publication 13.

NHMRC (1992) *Code of Practice for the Near-Surface Disposal of Radioactive Waste in Australia Publication 35.* Canberra: National Health and Medical Research Council, Radiation Health Committee.

NHMRC (1990) *Guidelines for Laboratory Personnel Working with Carcinogens or Highly Toxic Chemicals.* National Health and Medical Research Council/AGPS, Canberra.

NHMRC (1990) *Intervention in Emergency Situations Involving Radiation Publication 32.* Canberra: National Health and Medical Research Council, Radiation Health Committee.

NHMRC/NOHSC (1995) *Recommendations for Limiting Exposure to Ionising Radiation (Guidance note [NOHSC:3022(1995)])* and *National Standard for Limiting Occupational Exposure to Ionising Radiation* [NOHSC:1013(1995)]. *Radiation Health Series No. 39.* Canberra: Australian Radiation Laboratory. Australian Government Publishing Service.

NOHSC Australia (1991) *Control of Workplace Hazardous Substances.* National Occupational Health and Safety Commission/Australian Government Printing Service, Canberra.

NOHSC Australia (1995) *Exposure Standards for Atmospheric Contaminants in the Occupational Environment: Guidance Notes and National Exposure Standards.* National Occupational Health and Safety Commission/Australian Government Printing Service, Canberra.

NOHSC Australia (1994) *Model Regulation for the Control of Workplace Hazardous Substances: Code of Practice for the Control of Workplace Hazardous Substances.* National

Occupational Health and Safety Commission/Australian Government Printing Service, Canberra.

NOHSC Australia (1989) *National Strategy for the Management of Chemicals Used at Work*. National Occupational Health and Safety Commission/Australian Government Printing Service, Canberra.

NTP (1984) *Toxicity Testing Strategies to Determine Needs and Priorities*. National Toxicology Program/National Research Council, Washington.

Nichols, J.K. and Crawford, P.J. (1983) *Managing Chemicals in the 1980s*. Organisation for Economic Cooperation and Development, Paris.

OECD (1986) *Guidelines for Testing of Chemicals*. Organization for Economic Cooperation and Development Publications Service, Paris.

Parliament of Australia (1991) *Report of the Senate Select Committee on Agricultural and Veterinary Chemicals in Australia*. AGPS, Canberra.

PIAC (1992) *Toxic Maze*, parts 1 and 2. Public Interest Advocacy Centre, Sydney.

SCEC (1982) *Hazardous Chemicals: Second Report on the Enquiry into Hazardous Chemicals*. House of Representatives Standing Committee on Environment and Conservation. AGPS, Canberra.

Senate Select Committee on the Dangers of Radioactive Waste (1996) *No Time to Waste*. AGPS, Canberra.

Standards Australia (1995a) AS/NZS ISO 14000 Series *of Environmental Management Systems* standards. Standards Australia, Sydney.

Standards Australia (1995b). AS/NZS 4360 *Risk Management*. Standards Australia, Sydney.

Stewart, D. (1995) Attitudes to workplace hazardous substances. *J. Occup. Health Safety – Aust. N.Z.* 10: 375–376.

Strange, L. (1990) Right-to-Know-Right-to-Act: International Developments in Worker and Community Hazardous Substances Right-to-Know Laws. *A report to the Law Foundation*, Sydney.

UNCED (1992) *Agenda 21*. United Nations Conference on Environment and Development, New York.

Winder, C. (1997) Hazardous inconsistencies. *Aust. Safety News* 14–19.

Winder, C. (consulting ed.) (1985) *Managing Workplace Hazardous Substances*. CCH International, Sydney.

Winder, C. (1989) Government initiatives in the control of industrial chemicals in Australia. *Chem. Aust.* 56: 80–81.

Winder, C. (1993) The new legislation and national uniformity. *J. Occup. Health Safety Aust. N.Z.* 9: 299–301.

Woodruff, R.E. and Guest, C.S. (1997) *Developing the National Environmental Health Strategy*. Commonwealth Department of Health and Family Services, Canberra.

WorkCover. (1990) *NSW Chemicals Inquiry, Volumes I and II: Report to the Minister*. WorkCover Authority of NSW, Sydney.

WorkCover. (2001) *Occupational Health and Safety Regulation 2001*. WorkCover Authority of NSW, Sydney.

Wyatt, A. and Oxenburg, M. (eds) (1996) *Managing Occupational Health and Safety*. CCH Australia, Sydney.

Chapter 20

Managing workplace chemical safety

C. Winder

Introduction

There is a range of legislation relating to chemicals in workplaces. Employers must somehow comply with all this legislation.

With regard to the workplace, critical legislation has been introduced in the past decade relating to workplace safety and the environment. This has placed pressure on employers and workplace managers to manage chemicals better.

There are only two factors that affect safety:

- inherent properties (toxicity, flammability, corrosiveness, and so forth)
- factors relating to exposure such as the probability, duration, frequency and intensity of exposure.

Also, there are really only four approaches to chemical control (see Table 20.1). These approaches apply equally well at the national, industry, company or workplace levels.

Lastly, it is important to recognise that most, if not all, chemical substances can be handled and used safely, providing that:

- the hazards are known and understood
- correct handling and use procedures are in place and adhered to
- the correct equipment to handle and use chemicals is available, used and maintained
- workers are informed about hazards and trained in correct procedures and prompt action is taken to control and minimise problems that do arise.

This requires systems to ensure that hazards are identified, risk assessment procedures are in place, the right controls are used, and emergency procedures are in place. In turn, this requires good management. The risk management approach of

Table 20.1 Approaches to chemical safety

1   Prohibit chemicals
    Any risk unacceptable
    No risk
    No benefits
    Example: The US Delaney Amendment to the Food and Drugs Act (1976), which
    prohibits the introduction of any carcinogen into the USA

2   Check all chemicals
    Mandatory evaluation of hazard
    Resource intensive
    Risk minimisation
    Example: In Australia, all pharmaceutical substances require chemical, toxicological
    and clinical evaluation by law (Therapeutic Goods Act 1966)

3   Screen all chemicals, check all suspect chemicals
    Risk benefit analysis required
    Allows stepwise decision making
    Example: Assessment of the environmental effects of agricultural chemicals.

4   Allow chemicals
    Ignore the hazard
    Accept the risk
    Example: Occupational or domestic exposure to tobacco smoke (in situations where
    Approach 1 has not been implemented)

establishing organisational considerations, identifying chemical hazards, assessing risks, control risks and evaluating management activities has become the *de facto* means of managing workplace hazards. This approach outlines a range of activities as detailed below.

Risk management:

- establishing risk management systems outlining obligations of (among other things) suppliers of materials, occupiers, managers and employees
- establishing the boundaries of the risk management system (workplace safety only, or environmental impact, or both).

Identification of hazardous substances:

- criteria for determining a hazardous chemical
- requirements for hazard communication on chemical hazards, such as labels, material safety data sheets (MSDS), registers
- requirements for competency training of workers.

Assessment of the risks of hazardous substances:

- procedures for workplace assessment where chemicals are used.

Control of hazardous substances that are a risk to health:

- requirements for control of exposure or impact where risk assessments indicate that problems exist
- consideration of the need for workplace or environmental monitoring
- consideration of the need for health surveillance of workers
- systems for emergency response
- systems for environmental management.

Review of management systems and operational activities:

- internal audit procedures
- requirements for record keeping
- procedures for management review.

Importantly, risk management outlines an approach by which chemicals in the workplace can be managed.

Coming to grips with chemical safety management

Most senior managers can not be expected to be informed about every issue and decision within the organisation. How then can they meet their chemical safety responsibilities towards their employees? Some organisations nominate a person who is personally responsible for workplace safety, and presumably they are able to deal with chemical safety problems.

More commonly, senior management relies on an effective management system whose operation is guided by policy. A program for complying with legislation for chemicals should be built into this structure.

Principal concepts of a chemical safety management system

The management of occupational health and safety can be facilitated by:

- the workplace chemical safety policy
- the workplace chemical safety management system, containing a number of programs and subprograms
- procedures for those operations for which the information in the policy and program is not detailed or definitive enough
- strategic planning and integration.

The workplace chemical safety policy

A workplace health and safety policy, developed in conjunction with the employees, defines the organisation's goals and communicates senior management's commitment. It provides guidelines on how health and safety is to be managed. A prominent written policy assists in this process.

The content of a policy would typically include the following:

- purpose
- scope
- measurable targets and objectives
- actions and responsibilities.

Important features of the policy include:

- it is written in simple language
- it is an agreed set of principles between management and employees
- it outlines the responsibilities of relevant groups, such as management, employees, safety professionals (if any are available) and the occupational health and safety committee (if one exists)
- it is tailored to the specific needs of the workplace or organisation
- it can be integrated with normal processes in the workplace.
- it is workable.

The policy should be regularly reviewed. Conditions of the workplace change, as does legislation. The policy may need to be revised to reflect these changes.

The Chemical Safety Policy represents the first step in addressing chemical safety at work. It is a statement of intent. The next steps involve the establishment of a chemical safety management system.

The chemical safety management system

In line with contemporary management practice, organisations are increasingly devolving responsibility for responsibility and accountability to line management. For this to work effectively, strategic planning is becoming organisation-wide, and not the sole activity of senior management.

Framework of the management system

For the chemical safety system to work properly, it needs to be set in a conceptual and philosophical framework which allow its appropriate development. For this to occur, a number of fundamental issues need to be addressed:

- there needs to be full commitment from all stakeholders (especially employees and managers)
- the management system should be risk management in approach
- the management system should be proactive in nature
- the management system should provide the tools to allow change in safety culture
- there should be an expectation that there will be an initial quantal increase in chemical safety system effectiveness, followed by continual improvement.

Continual improvement

Continual improvement is crucial for sustaining the impact of any intervention, such as the development and introduction of a new management system. There is invariably a change in performance measures during the introduction of the system; often standards can slip after time. Therefore focus on the system, such as occurs through continual improvement, can ensure that any slip in standards is arrested (see Figure 20.1).

Basic elements of the chemical safety management system

Including chemical safety in the strategic planning process brings it in line with all other activities, incorporates it into those activities, and is one of the first steps in proactive chemical safety management.

The basic elements (subprograms) of a chemical safety program are those relating to the management of:

■ toxic or dangerous chemicals, especially those materials specific to the company
■ emergencies involving chemicals
■ 'near hits', incidents and accidents involving chemicals
■ rehabilitation of injured workers
■ workers' compensation.

With respect to 'near hits' the following should be noted: 'Quick thinking by the operator saved the workforce team from a near miss.' This is not quite true. The workforce was saved from a hit. A near hit is what they had. Near hits are

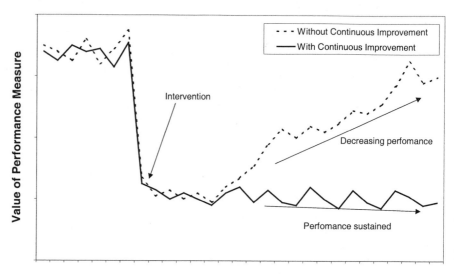

Perfomance Measure Over Time

Figure 20.1 Effect of intervention with and without continual improvement.

sometimes called near misses. This is an inapplicable designation, as a near miss is a *miss*. People don't seem to regard misses as something that need as much attention, so the term 'near hit' is more descriptive of the hazard, and the attention that needs to be given to it.

It is important to ensure the chemical safety management system is focused on the priority hazards within a workplace and sensitive to changes that may occur, including:

- design of new, or modification of existing, workplaces
- changes in work practices, such as the introduction of new materials, plant and processes
- changes in personnel, especially managers and supervisors
- changes in legislation, such as a new first aid regulation, or accident reporting.

Figure 20.2 outlines the components of the chemical safety management system.

Attributes of the management system

The chemical safety management system has a number of attributes:

- It is focused on priorities, with selection of strategies and activities of the organisation dependent on management imperatives, organisational priorities and resource availability. Policies and programs need to be identified, developed and implemented.
- It optimises the available resources. This is the operational side of managing the chemical safety program. Hazards and risks need to be identified through such things as workplace risk assessments, accident/incident investigation, review of rehabilitation and workers' compensation cases and emergencies.

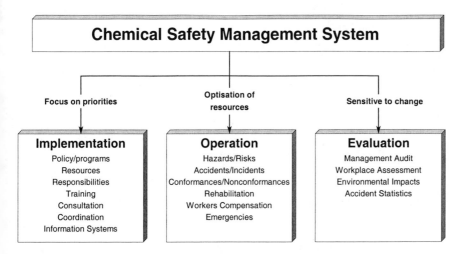

Figure 20.2 Components of the chemical safety management system.

- It is sensitive to changes, both in the internal and external environment, as changes in the external environment offer threats to the unprepared manager and opportunities to the visionary manager. Being aware of such changes will assist in positioning the management program to gain the opportunities. For example, information on pending regulatory changes, and the proactive and expeditious modification of existing programs can be exceptionally advantageous.

Further, changes in the internal environment will also assist in the effective management of the chemical safety management system. Management audits, regular assessment of the workplace, and accident statistics will aid in this process.

Chemical activities in workplaces can be very diverse. Unless proper systems are in place, they will not be managed safely, and can exacerbate accidents, injury or disease if they occur. Chemicals handling, use and disposal needs proper management.

Programs and subprograms within the management systems

The framework of the management system is one thing. What it deals with is another. Any matter that falls into the scope of a management system will need further systems for development, delivery and review. Of necessity, these will need to outline the ways in which the policy will be delivered, including the steps and organisational requirements which assist in that process. This is the program (or in some cases, the subprogram). For example, a planned maintenance program would outline the broad processes required for ensuring plant and equipment remains operational.

The content of any individual program may include the following:

- referral to the policy and overall commitment to the management system
- responsibilities and authorisations
- relevant skills/training
- relevant equipment
- operational requirements
- recommended sequence of activities
- other requirements (such as clean up or record keeping)
- referral to other programs where relevant.

The details of numbers, scope and range of such programs will vary from workplace to workplace, and can be extraordinarily intricate in their scope, complexity and outcomes. The types of individual programs could include:

Components common to some, if not all chemical safety programs include:

- hazard communication
- training
- workplace risk assessment
- operational control (including risk minimisation)
- workplace monitoring systems (for uncertain or unacceptable risks)

- environmental monitoring systems
- emergency preparedness (including fire and first aid)
- chemical safety auditing
- incident investigation and reporting
- waste minimisation
- record keeping.

Other, more specific, programs may include those for:

- procedures for suppliers
- purchasing and inventory control
- equipment, plant and process hazards
- emissions to air
- discharges to sewerage or stormwater
- 'first flush' rainwater control
- package specification
- decanting procedures
- contractor safety
- maintenance
- personal protective equipment
- product tampering
- recycling and reuse
- solid waste management
- treatment of hazardous wastes
- other trade wastes
- incidents in transit.

Where developed and implemented, each program within the chemical safety management system that has been identified as necessary in a particular organisation, will need procedures and work instructions (many, but not necessarily all, in written form) to ensure compliance with their requirements.

Taking the structures required to comply with the hazardous substances regulation, the management structures, systems and programs are shown in Figure 20.3.

A program for the management of chemicals at work (that also allows compliance with the requirements of the hazardous substances regulation) is shown in Figure 20.3.

One component in this program is the risk assessment subprogram. The process of risk assessment is absolutely critical to chemicals risk management. This process needs to be formal, open, and recorded. A sample risk assessment program is shown in Chapter 21 (Figure 21.3).

Specific instructions and specifications exist on how the programs are to be delivered. The procedures are detailed below.

Procedures

As noted above, the policy is a statement of intent and is generally worded in broad terms. The management system and the programs it contains specifies

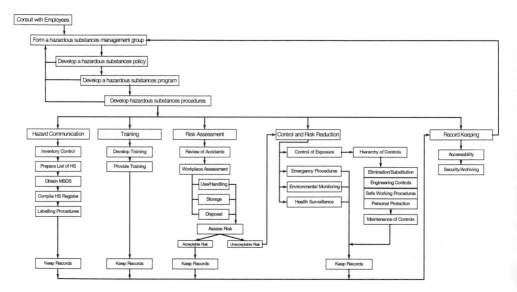

Figure 20.3 A compliance program for the hazardous substances regulation.

in more detail those aspects of individual activities to which the policy will apply (for example, the training program elements of the overall chemical safety management system).

Some activities, however, require more detail. Job specifications, safe working procedures, and the like require a detailed explanation, and in some cases, quite complicated instructions. For such tasks the procedure (normally written or delivered in specific training sessions) is required.

In some cases, a procedure can be performance-based (for example a lone-worker policy might specify that a worker should inform 'someone' that they are working alone, with specifying who that 'someone' is; it could be the worker's partner – the important thing is that 'someone' knows who will notice if the worker is absent for a long period of time.

However, most procedures (for example a batch card for the formulation of a paint in a paint factory) needs to be fairly specific in the materials, processes and steps to be followed in the task of making a particular batch of paint.

When developing procedures, or modifying procedures into the chemical safety management system (and this can be an extraordinarily complex task, as many organisations have thousands of procedures) a determination needs to be made about the safety content of such documents:

- where information safety or safe working procedures is absent, it can be added
- where there are many different varieties of procedures for the same basic process (one large organisation in New South Wales, Australia had 40 different confined space entry procedures), they can be harmonised
- where safety directions for a job may be in another document, these can be blended into working procedures so they become part of the job, not something outside it (which is invariably ignored).

In any event, the process of identifying which aspects and activities need procedures is a long-term job, and one that requires substantial effort. While a generic template can be developed for the structure and scope of the procedure, they may be quite complex once completed. For example, the specific requirements that all workers must follow in operational procedures for batch manufacture of a range of products.

The content of a technical procedure may include the following:

- responsibilities
- sequence of operations
- work methods
- safety and health precautions
- environmental safeguards
- characteristics and tolerances to be met
- equipment and materials to be used
- disposal procedures
- reference documents
- records to be kept.

Further development, modification or revision of such procedures requires input from the supervisors and knowledgeable workers who do those tasks on a regular basis.

The last stage adds a fourth element of planning.

The strategic plan

Strategic plans relate directly to corporate objectives, and indicate the forms of activity needed to achieve successful outcomes for the organisation. Strategies integrate the chemical safety management program into all the management systems of the organisation, and are directly identifiable with the achievement of one or more corporate objectives.

Policy, programs and procedures can be prepared for each aspect or problem identified in the risk assessment process, on an 'as needs' basis. These should aim to meet the chemical safety management system objectives:

- consistent with legislative compliance as a minimum
- consistent with the chemical safety policy
- documented
- aimed at significant workplace chemical risks
- includes options for control/prevention of chemical risks
- includes other requirements of the organisation (business, customers, shareholders, and the public).

Planning and the strategic plan

Strategic plans relate directly to corporate objectives, and indicate the forms of activity needed to achieve successful outcomes for the organisation. Strategies integrate the chemical safety management program into all the programs of the organisation, and are directly identifiable with the achievement of one or more corporate objectives.

In line with contemporary management practice, organisations are increasingly devolving responsibility for responsibility and accountability to line management. For this to work effectively, strategic planning is becoming organisation wide, and not the sole activity of senior management.

Including chemical safety in the strategic planning process brings it in line with all other activities, incorporates it into those activities, and is one of the first steps in proactive chemical safety management.

Once developed, the chemical safety management system should be integrated into everything the organisation does. This is best conducted by including the management system in the organisation's strategic plan.

Developing a strategic plan

A summary of the strategic planning process for chemicals and chemical safety follows.

Step 1 Identification of objectives

A major objective for chemical safety is to reduce losses related to injury and disease. Specific objectives that are linked with this are:

- prevention of injury and illness from exposure to chemicals
- provision of rehabilitation to return workers to their job safely and expeditiously
- management of claims to minimise loss to the organisation.

Step 2 Identification of strategies and programs to meet these objectives

Take prevention of injury and illness as an example. Strategies that deal with this objective include:

- a hazard control program
- emergency preparedness and an emergency action plan
- accident reduction plan through safety audits and incident investigation.

With regard to the hazard control program note that, depending on individual workplaces, specific programs may need to be developed to deal with high-risk hazards, such as personal protection, spill response, contractor training, and wastes.

Step 3 Identification of activities

This will be based on individual workplace processes, layouts and available chemical hazards. Essentially, now that the objectives (and programs to meet them) have been identified, the next step is to work out what has to be done. This can be assisted by setting goals or targets for each program. For instance, taking the hazard control program as an example, the activities will include:

- Encouragement of employees to report all accidents, incidents, spills or near hits involving chemicals. No penalty should proceed from this activity – hazard reporting is about identifying the causes of accidents, not ascribing blame.
- Regular hazard or safety inspections by supervisors and committees. While most good supervisors would regard inspection of the areas under their responsibility as an automatic, everyday occurrence, their inclusion in a chemical safety program means that the individual areas are inspected formally, with record keeping and follow-up of recommendations.
- Investigation of chemical hazards and the methods used for their control (including planned maintenance).

Targets or goals related to these activities might be:

- development of an incident reporting system and its effective use by workers
- hazard elimination (for example, of all cancer-causing chemicals from the workplace)
- successful action in the improved control of all hazards identified.

These targets may relate to the outcome of the strategy or the process required to achieve that outcome.

Step 4 Development of the action plan

The final step involves identifying the tasks that make up each activity listed in the strategic plan. Responsibility and time frames should be included so that senior management can review the plan to see what is (or is not) going on, who is doing it and by when. It should also be possible to understand what the plan will achieve in terms of the organisation's objectives.

What is not accountable for will not get done. What is not measured will not get done. Therefore, responsibility and time frames should be allocated to each task. Therefore, it is important to support the action plan with inclusion of chemical safety activities in job descriptions, contractor job specifications, performance appraisal and so on.

Basically, the plan should outline who should do what and when. This step is often overlooked in the chemical safety management process.

Translating the action plan into practice

As noted earlier, it is important to integrate chemical safety management into the normal management practices of the organisation. As with other organisational functions, this requires decisions to be made regarding:

- policy and programs
- resource allocation

- roles and responsibilities
- management information systems
- training
- coordination.

Systems need to be in place to allocate responsibility, provide information, training and supervision to those on the job, so that they know what to do, when to do it, and what happens if they don't.

Summary

The concept of legislation for the control of workplace chemical hazards is a good one. Only time will tell if recently enacted chemical safety legislation will make workplace use of chemicals safer. However, more effort is needed to ensure that the impact of legislation is both appropriate and enforced.

Chemical-induced injury and disease remains a significant problem in workers in industry. The responses of employers to the problems of chemical hazards in the workplace range from the authoritarian to the responsible, to the reactive, to the fragmented, to the negligent. These responses are made more complex by having to comply with a bewildering range of legislation and standards that are confusing and seemingly contradictory. As a result of this problem, a number of national and international initiatives have recommended the development of conventions, regulations and codes of practice to attempt to deal with the problems of chemical safety at work.

However, basic systems for a workplace program for the management of chemical safety are shown in Table 20.2. These programs differ in their needs, priorities, objectives and activities. However, one main point of this process is that it is not

---

Table 20.2 Workplace program for the management of chemical safety

Policy issues at the organisational level
Commitment from management for the program
Consultative processes between management and workers
Resourcing of chemical safety systems
Development of a chemical safety policy and program

The program
Assessment of workplace hazards
Hazard communication
Control of exposure
Procedures for the workforce to deal with minor emergencies, such as spills and splashes
Training and competency development
Requirements of special groups, such as the emergency services, contractors, visitors
Environmental issues, such as air pollution, water discharge and waste disposal.
Monitoring of the workplace
Health surveillance of workers

---

possible to control hazardous substances properly until some information on exist-ing hazards and controls is available.

Therefore the first activity is to conduct a workplace assessment. The workplace assessment requires the collection of information from chemical suppliers, so hazard communication is next. The workplace assessment and hazard communica-tion processes then drive the rest of the management activities, including:

- the need for training
- the need for control of exposure
- the need to control environmental impacts
- the need for emergency planning
- the need for monitoring and health assessment
- the need to keep records.

Risk management offers a means of dealing with these problems, by taking a more proactive and considered approach to develop a framework of measures to (i) consult with workers; (ii) identify, assess and control workplace chemical hazards; and (iii) review their effectiveness. Therefore, risk management offers a means of improving safety performance. In the area of chemical safety, this approach is aimed at both the suppliers of hazardous chemicals and at the employers who use them.

To be effective, chemical safety activities must be integrated into the manage-ment functions of planning, organising, leading, coordinating, controlling, direct-ing and evaluating workplace systems.

The management of problems that address workplace chemical safety should:

- be aimed to reduce workplace related occupational injury and disease
- be designed to provide an integrated structure to deal with all aspects of occu-pational hazards and adverse outcomes
- ensure a comprehensive approach to chemical safety management.

At the same time, internationally consistent ISO standards are in use, or are being developed, for quality systems, environmental management and occupa-tional health and safety. These standards outline a model for the management of quality, environment or safety, and the processes involved are applicable to the management of chemical safety. This process includes: obtaining commitment from senior management; instituting consultative mechanisms; developing a chem-ical safety policy; identifying components of the chemical safety management program; resourcing, implementing and reviewing the program; and integrating the program into the organisation's strategic plan.

Over the last decade, there has been significant activity in the way workplace issues should be managed, with international standards on quality being developed, such as the ISO 9000 series on quality. A new series on environmental manage-ment systems (the ISO 14000 series) is presently being finalised, and a third series on workplace safety and health is under development. These standards offer a model for the way a given issue (quality, environment, safety) can be managed. The chemical safety management program outlined above is consistent with this model, and is therefore capable of integration into management systems based on such standards.

Where workplace hazards are involved it is especially important to ensure that appropriate safety systems are incorporated into the workplace in a coherent management system. Only by blending a specific management program for chemical substances into the overall planning of an organisation will it be managed effectively and efficiently.

Where chemical hazards are involved it is especially important to ensure that appropriate safety systems are incorporated into the workplace in a coherent management system, and if appropriately developed, such a system should elicit 'yes' answers to the following questions:

- Do you have a materials inventory or chemicals register?
- Do you conduct routine safety audits (including chemical safety)?
- Do you inform employees regarding chemical hazards at your workplace?
- Do you train workers in dealing with chemical hazards?
- Do you routinely obtain material safety data sheets (MSDS) for the chemical materials in your workplace?
- Do you ensure workers are aware of what MSDS are and where they can be found?
- Do you have a system that evaluates the safety of all new products at your workplace?
- Is there adequate warning/placarding of chemical hazards and processes at your workplaces?
- Do you regularly maintain the technical measures (engineering controls, personal protection, and so on) used to control chemical hazards?
- Are there appropriate procedures for dealing with leaks, spills, fires and so on (including investigating and reporting)?
- Do you dispose of chemicals appropriately?
- Do you have procedures for investigating incidents involving chemicals?

If the response to any of these questions is negative, then a workplace chemicals safety management program may be needed.

Further reading

Eisner, H. (1988) Safety: the perils of self regulation. *Safety in Australia* May: 14.

HSE (1988) *Control of Substances Hazardous to Health Regulation.* Health and Safety Executive/Her Majesty's Stationary Office, London, UK.

ILO (1990) *Safety in the Use of Chemicals at Work. Convention No. 170.* International Labour Organisation, Geneva.

Leong, M. and Winder, C. (1998) Implementation of the NSW hazardous substances regulation at the local government level. *J. Occup. Health Safety Aust. N.Z.* 14: 389–394.

Lewis, S.M., Moore, H., Winder, C. and Mooney, C. (1993) *A Survey of Industrial Product Use in the Rockdale Area.* NOHSC Australia/NSW WorkCover Authority, Sydney.

Loullis, P.C., Braun, H., Malich, G. and Winder, C. (1998) Cross-national comparison of regulations concerning hazardous substances. *J. Hazard. Mater.* 62: 1143–1159.

NOHSC Australia (1994a) *Model Regulation for the Control of Workplace Hazardous*

Substances: Code of Practice for the Control of Workplace Hazardous Substances. National Occupational Health and Safety Commission/Australian Government Printing Service, Canberra.

NOHSC Australia (1994) *Guidance Note for the Control of Workplace Hazardous Substances in the Retail Sector*. National Occupational Health and Safety Commission/Australian Government Printing Service, Canberra.

Pearson, C., Game, W., Corbett, C., Jones, A. and St George, U. (1995) Hazardous substances in the construction industry. *J. Occup. Health Safety – Aust. N.Z.* 11: 510–512.

PIAC (1992) *Toxic Maze*, parts 1 and 2. Public Interest Advocacy Centre, Sydney.

Standards Australia (1995a) *AS/NZS ISO 14000 Series of Environmental Management Systems standards*. Standards Australia, Sydney.

Standards Australia (1995b) *AS/NZS 4360 Risk Management*. Standards Australia, Sydney.

Standards Australia (1997) *AS/NZS 4804 Occupational Health and Safety Management*. Standards Australia, Sydney.

Stewart, D. (1994) Attitudes to workplace hazardous substances. *J. Occup. Health Safety – Aust. N.Z.* 1995, 10: 375–376.

Valais, A. and Winder, C. (1998) We don't use chemicals: The ability of the NSW construction industry to comply with the hazardous substances regulation. *J. Occup. Health Safety – Aust. N.Z.* 14: 157–158.

Winder, C., Musry, A.H. and Tandon, R.K. (1994) Workplace incident notification: Results of a survey of sixteen factory sites in Western Sydney. *Qual. Assur. Good Pract. Law* 2: 372–385.

Winder, C. and Turner, P. (1992) Solvent exposure and related work practices amongst spray painters in automotive body repair shops. *Ann, Occup. Hygiene* 36: 384–394.

Winder, C. and Yeung, P. (1991) Solvent exposure in apprentice spray-painters: Workplace assessment and preliminary results. *J. Occup. Health Safety – Aust. N.Z.* 7: 151–153.

Winder, C., principal consultant (1995) *Managing Workplace Hazardous Substances*. CCH International, Sydney.

Winder, C. (1993) Chemical hazards and health effects of hairdressing. *J. Occup. Health Safety – Aust. N.Z.* 9: 359–371.

Winder, C. (1995) Best practice in workplace hazardous substances management. *Qual. Assur. Good Pract. Law* 4: 211–225.

_____

Chapter 21

Workplace assessment of toxic chemicals

C. Winder

Introduction

Assessment of the hazards and risks associated with the use of chemicals and the nature and cause of chemical-related injury and disease is an important factor in prevention. The purpose of assessment is to evaluate the risks to health arising from work involving the use of toxic chemicals and to determine appropriate measures to protect the health and safety of employees.

The process of workplace assessment

Before discussing workplace assessment and controls, it is necessary to look at total environmental exposure of the individual worker. Variations in temperature, radiation, atmospheric pollution, water and food contamination, viral and bacterial infections, and chemical insult are taken care of in the body largely by built-in protective mechanisms. However, extreme variations in some of these are not always able to be controlled and have to be reduced or protected by other means.

Some chemical exposure may be beneficial, in small amounts. Essential dietary trace elements such as copper, chromium, fluorine, iodine, molybdenum, manganese, selenium, and zinc are all toxic in high concentrations. Some of these have exposure standards and may even be carcinogenic. Hence some elements essential for survival can be hazardous in high doses.

Personal habits and voluntary exposure can add further stress to the individual. For example, synergistic effects of excessive tobacco smoking and exposures to chemicals such as asbestos, cotton, coal, grain, silica, and uranium are now well known. Add to this, social exposures such as alcohol, caffeine, other drugs and poor diet; and a picture emerges of a wide variety of additional stresses on the individual.

These can be further compromised by workplace exposures to a multitude of chemicals and agents.

Workplace assessment entails a process that identifies existing and potential hazards, evaluates the nature of the hazard and implementing appropriate measures for their control. Basic ways of controlling workplace hazards are shown in Figure 21.1 below, with the options that control at or close to the source being more desirable.

Workplace assessment can be of:

- Specific exposures
- Specific workplaces
- Specific processes
- Specific tasks
- Other handling issues.

While all workplaces are different and their assessment requires different expertise and knowledge, some generalisations can be made for assessing risks from exposure to toxic chemicals.

Approximately 36 million different chemical compounds have been identified: these are increasing by approximately 100,000–300,000 per year. Of these, less than 100,000 are in common use, and good-quality toxicological information is available for less than 10,000. Finally, less than 1000 substances have workplace or air quality standards in force. For example, the Australian list of exposure standards lists about 700 exposures, of which 582 are actual chemical entities. This makes the collection of information on chemical hazards problematic.

Workplace assessment of new chemicals and those already in use, and control of chemical inventories are important in controlling exposure.

Source	Source–worker link	Worker
■ Elimination	■ Housekeeping	■ Information
■ Substitution	■ Semi-automatic	■ Competency training
■ Process automation	control	■ Adequate supervision
■ Partial enclosure	■ Spray booths	■ Job rotation
■ Remote controls	■ Local exhaust	■ Worker enclosure
■ Isolation in space	ventilation	(such as crane cabs)
■ Isolation in time	■ General mechanical	■ Safe working
■ Enclosure	ventilation	procedures
■ Capture at source	■ Dilution ventilation	■ Personal monitoring
	■ Wet methods	■ Personal protective
■ Process change	■ Planned maintenance	equipment
	programs	■ Personal/general
	■ Continuous area	hygiene
	monitoring/alarms	

Figure 21.1 Basic processes of controlling workplace hazards.

Workplace chemicals risk assessment processes

One component of a workplace chemical safety program is the risk assessment sub-program. The process of risk assessment is absolutely critical to chemicals risk management. This process needs to be formal, open, and recorded. The notes below outline a suggested approach to risk management of chemicals.

The law and chemicals risk assessment

The basis of contemporary legislation for toxic chemicals is in line with contemporary occupational risk management approaches, and consists of identification, assessment and control of chemical risks in a coordinated, managed manner.

Assessment of the hazards and risks associated with the use of chemicals and the nature and cause of chemical-related injury and disease is an important factor in prevention.

The purpose of workplace assessment

The purpose of assessment is to evaluate the risks to health arising from work involving the use of toxic chemicals and to determine appropriate measures to protect the health and safety of employees. Assessment entails a process that identifies existing and potential hazards, evaluates the nature of the hazard and provides appropriate measures for their control. Workplace assessment of new chemicals and those already in use, and control of chemical inventories are important in controlling exposure.

Therefore, the purpose of the workplace assessment is to evaluate the risks to health arising from work involving the use of toxic chemicals and to determine appropriate measures to protect the health and safety of employees. Assessment entails a process that identifies existing and potential hazards, evaluates the nature of the hazard and provides appropriate measures for their control.

In many jurisdictions, a duty of the employer to carry out a workplace assessment where toxic chemicals are being used is mandatory. Under such circumstances, it is only necessary to assess work where there is potential for exposure.

The assessment focuses on work situation rather than individual substances. A practical way to carry out assessments is to divide the work up into jobs or tasks and assess the risks involved.

A suitable and sufficient assessment

The process of workplace assessment is outlined below.

Generic assessments

Generic assessments are workplace assessments where the assessment of one workplace is used for the assessment of another. Generic assessments are useful where one or more toxic chemicals are used or produced in work in the same or similar circumstances in more than one workplace.

Who should conduct workplace assessments?

There are a range of different types of people who could carry out a workplace assessment. These could include supervisors and managers, workers, health and safety representatives, unions, occupational health and safety personnel such as safety officers, nurses and doctors, the fire brigade, the council and outside experts.

Probably all these people could conduct a workplace assessment, and all could make useful observations about the risks of toxic chemicals. However, the simplest and most effective way to conduct a workplace assessment is a team approach, with knowledgeable workers and managers/supervisors working together.

This type of assessment may not accurately establish risks, but assesses their quality, that is, their ability to affect health. This type of assessment is therefore called a 'qualitative assessment'.

Workplace assessment is not something that any one person can cover entirely, and a team approach will provide a more objective assessment. The approach is not to accurately assess the problems of toxic chemicals, but to work out, to the best of available knowledge, which substances do not present a risk, and which substances may present a risk.

A competent assessor

The responsibility for the assessment lies with the employer. However, the person conducting the assessment should possess the ability to:

- interpret material safety data sheet (MSDS) and labels
- gather information in a systematic way
- openly discuss workplace issues with people and communicate effectively
- observe the conditions of work and to foresee potential problems
- if required, carry out simple diagnostic tests of control measures
- communicate effectively with employees, managers, contractors, experts and so forth
- be able to draw all the information in a systematic way to form valid conclusions about exposure and risks
- report the findings accurately (orally and in writing) to all relevant parties
- develop an action plan to eliminate or minimise risks
- know when to call in specialist help.

Identification of significant risks

If the assessment identifies significant risks to health, then steps must be taken to control the risk to health, although the word 'significant' should be defined. If the assessment process identifies unacceptable risks or unknown risks, then more expert advice may be sought.

If the assessment does not identify any major risks, or that any risk is already controlled in accordance with the MSDS (or equivalent information) then the assessment is a simple and obvious assessment, is complete, and no further assessment is needed.

Where assessments are found to considered inadequate, inaccurate or wrong, they should be repeated.

Review of the workplace assessment

For review of the workplace assessment, it should be revised whenever there is evidence to suggest that the workplace assessment is no longer valid or when there has been a significant change in the work to which the assessment relates. In any case, the validity of the assessment should be reviewed at intervals not exceeding 5 years. Again, the word 'significant' is not defined.

Record keeping

No risk to health

If the workplace assessment notes that the use of toxic chemicals does not constitute a risk to health, a short notification about the risk assessment and its findings can be attached to the product MSDS. The form of this notation is unclear, but should include the name of the assessor, the date of the assessment, and a statement outlining the conclusion of the assessment.

Significant risk to health

If the assessment process identifies significant risks to health, then such risks will have to be reduced. This is an important part of the assessment process, and a number of responsibilities are imposed on the employer. The first of these is to formally record the assessment, usually in some form of written assessment.

Where toxic chemicals are used in the workplace, the lack of assessment report or the lack of a record is a breach of an employer's duty of care.

Access to records

The assessment report is not a confidential document. It should be made available to workers. A step-by-step guide to a workplace toxic chemicals assessment is given below.

A step-by-step guide to workplace toxic chemicals assessment

Step 1 Identify all substances in the workplace

Materials already in use

The first step is to identify all substances (not just the toxic chemicals) in the workplace:

- refer to stock lists and so on
- check all locations and main sources

- check chemicals produced in work processes as intermediates, by-products or finished products
- maintenance/repair, research/testing activities
- compile inventory.

The end of this process should produce a list of all chemicals (the inventory).

Identifying the chemicals in work areas and in stores is not enough; there may be other sources of workplace chemical hazards which may form subsets of the materials inventory:

- all raw materials
- as some chemical hazards will only be produced during processing, it may also be necessary to create a process or reaction inventory
- all reaction intermediates
- any contaminants
- any by-products
- other hazard sources, such as ionising radiation, electrical, and so on.

Such an information base should contain at least a minimum amount of information. As a first stage this should indicate the name, manufacturer, whether the material has been classified by the supplier as hazardous, workplace location and the approximate amounts present in stock. The latter is often difficult: sometimes it is best to include maximum amounts if average amounts are too variable.

New materials

As part of the development of an information base, a process should be developed which identifies new materials entering the workplace, so that these can be targeted for assessment and action. Once in use, regular monitoring of exposure patterns may be required.

The inventory

The inventory is an important document. It must be at least a list of all toxic chemicals, but can be a list of all substances in the workplace. The inventory should contain at least a minimum amount of information and as a first stage, this should be the name, manufacturer, and approximate amounts in stock and workplace location. As some chemical hazards will only be produced during processing, it may also be necessary to create a process or reaction inventory as a subset of the materials inventory.

Options for creation of the information base include systems on computer, and a number of software systems are commercially available. However, sophisticated systems may not be necessary for workplaces where only a few chemicals are in use. Whatever system is used, a mechanism should be developed that allows review of the materials inventory on a periodic basis.

As part of the development of an inventory, a process should be developed which identifies new materials entering the workplace, so that these can be targeted for assessment and action. Once in use, regular monitoring of exposure patterns may be required.

The presence of an inventory does not necessarily mean that the assessment process is on a sound footing, unless links that allow identification of toxic chemicals are present (such as stores lists that contain fields listing individual items as toxic chemicals).

A sample inventory list is shown in Figure 21.2. This list has been expanded to include additional fields such as whether an MSDS exists and whether a risk exists. These fields are not needed in the identification process, but are useful later on in the assessment part of the process.

Step 2 Determine which substances are hazardous

The second step is to identify all the toxic chemicals on the inventory. This is often straightforward, for example the presence of a skull and crossbones or other hazard symbol on the label, or from specific risk entries in the MSDS. However, if there is any doubt, advice should be sought following contact with the manufacturer or importer.

From an operational perspective, it makes good sense to carry out chemicals risk assessments for any chemical with hazard properties, such as poisons, dangerous goods and toxic chemicals.

A flow chart that identifies chemicals with hazard properties is shown in Figure 21.3.

Workplace: Metal Cleaning Room				Approximate Number of Workers: 3-5							
Label? L / OK?	Trade Name	Makers Name and Address	Phone Number	MSDS available? Yes	No	Hazardous? Yes	No	?	A risk? Yes	No	?
1 ✓	Gluggo	Faultless Chemicals 87 Easy Street Driversnow		✓		✓				×	
2 ✓	Bung-up	Perfect Chemicals 666 Gravy Road Whitewash		✓			×			×	
3 ✓	Clogit	Clean Chemicals Box 99 Asawhistle			×		×			×	
4 ✓	Dipsi	So So Specialities Halfway house Middling		✓			×			×	
5 ✓	Strippa	Honestinjun Oil Locked Bag 13 Indasticks			×		×				×
6 ×	Thinno	Honestinjun Oil Locked Bag 13 Indasticks		✓		✓				×	
7 ✓	Bleach foam	Stodgy Chemicals Dead End Street Bokkaburk		✓		✓			✓		
8 ✓	Liquid scourer	Dodgy Chemicals Skid Row Woop Woop		✓		✓					×
10 ✓	Acidoff	Flybinite Minerals Dead End Street Dumpsville		✓		✓			✓		

Compiled by: Al K Hol, Team Leader Date: 17 October 2000
Use as many of these forms as you need to list all workplace hazardous substances

Figure 21.2 The workplace chemicals listing form (the index for the register).

Job Information

Location of job being looked at: _____

Day of activity: _____   Job being looked at: _____

Details of the job: _____

Names of the people looking at the job: _____

Chemical Information

Name of chemical(s) on container: _____

Is chemical(s) on the chemicals register? _____

Is a risk assessment needed? (Use flow chart below)

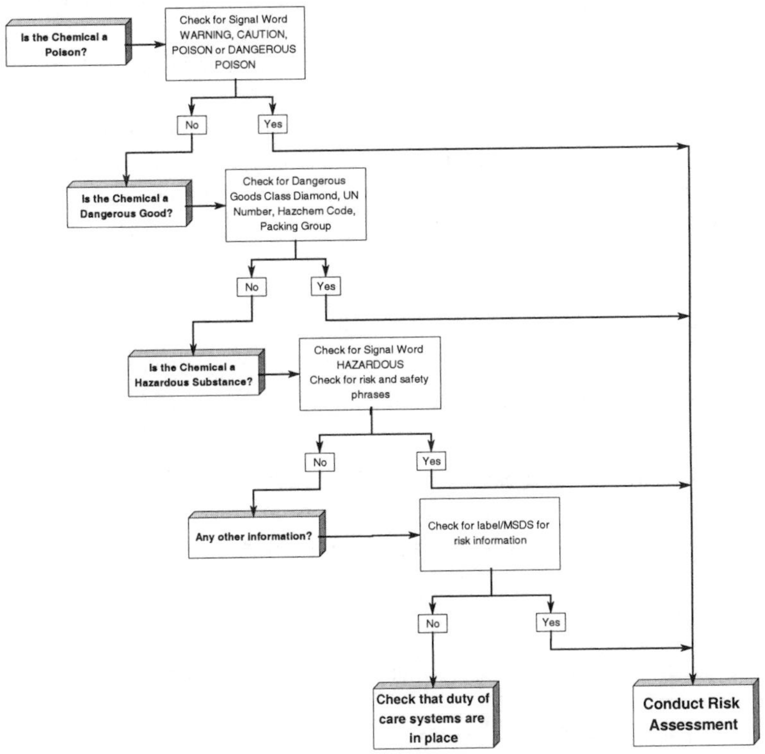

Risk Assessment Information

Is a risk assessment needed?   Yes ☐   No ☐

If risk assessment needed, fill out accompanying forms.

If risk assessment **not** needed, check that handling, use and disposal conditions are acceptable.

Figure 21.3 The process of hazardous chemical identification.

Step 3 Obtain information about the toxic chemical

The next step is to obtain information about the toxic chemicals identified on the inventory:

- from the label
- from the material data sheet
- from accident reports within the organisation
- from technical reference sources
- from practical experience.

In areas where chemicals are used, especially in manufacturing areas, it will also be necessary to consider the fate of materials. For example, a production process might use raw materials easily identified. Problem areas may be identified in the survey by observing work practices and making note of smells, dusts, smoke, waste materials and so on. This information should be sufficient to allow an accurate interpretation of the risks of the use of the material in normal (and reasonably expected abnormal) use.

In some workplaces, it may be appropriate to translate or summarise the information on the material safety data sheet to a one-page summary or index page that can be included in the chemical register. A hazard summary sheet is attached that allows an evaluation of MSDS and extraction of salient information that can be used in risk assessments (see Figure 21.4). Workers often find such summary sheets easy to use and are less reluctant to consult such sheets, rather than a 7-page safety data sheet.

The toxic chemicals register

The collection of material safety data sheets for each toxic chemical should be compiled into a toxic chemicals register. The register should be kept and maintained by a responsible person. It should contain as a minimum, a list of all toxic chemicals used or produced at the workplace, the material safety data sheets for at least all toxic chemicals, and relevant risk assessments. The register should be readily accessible to all employees with the potential for exposure to toxic chemicals.

The toxic chemicals register can be (appropriately indexed and catalogued):

- either a paper or computer database
- provide, as a minimum, a central listing of all substances which are used in the workplace together with basic information such as the material safety data sheet
- used as a tool to manage substances used at work
- a source of information.

There is great variability in the registers that already exist. However, the minimum requirements for such a register are the inventory and a material safety data sheet for every toxic chemical in the workplace.

The Hazardous Chemicals Summary (for inclusion with the MSDS in the Register)

Chemical Register Summary			
Chemical Name	*Paint-off Stripper*	**Manufacturer**	Fictitious Solvents Woop Woop
Dangerous Good?	Yes ☑ No ☐		Class 6.1
Poison?	Yes ☑ No ☐		Poison S6
Hazardous Substance?	Yes ☑ No ☐		Irritant
Hazardous Ingredients?	Chemical Name		Exposure Standard?
	Dichloromethane		50 ppm
	Phenol		5 ppm
	Ammonia		25 ppm
Storage	Solvent store No 6, level 3 Maintenance workshop		
Main Hazards	Flammable		Corrosive
	Irritant		Causes drowsiness
Controls	Ventilation		Yes ☑ No ☐
Precautions	Personal Protection	Overalls	Yes ☑ No ☐
		Gloves	Yes ☑ No ☐
		Eye Protection	Yes ☑ No ☐
		Respirator	Yes ☐ No ☑
		Other	Yes ☐ No ☐
Prepared by			
Date			

Figure 21.4 The hazardous chemicals summary (for inclusion with the MSDS in the register).

Obtaining information from the workforce

Individual workers or work teams are also useful sources of information.

An assessment planning sheet is attached that allows collection of information that can be used in the risk assessment process (see Figure 21.5).

Interim procedures

Once these materials have been identified, the level of information that they cover may be such an immediate problem that it may become necessary to develop priorities for action. Such priorities might include:

■   investigation of less hazardous alternatives

The purpose of this planning sheet is to identify potential chemical problems in your workplace. It is not a final assessment, but is used to gain an appreciation of chemicals problems in this workplace, as understood by the user.

Location of job being looked at: _____

  Day of activity: _____ Job being looked at: _____

Details of the job: _____

Names of the people looking at the job: _____

What chemicals are being used? For example: Solvent thinners Acid wash	How can they hurt you? For example: How are workers exposed? What are the health effects of the chemical? (Check MSDS)	How can you make them safer? For example: Use only in ventilated area Wear gloves and eye protection? (Check MSDS)	Done ✓

Use as many of these forms as you need. Hand the filled out forms to the Hazardous Substances Risk Assessor in your workplace for further assessment.

Based on *Hazpak!* NSW WorkCover Authority, Sydney, 1996.

Figure 21.5 The hazardous chemical risk assessment planner.

- action on chemicals that have poorly identified hazards
- development of policies for chemicals in use, where the present procedures are deemed inadequate
- requirements for monitoring.

In all other cases, the risk assessment process continues.

Step 4 Divide the work into units for assessment

The next step is to decide what is going to be assessed:

- by individual toxic chemical
- by groups of substances with the same hazards
- by process (such as manufacturer or research project)
- by activity (such as storage or disposal).

It is important to get the right level of assessment unit – the process of assessing the 200 or so chemicals in a laboratory would take a long time if conducted individually, when perhaps the units of assessment might be:

- assessment of storage of corrosives
- assessment of storage of flammables
- assessment of laboratory activities using toxic chemicals
- assessment of particular toxic chemicals
- assessment of disposal practices.

Information sources for the assessment

Information sources for such assessments include:

- description of company/location, working arrangements
- task description
- safe working practices
- accident/incident history
- any monitoring or health surveillance records
- previous assessment reports or external audits
- reports of unusual operating conditions
- permit to work procedures and records.

Steps 1–4 constitute the identification component in the workplace assessment process. Elements of this process (material safety data sheet collection, compilation of the register) also form part of the hazard communication subprogram of the toxic chemicals compliance program.

Step 5 Conduct a walk-through survey

The first four stages of the workplace risk assessment have been about finding out what hazards and hazardous processes and practices exist at the workplace. The next stage is about hazard assessment. The recognition of hazards depends on one fact: **A chemical must make contact, or be absorbed into, the human body to induce an effect.**

When identifying actual or potential problems, the following questions should be answered:

- How likely is it that a chemical will be released from the process or practice in which it is used?
- If release is likely, and control technology cannot be easily improved, what will be the nature, extent, intensity and duration of the resulting contamination?

The information collected during collation of the materials inventory and chemicals register is important in answering these questions.

The information obtained can then be used for evaluating the hazards associated with work involving hazardous chemicals. Risk should be evaluated by consideration of the following factors:

- assessing the magnitude of the hazard by: identifying the hazard(s), establishing the level of exposure (if any) which is not hazardous to health, determining the actual level of exposure to the hazard
- the possibility of safer alternatives
- the duration, intensity and frequency of exposure
- establishing pathways of hazard to the worker
- determining the numbers of workers at risk
- the nature and severity of potential health effects
- the likelihood of potential health effects at the exposures observed
- the economic and technical feasibility of alternate control measures.

As well as assessment of the workers involved directly in hazardous tasks, consideration must also be given to people who work nearby, who may come into contact inadvertently. This may occur through unplanned exposure or failure of control measures.

Once all the relevant information has been collected, the next step is to go out into the workplace and see how the toxic chemicals are being used. This part is called the assessment, and is best conducted as an initial walk-through with (if necessary) follow-up inspections.

The experienced professional can often evaluate, quite accurately and in some detail, the magnitude of hazards in a workplace without the benefit of any instrumentation. This is called professional judgement, and is frequently used in workplace assessment activities. A 'qualitative' evaluation can be made by individuals familiar with a process or operation. This requires familiarisation of workplace hazards, activities and controls. For a specific hazard, it also requires some judgement, and then investigation (for example, by a walk-through survey).

Qualitative or quantitative surveys

There are essentially two approaches to workplace assessment, relying on subjective (qualitative surveys), using professional judgement and knowledge of hazards or the workplace, or quantitative surveys, where the assessment is augmented by the results of monitoring.

Qualitative assessment

The process of workplace assessment essentially provides a framework for a qualitative preliminary subjective survey of workplace hazards. The recognition and evaluation activities were, in the main, based on the evidence that can be collected without taking any actual measurements. This type of evaluation is often very beneficial.

However, although the information obtained during a qualitative evaluation or walk-through inspection is always useful, only by measurement can there be a documentation of actual exposures.

Initial judgement

Estimation of the potential hazards to health requires that an initial determination be made of factors such as:

- the current or changing hazard parameters
- work operations normally conducted
- number and identity of workers
- any available information on likely exposures (from previous surveys, observations, estimates, and so on)
- attributable signs and symptoms
- pathways to the worker.

In such cases, it is appropriate to use 'worst case' scenarios in the estimation of exposure conditions.

The walk-through

An initial walk-through survey will then be necessary to support the findings of initial judgement and to reinforce the contents of the materials inventory. Such surveys are usually carried out by occupational health and safety professionals, such as occupational hygienists, but there is no reason why such surveys should not be carried out by informed personnel, such as occupational health and safety representatives, supervisors with a working knowledge of the area, or experienced workers. The main problem of workplace surveys by untrained personnel is that they may only look at one level of hazard, without seeing the different types of hazards that might occur. However, professionals sometimes miss things that the experienced operator understands implicitly. For this reason team assessment may be better than individual assessment. The assessor should:

- identify chemicals in use or being generated
- check routes of contact/exposure
- identify who may be exposed, define exposure characteristics (duration, intensity)
- identify which controls and work practices are in place
- identify other activities where exposure may occur (storage, transport, disposal).

There are a number of questions that can be asked on the walk-through.

- What materials are present?
- Are any hazardous?
- Where are they located?
- What handling procedures exist (storage, decanting)?
- On average, what quantities are used?
- What labels, material safety data sheets or other information is available on hazards?
- How are they used?
- How often are they used?
- What are the likely hazards?
- Is the substance released or emitted into the work area?
- Are workers exposed to the chemical? How often, how much and for how long?
- What controls are used to reduce exposure?
- Are there any risks associated with storage and transport of the substance?
- What disposal procedures are used?

This initial survey should be comprehensive, and take into account the whole workplace, including places where extensive chemical use is not expected (such as offices, photocopier chemicals) and lunch rooms (cleaning materials). It is usual to find materials missed in the identification stage.

In areas where chemicals are used, especially in manufacturing areas, it will also be necessary to consider the fate of materials. For example, a production process might use raw materials that are easily identified. Problem areas may be identified in the survey by observing work practices and making note of smells, dusts, smoke, waste materials and so on.

The minimum information required from the survey would include:

- the names and manufacturers of the chemicals
- what information is already available
- how the product is handled and used
- who is exposed
- what controls are used to control hazard.

Conditions in a workplace change with time. Inspection of all areas of the plant should be made on a regular basis to identify and eliminate foreseeable chemical (and other) problems. This should include review of all control measures in place and procedures for their maintenance.

Checklists

While nothing will be better that accurate knowledge of local conditions of knowledge and hazards, safety audits will be facilitated by the use of appropriately designed checklists. Checklists aid in auditing activities, assists in the identification of hazards, permit on-the-spot recording of findings and comments, and can provide a report of the audit. Checklists can also be tailored or developed for specific workplace applications. Whether the activity is called a 'walk-through survey', 'work-place inspection' or 'safety audit', many people feel more secure if they can use a checklist when required to critically assess the workplace from the health and safety viewpoint.

The checklist can be a vital monitoring aid and so also be an essential part of efficient health and safety system management.

However, simply applying a checklist from a handy source, such as a book or another organisation, invites checklist misuse. The indiscriminate use of checklists can actually hinder health and safety management.

Before any checklist is first used, some thought needs to be given to:

- how the checklist is to be applied
- the checklists' content.

Trialling its use, and improving the content on review, will further contribute to ensuring a good result. The advantages of using well-designed checklists are that they can:

- provide an aid to memory to ensure that important items of equipment or machinery, or aspects of work processes, are not overlooked
- enable a standardised approach to workplace inspection
- underpin a system of accountability for health and safety management by providing some of the records necessary for performance assessment
- increase the efficiency of recording, and standardise record-keeping (so that real comparisons and checks on progress are possible)
- reduce risk, for example, by providing written guidelines on the content and correct sequence for such procedures as plant shut-downs
- enable an efficient sharing of health and safety responsibility
- clarify inspection responsibilities (checklists should show who is responsible for performing the checks/collecting data, as well as who is responsible for following up on the findings of the check).

In some cases, the use of checklists can encourage complacency and half-measures, and promote a false sense of security. Some indicators are:

- Where there is no provision on a checklist for follow-up action. Checklist completion is not itself a goal, it is only an aid to a process of maintaining health and safety.
- Checklist items that are too general, too specific, or that invite subjective responses (such as 'Are the premises suited to the work being carried on there?') are inadequate tools for identifying problems.
- Checklists can substitute for training and information and/or disguise a need for training and information. For example, a user may be able to complete a checklist without understanding either what the check is directing his/her attention to, or why.
- Where the limitations of a checklist – that it is 'a check at a point in time' – are not appreciated, checklist use can reduce or replace an active sense of personal responsibility and vigilance.

Quantitative assessment

The other type of assessment is quantitative, involving collection and analysis of samples that represent actual exposures. These provide a more accurate picture of exposure conditions, but owing to their reliance on technology, are not used commonly, except where required for compliance with legislation or standards, or to form the basis for designing engineering controls.

Step 6 Evaluate the risks

The evaluation process builds on the assessment process, which identified which substances are hazardous. Here a systematic approach to assessing the risks involved in particular work needs to be established. This also forms part of the decision-making process resulting in an assessment of the degree of risk to health that exists from an occupational hazard. The next questions to answer are:

- How will the released chemical/contamination make contact with or be absorbed into the human body, by what routes and in what amounts?
- If absorption does occur, what are the consequences?

As well as assessment of the workers involved directly in hazardous tasks, consideration must also be given to people who work nearby, who may come into contact inadvertently. This may occur through unplanned exposure or failure of control measures.

The questions that should have been answered are:

- Do these toxic chemicals give unacceptable or unknown risks in use?
- Is the risk not significant?
- Is the risk significant?
- Is exposure high?
- Is the substance toxic?
- Do dangerous reactions occur?
- Can reasonably foreseeable leaks or spills occur?

The answers to these questions should assist in working out the nature of the risk to workers. There are a number of conclusions that can be made from these answers:

- Conclusion 1: risks insignificant and unlikely to increase
- Conclusion 2: risks are significant but are effectively controlled
- Conclusion 3: risks are significant but are not adequately controlled
- Conclusion 4: uncertain about risks
- Conclusion 5: need more information.

Obviously, for the first two conclusions no further action is needed except to record the conclusion. The remaining three conclusions need more information about further assessment or control.

Steps 5–6 constitute the assessment component of the workplace assessment process.

Conditions in a workplace change with time. Inspection of all areas of the plant should be made on a regular basis to identify and eliminate foreseeable chemical (and other) problems. This should include review of all control measures in place and procedures for their maintenance.

While nothing will be better than accurate knowledge of local conditions and hazards, safety audits will be facilitated by the use of appropriately designed checklists. Checklists aid in auditing activities, assist in the identification of hazards, permit on-the-spot recording of findings and comments, and can provide a report of the audit. Checklists can also be tailored or developed for specific workplace applications.

Step 7 Identify actions to control risks

Where the assessment identifies that an unacceptable risk exists, duty of care requires the risk to be controlled. The control process uses the assessment process to work out which risks have to be dealt with.

The information obtained can then be used for evaluating the hazards associated with work involving hazardous chemicals. Risk should be evaluated taking into account:

- the toxic chemicals involved and the possibility of safer alternates
- the routes of exposure
- the duration, intensity and frequency of exposure
- the nature and severity of potential health effects
- the likelihood of potential health effects at the exposures observed
- the economic and technical feasibility of alternate control measures.

There are many means by which these risks are controlled, including specific controls such as engineering controls, and general controls such as training. All have their place.

There are a range of actions that can produce control of the risk to acceptable levels:

- better hazard communication
- induction and training
- selection of appropriate control measures using the hierarchy of controls
- development of safe working procedures
- personal protection
- better storage
- disposal procedures
- emergency procedures
- first aid procedures
- ongoing monitoring
- health surveillance.

Step 7 constitutes the control component of the workplace assessment. These actions are outside the scope of these notes, and a control of exposure program would be required for a chemicals management program.

As part of contemporary occupational risk management, the identification, assessment and control process requires suitable records of the decisions made in the workplace assessment process. Therefore, the last steps of the workplace assessment process are about record keeping and review.

Step 8 Record the assessment

Record keeping is fundamental to the workplace assessment process. Even if the workplace assessment concluded that all toxic chemicals were under control, a notation is required in the register. A record is required for:

- simple and obvious assessments
- detailed assessments
- significant decisions arising out of the assessment.

A risk assessment form is shown (Figure 21.6) that allows

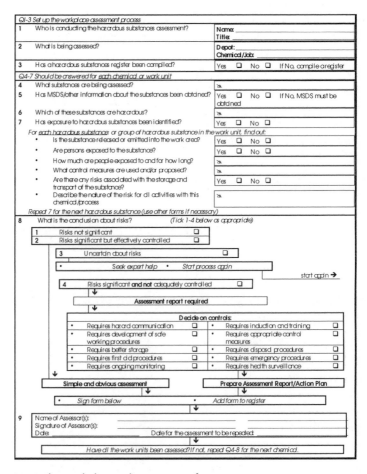

Figure 21.6 The workplace risk assessment form.

- decision to be made whether an assessment is a simple and obvious one
- an assessment needing a fuller evaluation, including identification of options for controlling risks that the assessor thinks would be useful for controlling risks
- records to be kept of the assessment.

Figure 21.7 shows a combination of the chemicals listing form and a rough risk assessment. This form works well in industries where chemical safety is poorly understood.

Workplace: Metal Cleaning Room											Approximate Number of Workers: 3-5

The hazardous substances listed below exist in this workplace. A copy of the MSDS has been obtained and a risk assessment conducted (see below for risk scores). Please ensure you comply with recommendations for controls of exposure.

Label?		Trade Name	Makers Name and Address	Phone Number	MSDS available?		Is the product							Controls (based on risk score) such as ventilation, gloves, respirator, boots
							Hazardous?			A risk?		Risk Score		
L	OK?				Yes	No	Yes	No	?	Yes	No			
1	✓	Gluggo	Faultless Chemicals 87 Easy Street Driversnow	1(-1/1()	✓		✓				✗	1		Wear impervious gloves
2	✓	Bung-up	Perfect Chemicals 666 Gravy Road Whitewash	(1((/)1(	✓			✗			✗	1		No special requirements
3	✓	Clogit	Clean Chemicals Box 99 Asawhistle	(1()(1()(1		✗		✗			✗	0		No special requirements
4	✓	Dipsi	So So Specialities Halfway house Middling	((1() ()<	✓			✗			✗	1		Wear impervious gloves
5	✓	Strippa	Honestinjun Oil Locked Bag 13 Indasticks	()()(1(-		✗			✗			1		Use in ventilated area, wear gloves and respirator
6	✗	Thinno	Honestinjun Oil Locked Bag 13 Indasticks	()()(1(-	✓		✓				✗	2		Use in ventilated area, wear gloves and respirator
7	✓	Bleach foam	Stodgy Chemicals Dead End Street Bokkaburk	(1((/)(1(	✓		✓			✓		3		Protect all skin. Wear gloves, safety glasses, apron
8	✓	Liquid scourer	Dodgy Chemicals Skid Row Woop Woop	(1()(-1(/	✓		✓		✗			1		Wear impervious gloves
9	✓	Unclog	Bodgy Chemicals Otherside Track Bokkaburk	(1((/)1(	✓			✗			✗	1		Wear impervious gloves, dust mask
10	✓	Acidoff	Flybinite Minerals Dead End Street Dumpsville	(1()(1(1(	✓		✓			✓		3		Protect all skin. Wear gloves, safety glasses, apron, respirator

Simple Risk Assessment

Class 3: (High Risk) Does the substance and its associated hazards have the potential to kill, or cause permanently disability, such as lung disease?

Class 2: (Medium Risk) Does the substance and its associated hazards have the potential to cause a serious injury, or illness, that will temporarily disable, such as Dermatitis?

Class 1: (Low Risk) Does the substance and its associated hazards have the potential to cause a minor injury that would not disable, such as mild skin rash?

Compiled by: Al K Hol, Team Leader _____	Date: 07 November 2000 _____
	Use as many of these forms as you need to list all workplace hazardous substances

NSW Occupational Health and Safety (Hazardous Substances) Regulation 1996, OHS Act 1983

Figure 21.7 The combined workplace chemicals listing/risk assessment form.

Step 9 Review the assessment

Workplaces change. Where a significant change has occurred, the workplace assessment may be rendered invalid. Reasons why an assessment should be repeated include:

- significant changes in volume or rate of production, plant, workplace layout, materials or control measures
- routine inspection that indicates standards are slipping
- accidents or near hits or losses of control
- new evidence becomes available about harmful effects of substances
- new materials, controls or technology become available.

Assessments should be reviewed at least once every 5 years.

Conclusion

A contemporary approach for dealing with (among other things) workplace hazards is risk management. This outlines a fairly standard approach of setting contexts, identifying hazards, assessing risks and controlling unacceptable risks, and monitoring and reviewing the system. This approach should work for all occupational health and safety hazards, and will work for hazardous chemicals.

This approach should be adopted for assessing toxic chemicals at work. The process while apparently problematic, is basically quite straightforward. A stepwise approach is recommended, and can be followed quite simply.

Further reading

HSE (1988) *Control of Substances Hazardous to Health Regulation*. UK Health and Safety Executive/Her Majesty's Stationery Office, London.

ILO (1990) *Safety in the Use of Chemicals at Work. Convention No. 170*. International Labour Organisation, Geneva.

Malich, G., Loullis, P.C., Braun, H. and Winder, C. (1998) Comparison of regulations concerning hazardous substances from an international perspective. *J. Hazard. Materials* 62: 143–159.

Standards Australia (1999) *AS/NZS 4360 Risk Management*. Standards Australia, Sydney.

Stewart, D. (1995)Attitudes to workplace hazardous substances. *J. Occup. Health Safety Aust. N.Z.* 10: 375–376.

Winder, C. (1999) Applying risk management approaches to chemical hazards. *J. Occup. Health Safety Aust. N.Z.* 15: 161–169.

Winder, C. (1995) Best practice in workplace hazardous substances management. *Qual. Assur. Good Pract. Law* 4: 211–225.

Winder, C. (1997) Hazardous inconsistencies. *Aust. Safety News*. March: pp. 14–19.

Winder, C. principal consultant (1995) *Managing Workplace Hazardous Substances*. CCH International, Sydney.

Chapter 22

Working examples on occupational toxicology

C. Winder and N.H. Stacey

Introduction

Toxicology and toxicological data have many uses relating to chemicals in the workplace. Occupational toxicologists and others using data of a toxicological nature will be called upon to put the information to particular purposes. Tasks will vary from the relatively simple to the complex. Questions may be asked that can be quickly answered from individual knowledge to those where extensive searching of the literature and interpretation of data are required.

At the same time such tasks can be learning experiences and can readily be used to illustrate how an understanding of the basics of toxicology are so important in enabling the occupational toxicologist to provide cogent responses to issues or tasks raised.

It is the intention of this chapter to provide examples of some of the tasks/situations that befall those using toxicological information so that it can be more readily appreciated just how the information is used and also to show how the tasks rely on the basics of the science.

- **Exercise 1** looks at the interface of chemistry and toxicology, and provides an indication of the numbers of molecules that exist in real terms.
- **Exercise 2** looks at estimating single-dose toxicity values such as LD_{50}s, and what such values can mean in the toxicological context.
- **Exercise 3** looks at evaluating primary irritancy data, and the implications of the toxicological assessment process.
- **Exercise 4** looks at evaluating repeated-dose data, and the implications of the toxicological assessment process.
- **Exercise 5** looks at evaluating toxicity data for the purposes of preparing a material safety data sheet (MSDS).
- **Exercise 6** looks at evaluating toxicity and occupational hygiene data for the purposes of establishing an exposure standard.

- **Exercises 7–11** investigate requests for information on chemical-related problems arising out of the workplace.
- **Exercise 12** is an exercise exploring issues relating to the management of chemicals in the workplace.

Exercises 2–6 look at the uses of toxicological data for more specific tasks, such as toxicity and irritancy classification, entries on product labels, MSDS and exposure standards. Exercises 7–12 look at wider issues of chemical exposures in workers, including adverse exposures, reproductive and carcinogenic risks, biological monitoring and management systems for chemicals. Four examples will be considered, from employers, OHS committees, unions and individual workers. Other case studies look at biological monitoring and developing management systems for chemicals. These situations are hypothetical although they are similar to the sorts of requests that the occupational toxicologist may expect to receive.

Exercise 1: Estimating the number of molecules of a toxicant in a drop of water

How many molecules of a toxicant (for example, a water-soluble pesticide) are there in a drop of water at a concentration of one part per million (1 ppm)?

Trainer's notes

A mole of any substance contains Avogadro's number (6.023×10^{22}) molecules. A mole of water (H_2O) weighs 18 g (two hydrogen atoms at 1 Dalton each, and one oxygen atom at 16 Daltons).

- How many molecules of H_2O in 1 ml?
  Number of molecules in a mole of water $= 6.023 \times 10^{22}$ (Avogadro's number)
  Weight of a mole of water $= 18$ g
  Number of molecules in a mole of water $= 6.023 \times 10^{22}$ in 18 g (or 18 ml)
  Number of molecules in 1 ml of water $= 6.023 \times 10^{22}/18$
  $$= 3.346 \times 10^{21}$$
- How many molecules of H_2O in a drop of water?
  Number of drops of water in 1 ml $= 20$
  Number of molecules in one drop $= 3.346 \times 10^{21}/20$
  $$= 1.673 \times 10^{20}$$
  $$\text{or } 16,730,000,000,000,000,000,000$$
- How many molecules of a toxicant at 1 ppm in a drop of water?
  Concentration of toxicant $= 1$ ppm.
  Number of molecules of a toxicant at 1 ppm.
  in a drop of water $= 1.673 \times 10^{20}/1,000,000$
  $$= 1.673 \times 10^{14}$$
  $$\text{or } 16,730,000,000,000,000$$

The point to be made here is that even though 1 ppm sounds small, there are still a large number of molecules involved.

Exercise 2: Classifying single-dose toxicity

Step 1

Table 22.1 lists some results of a single-dose study (estimating an LD_{50}) in the rat for a toxicant.

- What is the LD_{50}?

No control group is used in the study (a group of rats receiving 0 mg/kg). Why not? Because there is no point in *not* dosing animals.

- How toxic is this chemical?

Trainer's notes

There are a range of ways of calculating this value, such as eyeballing (the value is between 100 and 500 mg/kg), mathematically by interpolation, or by drawing a graph. The most effective way is to show the data graphically using a log dose plot (see Figure 22.1). This produces a value of about 400 mg/kg. This then leads into the second part of the exercise:

- How toxic is a chemical with an LD_{50} of 400 mg/kg?

This is a rhetorical question, because the toxicity of any chemical can only be made by reference to toxicity classification criteria (see for example, the toxicity rating scale in Chapter 2).

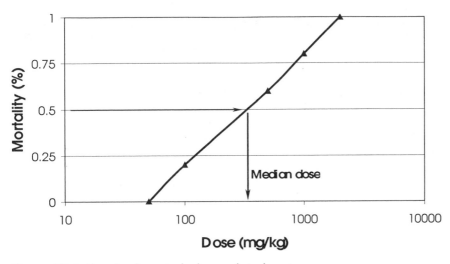

Mortality data: Single dose study in the rat

Figure 22.1 Mortality data: single-dose study in the rat.

Table 22.1 Single-dose study in the rat for a toxicant

Dose	Number of animals	Number dying
2000 mg/kg	5	5
1000 mg/kg	5	4
500 mg/kg	5	3
100 mg/kg	5	1
50 mg/kg	5	0

Step 2

The next part of this exercise is to look at LD_{50} data from more than one species.

One of the uses for the LD_{50} is to compare the toxic potentials of different compounds. Thus, if compound A has an LD_{50} that is twice that of compound B, some may claim that B is twice as potent as A. However, the LD_{50} is inadequate to describe either the slope of the linear portion of the log dose–response relationship or the behaviour in the lower and upper limits, and it becomes necessary to first see whether the log dose–response relationships for the two compounds are parallel.

The acute oral toxicity (LD_{50}) of 2,3,7,8-tetrachlorodibenzo-p-dioxin (TCDD) in various species shows significant variability (see Table 22.2).

These results indicate that the hamster is 5000 times less sensitive to 2,3,7,8-TCDD than the guinea pig. This species variability makes extrapolation of health effects to humans difficult.

- Which species should be used in human risk characterisation? Why?
- How toxic is 2,3,7,8-TCDD?

Table 22.2 LD_{50} data for 2,3,7,8-tetrachlorobenzo-p-dioxin (TCDD)

Species	LD_{50} (mg/kg body weight)
Guinea pig	0.001
Rat (male)	0.022
Rat (female)	0.045
Monkey	>0.07
Rabbit	0.115
Mouse	0.114
Dog	>0.3
Bullfrog	>0.5
Hamster	5.0

Trainer's notes

Discussion of these data is always challenging. The usual conclusions are the most sensitive species (erring on the side of caution), or the least sensitive species (because it allows more efficient commercial development), or the monkey (because it is phylogenetically closest to humans) or the rat (because it is usually the species specified in toxicity classification systems). The important point is that there is no 'correct' answer. Ultimately, no matter which species is used, TCDD falls into the most toxic category (see the toxicity rating chart in Chapter 2).

One point to introduce here is the concept of 'outlier species'. These are where some species specific aspect (for example biotransformation pathway anomalies) are not relevant to the human situation. If this is the case the use of inappropriate toxicological information may impair the process of human risk characterisation.

■ How toxic is 2,3,7,8-TCDD?

Again, this is a rhetorical question, because the toxicity of any chemical can only be made by reference to toxicity classification criteria.

Step 3

The next part of this exercise is to look at toxicity classification criteria.

Categories of toxicity have been devised on the basis of the wide variation in the dose of the chemical needed to produce death. These toxicity classifications, which use LD_{50} or ranges of lethal doses as an index, have immediate use. One such classification chart was shown in Chapter 2. An extract of this rating chart is show in Table 22.3.

Trainer's notes

By using Table 22.3 it can be seen that a chemical with an LD_{50} of 400 mg/kg can be classified as 'very toxic'. It can also be seen that no matter what species used, TCDD is classifiable as 'supertoxic'.

Discussion should then focus on the utility of such classification criteria, and what each classification means. Further, the classification table above is not used in any mandated classification system. Therefore, these data should be discussed in terms of

Table 22.3 Classification system for acute toxicity

Classification	Oral LD_{50} mg/kg	
Supertoxic	below 5	Less than 7 drops
Extremely toxic	5–50	Between 7 drops and 2.5 ml
Very toxic	50–500	Between 2.5 ml and 25 ml
Moderately toxic	500–5000	Between 25 ml and 500 ml
Slightly toxic	above 5000	Between 500 ml to 1000 ml

toxicity classification systems that exist nationally or internationally (for example, that of the Globally Harmonised System).

Such classification schemes serve to answer in lay terms the question: 'How toxic is this chemical?' However, they can only provide information for one route of entry into the body (the oral route) and in some exposure circumstances (such as occupational settings), dermal and inhalational routes are more important.

Step 4

This then leads to the last part of this exercise, about classification systems for acute toxicity from other routes of exposure.

Dermal and inhalational routes of entry into the body are important as these are the primary ways that people get exposed to chemicals in the occupational environment. However, most data on the acute toxicity of chemicals are available from oral toxicity studies. As already seen above, toxicity classification scales are based on oral toxicity.

Toxicity classifications for dermal and inhalational toxicity are not in widespread use. However, two exist. The first of these was the classification as outlined in the United Nations classification system for 'dangerous goods'. This classification system uses acute toxicity data to establish Packing Groups for toxic materials. The second classification system is that of the European Community 'hazardous substances'. A comparison of the two systems is made in Table 22.4.

A chemical is a powder with an oral LD_{50} of 23 mg/kg, a dermal LD_{50} of 575 mg/kg, and an inhalational LD_{50} of 0.33 mg/l.

What is its classification under the UN dangerous goods or the European Community hazardous substances toxicity classifications?

Table 22.4 Classification systems for classifying oral, dermal and inhalational toxicity

Classification system	Acute toxicity		
	Oral (mg/kg)	Dermal (mg/kg)	Inhalational (mg/l)
United Nations dangerous goods system Packing Group Risk			
PG I Severe	below 5	below 40	below 0.5
PG II Moderate	5–50	40–200	0.5–2
PG III Slight			
▪ Liquids:	50–200	200–1000	2–10
▪ Solids:	50–500	200–1000	2–10
European Community hazardous substances system Classification			
Very toxic	below 25	below 50	below 0.25
Toxic	25–200	50–400	0.25–1
Harmful	200–2000	400–2000	1–5
'Not hazardous'	above 2000	above 2000	above 5

Trainer's notes

The best way to discuss this data is to draw a table:

Classification system	Toxicity	UN	EC
Oral	23 mg/kg	Toxic 2	Very toxic
Dermal	575 mg/kg	Toxic 3	Harmful
Inhalational	0.33 mg/L	Toxic 1	Toxic

Discussion should then focus on which value should be used for classification of the product. While these systems do provide some measure of equivalence between the different routes of exposure, they are very simple in approach. And they will allow a classification to be made.

Further, while perhaps oral toxicity data are not completely relevant to the workplace, they cannot be dismissed. Therefore, this chemical is categorised as Toxic 1 in the UN system and Very Toxic in the EC (where the label would be required to include a skull and crossbones symbol):

Figure 22.2 EC pictogram for toxic and very toxic classifications

With regard to providing advice on the toxicity of this material, it is apparent that it is toxic orally and by inhalation, and at least harmful through the skin. If worker exposure to the chemical is possible, then the occupational toxicologist should recommend that workplace hygiene methods should concentrate on minimising inhalational exposure, with skin protection a second priority.

Discussion of this part of the exercise should also cover the transnational problems of different classification systems, and international moves to introduce a globally harmonised system (GHS) for the classification of chemicals.

Exercise 3: Classifying irritancy

Step 1

The irritation scores for the three rabbits used in a primary irritation study for a proprietary chemical are given in Table 22.5.

- What is the final primary irritation score?
- What does this score indicate?
- How should users handle this chemical?

Table 22.5 Irritancy scores

Time	Skin redness (erythema)			Skin swelling (oedema)		
	Rabbit 1	Rabbit 2	Rabbit 3	Rabbit 1	Rabbit 2	Rabbit 3
12 hours	4	3	4	4	3	3
24 hours	3	2	4	3	2	3
48 hours	3	2	2	2	1	3
72 hours	2	1	1	1	0	1
96 hours	1	0	0	0	0	0
10 days	0	0	0	0	0	0

Trainer's notes

The usual approach with data of this nature is to average all scores at 24, 48 and 72 hours, giving a score of 36/18 or 2.0.

A score of 2 or more in the classification system used in the European Community indicates that a substance is an irritant, and the label of such a material should be labelled with the 'Xi' irritancy symbol:

Figure 22.3 EC pictogram for irritant or harmful substance

What does it mean if a chemical product is labelled with this symbol?

Step 2

The results for irritancy studies for a second chemical product are available for both skin and eye irritation, and are summarised below.

Skin irritancy results

With an assessment score of 2.2, the chemical is to be designated under the EC Hazard Classification System as irritant (Xi). All findings were reversible within 10 days.

Ocular irritancy results

With an assessment score of 0.0, the Chemical was not irritant to the rabbit eye.

Trainer's notes

It is quite rare, but not impossible, for a chemical to be a skin irritant, but not an eye irritant (for example, the chemical is an oil, and not soluble in water). The Eye Irritancy Study looks at effects on conjunctiva, the iris and the cornea, and if the chemical does not contact the eye, it may be difficult to reach an irritating dose.

With a skin irritancy score of 2 or more in the classification system used in the European Community the chemical will be classifiable as an irritant, and the label of such a material will contain the 'Xi' irritancy symbol (see Figure 22.3).

■   So what is the utility of the eye study?

The utility of eye irritancy studies for skin irritants is questionable. In most cases, chemicals that are skin irritants are also usually eye irritants (the chemical in the exercise above is a rare exception to the rule). Indeed, the reasons for carrying out an eye irritancy study for chemicals that have been established as skin irritants is dubious, as it could be concluded that a skin irritant is an eye irritant, without the need for additional testing. This has become common practice.

■   What recommendations should be given to workers using this product?

The chemical is a skin irritant, so skin protection is probably necessary.

■   But the chemical is not an eye irritant so is eye protection necessary?

In any eye exposure, there will always be skin exposure, therefore, eye protection should be worn where there is potential exposure to any irritant.

Exercise 4: Classifying repeated-dose toxicity

Step 1

The results in Figure 22.4 below are data on osteosarcoma (a type of bone cancer) taken from a 2-year feeding study conducted by a US government-sponsored toxicity testing facility in the USA in rats drinking sodium fluoride. Between 50 and 80 animals were used for the data points at 0 (control animals drinking deionised water), 10, 45 and 78 ppm. One other data point (the second on the curve) is not available from the study, but was taken using historical control data. That is, the toxicity testing facility used tap water for control animals in their other carcinogenicity bioassays, and this tap water contained an average of 27 ppm fluoride. Therefore, historical control data was available from 6000 control animals (this is the basis of the second data point). This one data point showed that 0.5% of 6000 control animals consuming 27 ppm had bone cancer (this is more than enough to establish a background incidence rate).

■   Is fluoride a carcinogen?
■   Is it possible to set a lowest observable effect level (LOEL) and a no effect level (NOEL) for fluoride?
■   If so, what are the LOEL and NOEL?

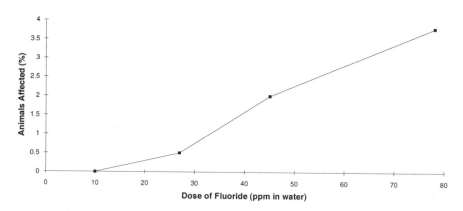

Figure 22.4 Incidence of osteosarcomas in long-term repeated-dose study of sodium fluoride.

Trainer's notes

Firstly, these data show that the dose–response relationship can apply to repeated dose situations.

Secondly, discussion about whether sodium fluoride is a carcinogen should focus on the point that although there does appear to be a NOEL in these data, the question was: 'Is sodium fluoride a carcinogen?' *not* 'Is there a dose of sodium fluoride where it is (or is not) a carcinogen?'

This is an important point, because most carcinogen classification systems do not consider the impact of dose or exposure when making a recommendation that a given exposure is carcinogenic. This then leads to the conclusion that (in rats) sodium fluoride is carcinogenic.

Discussion can consider possible mechanisms of fluoride on bone metabolism, the possibility of feasible interactions and whether there is bone toxicity at higher doses.

Discussion can then move to rat–human extrapolations, and the uncertainties in this process.

Comment can also be sought on whether it is possible to establish NOELs for carcinogens (for example, is 10 ppm a NOEL, and 45 ppm a LOEL (27 ppm if the historical control data is used)).

Step 2

The European Community carcinogen classification criteria state:

Category 1 carcinogenicity: Sufficient evidence to establish a causal relationship between human exposure and the development of cancer on the basis of epidemiological data.

Category 2 carcinogenicity: Sufficient evidence on the basis of long-term animal studies or other relevant information, to provide a strong assumption that human exposure can result in the development of cancer.

Category 3 carcinogenicity: Some evidence from appropriate animal studies that human exposure can result in the development of cancer, but that this evidence is insufficient to place the substance in Category 2.

- How should sodium fluoride be classified?

Trainer's notes

The data above is from a study in rats, so Category 1 (a causal relationship in humans) does not apply.

If (as already agreed) sodium fluoride is considered a carcinogen in animals, then discussion should focus on whether it is a Category 2 or 3 carcinogen. Once this is decided, it should be noted that in the EC, a Category 2 carcinogen is required to carry the risk phrase 'May cause cancer'; a Category 3 carcinogen is required to carry the risk phrase 'Possible risk of irreversible damage'.

Lastly, comment on the health effects of diet totally devoid of fluoride.

Step 3

- What is the likely impact of sacks of sodium fluoride containing the warning 'May cause cancer' on a group of workers handling such a material?
- What advice should the occupational toxicologist offer in such circumstances?

Trainer's notes

A carcinogen warning creates significant concern in working populations. Sodium fluoride has extensive use worldwide in water fluoridation, and therefore water utility workers are likely to handle this chemical.

Discussion should focus on the risk assessment process for the tasks involving this process, and the possibility of elimination of this chemical from the workplace. While elimination is possible, it is unlikely, and discussion can also focus on the public health benefits of water fluoridation as opposed to the carcinogenic risk revealed in these data.

Therefore discussion should then focus on minimising exposure and the use of control measures using the hierarchy of controls.

Exercise 5: Preparing material safety data sheets

Step 1

Below is the information given by a printout of a computer bibliographic database for a general solvent and chemical intermediate used in the chemical industry. It is sufficient for the purpose of this exercise to know that it is a chlorinated solvent.

COMPUTER DATABASE PRINTOUT
IDENTITY:

Name:	#######
CAS Number:	############ UN Number: ####
Use(s):	Chemical intermediate, Solvent, Agricultural excipient, Industrial chemical; Vapour degreasing

CHEMICO-PHYSICAL PROPERTIES:

Appearance:	Clear, colourless liquid with a sweetish chloroform smell
Odour Threshold:	21 ppm
Specific Gravity:	1.5 at 20°C
Melting Point:	−85°C
Boiling Point:	87°C
Refractive Index:	1.48
Vapour Pressure:	58 mm Hg at 20°C; 77 mm Hg at 25°C; 100 mm Hg at 32°C
Vapour Density:	4.53 (Air = 1)
Flash Point:	60°C
Flammability Limits:	Upper limit: 90% Lower limit: 12.5%
Autoignition Point:	410°C
Reactivity:	Reacts violently with Al, Ba, N_2O_4, Li, Mg, Na, Ti, alkalis, nitrates, liquid oxygen, ozone, anhydrous perchloric acid.
Combustion Products:	Hydrogen chloride, phosgene, carbon dioxide in the presence of complete combustion (carbon monoxide if oxygen supply limited), toxic smoke. Hazardous polymerisation will not occur.
Water solubility:	Practically insoluble.
Solvent Solubility:	Soluble in most solvents.

TOXICITY REPORTS:

Species	Route	Test	Protocol	Result
Toxic Effects:				
Rat	Oral	Single dose toxicity: TD_{Lo}		2688 mg/kg
Rat	Oral	Single dose toxicity: TD_{Lo}		1140 mg/kg
Rat	Oral	Single dose toxicity: TD_{Lo}		36,000 mg/kg
Mouse	Oral	Chronic toxicity: TD_{Lo}	77 wks	455,000 mg/kg
Human	Oral	Case study: TD_{Lo}		2143 mg/kg
Rabbit	Skin	Primary irritation	2 mg, 24 hrs	Severe irritant
Rabbit	Eye	Primary irritation	20 mg, 24 hrs	Moderate irritant
Rat	Inhalational	STRD toxicity: TC_{Lo}	24h/20 days	10,872 mg/m³
Rat	Inhalational	STRD toxicity: TC_{Lo}	4h/17 days	604 mg/m³
Rat	Inhalational	STRD toxicity: TC_{Lo}	6h/20 days	10,872 mg/m³
Rat	Inhalational	STRD toxicity: TC_{Lo}	4h/14 days	604 mg/m³
Mouse	Inhalational	STRD toxicity: TC_{Lo}	7h/5 days	604 mg/m³
Mouse	Inhalational	STRD toxicity: TC_{Lo}	24h/4 wks	906 mg/m³
Rat	Inhalational	Chronic toxicity: TC_{Lo}	7h/2 yrs	906 mg/m³
Mouse	Inhalational	Chronic toxicity: TC_{Lo}	7h/2 yrs	906 mg/m³
Hamster	Inhalational	Chronic toxicity: TC_{Lo}	6h/77 wks	604 mg/m³
Rat	Inhalational	Single dose toxicity: TC_{Lo}		2688 mg/m³
Human	Inhalational	Case study: TC_{Lo}	10 mins	6900 mg/m³

Human	Inhalational	Case study: TC_{Lo}	83 mins	$966\,mg/m^3$
Human	Inhalational	Case study: TC_{Lo}		$812\,mg/m^3$
Human	Inhalational	Case study: TC_{Lo}	8 h	$664\,mg/m^3$

Lowest Lethal Effects:

Rabbit	Oral	Single dose toxicity: LD_{Lo}		$7330\,mg/kg$
Human	Oral	Case study: LD_{Lo}		$7000\,mg/kg$
Human	Inhalational	Case study: LC_{Lo}		$17,516\,mg/m^3$
Rabbit	Skin	Single dose toxicity: LD_{Lo}		$1800\,mg/kg$
Dog	Skin	Single dose toxicity: LD_{Lo}		$150\,mg/kg$
Rat	Inhalational	Single dose toxicity: LC_{Lo}	4 h	$48,321\,mg/m^3$
Rabbit	Inhalational	Single dose toxicity: LC_{Lo}		$66,442\,mg/m^3$
Guinea Pig	Inhalational	Single dose toxicity: LC_{Lo}	40 mins	$224,694\,mg/m^3$
Cat	Inhalational	Single dose toxicity: LC_{Lo}	2 h	$32,500\,mg/m^3$
Human	Inhalational	Case study: LC_{Lo}		$17,516\,mg/m^3$
Dog	Intravenous	Single dose toxicity: LD_{Lo}		$150\,mg/kg$
Cat	Intravenous	Single dose toxicity: LD_{Lo}		$5864\,mg/kg$

Lethal Effects:

Mouse	Oral	Single dose toxicity: LD_{50}		$2402\,mg/kg$
Dog	Oral	Single dose toxicity: LD_{50}		$1900\,mg/kg$
Mouse	Skin	Single dose toxicity: LD_{50}		$16,000\,mg/kg$
Rabbit	Skin	Single dose toxicity: LD_{50}		above $20,000\,mg/kg$
Rat	Intraperitoneal	Single dose toxicity: LD_{50}		$1282\,mg/kg$
Mouse	Intravenous	Single dose toxicity: LD_{50}		$33.9\,mg/kg$
Rat	Inhalational	Single dose toxicity: LC_{50}	4 h	$51,039\,mg/m^3$

Genetic Toxicity:

Genotoxicity Assay	Genetic endpoint	Result
Escherichia coli	Point mutations	Weakly positive
Saccharomyces cerevisiae	Gene conversion	Positive
Saccharomyces cerevisiae	Reversion	Positive
Drosophila melanogaster	Sex-linked lethal	Negative
RLV F344 rat embryo	Cell transformation	Positive
Rat	Dominant lethal assay	Negative
Mouse	Dominant lethal assay	Negative
Mouse spot test	Sperm morphology	Positive

Carcinogenicity Bioassays:

Species	Agency	Route	Conclusions
Rat	NCI, 1976	Gavage	No evidence
Mouse	NCI, 1976	Gavage	Clear evidence
Rat	–, 1980	Inhalational	No evidence
Mouse	–, 1980	Inhalational	No evidence
Hamster	–, 1980	Inhalational	No evidence
Rat	NTP, 1983	Feed	Inadequate study
Mouse	NTP, 1983	Gavage	Clear evidence
Rat	NTP, 1988	Gavage	Inadequate study
IARC Evaluation:	Human:	Inadequate (1987)	
	Animal:	Limited (1987)	

Before the exercise of preparing a material safety data sheet, it is necessary to consider the significance of these data.

- What is the significance of TD, LD_{Lo}, and LD_{50} data?
- What is the significance of data from nonoccupational routes of exposure (for example, intraperitoneal, intravenous)?
- How should inconsistent data for the same species and route of administration (for example, rat/inhalational) be handled?
- How should variable data for the same route but from different species be evaluated?
- Do the results from repeated-dose studies need to be assessed differently from single-dose studies?
- What is the significance of human data?
- How should the genotoxicity data be categorised?
- What are the implications of the carcinogenicity data?
- What is the final toxicity rating of this material?

Notes:
LDL_o Lowest lethal dose; TDL_o Lowest toxic dose
LCL_o Lowest lethal concentration; TCL_o Lowest toxic concentration
IV Intravenous; IP Intraperitoneal.

Step 2

The second part of this exercise looks at the process of preparing an MSDS for this product. A company is marketing this chemical in a new product called 'Spot-Splott'. This is a formulated product containing 98.5% of this ingredient, and 1.5% of an organic surfactant, to be used as a degreaser and cleaner for metal parts.

Write an MSDS for this product as part of the company's commitment to hazard communication to workers and customers.

If a recommended format is being used, reasons should be given for its selection.

- Discuss which of the data in the printout is suitable for conversion to information that can be included on an MSDS.
- How should contentious or unequivocal information (such as carcinogenicity evaluation) be included?
- A competitor's MSDS for the same product is silent on this issue. What should this MSDS say?
- What sorts of controls will be recommended?
- What sorts of emergency measures will be recommended?
- What sort of response do you think this MSDS will receive from your marketing people?
- What sort of response do you think this MSDS will receive from your customers?

Exercise 6: Setting exposure standards

This exercise is aimed at developing skills in evaluation of material and being critical. The important points to bring out are:

- how much weight is given to toxicity data in the exposure standards setting process
- how empirical data are used in the standards setting process
- the contribution that each piece of data makes in the final decision.

Below is an extract from an occupational toxicological hygiene abstract relating to the chemical used in the previous exercise.

Toxicity Hygiene Overview

####### (CAS Number: ############) is a liquid with an appreciable vapour. This material is a skin, eye and respiratory irritant to single exposures. It may be absorbed through the respiratory system or across the skin to produce conventional symptoms of solvent-induced central nervous system depression.

Organ systems affected by prolonged and/or repeated exposure of rats and mice to this material include: (i) the central nervous system – (euphoria, drowsiness, headaches, analgesia, anaesthesia); (ii) the liver – hepatocyte degeneration, hepatocellular carcinomas (in mice only); (iii) kidney – glomerulonephritic degeneration; (iv) cardiovascular – skin redness and flushing, ventricular arrythmias (at high exposures); (v) skin – irritation, vesication, paralysis of finger (following immersion); (vi) respiratory system – tachypnea. Some of these effects are exacerbated by alcohol consumption.

In workers, the main hazard is from inhalation of vapour, although significant exposure is possible if skin contact is substantial. Once absorbed, the chemical is rapidly partitioned into adipose tissue. Effects reported in different epidemiological studies of chronically exposed workers include:

- subjective symptoms (vertigo, fatigue, headache) and experimental findings (short-term memory loss and increased misunderstanding) at 85 ppm (exposure data not expressed as a time-weighted average)
- intolerance to alcohol, tremors, giddiness and anxiety above 40 ppm.

Evidence is also available from controlled exposures of volunteers under laboratory conditions:

- significantly decreased performance in visual perception and motor skill in volunteers at 1000 ppm, but not 300 or 100 ppm (after 2 hours' exposure)
- significantly decreased performance in numerical ability in volunteers exposed to 100 ppm for 70 minutes
- drowsiness and upper airway irritation at 27 ppm, headache at 81 ppm, dizziness at 201 ppm (4-hour exposure)

- significantly decreased performance on a perception test, a complex reaction time test and manual dexterity test was found following two 4-hour exposures separated by a 1.5-hour interval at 110 ppm.
- in a repeated exposure experiment (7 hours/day), slight fatigue and sleepiness were reported on day 5 above 20 ppm.

Based on the data in this extract, and on the data in the MSDS preparation exercise, establish a time-weighted average (TWA) exposure standard for this material.

Trainer's notes

Provide justification for the exposure standard level established.

- What information was important in reaching the conclusion for the exposure standard?
- Was it health based?
- Was it based on studies where there were constant or variable exposures?
- What might be other reasons for establishing an exposure standard?

Typical exposures for this chemical are below 10 ppm for most exposures, but 30–80 ppm in use as a solvent and 50–200 ppm in vapour degreasing.

- What is the significance of these data and how did it relate to the exposure standard that has been set?
- It may be that controlling exposures to the exposure standard is not technically feasible in those industries.
- What does this mean for the standard that has been set?
- Should a lower but realistically practical exposure standard be set that has some likelihood of being met in workplaces?
- A similar chemical has a TWA exposure standard of 75 ppm, based on its ability to cause drowsiness and narcosis; should exposure standards for other compounds be used for comparison purposes?
- Is it possible to establish any short-term exposure limits?
- Does the fact that this chemical causes cancer in mice have any bearing on the standard setting process?

Exercise 7: Request for assistance from an employer, degreasing workers

A letter is received from a company with the following text:

METALPREP METALCLEAN PTY LTD

OccTox Consulting and Advisory Services   (a subsidiary of Multinatio, Inc)
PO Box 100
Erewhon 12345   March 6, 2003

Dear Sir/Madam

LIVER FUNCTION TESTS IN DEGREASING WORKERS

I write to seek advice on the possible causes of liver dysfunction found in one of our workers. Two weeks after returning to work after the Christmas/New Year holiday break, one worker was found to have elevated liver function tests, from his regular six months screening test. In particular:

- transaminanses – moderately elevated;
- alkaline phosphatase – essentially normal;
- gamma-glutamyl transpeptidase – essentially normal.

The worker concerned has been with our company for three years and has not previously been found to have abnormal liver function tests. He has reported alcohol intake to be minimal – essentially social consumption over the weekend. The worker is one of six employed in our metal degreasing operation where the main chemical used is trichloroethylene. With regard to his co-workers, three had no abnormality in their liver function tests while one had mildly elevated gamma-glutamyl transpeptidase and the other had borderline aspartate aminotransferase.

All six workers carry out essentially the same tasks and get approximately equivalent exposures. Periodic air monitoring carried out by an occupational hygienist from corporate headquarters about eighteen months ago indicated no extraordinary excursions above the time-weighted average exposure standard for trichloroethylene (50 ppm) in the degreasing unit during normal operations.

There was one minor incident a few days prior to routine blood sampling last year. Some trichloroethylene was spilled during planned maintenance operations. About 15 L was spilled, but all workers were evacuated immediately. Clean up and decontamination was conducted by our spill team (including two workers from the degreasing unit who had normal liver function tests). Rough estimates suggest a likely peak air concentration of 150 to 250 ppm, for a period of about 20 minutes.

Please do not hesitate to contact me for further information if necessary. We look forward to receiving your advice at your earliest convenience.

Yours sincerely

Sol Vent

MR SOL VENT
MANAGER, METAL CLEANING OPERATIONS

Trainer's notes

There is no 'correct' answer, necessarily, to this problem. However, the steps taken and the kind of information sought are the important aspects. It should be appreciated that different individuals will approach this problem in different ways. For instance many will want to seek more information on a variety of points before responding. This is laudable and probably necessary but for the purposes of this exercise where this is not possible, information can still be identified that must be sought to provide a response to the request.

There are a range of different toxicological and occupational health principles that this exercise can use to illustrate:

- Dose–effect
- Extrapolation from animal data
- Route of exposure
- Duration of exposure
- Disposition: absorption, distribution and excretion
- Biotransformation
- Toxicity versus risk
- Target organ toxicity
- Toxicity by class of chemical
- Use of information sources
- Exposure standards: values and pitfalls
- Use of regulatory control
- Consideration of extraneous factors
- Integration of toxicology, medicine and hygiene
- Data interpretation and the importance of a thoughtful and considered approach.

Ultimately, the information that needs to be sought in formulating a reply to the letter in this exercise includes:

1 What are the tasks of a degreaser with regard to likely exposures?
2 Is trichloroethylene hepatotoxic? Under what circumstances?
3 How are the liver function tests to be interpreted with regard to chemical (and especially trichloroethylene) exposure?
4 What is the basis for the exposure standard for trichloroethylene? Is it related to liver effects or to some other toxic response?
5 Are there other potential hepatotoxicants to be found in this type of workplace?
6 Could chemicals have been mislabelled, contaminated or the like?
7 Is there a possibility of the toxicity being due to interaction with other chemicals, at work or otherwise?

Once information is obtained on the above a letter of reply can be drafted. It should be remembered that it is going to the manager and should be succinct and to the point. If it is felt that more information or investigations are required these should be stated but at the same time provide as much information as possible. Recommendations should be clearly provided for an appropriate course of action. For example, one recommendation might be that the liver function tests be repeated. The reader may care to address the problem and draft a letter of reply.

Exercise 8: Request for information from an OHS Committee, reproductive hazards

Your workplace safety committee notes that new evidence from Scandinavia has suggested that a material used extensively in your workplace may cause embryroxicity in pregnant animals at high doses and spontaneous abortions in workers.

- Workplace: A factory manufacturing textiles/furniture.
- Workers: 78 in production areas (about two-thirds men); 32 in office areas (12 men and 20 women).
- Process: Thick felt is fixed to surface covers of furniture and mattresses with a flexible adhesive called Mutoglue. To make the adhesive stick better and penetrate the material, the felt is immersed in a vat containing a watery solvent mixture called Mutosolv for about 15 minutes before fixing. All processes are manual.
- Chemical products: Mutofix (and Mutosolv) was introduced about 2.5 years ago, as a replacement for an adhesive which contained n-hexane. The agent of concern is found in both Mutofix (at a concentration of 0.4%) and Mutosolv (4%). Neither the label nor the MSDS mention anything about reproductive problems.
- Exposed workers: The number of workers involved varies, depending on the size of the job (mattresses need more workers than chair covers), the number of items being made (the size of the order) and the amount of the material needing fixing. However, not more than 20 workers are used for any particular order, although probably most if not all manufacturing workers have been involved in fixing at some stage. Three factory workers and one office worker are known to be pregnant, and there are also at least two expectant fathers working in the factory.
- OHS procedures: The factory has roof vents and ceiling fans that are used as dilution ventilation. Dust masks are used extensively by workers handling felt as it releases fibres and dust. All workers also wear gloves (bought from a local supermarket) when dipping felt into Mutosolv and when fixing with Mutofix. No monitoring has been carried out in the factory since before the solvent adhesive was replaced, as it is considered that there is no need for it (the solvent has a faint chemical smell).

What are the necessary steps to be taken to control the risk by: (i) the safety officer; (ii) the visiting occupational physician; (iii) a consultant occupational hygienist bought in to investigate the problem?

Trainer's notes

It is common in situations such as this to make an assumption that this is an 'industrial relations' issue. Many occupational health issues arise because of the actions of a group of workers and they may have other outcomes in mind. However, the occupational toxicologist should not be distracted by such issues, and stick to the facts of the case.

Reproductive hazards are a particular problem in workplaces when they arise. They need to be dealt with openly and with sensitivity.

While it is reasonable to accept such assertions at face value, the first step is to check the *bona fide* facts of the source – authoritative information is more useful than second-hand or subjective reports.

This particular problem can be viewed from two perspectives: what to tell the workers and what to do in the workplace (see the chapter on reproductive hazards for more information).

Exercise 9: Request for information from a union, plastics workers

A letter is received from a union, requesting information as follows:

	CHEMICALS WORKERS UNION
OccTox Consulting and Advisory Services	Plastics and Elastomers Division
PO Box 100	
Erewhon 12345	March 6, 2003

Dear Sir/Madam

MOCA EXPOSURE IN WORKERS

It has been brought to our attention that some of our members might be exposed to the chemical MOCA (methylene-*bis*-2-chloroaniline) in the course of their work in polyurethane manufacture. We understand that this chemical may have cancer-producing effects.

We therefore request that you provide a report to indicate whether or not these workers' health may be at risk and what steps might be taken to rectify the situation.

Yours sincerely

Polly Vinel

MS POLLY VINEL
INDUSTRIAL OFFICER

Trainer's notes

Again, the steps to be taken in order that a satisfactory response might be provided will be outlined but an actual response will not given.

Again, there are a range of different toxicological and occupational health principles that this exercise can use to illustrate. Indeed, many of the basic aspects relevant to Exercise 8 are also relevant here. However, this exercise should highlight some additional matters, such as:

- specific regulations exist covering particular chemicals in some jurisdictions
- consideration of product performance when addressing the use of chemicals

- the importance of dermal absorption of some chemicals
- the advantage of biological monitoring over environmental monitoring in some circumstances
- the relevance of atmospheric standards for some chemicals as compared to biological exposure indices
- the importance of considering all relevant toxicity data when deliberating on the use of a chemical
- the necessity to consider various ways of dealing with a hazardous chemical.

Information to be sought:
How is methylene-*bis*-2-chloroalanine (MOCA) used in this industry?
Are there physicochemical properties of MOCA that are relevant to potential exposure?
What is the likelihood of exposure? Is there any evidence to suggest that it is a particular problem for this chemical?
Is MOCA a carcinogen? What is the strength of evidence? Is there cause for concern?
What might be done to rectify the situation if it is a problem? What are the options for:
    Elimination
    Substitution
    Containment
    Safe working procedures
    Ventilation and monitoring
    Personal protective equipment.
What regulations cover the use of the chemical, if any?

    This exercise has two parts, the first being gathering, collating and interpreting toxicological information; the second being preparation of a reply to the request.

    Again, the letter of reply needs to be concise and address the issues raised in the request. Important aspects relevant to the use of this chemical should be raised. For example, in New South Wales, Australia, a licence is required for its use and there are conditions attached to this licence – this should be included.

Exercise 10: Request for information from an individual worker, an operating-theatre nurse

Occasionally, individuals with chemically related problems contact the occupational toxicologist. For example, an ex-nurse originally worked for more than 10 years on the wards of a general hospital, and 5 years ago she undertook postgraduate training to become a theatre nurse. Once qualified, the nurse began working in a suite of operating theatres in a large city hospital. Three years ago she was transferred full time to the operating theatres where gastroscopy, bronchoscopy and endoscopy procedures were being carried out. The fibre optic devices used in such procedures are too fragile to be sterilised using conventional sterilisation techniques, and after each procedure, had to be washed, sterilised with a 0.2% solution of the chemical sterilant glutaraldehyde, and rinsed prior to re-use. This was part of the nurse's daily duties. The nurse began to show signs and symptoms possibly related to exposure to this chemical, and contacted an occupational toxicologist for information.

Glutaraldehyde is a hazardous material. Searching for the health effects of glutaraldehyde can form an initial step in this exercise. This includes its well-established irritant and sensitising properties. However, glutaraldehyde is known to cause headaches, light-headedness and nausea, suggesting nonirritant effects common to many organic substances. The health hazards of glutaraldehyde are also based on the concentration of glutaraldehyde in the product being used:

- above 25% corrosive
- between 1% and 25% a skin irritant
- above 5% capable of causing serious eye damage
- between 1% and 5% an eye irritant
- above 1% respiratory irritant; a skin sensitiser.

Within 6 months of her new position, the nurse found herself with increasing levels of exhaustion. This was accompanied by a headache which was different from the mild headaches of low incidence that she got from time to time previously, in terms of localisation, intensity, duration and inability to be controlled with over-the-counter analgesics.

After a year, the exhaustion had been diagnosed as chronic fatigue, and the intense headache was accompanied by neuropsychological symptoms such as forgetfulness, inability to concentrate and anxiousness. The nurse also found other symptoms were occurring, such as gastric problems (nausea and diarrhoea), respiratory problems (wheeze developing, symptoms of upper airway irritation) and skin rash.

In her second year in the position, neuropsychological symptoms were characterised by a clinical psychologist as 'organic brain disease', and the wheeze developed into asthma (diagnosed by a respiratory specialist). This specialist also noted that the likely cause of asthma was from occupational exposure to glutaraldehyde.

On return to work, the nurse requested job transfer or retraining, and for advice on chemical exposures in the operating theatres. She was informed that no other jobs existed, and that she must continue to work in her current location. Also, following a complaint about smells in the operating theatres by a patient, occupational hygiene surveys carried out in the theatre complex overall (but not specifically in the theatres where the nurse is working) showed that all chemical exposures were 'within acceptable levels'.

On her return to work that afternoon, the symptoms of headache, nausea and wheeze arose as soon as she smelled the residual glutaraldehyde smell that was always present in the operating theatres. She refused to work and was suspended from duties.

The nurse is considering initiating a workers' compensation claim for damages through the courts, and contacted an occupational toxicologist for advice.

There are a number of critical toxicological concepts to be covered in this case study.
Toxicity of glutaraldehyde:

■   Describe the hazards of glutaraldehyde.
■   Outline the health problems that arise following exposure to glutaraldehyde.

Occupational hygiene aspects of glutaraldehyde:

■   What are the concentrations of glutaraldehyde found in workplace air that are
associated with health problems?
■   Does the exposure standard for glutaraldehyde provide adequate protection?
■   Are any of the symptoms in this case specifically related to the toxicity of
glutaraldehyde?
■   Was the occupational hygiene survey carried out in such a way that it can be used
as a measure of 'acceptable risk'?

Working conditions:

■   Was the nurse given appropriate information, instruction and supervision to carry
out her duties safely?
■   Was specific information about the hazards of glutaraldehyde provided?
■   Was specific instruction given about how to minimise exposure to glutaraldehyde
provided?
■   Did the nurse comply with any controls given to minimise exposures?

Other issues can be included in the discussion, such as whether workers'
compensation or common law claims could be initiated (this varies from state to state
and nation to nation).

Inevitably, cases such as this, which have a 'psychological' component are complex
and can be difficult to establish, although there is no reason why psychological
symptoms do not follow toxic exposures. Certainly, concepts such as dose–response,
acceptable exposure and what constitutes suitable working conditions are important in
this exercise. However, one question that should be asked is

■   Would any of the symptoms have arisen if this nurse had not been exposed to
glutaraldehyde?

Exercise 11: Biological monitoring – processes and pitfalls

A program for the biological monitoring of chlordimeform was used as the basis for
this exercise. It is provided to stimulate the reader's thinking about biological
monitoring programs and issues that arise from these. The reader is invited to
proceed through the exercise in a step by step manner and think about each step
before proceeding to the next.

Step 1

Chlordimeform is an insecticide that had extensive use in the cotton industry from about 1978 until 1986. Its use was discontinued due to association of a major metabolite with human bladder cancer which was consistent with the class of chemical and positive carcinogenicity results in experimental animals.

During the period of its use it was required that workers participate in a program to monitor exposure to the chemical.

Question 1: What possibilities can you think of for methods of monitoring exposure?

Question 2: What do we need to know about the chemical to decide if such suggestions are feasible?

Step 2

Consider the information on the disposition and detection of chlordimeform as follows:

- Absorption: rapidly absorbed across skin, lung and gut.
- Distribution: essentially throughout body including the central nervous system. No significant bioaccumulation.
- Biotransformation: several metabolites.
- Excretion: chlordimeform and metabolites are rapidly excreted by mammals.
- In humans, using a method that detects only those compounds that are hydrolysed to 4-chloro-o-toluidine, mean recovery from urine 72 hours after absorption was 38%.
- Detectable residues were approaching zero in most subjects 36 hours after absorption. Mean half-life was estimated to be 8.8 hours.

Step 3

Consider the information on the urine monitoring scheme as follows:

- All chlordimeform workers were required to submit urine specimens for analysis. A spot sample was collected between 24 and 48 hours after exposure ceased. The allowable action level was 0.2 mg/l, whereby workers were suspended from exposure until levels had returned to normal.
- Supervisors were responsible for collecting samples which were analysed within 96 hours.
- Actions following detection of levels over the limit included:

  - Immediate removal of the worker from further exposure to chlordimeform.
  - Completion of an incident report detailing the circumstances causing the high exposure, if possible (this is the responsibility of the marketing companies but sometimes a pesticides inspector is also involved in the investigation).

- Appropriate remedial actions, including retraining of the offending worker if necessary.

- During the operation of the scheme some changes were introduced, including in the initial growing season:

  - Incentive payments were offered by the marketing company with the intent of encouraging timely submission of correctly collected and labelled urine samples and to foster safe working habits. These incentive payments were $2 for each correctly filled and labelled urine sample collected 24–48 hours after exposure; and $5 if the sample result was below 0.1 mg/l.
  - Supervisors were also paid 10 cents per litre of product applied if samples were promptly collected and submitted, and work logbooks correctly maintained.

Question 3: Do you consider the incentive scheme a useful means of collecting samples? Is it possible that such a system might be open to abuse? Would you suggest any changes to the scheme?

Question 4: Does the information in Table 22.6 support your responses to Question 3.

Step 4

Look at Tables 22.7 and 22.8.

Question 5: What do they tell you about exposure?
Question 6: What are possible reasons for overexposure?
  For information, the main causes stated were:

- Exposures obtained from post-application maintenance. This was the most commonly reported cause in the subsequent season (51 reports), and was commonly attributed to not wearing waterproof clothing when washing down

Table 22.6 Urine analysis results corrected to standard reference times

Worker number	Time after exposure (hours)	Chlordimeform metabolites (mg/L)	Concentration (mg/l)	
			At 8 hours	At 24 hours
1	31	0.73	5.35	1.13
2	34	0.85	8.08	2.02
3	36	0.27	3.05	0.76
4	38	1.20	16.12	4.03
5	48	0.31	9.9	2.48
6	15	0.26	0.48	0.12
7	3	0.79	0.51	0.13
8	46	0.16	4.30	1.07

Table 22.7 Exposure level versus worker category

Category	Mean urine chlordimeform (mg/L)	
	1984–1985	1985–1986
Loader/mixer	0.62	0.52
Supervisor	0.41	0.14
Marker	0.13	0.17
Driver	0.17	0.13
Pilot	0.16	0.17

Table 22.8 Multiple exposures greater than 0.2 mg/L versus worker category in a subsequent cotton season

Operator number	Workers and their number of exposures above 0.2 mg/L		
	Loader/mixer/supervisor	Pilots	Driver/marker
1	2, 4, 8, 9	3, 7	2, 2, 2, 3, 3, 8
2	2, 2, 3, 8, 11	3	2, 2, 2, 4
3	2, 2, 3, 4, 6	2, 2	2, 2, 4
4	3, 6	2	3
5	4	2, 2	
6	13	3, 8	2, 2, 4
7	2, 2, 2		
8		5	

vehicles and equipment. It appears that workers did not fully appreciate the degree to which aircraft and marker vehicles were contaminated and the ease with which chlordimeform was absorbed through the skin. This was despite emphasis on these facts in the training sessions. Many workers do not expect exposure to occur when they are not directly involved in mixing or spraying pesticides.

- Spray drift while marking was the second most common cause reported. This seemed to be a somewhat standard excuse and the 45 reports in the subsequent season accounted for almost all the multiple high exposures detected in markers and drivers.

- No protective clothing worn (33 reports in the subsequent season) does not appear to be an acceptable reason. These situations should have been prevented by proper supervision.

- Accidental spillage on body. This reason accounted for a large proportion of the sample results exceeding 2.0 mg/l. Results of 2.3, 2.5, 2.5, 4.4, 11.1 and 12.3 mg/l were attributed to incidents which involved skin contact while uncoupling hoses, changing drum probes or removing a blocked filter (1 instance).

- Exposure while repairing faulty equipment was reported 20 times in the subsequent season. It seems difficult to adopt adequate protective measures while repairing equipment in the work environment.

Less commonly reported reasons for high exposures were torn protective clothing, poor quality equipment, failing to wash after work and extended periods of general exposure. It is obvious that many of these high exposures could have been avoided by better attention to work hygiene and adherence to conditions stipulated in permits. It is disturbing that sporadic, relatively high exposures result from accidents.

Trainer's notes

1 The toxicity of chemical obviously important – but is comprehensive monitoring appropriate?
2 How best to monitor exposure?
3 The importance of skin absorption.
4 The importance of biotransformation processes.
5 The importance of toxicokinetics (of excretion).
6 Importance of assay procedure.
7 Issues of the 'penalty' of removing workers with excessive exposures.
8 The issue of incentive payments in the initial season.
9 Interpretation of results
    targeting of particular groups
    targeting of particular individuals.
10 Reasons given for overexposures.

Exercise 12: Workplace chemicals management

A company in the manufacturing sector uses a wide variety of hazardous chemicals in the production of its products. The company employs about 500 workers, including 150 office staff, 300 production workers (split into three 8.5-hour shifts), 25 maintenance staff and the remainder as cleaners, laundry staff, canteen personnel and security officers. Other hazardous chemicals are also used by seven staff in a Quality Control Laboratory.

Systems for the control of chemicals have been developed in a completely reactive manner over the years, and include:

- A dangerous goods licence for flammable liquids. Solvents are stored in a placarded flammable goods store, but other dangerous goods (including corrosives and poisons) are also present on site.
- An occupational health and safety committee, which has developed a health and safety policy, and spends a lot of time looking at accident reports and rehabilitation cases. While there is some concern about chemicals voiced by some workers, the committee doesn't know what to do about it, except think about developing some training.
- Information on hazards of some products (material safety data sheets) is available in the production manager's office, the laboratory and the occupational health clinic. The responsibility for the purchase of materials below $500 resides with line management.

- Safety equipment is available:

    - some access to respirators for solvent exposure tasks (most workers don't use them)
    - all workers are issued with eye protection, although compliance is virtually nonexistent
    - first aid boxes, and trained first-aiders
    - fire extinguishers, fire blankets, safety showers and eyewash stations at selected locations throughout the factory.

- Visits by a visiting occupational health physician 4 hours a week
- Visits by a waste contractor who removes

    - two dumpsters of industrial waste per week, and
    - large numbers of 2001 drums of liquid waste (grease trap waste, acid waste and spent solvents in approximately equal amounts) about once every 6 months.

- Chemical safety was added to the duties of the quality control laboratory manager about 5 years ago, but she is usually too busy to deal with anything other than urgent jobs. Overall, attitudes to safety are quite good, but most workers have little idea of the hazards of the materials they are using.

The purpose of this exercise: The company has just been merged with a multi-national company, which is introducing changes in the way the company is being run. As part of this process, new managers have been appointed in a number of areas, including personnel, production and safety. However, the workforce is largely unchanged, although concern is being expressed at the new changes.

You are the new occupational health and safety manager. You have inherited a department comprising a rehabilitation coordinator, a safety officer (who works part-time in the laboratory, and spends the rest of his time dealing with hazardous wastes) and two part-time occupational health nurses (one from an agency). The possibility of new staff appointments in your department disappeared with the introduction of your position.

- What activities do you need to undertake to address the problems of chemical hazards in the short and long term?
- What systems, policies and programs do you need to put in place to ensure that chemical safety issues are being effectively managed?

Trainer's notes

This particular exercise can be viewed from two perspectives:

- what are the policy implications, and what should be included in a workplace chemicals management system?

    Some of the policy implications involve the following:

- What is the level of senior management commitment?

- What consultative mechanisms exist (management–workers, management–safety unit and so on)?
- What is the present level of compliance with relevant legislation?
- What chemicals are used at the factory?
- What controls are in place (and which are being used)?
- Have there been any incidents/accidents involving chemicals?

Some of the subprograms that seem to be included in a chemicals management system include:

- Policy, program, procedures
- Chemicals inventory control
- A workplace assessment program for chemicals
- A hazard communication program
- Training for hazardous chemicals
- Proper use and maintenance of controls (including PPE – personal protective equipment)
- Contractor management
- Waste management program
- Emergency response for chemicals (as part of a larger emergency planning system)
- Health surveillance for selected workers
- Environmental monitoring (possibly).

Note: See also the chapter on managing workplace chemicals (Chapter 20).

Index